WORK, INDUSTRY & Canadian Society

WORK, INDUSTRY & Canadian Society

5TH EDITION

HARVEY J. KRAHN
University of Alberta

GRAHAM S. LOWE
The Graham Lowe Group Inc. and University of Alberta

KAREN D. HUGHES
University of Alberta

THOMSON
™
NELSON

Australia   Canada   Mexico   Singapore   Spain   United Kingdom   United States

THOMSON

NELSON

Work, Industry, and Canadian Society, Fifth Edition

by Harvey J. Krahn, Graham S. Lowe, and Karen D. Hughes

Associate Vice President, Editorial Director:
Evelyn Veitch

Senior Acquisitions Editor:
Cara Yarzab

Marketing Manager:
Lenore Taylor

Developmental Editor:
Sandra Green

Permissions Coordinator:
Indu Ghuman

Senior Production Editor:
Bob Kohlmeier

Copy Editor:
Margaret Crammond

Proofreader:
Margaret Crammond

Indexer:
Dennis A. Mills

Production Coordinator:
Hedy Sellers

Design Director:
Ken Phipps

Interior Design:
Katherine Strain

Cover Design:
Johanna Liburd

Cover Image:
Interchange Suite (Wondrous Country), Roads and Bridges, by Yael Brotman; drypoint on paper

Compositor:
Interactive Composition Corporation

Printer:
Transcontinental

Library and Archives Canada Cataloguing in Publication

Krahn, Harvey
Work, industry, and Canadian society / Harvey J. Krahn, Graham S. Lowe, and Karen D. Hughes. — 5th ed.

Includes bibliographical references and index.
ISBN 0-17-640610-7

1. Work—Social aspects—Canada. 2. Industrial sociology—Canada. I. Hughes, Karen D., 1960– II. Lowe, Graham S. III. Title.

HD6957.C3K72 2006
306.3'6'0971
C2005-905638-X

BRIEF TABLE OF CONTENTS

DETAILED TABLE OF CONTENTS

Contents

Contents

Contents

Contents

Contents

Contents

LIST OF TABLES AND FIGURES

PREFACE

In this fifth edition of *Work, Industry, and Canadian Society* we have drawn on the ever expanding literature on work and employment to incorporate new empirical findings, consider ongoing theoretical and policy debates, and provide a more international perspective on employment trends. The most significant change from the fourth edition, however, is the updating of almost every table and figure, and most text discussions of data, using the most current information available. We have also incorporated approximately 400 new references (and deleted almost as many old ones) to reflect the most up-to-date findings and debates in the research literature. We have not changed the basic structure of the text. Rather, we have modified each chapter, in some cases adding new sections and deleting old ones. Our vantage points have shifted, as one of us (Graham Lowe) has left the academic world to run a private-sector workplace consulting business. In turn, Karen Hughes has joined the author team, contributing extensive and complementary expertise. The two original authors are delighted to welcome Professor Hughes as a collaborator to this fifth edition.

In the four years since we last revised the text, there have been major changes in the world of work. The dramatic drop in national unemployment rates to a 25-year low, a booming North American economy, the rise of China as a new powerhouse in the world economy, heightened concerns about terrorism and security, employers' anticipation of baby-boomer retirements, continuing labour market polarization—these are some of the workplace-defining trends since we prepared the last edition of this text. While the fast-changing and fluid nature of the national and international contexts of work is part of what makes this research area so interesting, this continual change also presents a large challenge to textbook authors. Students using this text should be encouraged to consider the patterns of change and continuity since the summer of 2005, when we prepared this fifth edition.

As with previous editions, the author team worked as equal partners on this textbook. Even if we might want to take individual credit for specific ideas, by now it is difficult to remember where an idea originated. We are very grateful to Nickela Anderson who served as an excellent research assistant, hunting down new sources, assisting with the updating of information in the figures and tables, and compiling the references. We also thank Statistics

Canada for making available the high-quality data that we have mined extensively in all five editions of the text. In planning this edition, we received helpful comments from various colleagues across the country who have used the book in their teaching, including Lee Chalmers at the University of New Brunswick, Charlene Gannage at the University of Windsor, Jan Kainer at York University, Wolfgang Lehmann of the University of Western Ontario, Greg McElligott at McMaster University, Vicki Nygaard at the University of Victoria, Marianne Sorenson of Athabasca University, and Phillip Vannini of the University of Victoria. Equally important, feedback from undergraduate students at the University of Alberta has led to other modifications. At Nelson, Cara Yarzab encouraged us to take on another revision and agreed to a deadline that fit our schedules, and Sandra Green managed the editorial process. We would also like to thank Bob Kohlmeier, the production editor, and Margaret Crammond, our copy editor, for many helpful editing suggestions which improved the text. Any errors and omissions are, of course, our own responsibility.

Harvey J. Krahn and Karen Hughes, *University of Alberta*
Graham S. Lowe, *The Graham Lowe Group Inc. and University of Alberta*
Edmonton and Kelowna, August 2005

INTRODUCTION

Work is undeniably an essential human activity. It is the basis for the economic survival of individuals and society. Beyond this, an individual's work activity structures much of her or his time and, one hopes, provides a source of personal fulfillment. An occupation also shapes identity and, in the eyes of others, largely determines an individual's status or position in society. Stepping back to gain a wider perspective, one also sees that Canadian society and the quality of life we have come to enjoy are the historical products of the collective work efforts of millions of women and men. Thus, while the *sociology of work and industry* may at first glance seem as potentially boring as the part-time job you just quit at your local shopping mall, it is not difficult to justify the subject's social relevance.

When we wrote the third and fourth editions of this book we began by noting that the pace and scope of change in Canadian workplaces, the national labour market, and the global economy had increased in the 1990s. Halfway into the opening decade of the 21st century, we still stand by this conclusion. But despite some truly remarkable developments in the local, national, and global contexts, the overall pattern of change is still best called evolutionary rather than revolutionary. As we note in various places in the text, debates continue about the extent to which "postmodern" or post-bureaucratic work systems, jobs, or organizations have swept away older forms of work. But many of the hierarchical, inflexible, and conflict-prone features of the 20th-century workplace and labour market remain, albeit in somewhat different forms.

At the same time, some of the prominent trends documented in the fourth edition have clearly shifted direction. For example, global economic integration continues, but the grassroots opposition so evident at G8 or other "summits" of the leaders of industrial nations has placed the actions of transnational corporations and national governments under greater scrutiny. New information and communication technologies continue to push the frontiers of work, but the shakedown in the high tech sector after the dot-com stock-market bubble burst in 2001 raises a host of questions about the sustainability of the so-called "new economy," the future economic potential of the Internet, and the implications for the livelihood of workers in this sector. In previous editions, we described the new, powerful national economies in East Asia; yet soon thereafter, they were shaken by a deep financial crisis.

China is now the emerging global economic powerhouse, flooding world markets with inexpensive—but increasingly high quality—goods and showing an insatiable appetite for natural resources. India is also quickly establishing a global presence, hosting a growing number of "offshore" call centres and information technology development centres for North American and European multinational companies. We also described how a North American continental free trade agreement had begun to leave its mark. The debates on continental integration continue, and the political agenda is now broadened with discussions of a free trade area that will extend throughout Central and South America. Unemployment was a leading public policy issue in the mid-1990s, but by the end of the century robust economic growth in many, but not all, parts of Canada had pushed down the national jobless rate. By 2005, Canada was experiencing its lowest level of unemployment in 25 years. Employers and policymakers were voicing concerns about labour shortages caused by the combination of a hot economy and baby-boomer retirements. Yet corporate downsizing, a trend widely associated with the retrenchment of the 1990s, has continued, often due to mergers and acquisitions. Relentless change in work organizations has become "the new normal." While the trend toward more nonstandard employment continues, it does seem to have levelled off in some respects. However, in other ways described in the following chapters, employment is becoming still more fragmented and tenuous. Overall, labour market polarization continues and social inequality has increased.

In this fifth edition we examine these and other trends that continue to influence the work opportunities and experiences of Canadians. In particular, we are concerned with the consequences of different types of work arrangements for both the individual and society. Diversity and flexibility in jobs and work organization are even more prominent themes. As the book's title suggests, a sociological analysis of work must be grounded in its industrial context—hence, our emphasis on the underlying economic forces that have shaped and continue to shape work opportunities. But a comprehensive understanding of the world of work also requires us to consider other features of society. For example, there are basic connections between a person's paid work and her or his family responsibilities. Society-wide value systems influence the work expectations of employees, as well as the behaviour of employers. The state plays a pivotal role in determining the nature and rewards of work through its employment standards and labour legislation, various labour market programs, and the educational system. Consequently, our discussion of work, industry, and Canadian society frequently ranges well beyond the work site.

The sociology of work and industry is not a neatly defined area of scholarship. This is an advantage, in our view, because, more than is the case for most sociological specialties, we are forced to incorporate the insights of related disciplines. Classic sociological theory, political economy, organization and management studies, industrial relations, labour history, labour economics, macroeconomics, women's studies and feminist theory, stratification, race and ethnic relations, and social policy are some of the diverse literatures upon which we draw. Because of this scope, our coverage of the literature is far from exhaustive. Instead, we selectively examine theoretical discussions and research findings in an attempt to highlight key themes and debates in the sociology of work and industry.

A few brief definitions may be useful. First, *work* refers to activity that provides a socially valued product or service. In other words, through our work we transform raw materials as diverse as information, iron ore, and wheat into something that is needed or desired by all or part of society. The emphasis in this definition is on activity; action verbs such as *cook, hammer, clean, drive, type, teach, sculpt,* or *serve* are typically used to describe work. Obviously, such a definition is very broad, including both paid and unpaid work, activities ranging from the legal to the illegal, and from the highly esteemed to the undesirable and despised.

But, with a few exceptions, we have restricted our focus to paid work. This approach recognizes the centrality of the employment relationship in modern capitalist societies, both in terms of the production of wealth and the quality of life. Hence, we discuss housework and childcare—the major forms of unpaid work in our society—only briefly. Similarly, volunteer work receives only a few comments, as does work in the expanding "informal economy," where goods and services may be exchanged without cash transactions, and work in the traditional hunting and gathering economies that still exist among some of Canada's Aboriginal peoples. By spending less time on these non-paid forms of work we are not suggesting that they are unimportant, but that they warrant careful study in their own right. But we do recognize that, throughout the life course, paid work is variously combined with school work, child rearing, care of dependent relatives, household work, and community volunteering. Therefore, at various points in the following chapters, we examine the interaction of paid employment with some of these other types of work.

The relationship between employer and employee forms the cornerstone of the sociological study of work. We highlight throughout the book how the nature of employment relationships has been significantly altered during the past quarter century. By centring our analysis on workplace relations and the organizational structures in which these are embedded, we are able to address many of the core themes of classical sociological theory, including *inequality* and the distribution of scarce resources, power and how it takes shape in authority relations, and the shaky balance between the ever present potential for *conflict* and the need for *cooperation* and *consensus*. A basic problem in all societies involves the distribution of wealth, power, and prestige. Paid work is central to this dilemma of distribution. Put more concretely, who gets the well-paying, interesting, and high-status jobs? Who is most vulnerable to unemployment? How does cooperation among employees and between management and employees occur? To what extent are employees integrated into their work organizations? Are there inherent conflicts of interest between employers and employees?

Another key theoretical concern in research on the sociology of work and industry is the interplay between the actions of individuals and the constraints exerted on them by social structures. For us, the *agency–structure debate* translates into some very basic and practical questions: How limiting and restrictive are work organizations? How much can employee actions, individually or collectively, re-shape these structures? It is easy to detect the themes of *stability* and *change* just beneath the surface of these questions. In addressing these obvious tensions between individual action and social structures, it is helpful to distinguish between *macro* (social structure) and *micro* (individual) *levels of analysis.* At the macro level, we focus on the changing global economy, labour markets, industries and occupations, bureaucracies, unions and professional associations, and a variety of institutions that *structure,* or establish, regular and predictable patterns of work activity in society. We must be careful, however, not to place too much emphasis on social structures, because in doing so we risk losing sight of the individuals who are engaged in performing the work and, sometimes, re-shaping these structures. And, in the Internet age, more work relationships and business arrangements can be "virtual," which gives a whole new meaning to the concept of social structure.

As a counterbalance to this macro view, an individual-level micro perspective on work in Canadian society is needed. This we provide by asking questions such as: How do individuals experience their work situation? What are the ingredients of job satisfaction? Or, looking at the negative side, what causes job dissatisfaction, alienation, and work-related stress? What do people do if they don't like their work situation? What types of work values are held by Canadians? Overall, what factors contribute most to overall quality of working life? In short, a micro-level analysis investigates how people experience their jobs, how they adapt to working conditions that are often far from ideal, and how, in some instances, they attempt to improve these conditions within the confines of what is possible, given existing work institutions.

To elaborate this point, while an individual's work options may be limited by social structures, there is nothing inevitable about the daily work routines, employment relationships, and legislative frameworks found in capitalist societies. Thus, we will argue that work can be demeaning or rewarding, depending in large part on how it is organized and who has authority. Further, we will highlight some of the possibilities offered by more egalitarian and humanized forms of work. However, we stop well short of advocating a utopian vision of work. Instead, we approach the subject of work from a *reformist perspective,* arguing that even within the confines of postindustrial capitalism the quality of working life can be improved for many. We believe that through careful theoretical reasoning and rigorous empirical research, sociologists can further the development of enlightened public policy and help individuals create better workplaces for themselves.

It will quickly become clear that we are not presenting a strictly Marxist analysis of work. Similarly, we go beyond a simplistic Weberian approach that gives undue emphasis to employees' understandings of their work situations. We certainly do not advocate a functionalist or Durkheimian model of the work world that allows the need for social integration and stability to take on a larger-than-life presence. To varying degrees we have been influenced by these classical theoretical orientations, and less influenced by current trends in postmodernist thinking. More than anything, our starting point is a *conflict perspective* of the social world. Thus, a Marxist perspective has informed the book's emphases on inequality, power, conflict, and control. The Weberian approach sensitizes us to the centrality of individual actors, and their beliefs and behaviours, in sociological analysis. The Durkheimian tradition reminds us that social integration, consensus, and stability within work organizations are the motivation for much management theory and practice.

Throughout the text we also strive to provide a comprehensive overview of the sociology of work and industry from a Canadian perspective. At the same time, we have incorporated more comparative material than in previous editions, recognizing that our understanding of work in Canada is enriched by comparisons with other nations, especially in today's global economy. Thus, this fifth edition of *Work, Industry, and Canadian Society* retains the basic themes and guiding questions addressed in the four earlier editions. However, a considerable amount of new material has been added since we last revised the book in 2001, while a few sections have been eliminated. As much as possible, empirical trend data have been updated to the most recent available. While the revisions are far too numerous to list here, they include enriched or new discussions of immigration and ethnicity, the implications of an aging workforce, outsourcing and offshoring, knowledge workers and professions, vulnerable workers, business ethics and social responsibility, and human rights. We also have included the latest debates and research findings in the major topic areas covered in the book. A new feature at the end of each chapter is a list of questions to stimulate class discussion and help students in studying the material.

As with the previous editions, the book is written with university undergraduates in mind. But it could also be used profitably by graduate students seeking a general assessment of the literature from a Canadian perspective, and by non-students seeking a better understanding of contemporary work trends. We have provided an extensive guide to sources, including relevant websites, allowing readers to directly access the printed literature and electronic information they require to find answers to their own research questions.

Chapter 1 of this book begins with an overview of the industrialization process and the rise of capitalism in Europe. This short history lesson provides a backdrop against which we can compare the somewhat later Canadian industrialization experience and the even more recent process of industrialization in a number of Asian countries. This first chapter also introduces some aspects of the theoretical writings of Karl Marx, Adam Smith, Max Weber, and Émile Durkheim. These social philosophers developed their assessments of the problems and prospects of work in industrial capitalist societies from their observations of how Europe was transformed during the Industrial Revolution. But the global economy, the industrial structures of advanced capitalist societies, and the employment opportunities found within them have all changed dramatically over the past century. Thus, we also examine a number of more recent theoretical perspectives on work in postindustrial society. Both the classical and contemporary theories resurface throughout the text as we focus on more specific topics within the sociology of work and industry.

Chapter 2 offers a detailed overview of the major industrial, labour market, and labour force trends in Canada. We begin by examining changes in the composition of the labour force over the past few decades, and then focus on how the industrial and occupational distribution of employment has also changed. Labour force participation patterns and unemployment rates are then examined, along with the rise in nonstandard forms of work (temporary and part-time, for example).

Labour markets are central institutions in the contemporary work world, basically determining who gets the good jobs and who ends up in the less desirable ones. Chapter 3 examines how the Canadian labour market operates by comparing recent findings with the predictions of two competing theories, the human capital model and the labour market segmentation approach. This chapter also focuses directly on the problem of growing labour market polarization and increasing social inequality.

The growing labour force participation of women is clearly one of the most remarkable social changes of the 20th century. The question of how gender affects labour market outcomes is woven into the discussions in Chapters 2 and 3. But, given the importance of gender issues in the workplace, Chapter 4 provides an in-depth analysis of women's employment. This chapter outlines the transformation of women's economic roles, emphasizing the ways in which gender has become a source of entrenched divisions both within the labour market and within work organizations. The current push for gender equality is also highlighted, as are the challenges of work–life balance.

Chapter 5 shifts the focus to how work activities are typically organized and managed in large bureaucracies. Here we discuss organizational theories that attempt to explain the structure of large organizations, as well as the way change occurs within them. Equally important, we introduce the range of management theories that have been developed over the past decades. We examine these models of management with particular attention to whether they counter some of the problems of bureaucracy and attempt to humanize working conditions, and whether they offer workers more autonomy or managers more control.

Chapter 6 is, in some ways, a debate with the mainstream management approaches. Our starting point is the labour process perspective, a much more critical assessment of employer–employee relationships in capitalist society. We examine different methods managers have used to control and gain compliance from workers, and ask whether new computer-based technologies, as well as the new management approaches, really provide workers with more autonomy and a chance to enhance their skills. Then, looking for alternatives, we discuss current approaches to addressing health and safety concerns in the workplace since, to some extent, they require co-management by workers and management. We conclude by discussing different forms of industrial democracy as well as examples of worker ownership, arguing that these approaches to organizing and managing work do have something to offer.

Unions receive some comment in earlier chapters, but Chapter 7 is devoted to this subject. We discuss various theories about the origins and functions of unions and document the development of the organized labour movement in Canada. We also describe the legal framework that has emerged for regulating union–management relations. This chapter concludes by examining worker militancy, particularly strike patterns, and by asking whether unions represent a revolutionary or reformist collective response by workers to the greater power of employers.

We conclude, in Chapter 8, by examining the individual experience of work. After commenting on society-wide work values and how they vary across time and culture, we then discuss work orientations, asking what individual employees seek to obtain from their work. Are their work orientations (or preferences) perhaps shaped by the work opportunities available to them? Equally important, are we seeing the emergence of new work orientations in response to growing employment insecurity and income inequality? From here we go on to discuss a wide array of research findings on job satisfaction, alienation, and workplace stress. A major theme developed in this chapter is that negative work experiences can have serious and lasting effects on individuals.

Finally, in a short Conclusion, we review the trends described and the main conclusions drawn in the previous chapters. We also raise some important questions about the future of work in Canada. In the early 21st century, we can see more paradoxes and contradictory trends in labour markets and workplaces. Forms of work are becoming more varied. Employers continue to downsize amid concerns about a scarcity of "talent." Polarization in the quality of working life is increasing. New human resource management strategies, job redesign, industrial democracy, healthy workplace initiatives, and information technology hold out possibilities for creating better workplaces. But these trends could just as easily have negative effects on the quantity and quality of work. We do, however, have some control and room to manoeuvre. In our opinion, what Canada needs is more public debate among employers, unions, professional associations, governments, and other labour market stakeholders to determine how best to balance the interests of workers, who want both a decent standard of living and an improved quality of working life, with the economic growth, profits, and labour market flexibility sought by employers and government.

CAPITALISM, INDUSTRIALIZATION, AND POSTINDUSTRIAL SOCIETY

INTRODUCTION

In the early 1870s, John Henry Lincoln, known as Harry, travelled from his home and birthplace in Huntingdonshire in England to New Haven, Sussex, to attend the wedding of his sister to Fred Waters. There he met Fred's sister, Anne, and that was the beginning of it all, for they soon married. Harry was 23 and a master boot maker by trade. Anne was from a well-to-do family, and her desire to marry a craftsman caused something of a stir. After marrying, Harry set up his own boot-making and repair shop, while Anne devoted herself to raising their family. Together they would have 11 children.

In the early 1900s the family left to homestead in Canada. Packing their possessions into 10 trunks, they travelled two weeks by sea to Montreal, then 1,700 miles by train to Manor, Saskatchewan. Once settled, one of the younger sons, Arthur, moved farther west to Alberta, where he homesteaded, married, and raised three children. A resourceful man, he survived two World Wars and the Great Depression, working in farming, sales, and then his own business. In the postwar prosperity of the 1950s, he saw his daughter move to the big city of Edmonton and juggle a career, marriage, and motherhood at a time when few women did. Her three "baby-boom" children would all go on to university, the youngest eventually teaching there and collaborating on the book that you are now reading.[1]

Reflecting on this story, it is hard not be struck by the huge change that has occurred in just a few generations. Eleven-children families are not the norm. Farming has been replaced by factory work, and then again by white-collar, retail, and knowledge jobs. Canada still relies on immigrants to fuel its economy, but not primarily those from Britain as was once the case. And while people still migrate for work, work itself is increasingly mobile, with production being spread across the globe. This book examines these changes and the types of paid work currently done by Canadians. Today, most work for wages or a salary in bureaucratic organizations. In two out of three two-adult households, both partners are employed outside the home. In a labour force of just under 16 million individuals, almost three in four workers are in the service sector. Compared to the past, fewer workers are full-time employees. More than one in four are working either in temporary or part-time jobs; almost one in six are self-employed. For every 13 employed Canadians, one is unemployed.

These present labour market realities become more interesting when we realize that only a generation ago some of these trends had just begun to emerge. And going back a century, as we have seen, the differences in work patterns are huge. What social and economic forces led to the shifts from agriculture to manufacturing, then to services? Why has education become so important? Why have two-earner households become the norm? Why are women, Aboriginal people, and visible minorities underrepresented in better-paying jobs, and what are the social consequences? To answer these and many related questions, we examine the complex process of industrialization by looking back in history and at other societies. This historical and comparative approach will help us to understand present patterns and trends, assess their significance for individuals and society, and respond to the future challenges they pose.

Industrialization refers to the technical aspects of the accumulation and processing of a society's resources. *Capitalism* is a term used to describe key aspects of the economic and social organization of the productive enterprise. An industrial society is one in which inanimate sources of energy such as coal or electricity fuel a production system that uses technology to process raw materials. But labelling a society "industrial" tells us little about the relationships among the individuals involved in the productive process. A capitalist system of production is one in which a relatively small number of individuals own and control the means for creating goods and services, while the majority have no direct ownership stake in the economy and are paid a wage to work for those who do.

Studying the rise of capitalism requires us to make sense of large, complex processes spanning generations and leaving no aspect of daily life unaffected. The theoretical writings of Adam Smith, Karl Marx, Émile Durkheim, and Max Weber help us to explain, not just describe, the causes and consequences of capitalist development. But as we will see, the sociological concepts of *power, control, inequality,* and *conflict* that these early social scientists used to analyze changes in their times continue to be central to today's debates about the nature of 21st-century postindustrial society. Therefore, in this chapter and later in the book, we address questions such as: Is the rise of a knowledge-based, high-tech economy creating more flexible ways of organizing economic activity that benefits both workers and employers? How are new information technologies transforming the content and organization of work? Are there more losers than winners as a result of the economic restructuring of the past two decades? Are well-educated knowledge workers becoming the new elite in society? Is economic globalization increasing the power of transnational corporations at the expense of national governments? And with the rapid economic development of East Asia, the collapse of the Soviet Union, and the uneven transitions of Eastern European countries toward market-based economies, do we need to rethink long-accepted explanations of capitalist development?

THE ORIGINS OF INDUSTRIAL CAPITALISM

Capitalism and industrialization dramatically re-shaped the structure of European society, economically, socially, and spatially. These changes occurred over centuries, with a different pace and pattern in each country. While the details of these changes differed internationally and regionally within countries, the result was profound change in how, where, for whom, and under what conditions individuals worked.

The emergence of capitalism in Europe consisted of two basic periods: *mercantile* or *commercial capitalism,* which began in the 1500s, and *industrial capitalism,* which evolved somewhat later.[2] In the mercantile period, merchants and royalty in Spain, Holland, England, and France accumulated huge fortunes by trading internationally in a variety of goods, including spices, precious stones and metals, sugar, cotton, and slaves. An elaborate trading network evolved, linking Africa, Asia, and the American colonies with Europe. This global trade and the pillage of cultures (slaves from Africa, for example, and vast amounts of gold and silver from Central and South America) provided wealth that would subsequently fuel the growth of industrial capitalism in Europe (Beaud 1983).

These early signs of capitalist commercial activity emerged out of a *feudal society;* the *Industrial Revolution* had not yet begun. Most people still lived in the countryside. The class structure of these agrarian societies consisted of a relatively small aristocracy and merchant class, most of whom lived in the cities, a rural landowning class, and a large rural peasantry. Work typically involved peasants farming small plots of land they did not own. Landowners received rent, usually in the form of agricultural produce, little of which was sold for cash. Generations of peasant families lived and died on the same feudal estates.

Thus, feudal Europe was predominantly a *pre-market* economy in which the producer was also the consumer. It was also a *pre-capitalist* economy because wage labour was rare and a business class had not yet become dominant. Feudal lords accepted rent and expected services in the form of manual labour required for the upkeep of the estate. In return, they allowed historical tenancy relationships to continue and provided some protection, if necessary, for their tenants. Thus, feudalism was built upon a system of mutual rights and obligations, reinforced by tradition. One's social position was inherited. But this relatively stable society also stifled economic progress.[3]

Did the decay of feudalism lead to the rise of capitalism, or was it the other way around? Scholars are divided on this question. Some argue that factors internal to feudal society, such as growing rural populations, deterioration of land, and landlords demanding more rent, forced people off the land and into the cities where they could form an urban working class. Others counter that as mercantile capitalism developed in urban areas, and as the market economy slowly began to make an impact on rural life, cities began to attract landless serfs. This debate is difficult to resolve, since the two processes influenced each other (Hilton 1976; Aston and Philpin 1985). What is undisputed is that capitalism brought with it an entirely new social order.

Early Capitalism

Industrial capitalism began to emerge in the early 1700s. The production of goods by artisans, or by the home-based *putting out system* in which merchants distributed work to peasant households, led to larger workshops ("manufactories") that made metal, cloth, glass, and other finished goods (Beaud 1983). By the late 1700s, various inventions were revolutionizing production techniques. James Hargreaves's spinning jenny transformed work in textile industries. Growth in trade and transportation, the construction of railways, and military demand for improved weapons encouraged new techniques for processing iron and other metals. Inventors also were devising ways of harnessing water and steam, as exemplified in James Watt's steam engine.

These early inventions facilitated a new form of work organization: the *industrial mill*. A technical breakthrough that involved harnessing many machines to a single inanimate energy source, the mill also had immense social implications, consolidating many workers under one roof and the control of managers (Beaud 1983: 66–67; Burawoy 1984). A growing class of impoverished urban wage-labourers endured horrific working conditions in early industrial mills. Some workers resisted this trend, particularly artisans who previously had controlled their own labour. Episodes of destroying textile machinery occurred between 1811 and 1816 in a number of British communities. The unemployed craftsmen involved (called *Luddites*) were not unthinking opponents of technological innovations, but skilled workers frustrated by changes that were making their skills obsolete. These uprisings were quashed by the state; some participants were jailed or deported, others were hanged (Beaud 1983: 65; Grint 1991: 55).

The emergence of industrial capitalism also changed the *gender-based division of labour*. While women, men, and children had done different types of work in feudal times, much of it at home, they were often engaged in parallel work activities (Middleton 1988). The putting out system, particularly in the textile industries, brought many women into the paid labour force along with men, since they could work out of their home and still carry out domestic work and childcare. Early textile factories also employed men, women, and children (Berg 1988). But as manufacturing developed and expanded, it became the preserve of men. While a gendered division of labour had long existed, it became more pronounced with the rise of industrial capitalism.[4]

The Great Transformation

In only a matter of decades, factory production dominated capitalist societies. The urban landscape also changed as manufacturing cities grew to accommodate the new wage-labour force. Mechanization and the movement to factory-based production proceeded even faster in the 1800s than it had in the previous century. Manufacturing surpassed agriculture in its annual output. Industrial production in Britain, for example, increased by 300 percent between 1820 and 1860. The portion of the labour force employed in agriculture in Britain, France, Germany, the United States, and other industrializing countries declined, while employment in manufacturing and services rose rapidly. By the end of the 19th century, industrial capitalism was clearly the dominant system of production in Western nations.

According to Karl Polanyi, the *great transformation* that swept Europe with the growth and integration of capital, commodity, and labour markets—the foundation of capitalism—left no aspect of social life untouched.[5] The struggle for democratic forms of government, the emergence of the modern nation-state, and the rapid growth of cities are all directly linked to these economic changes.

Along with new technologies, the replacement of human and animal sources of energy with inanimate sources, and the emergence of an integrated market system for finance, commodities, and labour, this era also saw dramatic changes in how work was organized. In time, the relatively stable landlord–serf relationships of feudalism were replaced by wage-labour relationships between capitalists and labourers. Employers paid for a set amount of work, but also determined exactly how, and under what conditions, work would be done. Previously independent artisans lost out to the factory system. In the end, the result was a higher standard of living for most residents of the industrialized countries. But the interrelated processes of change that created a market economy also led to new problems of *controlling, coordinating, and managing* work—central themes in this book.

CANADA'S INDUSTRIALIZATION

The process of industrialization in Canada lagged behind that in Britain and the United States, and can be traced back to the mid-1800s.[6] As a British colony, Canada's role had been to provide raw materials, rather than to produce finished goods that would compete on world markets with those of the mother country. Canadian economic elites focused on traditional activities such as exporting *staple products* like timber and fur to sell on world markets, and developing transportation networks (particularly railways) that could link the resource-producing regions of the country with the port cities involved in export trade.

Work in Pre-industrial Canada

The first half of the 19th century, then, was a pre-industrial economic era in Canada. Canada still had a pre-market economy, since most production and consumption took place in households. In fact, given the peculiarities of a colonial economy dependent on Britain, even land was not really a marketable commodity. By the mid-1830s, less than one-tenth of the vast tracts of land that had been given by the French, and later British, monarchy to favoured individuals and companies had been developed for agriculture.

At the same time, immigration from Europe was increasing. Shortages of land for small farmers, potato famines in Ireland, and dreadful working conditions in many British factories fuelled immigration to the New World. Large numbers of immigrants landed in Canada, only to find shortages of urban factory jobs and little available agricultural land. Thus, the majority of these immigrants sought employment in the United States, where factory jobs and land were more plentiful (Teeple 1972).

Some of the immigrants who stayed in Canada were employed in building the Welland and Rideau canals—the first of many transportation megaprojects—in the first half of the 19th century. The influx of unskilled workers created a great demand for such seasonal jobs, which often involved 14 to 16 hours a day of very hard and poorly paid work. Consequently, poverty was widespread. In the winter of 1844, the *St. Catharines Journal* reported that

> the greatest distress imaginable has been, and still is, existing throughout the entire line of the Welland Canal, in consequence of the vast accumulation of unemployed labourers. There are, at this moment, many hundreds of men, women, and children, apparently in the last stages of starvation, and instead . . . of any relief for them . . . in the spring . . . more than one half of those who are now employed must be discharged.[7]

The Industrial Era

By the 1840s, Canada's economy was still largely agrarian, even though the two key ingredients for industrialization—an available labour force and a transportation infrastructure—were in place. Prior to Confederation in 1867, some of Canada's first factories were set up not in Ontario or Quebec, as we might expect, but in Nova Scotia. Shipbuilding, glass, and clothing enterprises were operating profitably in this region before the Maritime provinces entered Confederation (Veltmeyer 1983: 103). After 1867, manufacturing became centralized in Ontario and Quebec, resulting in the *deindustrialization* of the Maritimes. A larger population base, easy access to U.S. markets, and railway links to both eastern and western Canada ensured that the regions around Montreal and Toronto would remain the industrial heartland of the country.

At the time of Confederation, half of the Canadian labour force was in agriculture. This changed rapidly with the advance of industrialization. By 1900, Canada ranked seventh in production output among the manufacturing countries of the world (Rinehart 2006; Laxer 1989). The large factories that had begun to appear decades earlier in the United States were now springing up in Toronto, Hamilton, Montreal, and other central Canadian cities. American firms built many of these factories to avoid Canadian tariffs on goods imported from the United States. This began a pattern of direct U.S. foreign investment in Canada that continues today.

These economic changes brought rapid urban growth and accompanying social problems. Worker exploitation was widespread, as labour laws and unions were still largely absent. Low pay, long hours, and unsafe and unhealthy conditions were typical. Workers lived in crowded and unsanitary housing. Health care and social services were largely nonexistent. In short, despite economic development in the decades following Confederation, poverty remained the norm for much of the working class in major manufacturing centres (Copp 1974; Piva 1979).

The Decline of Craftwork

Traditionally, *skilled craftworkers* had the advantages of being able to determine their own working conditions, hire their own apprentices, and frequently set their pay. Some worked individually, while others arranged themselves into small groups in the manner of European craft guilds. But this craft control declined as Canada moved into the industrial era. Factory owners were conscious of the increased productivity in American factories obtained through new technology and "modern" systems of management. Dividing craft jobs into many simple tasks allowed work to be performed by less skilled and lower-paid employees. Mechanization further cut costs while increasing productivity.

As these systems entered Canada, the job autonomy of craftworkers was reduced, resulting in considerable labour unrest. Between 1901 and 1914, for example, more than 400 strikes and lockouts occurred in the 10 most industrialized cities of southern Ontario (Heron 1980; Heron and Palmer 1977). While these conflicts may look like working-class revolt, they are more accurately seen as a somewhat privileged group of workers resisting attempts to reduce their occupational power. While there were large numbers of skilled workers who experienced the "crisis of the craftsmen" (Heron 1980; also see Rinehart 2006), there were larger numbers of unskilled manual labourers whose only alternative to arduous factory work was seasonal labour in the resource or transportation industries, or unemployment.

There were, in fact, thousands of workers employed in the resource extraction industries throughout Canada, and many others worked on constructing the railways. In *The Bunkhouse Man,* Edmund Bradwin estimates that up to 200,000 men living in some 3,000 work camps were employed in railway construction, mining, and the lumber industry during the early 20th century. These workers were from English Canada and Quebec, as well as from Europe and China. Employers considered immigrants to be good candidates for such manual work, as they were unlikely to oppose their bosses. This hiring strategy often did ward off collective action, although immigrants sometimes were the most radical members of the working class.[8]

The creation of a transcontinental railway led to a high demand for coal. Mines were opened on Vancouver Island and in the Alberta Rockies, with immigrants quickly taking the new jobs. Mine owners attempted to extract a lot of work for little pay, knowing they could rely on the military to control unruly workers. It has been estimated that in the early 1900s, every 1 million tons of coal produced in Alberta took the lives of 10 miners, while in British Columbia the rate was 23 dead for the same amount of coal.[9] These dangerous conditions led to strikes, union organization, and even political action. In 1909, Donald McNab, a miner and socialist, was elected to represent Lethbridge in the Alberta legislature. The same year, the Revolutionary Socialist Party of Canada elected several members to the British Columbia legislature (Marchak 1981: 106; Seager 1985). But, although the labour movement did take root in resource industries, it never had the revolutionary spark that some of its radical leaders envisioned.

RETHINKING INDUSTRIALIZATION

While countries differ in the timing, pace, and form of industrialization, there still appear to be some similar underlying dynamics and processes. Industrialized countries tend to be highly urbanized; production typically takes place on a big scale using complex technologies; workplaces tend to be organized bureaucratically; and white-collar workers make up most of the workforce. Citizens are reasonably well educated, and generally an individual's level of education and training is related to her or his occupation. Such similarities led American social scientists in the 1950s to claim

> The world is entering a new age—the age of total industrialism. Some countries are far along the road, many more are just beginning the journey. But everywhere, at a faster pace or a slower pace, the peoples of the world are on the march towards industrialism. (Kerr et al. 1973: 29)

This is the *logic of industrialism* thesis, a deterministic and linear argument about the immensity and inevitability of industrial technology. It contends that industrialism is such a powerful force that any country, whatever its original characteristics, will eventually come to resemble other industrialized countries. In current debates about globalization, the same argument has been restated as a *thesis of economic, political, and cultural convergence.* Yet, as we will see in later chapters, comparisons of work patterns in various industrialized countries do not support this prediction. For example, there are pronounced cross-national variations in unemployment rates, innovative forms of work organization, unionization and industrial relations systems, and education and training. And as noted below, globalization also is not a standardized process, but rather is playing out quite differently at national, regional, and community levels.

Our historical overview also shows that Canada's industrialization occurred later and was shaped by its colonial status; that immigration was a major factor in creating a workforce; and that resource industries played a central role. Other countries such as Sweden and Japan also industrialized later, but their experiences differed from Canada's. In recent decades, Central and South American nations such as Mexico and Brazil have undergone much more rapid industrialization. Again, the process differed. In particular, inequality has become even more pronounced in these countries (Chirot 1986: 251–55). By reflecting on the experiences of the recently industrialized nations of East Asia, and on the stirrings of capitalism in China, India, and the former Soviet Union, we will come to appreciate even greater diversity in this process. Other than at the broadest level of analysis, there does not appear to be an inherent "logic of industrialism."

This becomes clear when we compare the 19 countries in North and South America, Asia, and Europe presented in Table 1.1. Keep in mind that comparing nations at different phases and levels of development presents methodological challenges, mainly because perfectly comparable data are

TABLE 1.1 Social and Economic Indicators for Selected Countries, 2003–2004

	Population 2003 (millions)	Urban Population as % of Total (2004)	PPP Gross National Income[a] (per capita) 2003	Average Annual GDP % Growth 1990–2003[b]	Net % of Relevant Age Group Enrolled in Secondary Education (2002/03)	Labour Force: % Female (2003)	% Share of Income or Consumption in Lowest 20% of Income Earners	% Share of Income or Consumption in Highest 20% of Income Earners
Canada	32	81	30,040	3.3	98	46.2	7.0	40.4
United States	291	80	37,750	3.3	85	46.6	5.4	48.5
Brazil	177	84	7,510	2.6	72	35.5	2.4	63.2
Mexico	102	76	8,980	3.0	60	34.4	3.1	59.1
Chile	16	87	9,810	5.6	79	35.1	3.3	62.2
India	1,064	28	2,880[c]	5.9	—	32.6	8.9	43.3
China	1,288	40	4,980[d]	9.6	—	45.0	4.7	50.0
Indonesia	215	47	3,210	3.5	—	41.2	8.4	43.3
Thailand	62	32	7,450	3.7	—	47.0	6.1	50.0
Korea (Republic)	23	81	18,000	5.5	87	40.7	7.9	37.5
Singapore	44	100	24,180	6.3	—	38.6	5.0	49.0
Japan	128	66	28,450	1.2	100	41.8	10.6	35.7
Russian Federation	143	73	8,950	−1.8	—	49.1	8.2	39.3

Ukraine	48	67	5,430	−5.3	48.8	8.8	37.8
United Kingdom	59	89	27,690	2.7	44.1	6.1	44.0
Germany	83	88	27,610	1.5	42.6	8.5	36.9
Sweden	9	83	26,710	2.3	48.0	9.1	36.6
France	60	76	27,460	1.9	45.5	7.2	40.2
Switzerland	7	38	32,220	1.2	40.7	6.9	40.3

[a] PPP GNI is gross national income converted to international dollars using purchasing power parity rates. An international dollar has the same purchasing power over GNI as a U.S. dollar has in the United States.

[b] Gross Domestic Product per capita is based on GDP measured in constant prices. Growth in GDP is considered a broad measure of the growth of an economy. GDP in constant prices can be estimated by measuring the total quantity of goods and services produced in a period, valuing them at an agreed set of base year prices, and subtracting the cost of intermediate inputs, also in constant prices.

[c] The estimate is based on regression; others are extrapolated from the latest International Comparison Programme benchmark estimates.

[d] Estimate based on bilateral comparison between China and the United States.

Source: Compiled from The World Bank Group, *World Development Indicators 2005*, available online at http://www.worldbank.org/data/wdi2005/wditext/Cover.htm; World Bank, *World Development Report 2005: A Better Investment Climate for Everyone* (Oxford: Oxford University Press, 2004), available online at http://econbeta.worldbank.org/WBSITE/EXTERNAL/EXTDEC/EXTRESEARCH/EXTWDRS/EXTWDR2005/0,,contentMDK:20259914~menuPK:477673~pagePK:64167689~piPK:64167673~theSitePK:477665,00.html; and United Nations Statistics Division, *Social Indicators: Indicators on Human Settlements*, available online at http://unstats.un.org/unsd/demographic/products/socind/hum-sets.htm.

lacking. However, the World Bank has constructed basic comparative measures that perhaps are as precise as we can get, so we use these in Table 1.1. Canada provides a useful benchmark for interpreting the relative socioeconomic conditions in other countries. After noting which countries are included in Table 1.1, scan each column to get a sense of the huge differences in population, urbanization, per capita income, annual economic growth, women as a proportion of the total labour force, and inequality (share of all income or consumption accounted for by the top and bottom *quintiles,* or equal groupings of 20%). Keep in mind that in many developing nations, significant economic activity still occurs in the informal sectors of the economy, which makes it difficult to capture in the official government statistics presented in Table 1.1 (Martin 2000). Still, these basic social and economic development indicators raise a host of questions about the underlying historical, cultural, and political contexts in which countries industrialize.

The East Asian Tigers

The *Four Tigers* of East Asia—Singapore, Hong Kong, Taiwan, and South Korea—have attracted much attention from scholars, businesses, and politicians. In a few decades, these economies have become highly industrialized, either surpassing or fast approaching per capita incomes in the major Western industrial nations. While it took Britain 58 years (1780 to 1838) and the United States 47 years (1839 to 1886) for per capita gross domestic product (the total output of the economy) to double, it took South Korea only 11 years (1966 to 1977) (*The Economist* 16 October 1993: 80). In the same region, another set of *newly industrializing countries* (NICs)—Thailand, Indonesia, and Malaysia—are following just behind the Four Tigers. While the Four Tigers and the NICs exhibited impressive economic growth rates through most of the 1990s, a major financial crisis in 1997–98 shook the foundations of many of these economies. The 2001 dot-com crash and global downturn slowed recovery, but recent growth in the region has been strong. Even with the devastation of the 2004 tsunami that affected parts of Indonesia and Thailand, analysts predict renewed expansion, with growth the highest it has been since the 1997 financial crisis (World Bank 2005b).

The World Bank (1993: 277) considers rapid economic growth and greater reductions in income inequality than in other developing nations to be the signal features of the "East Asian miracle." But other factors are also important, such as rising output and productivity in agriculture at the same time that a manufacturing export sector was developing, steep declines in birthrates so that population growth was constrained, heavy investment in human resources through the expansion and improvement of educational systems, and a bureaucratic state run by a career civil service.

But this overstates the similarities among these nations; in fact, there are striking differences. For example, capitalism has taken distinctive forms in the Four Tigers.[10] A few giant diversified corporations (called *chaebol*) that have close ties to the state dominate the South Korean economy. Taiwan's dynamic export sector is sustained by a network of highly flexible small and medium-sized manufacturing firms with extensive *subcontracting systems* that often extend into the informal sector of the economy. In Hong Kong (and also Taiwan) the main organizing unit of business is the Chinese *family enterprise.* These firms give utmost priority to the long-term prosperity and reputation of the family—a very different concept than the rugged individualism of North American capitalism. Singapore is set apart by the prominent role of the state in setting the direction of industrialization, which has involved much greater reliance on direct investment by foreign multinational corporations. These variations in business systems account for diverse industrialization paths and approaches to work organization within firms.

Signs of Capitalism in China, India, and the Former Soviet Union

We've all seen media reports of rich "peasants" in China buying German luxury cars, India's "Silicon Valley," and the garishly lavish lifestyles of Russia's new entrepreneurs. How is economic change transforming these countries? Social scientists have asked this question, attempting to understand the complex change occurring in "transitional economies." Recent research looks beyond "markets" in a narrow economic sense, attempting to present a more comprehensive explanation of what influences the economic behaviour of individuals in these societies. Factors such as a society's culture, norms, and customs, as well as existing social networks, state policies, and work organization, all influence how markets take shape and how individuals pursue new opportunities within them (Basu 2004; Jha 2002; Nee and Matthews 1996: 403).

China is a case in point. Since introducing market reforms in the late 1970s, the Chinese economy has grown at an astonishing rate of nearly 10 percent per year (World Bank 2005a). As Table 1.1 shows, China's growth from 1990 to 2003 far outstripped that of Canada, the United States, and many other industrialized countries. As the world's most populous nation, with nearly 1.3 billion inhabitants, China is seen by corporations as a huge market to conquer. Its workers are also being integrated into the global economy, especially in its "special economic zones," which attract foreign manufacturers such as Wal-Mart and The Gap. "Made in China" now appears on a growing number of consumer items that Canadians buy. Indeed, China is a leading recipient of foreign direct investment by multinational corporations, with more than half of its trade now controlled by foreign firms (Fishman 2005: 15).[11] It is also making significant investments in the West, in particular in the oil industry including Alberta's oil sands.

Is communist China being overtaken by capitalist free enterprise? The short answer is no, but there are remarkable transformations occurring, especially in coastal industrial regions in the south. But the Chinese version of an entrepreneurial spirit is deeply rooted, and can be seen in the rich and powerful dynasties established by mainland Chinese families who migrated throughout Asia around a century ago (*The Economist* 9 March 1996: 10). Chinese-style capitalism, influenced of course by the Western version, built Singapore, Taiwan, and the former British colony of Hong Kong that has now reunited with mainland China.

Unlike the former Soviet Union, where privatization is now widespread, state ownership of property remains dominant in mainland China. Furthermore, the bureaucratic systems of state-run work units and the Communist Party remain important influences on income levels and access to housing for many workers (Adamchak, Chen, and Li 1999). But the rules have changed: farming has shifted from collective farms to households, giving farmers full return on anything they produce over a set quota. And while state-owned enterprises are still prominent, at the local level hybrid corporations straddling the state and private sector are permitted. Though the Communist Party and state institutions continue to strongly influence individuals' occupational rewards, market reforms have coincided with rising social inequality, sparking debate over the specific causes of such change (Wu and Xie 2002). Another much-debated question is whether economic reform may lead to political reform and the eventual emergence of a democratic state.

India is another country that has received enormous attention lately. Like China, it is a huge country, with a population of over 1 billion. With its higher birthrate it will actually surpass China in size by 2035 (Word Bank 2005a; *The Economist* 3 March 2005). Despite high levels of poverty and illiteracy, India has invested heavily in computer engineering and science. Its growing technological savvy has attracted great interest, fuelling rapid economic growth of nearly 6 percent per annum (see Table 1.1). White-collar outsourcing has been a key source of growth in India. In *The World Is Flat* (2005), Thomas Friedman cites a litany of jobs now being done there for North American and European firms—from call-centre work to software engineering, income tax preparation, and the reading of CAT scans. India's attraction is not simply cheaper labour costs and time differences that facilitate 24/7 production, but a deep pool of highly skilled technology workers and fluency in English as a legacy of British colonialism. Recent forecasts suggest that IT workers will grow from 160,000 in 2003 to 2.7 million by 2012, and back office processing (BPO), now employing a quarter of a million workers, will quadruple by 2012 (*The Economist* 3 March 2005).

As in China, India's economic growth has come through conscious state planning, especially in the mid-1980s under the government of Rajiv Gandhi, who sought to make India highly competitive in the high tech sector. But because India is a democratic society, with an increasingly fragmentary political system, there has been difficulty furthering reform, especially at the local level (Jha 2002; Basu 2004; Parthasarathy 2004). Thus, companies such as Dell, Microsoft, and IBM employ an elite group of knowledge workers in specially created "campuses" or STPs (software technology parks). For example, at InfoSys campus, the "jewel" of India's IT sector, employees enjoy a resort-size swimming pool, putting greens, restaurants, and a health club (Friedman 2005: 4). Outside the gates of Infosys, however, other parts of the economy remain untouched. Today 35 percent of Indians live on less than $1 a day, and rates of adult literacy and life expectancy remain low.

Compared to China and India, the former Soviet Union has taken a much different course. As Robert Brym observes, a "strange hybrid of organized crime, communism and capitalism grew with enormous speed in the post-Soviet era" (1996: 396). Private enterprise, or market liberalization, was introduced in the mid-1980s, but competitive capitalism as we know it has failed to emerge. The reason, in Brym's view, is that the old Communist Party elite, along with organized criminal gangs that had prospered in the corruption of the Soviet economy, were well placed to take advantage of any new opportunities. The result is a predatory variety of capitalism that enriches a small elite while breeding corruption and making life worse for ordinary Russians. Thus, despite the emergence of a private sector and state policies aimed at economic reform, and contrary to *market transition theory*—which has been developed to explain recent social and economic change in China and the former Soviet Union—Russia has not really experienced increased entrepreneurial activity, greater labour market returns to education, or reductions in inequality (Gerber and Hout 1998).

A broadly similar pattern is typical of other Eastern bloc nations. Former Communist Party elites, who still control state resources and have good business networks, continue to benefit in the post-Soviet era, while living standards for the majority have declined (Rónas-Tas 1994; Szelényi and Kostello 1996). Unlike earlier transitions to capitalism in western Europe and North America, what is occurring in the former Soviet bloc does not involve the rise of a new economic elite or the creation of a working class—both of which existed in different forms under communism. Thus, Polanyi's "great transformation" from feudalism to a market economy in 19th-century Europe represented a unique experience of social and economic change. Subsequent transformations to market-based capitalism have taken some unexpected twists and turns.

THEORETICAL PERSPECTIVES

Now that we have some historical and comparative background on industrialization and capitalism, we can examine major explanations of the causes and consequences of these changes. The following sections present a range of classical and contemporary theories of social and economic change. After acquainting ourselves with these theories and their analytic concepts, we can apply their insights throughout the book when considering specific work issues.

Karl Marx on Worker Exploitation and Class Conflict

Karl Marx spent a lifetime critically examining the phenomenon of industrial capitalism. His assessment of this new type of society was presented within a very broad theoretical framework. He called the overall system of economic activity within a society a *mode of production,* and he identified its major components as the *means of production* (the technology, capital investments, and raw materials) and the *social relations of production* (the relationships between the major social groups or classes involved in production).

Marx focused on the manner in which the ruling class controlled and exploited the working class. His close colleague, Friedrich Engels, documented this exploitation in his 1845 book, *The Condition of the Working Class in England.* Engels (1971: 63) described one of London's many slum districts, noting that other industrial cities were much the same:

> St. Giles is in the midst of the most populous part of the town, surrounded by broad, splendid avenues in which the gay world of London idles about. . . . The houses are occupied from cellar to garret, filthy within and without, and their appearance is such that no human being could possibly wish to live in them. But this is nothing in comparison with the dwellings in the narrow courts and alleys between the streets, entered by covered passages between the houses, in which the filth and tottering ruin surpass all description. . . . Heaps of garbage and ashes lie in all directions, and the foul liquids emptied before the doors gather in stinking pools. Here live the poorest of the poor, the worst paid workers with thieves and the victims of prostitution indiscriminately huddled together.

It was from such firsthand observations of industrializing Europe that Marx developed his critique of capitalism.

Class conflict was central to Marx's theory of social change. He argued that previous modes of production had collapsed and been replaced because of conflicts among class groups within them. Feudalism was supplanted by capitalism as a result of the growing power of the merchant class, the decline of the traditional alliance of landowners and aristocracy, and the deteriorating relationship between landowners and peasants. Marx identified two major classes in capitalism: the capitalist class, or *bourgeoisie,* which owned the means of production, and the working class, or *proletariat,* which exchanged its labour for wages. A third class—the *petite bourgeoisie*—comprising independent producers and small business owners, would eventually disappear, according to Marx, as it was incorporated into one of the two major classes. Marx argued that capitalism would eventually be replaced by a socialist mode of production.

The catalyst would be revolutionary class conflict, in which the oppressed working class would destroy the institutions of capitalism and replace them with a *socialist society* based on collective ownership of the means of production.

Marx sparked ongoing debate regarding the nature of work and of class conflict in capitalist societies (Zeitlin 1968; Coser 1971). Ever since, social, political, and economic analysts have attempted to reinterpret his predictions of a future worker-run socialist society. No capitalist society has experienced the revolutionary upheavals Marx foresaw. And the collapse of the Soviet communist system in Eastern Europe ended speculation that communism would evolve into true socialism. In fact, Marx probably would have been an outspoken critic of the Soviet communist system, given its extreme inequalities in power distribution and harsh treatment of workers. He would also have condemned the inequalities in today's capitalist Russia.

Marx's critique of capitalism has also shaped research in the sociology of work and industry, as analysts attempt to refine or refute his ideas. First, Marx emphasized how the capitalist profit motive is usually in conflict with workers' desires for better wages, working conditions, and standards of living. Second, Marx argued that the worker–owner relationship led to workers losing control over how they did their work and, hence, to the dehumanization of work. Third, Marx predicted that the working class would eventually organize to more actively oppose the ruling capitalist class. In short, Marx wrote about inequality, power, control, and conflict. His enduring legacy for sociology was this more general *conflict perspective.* He recognized that the relations of production in industrial capitalist society typically are exploitative, with owners (and their representatives) having more power, status, and wealth than those who are hired to do the work. Almost all of the debates about better ways of organizing workplaces and managing employees, the need for unions and labour legislation, and the future of work in our society stem from this basic inequality.

Adam Smith: Competition, Not Conflict

Adam Smith wrote *The Wealth of Nations* in 1776, during the early Industrial Revolution in England, extolling the wealth-producing benefits of capitalism. Thus, he is often portrayed as the economic theorist whose ideas outlasted those of Marx who, writing some time later, predicted the eventual downfall of capitalism. Even today, Adam Smith's ideas are frequently used to call for less government intervention in the economy, the argument being that "the unseen hand of the market" is best left alone.

It is important to note, however, that Adam Smith did not condone the exploitation of workers. He recognized that working conditions in the industrializing British economy were far from satisfactory, and actually argued that higher wages would increase the productivity of workers and the economy as a whole (Weiss 1976; Saul 1995: 150). But Smith was very clear about what he saw as a key underlying principle of capitalism. *Competition among individuals and enterprises,* each trying to improve their own position, led to growth and the creation of wealth. As he put it: "It is not from the benevolence of the butcher, the brewer, or the baker, that we expect our dinner, but from their regard to their own self-interest" (Smith 1976: Book 2, Chapter 2: 14).

Thus, for Adam Smith, the "profit motive" was the driving force of capitalism. Individuals and firms in aggressive competition with each other produced the "wealth of nations." Where Marx (some decades later) saw conflict, exploitation, and growing inequality, Adam Smith saw competition leading to greater wealth.

These different perspectives continue to underpin arguments for, on the one hand, restructuring employment relationships to reduce inequality and, on the other, eliminating barriers to competition that are perceived to be holding back the creation of wealth. These counter-positions highlight a central contradiction of capitalist economies. Many people like to believe that the "marketplace" is best left unchecked if greater wealth is to be produced. At the same time, it is clear that social inequality increases without labour legislation, employment insurance, and other programs that smooth the rough edges of capitalism. The question, then, is what kind of society do we want?

Diverse Perspectives on the Division of Labour

Human societies have always been characterized by a basic *division of labour*— essentially, how tasks are organized and distributed among workers. In primitive societies, work roles were assigned mainly according to age and gender. But with economic development these roles became more specialized, and the arrival of industrial capitalism further intensified this process. Once a certain scale of production was reached, it was much more efficient to break complex jobs into their component tasks.

Pre-industrial skilled craftworkers are often idealized. C. Wright Mills, for example, emphasized the personal satisfaction derived from being involved in all aspects of the creation of some product, being free to make decisions about how the work should be done and to develop one's skills and abilities (Mills 1956: 220). Obviously not all pre-industrial workers were fortunate enough to be craftworkers, but it is clear that such opportunities were reduced with the growth of industrial capitalism. Part of the change was due to the loss of control over work that accompanied the spread of wage-labour. Equally significant was the division of work processes into simpler and smaller tasks, each done by an individual worker.

In *The Wealth of Nations,* Adam Smith identified the division of labour as a key to capitalism's success. Using the example of a pin factory, he described how productivity could be greatly increased by assigning workers to specific tasks such as stretching wire, cutting it, and sharpening it. Whereas individual workers might produce 20 pins a day each by doing all of the operations themselves, with a well-defined division of labour, 10 people could make 48,000 pins a day. The greater productivity, Smith reasoned, came from the increased dexterity a worker could master in repeating a single task over and over again, the time saved in not having to change tasks and shift tools, and the added savings obtained from designing machines that workers could use to repeat the single task (Smith 1976; Braverman 1974: 76). The real advantage of this form of work organization would only be realized when factories made large quantities of a product—precisely the goal of industrial capitalism.

In 1832, Charles Babbage translated Smith's principles into practical cost-cutting advice for entrepreneurs. By subdividing tasks, he argued, less skill was required of any individual worker. Consequently, employers could pay less for this labour. Workers with fewer skills simply cannot demand as high a reward for their work (Braverman 1974: 79–83). The early history of industrialization is full of examples of this basic economic principle at work. The advent of factories with detailed divisions of labour invariably replaced skilled craftworkers with unskilled factory workers who were paid less—just the sort of outcome that Marx criticized.

Returning to Mills's description of the craft ideal, we can easily see some of the problems. A central feature of *craftwork* was the degree to which one individual was involved in all aspects of the creation of some product. The resulting sense of pride and self-fulfillment was an early casualty of industrialization. Furthermore, the minute subdividing of tasks in an efficient mass-production factory or administrative system inevitably led to repetitious, boring work. Little wonder, then, that many craftworkers actively resisted factory-based production.

Marx's discussions of work in capitalist society focused on the negative consequences of an excessive division of labour. For him, capitalism itself was the source of the problem. The division of labour was simply a means to create greater profits from the labour of the working class. The development of huge *assembly-line factories* in the early 20th century epitomized this trend. Henry Ford, the inventor of the assembly line, took considerable pride in recounting how his Model T factory had 7,882 specific jobs. Ford calculated that about half the jobs required only "ordinary men," 949 required "strong, able-bodied men," while the rest could be done by women, older children, or men who were physically disabled, he reasoned. These observations do not reflect a concern for disabled workers, but instead highlight the extreme fragmentation of the labour process, to the point that even the simplest repetitions became a job.[12]

Émile Durkheim, an early 20th-century French sociologist, provided the alternative, conservative assessment of capitalist employment relations, particularly the division of labour. He noted that industrial societies contained diverse populations in terms of race, ethnicity, religion, occupation, and education—not to mention differences in beliefs and values. Durkheim pointed to evidence in European industrialization of group differences creating conflict over how scarce resources should be distributed, over rights and privileges, and over which beliefs and values set the standard. Durkheim saw the division of labour as a source of *social cohesion* that reduced this potential for conflict.[13] He reasoned that individuals and groups engaged in different tasks in a complex division of labour would recognize their mutual interdependence. In turn, this would generate tolerance and social harmony.

Durkheim believed that individuals in modern society are forced to rely on each other because of the different occupational positions they fill. In simple terms, lawyers need plumbers to fix their sinks while plumbers need teachers to educate their children. By the same logic, capitalists and their employees are interdependent. Without cooperation between the two groups, the economy would grind to a halt. We will see in later chapters how Durkheim's positive assessment of the division of labour shaped management theories that assume shared interests in the workplace. While Marx has influenced conflict perspectives on work in modern society, the conservative assumptions of Durkheim's general model are the backbone of the *consensus approach.*

Max Weber on Bureaucratic Organizations

Max Weber, a German sociologist also writing in the early 20th century, addressed yet another major change accompanying capitalist industrialization— *bureaucracy.* Weber noted that Western societies were becoming more rational, a trend most visible in the bureaucratic organization of work. Informal relationships among small groups of workers, and between workers and employers, increasingly were being replaced by more formal, impersonal work relations in large bureaucracies. Rules and regulations were now determining workers' behaviour. While Weber was concerned about the resulting loss of personal work relationships, he believed that this was far outweighed by greater organizational efficiency. For Weber, bureaucracy and capitalism went hand in hand. Industrial capitalism was a system of rationally organized economic activities; bureaucracies provided the most appropriate organizational framework for such activities.

What defined Weber's "ideal-type" bureaucracy was a precise division of labour within a hierarchy of authority (Weber 1946: 196–98). Each job had its own duties and responsibilities, and each was part of a chain of command in which orders could be passed down and rewards and punishments used to ensure that the orders were followed. But the power of the employer could not extend beyond the bureaucracy. The contract linking employer and employee was binding only within the work relationship. Also necessary for efficiency were extensive written records of decisions made and transactions completed.

Recruitment into and promotion within the bureaucratic work organization were based on competence, performance skills, and certifications such as educational credentials. Individual employees could make careers within the organization as they moved as far up the hierarchy as their skills and initiative would carry them. Employment contracts assured workers a position so long as they were needed and competently performed the functions of the office. In short, rationality, impersonality, and formal contractual relationships defined the bureaucratic work organization.

Yet bureaucracies were not unique to 19th-century capitalism. A somewhat similar form of centralized government had existed in ancient China, and European societies had been organizing their armies in this manner for centuries. What was unique, however, was the extent to which workplaces became bureaucratized under capitalism. At the beginning of the 20th century, increased competition and the development of big, complex industrial systems demanded even more rationalized production techniques and worker control systems. Large bureaucratic work organizations would become the norm throughout the industrial capitalist world. But bureaucracy also became the norm during the 20th century in industrialized communist countries of the former Soviet bloc. So, while closely intertwined, bureaucracy and capitalism are, in fact, separate phenomena.

The Managerial Revolution

Large corporations had begun to dominate the Canadian economy by the early 20th century. Formerly, most manufacturing enterprises had been owned and controlled by individuals or families. Then came the joint-stock companies in which hundreds, sometimes thousands, of investors shared ownership and profits. Such a diverse group of owners obviously could not directly control the giant corporation in which they had invested. Consequently, a class of managers who could run the enterprise became essential (Berle and Means 1968).

As this pattern spread, observers began to question the Marxist model of industrial capitalism that portrayed the relations of production as a simple two-class system: capitalists who owned and controlled the means of production versus workers who had little choice but to exchange their labour for a wage. An alternative model became popular. The *managerial revolution* theory predicted a new era of reduced conflict and greater harmony in the workplace (Burnham 1941). The theory held that *managers*, who were salaried workers and not owners, would look beyond profits when making decisions; the good of both the company and the workers would be equally important. Since ownership of the firm was now diffused among many individuals, power and control of the enterprise had essentially shifted to a new class of professional managers.

Several decades of debate and research later, it is generally agreed that this perspective on industrial relations in capitalist society exaggerated the degree of change (Zeitlin 1974; Hill 1981: 71–76; Scott 1988). First, family owner-ship patterns may be less common, but they certainly have not disappeared. In Canada, the prominence of names like Molson, McCain, and Bronfman demonstrate the continuing role of powerful families in the corporate sector. Furthermore, while ownership of corporations involves more individuals, many corporations are still controlled by small groups of minority share-holders. Individuals who serve as directors of major corporations are linked in a tight network of overlapping relationships, with many sitting on the boards of several major corporations simultaneously. Another postwar trend has been the corporate ownership of shares. Concentration of ownership increased as a few large holding companies replaced individual shareholders. Thus, the belief that the relatively small and very powerful capitalist class described by Marx has virtually disappeared is not supported by current evidence.[14]

Advocates of the managerial revolution theory also must demonstrate that, compared with earlier capitalists, the new breed of managers is less influenced by the bottom line of profit. But research continues to show that senior managers and corporate executives think and act in much the same way as capitalist owners. They share similar worldviews, often come from the same social backgrounds as owners, and hold large blocks of shares in the corporation, where they frequently serve as directors as well. In brief, the optimistic predictions of an era of industrial harmony brought on by new patterns of corporate ownership and management are largely unfounded.

POSTINDUSTRIALISM AND GLOBALIZATION
Postindustrial Society

Continuing social and economic change has led some social scientists to argue that we have moved out of the industrial era into a *postindustrial society.* Daniel Bell, writing in the early 1970s, was among the first to note these transformations in the U.S. occupational structure.[15] The Industrial Revolution had seen jobs in the manufacturing and processing sectors replace agricultural jobs. After World War II, jobs in the service sector had become much more prominent. The number of factory workers was decreasing, while employment in the areas of education, health, social welfare, entertainment, government, trade, finance, and a variety of other business sectors was rising. White-collar workers were beginning to outnumber blue-collar workers.

Bell argued that postindustrial societies would engage most workers in the production and dissemination of knowledge, rather than in goods production as in industrial capitalism. While industrialization had brought increased productivity and living standards, postindustrial society would usher in an era of reduced concentration of power (Bell 1973: 358–67). Power would no longer merely reside in the ownership of property, but also in access to knowledge and in the ability to think and to solve problems. *Knowledge workers*—technicians, professionals, and scientists—would become a large and important class. Their presence would begin to reduce the polarization of classes that had typified the industrial age. In contrast to the managerial revolution thesis, which envisioned a new dominant class of managers, Bell felt that knowledge workers would become the elite of the postindustrial age.

More recently, Richard Florida (2002) has added to this line of thinking in *The Rise of the Creative Class*. Like Bell, Florida sees the shift from goods to knowledge production as a crucial change. But for Florida, knowledge and information are simply the tools and materials for what really drives growth in a postindustrial age—*creativity*. Mapping the U.S. economy, Florida suggests there is an emerging *creative class* that accounts for roughly one-third of workers. At the top is a *super creative core* of scientists and engineers, poets, professors, novelists, entertainers, and artists. A second layer includes *creative professionals,* who possess high levels of human capital and formal education, and work in a range of fields such as business, finance, and law. According to Florida, these workers are transforming the economy through a new work ethic that places high priority on interesting work, flexible forms of organization, and dynamic places in which to live and work.

But what about those left out of this elite group? Writing about this more recently in *The Flight of the Creative Class,* Florida (2005) acknowledges a growing divide between creative workers and a "service class" who provide cleaning, childcare, and other services for them. Other writers such as Robert Reich (2000) in *The Future of Success* have noted that the knowledge economy offers great opportunities for well-educated, creative workers. But at the other end of the economic spectrum, job quality has eroded sharply through declining pay and job security and increased demands for continual effort. Reflecting on this economic divide, Reich remarks "Not for a century has America endured, or tolerated, this degree of inequality" (2000: 107). Likewise Florida notes that today the United States is the most unequal of all industrialized nations, with inequality rates nearly double those of Sweden and Japan (2005: 186).

Why the optimism in the early theories of postindustrial society? These explanations of social and economic change were developed in the decades following World War II, a time of significant economic growth in North America.[16] White-collar occupational opportunities were increasing, educational institutions were expanding, and the overall standard of living was rising. Hence the optimistic tone of the social theories being developed. Yet, as we discuss at various points in the book, other commentators paint a negative picture of the rise of service industries and an expanding white-collar workforce. These critical perspectives point to job deskilling, reduced economic security, the dehumanizing impact of computers, and widening labour market polarization—trends that seem more pronounced since the early 1980s.

Today, these themes are being revisited in debates about the so-called *new economy*, which essentially refers to industries based on new information and communication technology, especially the Internet. As Don Tapscott (1996: 43) argues, what distinguished the old industrial economy from the new "digital economy" is that the latter is "built on silicon, computers, and networks" and its main inputs and outputs are information. Thus the new economy is also a knowledge-based economy. Taking stock of these trends, sociologist Manuel Castells (1996, 1998) concludes that a new technical-economic paradigm, or model, has emerged—a networked society. As old boundaries imposed by time and space have broken down, opportunities to organize our collective and individual lives in more flexible ways emerge—with profound implications for society. The big unanswered question, of course, is who the winners and losers will be. There are growing concerns about a "digital divide," based on social groups or communities that are included in or excluded from the technology and knowledge of the new economy. However, more research is required to assess the many ways that access to and use of information technology can affect working conditions and work rewards (Hughes and Lowe 2000).

Industrial Restructuring

Technological change is accelerating at a time when the international economy is in upheaval. The deep recessions of the early 1980s and early 1990s, fierce international competition, multinational free trade arrangements, and the spectacular growth of Asian economies have had a major impact on Canada. In fact, it is becoming more difficult to think in terms of discrete national economies. In the words of Robert Reich (1991: 6), "[m]oney, technology, information, and goods are flowing across national borders with unprecedented rapidity and ease." This *globalization* of economic activity continues to bring about fundamental readjustments in the Canadian economy and labour market, including plant shutdowns, job loss through downsizing, corporate reorganization and mergers, and the relocation or expansion of company operations outside Canada.[17]

But are these trends new? Writing in the early 20th century, economist Joseph Schumpeter considered *industrial restructuring* a basic feature of capitalism. According to Schumpeter, this process of "creative destruction" involved breaking down old ways of running industry and building up more competitive, efficient, and high-technology alternatives.[18] North American industry clearly is engaged in this process today. But while necessary for the economy as a whole, industrial restructuring can also diminish the quality and quantity of work for individuals. Job losses may be part of the process for some, while others may find themselves with much less job security.

Industrial restructuring involves interrelated social, economic, and technological trends. Crucial is the shift from manufacturing to services. Canada's service industries have rapidly grown in recent decades, compared with declining employment in agriculture, resource, and manufacturing industries. Indeed, both Canada and the United States have experienced *deindustrialization*. This concept refers to declining employment due to factory closures or relocation, typically in once-prominent manufacturing industries: steel, automotives, textiles, clothing, chemicals, and plastics. Once mainstays of the Canadian and U.S. economies, these industries are now sometimes referred to as "sunset industries," because they have failed to adapt quickly to shifting consumer demands. Factories have been sold off, shut down, or relocated to areas such as Mexico, China, or other developing nations where labour is cheap and employment rights and environmental standards are lax.

Canada has been more vulnerable than the United States to deindustrialization. By the early 1990s dozens of large multinationals, encouraged by the Canada–U.S. free trade agreement, had announced or implemented plant closures in central Canada, seeking cheaper labour and fewer regulations. For example, after purchasing the Bauer hockey equipment company, Nike announced in 1997 that it was closing the Ontario factory.[19] But Canadian-owned firms have also been making similar moves. As Canadian corporate giants like Bombardier and Nortel have become global competitors, they have shifted more of their operations, and therefore job opportunities, out of the country.

Globalization

The term *globalization* has become part of everyday language, but what does it really mean, and can evidence document such a trend? The basic idea is not new; in Canada's colonial past, the masters of the British Empire no doubt envisioned their reach as global. Yet searching for a definition of globalization, one is struck by the great many meanings it conveys. Advocates of globalization echo earlier themes in this chapter: the "logic of industrialism" view that predicts the inevitable spread of capitalist markets and national convergence, and the postindustrial vision of economic progress through technology and information. As Gordon Laxer (1995: 287–88) explains, globalization typically refers to four interrelated changes:

> Economic changes include the internationalization of production, the harmonization of tastes and standards and the greatly increased mobility of capital and of transnational corporations. Ideological changes emphasize investment and trade liberalization, deregulation and private enterprise. New information and communications technologies that shrink the globe signal a shift from goods to services. Finally, cultural changes involve trends toward a universal world culture and the erosion of the nation-state.

Corporations, and often governments, promote globalization as a means by which expanding "free markets" will generate economic growth and elevate living standards. Signs of an increasingly global economy are visible in multinational trade agreements, financial markets, and economic treaties. Examples include NAFTA, the European Union, and the Association of South East Asian Nations; international regulatory frameworks such as the World Trade Organization (WTO); and the integration of financial markets through information technology. Critics—and there are many—detect more sinister aspects of globalization. For example, the protesters at the Summit of the Americas in Quebec City in April 2001 (where government leaders from Canada, the United States, and Central and South America discussed creating a Free Trade Area of the Americas by 2005) raised concerns that such treaties threaten national cultures, workers' rights, the environment—and ultimately, democracy. Even champions of globalization recognize these basic challenges. For example, as Thomas Friedman (2000: 42) writes in his book on globalization, *The Lexus and the Olive Tree,*

> Any society that wants to thrive economically today must constantly be trying to build a better Lexus and driving it out into the world. But no one should have any illusions that merely participating in this global

economy will make a society healthy. If that participation comes at the price of a country's identity, if individuals feel their olive tree roots crushed, or washed out, by this global system, those olive tree roots will rebel.

Community-based opposition is only one of the reasons why a globally-integrated economic system remains largely an ideal. In a truly global economy, corporations would operate in a completely transnational way, not rooted in a specific national economy. But research suggests that there are few such corporations (Hirst and Thompson 1995). While McDonald's, IBM, and General Motors may do business in many countries, operating vast networks of subcontractors and suppliers, they remain U.S.-based. The United States, Japan, and Europe still account for most of the world's trade, largely through corporations located in these countries. While multinational corporations account for 25 percent of the world's production, they employ only 3 percent of the world's labour force (Giles 1996: 6). Even so, computers, software, clothes, shoes, cars, and a vast range of other goods are produced by global business networks that span several continents (Rinehart 2006: Chapter 6). Some observers argue that large corporations' sustained power will depend on creating and controlling these networks or "business alliances" (Harrison 1997). However, Anthony Giles (1996: 6) cautions, "Beyond the obvious technological, economic and logistical hurdles, there are a host of cultural, legal, political and linguistic factors which complicate the development of genuine globally integrated production systems."

It is much easier to imagine production systems spanning several continents than it is to envision a largely global labour force. Despite expanding international markets for some goods and services, labour is still a local resource. According to Hirst and Thompson (1995: 420), "Apart from a 'club-class' of internationally mobile, highly skilled professionals, and the desperate, poor migrants and refugees who will suffer almost any hardship to leave intolerable conditions, the bulk of the world's populations now cannot easily move." In fact, Canada plays an important role in this regard, being one of the few nations to accept relatively large numbers of immigrants annually.

Picking up on this theme, Saskia Sassen (2002) argues that mainstream accounts of globalization have overemphasized the export of low-wage work to developing nations while ignoring the growing role migrant workers play in the industrialized North. As Sassen points out, the managerial, professional, and technical elite who live in global cities are increasingly dependent on low-paid service workers to maintain the infrastructure that allows them to do their work. Office buildings need to be cleaned, young children and elderly parents need to be cared for, groceries purchased, meals made, dishes done. Much of this work, along with other vital services, is done by migrant women, many of whom have moved in search of work to support their own families back home (see also Stasiulis and Bakan 2003).

Although labour might not become part of a truly global market, globalization may still have an impact on labour practices, notably through the public's growing concern about the labour practices of nationally based firms operating in developing countries (one could say the same about environmental practices). Global media coverage has helped to raise North Americans' awareness. Recently, multinational corporations like Nike have been the targets of public campaigns because of their labour practices in Asian countries (Heinzl 1997). Reflecting these concerns, a 1997 survey of 98 of Canada's largest corporations by the Montreal-based International Centre for Human Rights and Democratic Development found that while 42 percent of the responding firms agreed that international business has a role to play in promoting human rights and sustainable development, only 14 percent had a code of labour standards that protects basic human rights—freedom of association, non-discrimination, and elimination of child labour and forced labour.[20]

Canada and Free Trade

The 1989 Free Trade Agreement (FTA) with the United States and the 1994 North American Free Trade Agreement (NAFTA), which included Mexico, committed Canada to a policy of more open, less regulated markets. While claims of massive job losses directly from the FTA and the NAFTA have not been borne out, the permanent factory closures and job losses that occurred in central Canada in the early 1990s have partly been linked to free trade. Nevertheless, strong economic growth and job creation in Canada and the United States have offset these losses. Significant industrial restructuring was already taking place, but the FTA probably accelerated the process, allowing the individuals and communities negatively affected less time to respond. Between the first quarter of 1990 and the second quarter of 1991, Canada lost 273,000 manufacturing jobs, a decline of 13 percent in the industry as a whole. As the Economic Council of Canada (1992: 4) concluded, "Many of the jobs that were lost will not return." These issues are presumably being monitored, since NAFTA includes both an accord on labour and a commission for labour cooperation, which is mandated to track labour market trends in the three NAFTA countries and to address complaints about workers' rights, collective bargaining, and labour standards.[21]

To put NAFTA in perspective, in 2003 the three participating countries had a combined labor force of 210 million workers: 71 percent in the United States, 21 percent in Mexico, and 8 percent in Canada (World Bank 2005a: Table 2.2). Mexico's labour force has been growing at about twice the rate of the labour forces of its two northern neighbours. While its manufacturing sector declined in the early 1990s and was hit hard by the peso crash of 1994, it has since rebounded, adding 1.5 million new workers from 1996 to 1999 (Commission for Labor Cooperation 2003: 81). The migration of jobs from Canada or the United States to Mexico was a major concern of NAFTA opponents. There is little doubt that the *maquiladora* factories along Mexico's northern border have been booming. From 1996 to 1999 the number of *maquila* plants rose by 39 percent, bringing employment to 1.14 million (Commission for Labor Cooperation 2003: 81). Were these "new" jobs, or jobs relocated from high-wage countries like Canada or Japan? From available evidence, it seems that fewer jobs than expected have migrated south.[22] And much of the new foreign investment in Mexico is from firms based in Japan, East Asia, and Europe seeking a base from which to supply the North American consumer market. Maquilas have also become more diverse, as auto and pharmaceutical manufacturers such as Volkswagen, Honda, Eli Lilly, and Mercedes Benz have set up operations.

Jobs may be migrating south, but only in some manufacturing industries, and if firms seek low wages, they will likely relocate to China, not Mexico. Actually, there is as much concern that competitive pressures from NAFTA in some manufacturing industries, such as auto and auto parts manufacturing, have accelerated the trend toward *nonstandard* or *contingent* work (part-time, temporary, contract) in the United States and Canada (Roberts, Hyatt, and Dorman 1996). Workers in Canada's resource industries have also benefited less than they could given U.S. refusals to comply with NAFTA rulings on ongoing softwood lumber disputes. Others would argue that the group most negatively affected by NAFTA is Mexican workers. While maquila workers' dollar-an-hour pay, lack of rights, and working conditions may be deplorable by Canadian standards, they are part of an emerging middle class in Mexico and relatively advantaged compared to the majority of Mexicans—the rural poor. The Zapatista uprising in the rural state of Chiapas briefly drew the world's attention to those who will not benefit from NAFTA. Though maquila workers face an uphill battle, they have had some success organizing for better pay and benefits, as in the 1994 peso devaluation (Kopinak 1996: 199). More recently they have worked with Canadian and U.S. labour and student groups to bring complaints about labour rights violations under NAFTA's labour side agreement. The Maquiladora Solidarity Network (MSN) in Canada and the United Students Against Sweatshops (USAS) are just two examples of emerging global solidarity networks that help monitor and try to challenge labour violations (see the MSN website at http://www.maquilasolidarity.org).

THE ROLE OF THE STATE IN TODAY'S ECONOMY

Clearly, the information technology revolution, industrial and labour market restructuring, and economic globalization are proceeding in a political context dominated by *laissez-faire* beliefs that advocate free markets with little or no government interference.[23] In a sense, this political environment is also a key determinant of Canadians' future employment prospects. These three large trends have the potential to either improve employment opportunities for Canadians or lead to further labour market disruptions. It is difficult to predict just what their ultimate impact might be, but it is important to consider how these forces could be shaped to our collective advantage. Public policy in other industrial countries in Europe and Asia has been more proactive in attempting to influence the course of technological, labour market, and economic change.

The FTA and NAFTA were negotiated by governments, and have been hotly contested political issues in all three countries affected. These trade agreements underscore the evolving role of the state (or government) in the economy and the labour market. The history of industrial capitalism is replete with instances of different forms of state intervention. For example, in the early Industrial Revolution, the French government forced unemployed workers into factories in an attempt to give manufacturing a boost. We noted above that in the early 19th century the British government dealt harshly with the Luddite protests against new technologies. And the Canadian government adopted its National Policy in 1879 to promote a transcontinental railway and settlement of the West.

In fact, as Canada industrialized, the government heavily subsidized the construction of railways in order to promote economic development. It also actively encouraged immigration to increase skilled labour for factory-based production and unskilled labour for railway construction. At times, it provided military assistance to employers combating trade unionists and introduced laws discriminating against Chinese and other non-white workers. But the Canadian government also passed legislation that provided greater rights, unemployment insurance, pensions, and compensation for workplace injuries. While some might argue that these initiatives were designed mainly to ensure industrial harmony and to create an environment conducive to business, it is nonetheless true that these labour market interventions benefited workers.

A consensus, or compromise, was reached between employers and workers (mainly organized labour) in the prosperous post–World War II period. Acting on *Keynesian economic principles* that advocated an active economic role for the state, the Canadian government attempted to promote economic development, regulate the labour market, keep unemployment down, and assist disadvantaged groups—in short, to develop a "welfare state." While the Canadian state was never as actively involved in the economy and the labour market as some European governments, it nevertheless saw itself playing an important role.[24]

But industrial restructuring in the 1980s was accompanied (some would say facilitated) by new conservative political doctrines based on free-market economics (Marchak 1981; Saul 1995; Kapstein 1996). Ronald Reagan in the United States, Margaret Thatcher in Britain, and Brian Mulroney in Canada argued that economies would become more productive and competitive with less state regulation and intervention. Public policy was guided by the assumption that free markets can best determine who benefits and who loses from economic restructuring. The ideas of Adam Smith were used to justify the inevitable increases in inequality. High unemployment came to be viewed as normal. In some jurisdictions, labour rights were diluted, social programs of the welfare state came to be seen as a hindrance to balanced budgets and economic competitiveness, and government-run services were privatized—changes that directly or indirectly affected employment relationships.[25] As John Ralston Saul (1995) argues, these free-market ideologies erode democratic freedoms, threatening to bring about what he calls "the great leap backward." Similarly, with the rapidly developing economies of East Asia, governments routinely use economic imperatives to restrict individual and collective rights.

Thus, in today's economic climate of globalization and restructuring, proponents of free-market economics seem to have more influence. But as we noted when discussing the ideas of Adam Smith, an unchecked marketplace generally leads to greater inequality. So, when answering the question "what kind of society do we want?" it will be important to determine the role of the state in achieving the desired society. Looking back, we can see that differing government policies have interacted with economic globalization to produce different effects at the local level. Looking forward, there is no fixed trajectory for globalization. Political choices by citizens and their governments undoubtedly will continue to shape the process in diverse ways (Cameron and Stein 2000). Canadians are concerned about labour market inequalities, employment, workers' rights, and employers' responsibilities to communities. The extent to which governments are held accountable for these issues will determine the impact of global economic forces on our daily economic lives.

CONCLUSION

From feudalism to NAFTA and globalization—we have covered a lot of territory in this chapter. We began with a historical overview of the origins and development of industrial capitalism, then discussed various theories that explain and evaluate the causes and consequences of this complex process. As economies and societies continually evolve and the nature of work is transformed, new social theories have been constructed accordingly.

We have highlighted important social and economic changes initially unleashed by the Industrial Revolution in Europe. As feudalism gave way to capitalism, markets grew in importance. A new class structure evolved and a predominantly rural society became urban. Factory-based wage labour became the norm, while craftwork declined. Larger workplaces demanded new organizational forms and, in time, bureaucracies evolved to fill this need. And while industrial innovations led to substantial increases in productivity, it was some time before the standard of living of the working class began to reflect this increase. These far-reaching social and economic changes were the focus of early sociologists like Marx, Durkheim, and Weber, and economists like Adam Smith. Having had a brief look at the changing worlds that these social theorists were observing, we are now better equipped to understand their concerns and conclusions.

Our overview of the industrialization experience in Canada and several other countries takes us into the present. Much larger workplaces, new technologies, a more complex division of labour, growth in white-collar occupations, and a new class of managers were among the changes observed as industrial capitalism matured in 20th-century Canada. Once again, these changes generated a variety of new social theories dealing with technology, social inequality, skill, knowledge workers, increasing convergence, and labour–management cooperation. But the optimistic predictions of reduced inequality and conflict in a postindustrial era are not well supported by the data. Instead, as we will see in the next chapter, employment trends suggest a growing gap between more and less advantaged workers.

We have also highlighted a number of global economic and technological forces now re-shaping employment patterns. Like other advanced capitalist societies, Canada has become a service-dominated economy. The information technology revolution is having a major impact on both the quantity and quality of work in the Canadian labour market, as are processes of industrial and labour market restructuring. The eventual outcomes of these trends are still unclear. We will also have to confront the limits to growth, as environmental sustainability becomes a more pressing global problem (World Commission on Environment and Development 1987). We may see a general improvement in the standard of living and the quality of working life, or the benefits may go primarily to those who already have better jobs, thus contributing to increased polarization in the labour market and in society as a whole. In subsequent chapters, we will return frequently to this basic question about increasing or declining inequalities.

DISCUSSION QUESTIONS

1. What is industrial capitalism and how did it develop in Canada?
2. In what ways did the emergence of industrial capitalism in Canada follow the path of other countries, such as Britain? In what ways was the Canadian experience unique?
3. What is meant by the "logic of industrialism" thesis? Using three countries discussed in the text, how would you evaluate the merits of this idea?
4. How did Karl Marx, Adam Smith, and Max Weber believe industrial capitalism would affect workers? What are some of the main differences in their ideas?

5. In your opinion, do any of the "grand theories" of Marx, Weber, or Durkheim have relevance to explaining the emergence of industrial capitalism in Canada?

6. What is a postindustrial society? Based on material you have read in Chapter 1, would you characterize Canada as postindustrial? Why? Why not?

7. Drawing on your reading in Chapter 1, discuss how free trade, globalization, and industrial restructuring are re-shaping the Canadian economy. What do you see as the most important changes taking place?

8. In your view, how has the quality of work and social inequality changed over the past several decades in Canada? Have economic prospects improved, declined, or diverged?

NOTES

1. Abridged from Joyce Doyle's (1980) *Our Family History*.

2. Beaud's (1983) history of capitalism is a major source for our brief discussion; also see Grint (1991: 48–83) on pre-industrial and early industrial work patterns.

3. Over the centuries, however, technological innovations did lead to important changes in work patterns. See, for example, White's (1962) detailed historical analysis of the impact of draught horses and the wheeled plough on agriculture, and of the effects of the invention of the stirrup on the practice of warfare.

4. Grint (1991: 69–73). Cohen (1988: 24) concludes that industrialization in Canada did not lead to a sharper gender-based division of labour since, as was not the case in England, such segmentation already existed prior to industrialization.

5. Polanyi (1957). See Boyer and Drache (1996: 8–12) for an application of Polanyi's ideas to the current era.

6. This discussion is drawn from a variety of sources, many cited individually below. For a useful overview of this period in Canadian history, see Ryerson (1968), Laxer (1989), Kealey (1995: Part One), Palmer (1992), and the Canadian labour studies journal *Labour/Le Travail*.

7. Bleasdale (1981: 13); see Wylie (1983) on the building of the Rideau Canal.

8. Bradwin (1972); see Avery (1995) and Creese (1988–89) on strikes and radical behaviour among immigrant workers.

9. Mine accidents that took the lives of many miners at one time are mainly responsible for these high averages. For example, 189 miners died in a mine explosion in Hillcrest, Alberta, in 1914. A memorial to these miners and background on the disaster can be found at http://members.tripod .com/~coalminersmemorial/hillcrestminedisaster.html. See McCormack (1978: 9) on British Columbia, and Caragata (1979: 16–21) on Alberta coal miners during this era.

10. See World Bank (2001), Holzer (2000), Whitley (1992), Shieh (1992), and Hsiung (1996) for accounts of economic development in the region. The argument about the critical role of a "Weberian" form of government bureaucracy is presented by Evans and Rauch (1999).

11. In 2003, China received $53 million in direct foreign investment, more than any other country in the world. Estimates suggest that since starting reforms, China has received approximately half a trillion dollars in foreign investment. (Fishman 2005: 15).

12. Toffler (1980: 50) provides the quote from Henry Ford.

13. Durkheim (1960) did allow that a "forced" division of labour, where individuals have no choice over how they participate in the productive system, would not lead to increased social solidarity. But he argued that this and other abnormal forms of the division of labour would disappear as industrial capitalism matured further.

14. See Antoniou and Rowley (1986) on family ownership patterns, Grabb (2004) on concentration of ownership, and Carroll (2004).

15. Bell (1973); current and more critical perspectives on postindustrial society are reviewed by Krishnan Kumar (1995), Clement and Myles (1994), and Nelson (1995).

16. The managerial revolution perspective (Burnham 1941; Berle and Means 1968) was developed earlier, during an era when corporate concentration in North America was proceeding rapidly and when concerns about the excessive power of the corporate elite were being publicly debated (Reich 1991: 38).

17. See Lowe (2000: Chapter 4) for a discussion of industrial restructuring and the rise of the "new economy" in Canada. Also see Boyer and Drache (1996) and Barnet and Cavanagh (1994: Part Three) for critical perspectives on restructuring, globalization, and work.

18. Schumpeter's views are discussed in Bluestone and Harrison (1982: 9).

19. Drache and Gertler (1991: 12–13). See Heinzl (1997) on Nike's closure of the Bauer factory. Also see Mahon's (1984) analysis of restructuring in the Canadian textile industry.

20. International Centre for Human Rights and Democratic Development (1997). The organization's website (http://www.ichrdd.ca) has a number of reports on globalization and labour rights. On human rights at work, and the goal of promoting "decent work" for all, see International Labour Office (1999) and the International Labour Organization website (http://www.ilo.org/public/english/). Also see Adams and Hallock (2001) on the anti-sweatshop movement and its impact on corporate codes of conduct.

21. Commission for Labor Cooperation (2000, 2003). For current information, see the North American Institute's website (http://www.northamericaninstitute.org) and the Commission for Labor Cooperation website (http://www.naalc.org). The latter monitors the North American Agreement on Labor Cooperation. The NAALC's study of adjustments in the North American garment industry under NAFTA focuses on the importance of labour standards (http://www.naalc.org/english/publications/garment_sum.htm/). See Adams and Singh (1997) and Gunderson (1998) for an assessment of the impact of NAFTA on labour policies and worker rights. Vosko (1998) contrasts the different regulatory responses under NAFTA and the European Community to the spread of temporary employment.

22. Orme (1996: 13); Arsen, Wilson, and Zoninsein (1996); for different perspectives on NAFTA, see Bognanno and Ready (1993).

23. See Saul (1995) for a critique of this ideology.

24. See Olsen's (1999) comparative review of recent changes to Sweden's welfare state.

25. See Bamber, Landsbury, and Wailes (2004) for a comparative analysis of how employment relations and labour–management relations have changed recently in industrialized nations.

2

CANADIAN EMPLOYMENT PATTERNS AND TRENDS

INTRODUCTION

Pick up a copy of any Canadian newspaper and flip through it, looking for feature stories on work-related topics. You'll likely come across articles based on labour force statistics collected by Statistics Canada, or by some other public- or private-sector data collection agency. Most common are the unemployment rates, updated monthly. But the news media also carry stories on a wide array of themes, such as income patterns, industrial change, part-time work, self-employment, strikes, and workplace health and safety.

Understanding these labour force statistics is important both for learning more about the Canadian labour market and for developing statistical literacy. In a society in which the media constantly bombard us with the latest figures on one social trend or another, it is helpful to have some understanding of the definition and source of these numbers, and even more important, to be able to interpret them critically. Hence, we will discuss these trends with reference to some of the broad theories of social and economic change discussed in Chapter 1. At the same time, this statistical overview will provide us with the background needed to evaluate, in Chapter 3, competing theories of how labour markets operate.

We begin by outlining key demographic factors shaping the workforce. We then go on to discuss labour force participation trends and Canada's occupational and industrial structures, emphasizing the rise of the service economy. Canada's unemployment trends and the growth in nonstandard jobs in this country are other focuses of this chapter. We conclude with an analysis of where current labour market trends may be taking us, raising important issues about the future of work.

DATA SOURCES

A major data source is the Canadian census, a survey of the entire population of the country conducted by Statistics Canada. The most recent national census was completed in 2001, and we draw on those results in this edition of the book. The huge effort and cost involved in collecting information about 31 million Canadians means that only a limited number of questions can be asked about each household and the individuals within it. Furthermore, because a census takes place every five years, results are rarely up-to-date. Despite this, a census provides the most complete picture of the Canadian labour market at specific points in time, and allows us to accurately plot historical changes.

A second useful source is the monthly Labour Force Survey, also conducted by Statistics Canada. Unlike the census, which attempts to cover a range of topics, this random sample survey is designed to collect only work-related information and so provides much more detail. Because the survey is done every month, information is always updated. In order to get precise estimates of the labour market activities of Canadians aged 15 and older, a large sample is needed. Approximately 54,000 households (most providing information on more than one individual) are included in the sample.[1] Households remain part of the sample for six months before being replaced.

The Labour Force Survey provides, for example, monthly estimates of unemployment and labour force participation rates and descriptions of industries and occupations experiencing job growth or decline.[2] Frequently, more detailed surveys, designed to study specific work-related topics, are "piggy-backed" onto the Labour Force Survey. Examples of such topics include student summer employment, self-employment, involuntary part-time work, and job search behaviour. Many of the statistics cited in this chapter are obtained from this useful source.[3]

THE DEMOGRAPHIC CONTEXT OF LABOUR MARKET CHANGE

Whether our focus is on the labour market today or 10 to 20 years in the future, there is no denying that *demographic shifts* (population changes) under way in Canadian society set some basic parameters. Of all the employment trends we will consider in this chapter, those related to the demographic composition of the labour force are perhaps the only ones we can confidently project into the future. So while economists debate whether the unemployment rate may rise or decline next year, demographers (who specialize in studying the structure and dynamics of population) can quite accurately predict birthrates, life expectancy, population growth, age distributions of the population, and related trends. Three of these trends deserve particular attention: aging, cultural diversity, and educational attainment.

Workforce Aging

While we might dispute David Foot's claim in his best-selling book, *Boom, Bust and Echo,* that "[d]emographics explain about two-thirds of everything," we do agree that Canada's demographic trends influence many economic and social changes.[4] Foremost among these demographic factors is *population aging,* which has significant implications for job opportunities, pensions, work values, and organizational structures. The *baby-boom generation,* born between 1946 and 1964, is the largest generation in Canadian history. In the words of historian Doug Owram (1996: xiv), "Economics, politics, education, and family life would all have been considerably different without the vast demographic upsurge of births after the Second World War." As baby boomers move through the life course, they are leaving few institutions unchanged—from a revolution in popular music in the 1960s to the rapid expansion of postsecondary institutions in the 1970s to debates over mandatory retirement in the 2000s. Coupled with smaller birth cohorts following in their wake, due to sharply declining birthrates, the aging of the baby boom has fundamentally altered the demographic shape and social structure of Canada.

This huge and slowly aging bulge in the workforce has affected the career opportunities of many Canadian workers. The baby-boom generation entered the workforce when the economy was still expanding, and many obtained good jobs. Some of the smaller generations that followed them into the labour force have been less fortunate, since higher unemployment and global economic uncertainties have frequently led to layoffs, not hiring. Yet many baby boomers found their careers beginning to *plateau* in the 1990s. Because most work organizations are pyramids, career success has been defined in terms of climbing a ladder that has room for fewer and fewer people on each higher rung. But with more older workers, the competition for the few top jobs has intensified. For younger workers experiencing difficulty finding satisfactory entry-level jobs, this may not seem a serious problem. But, for long-term employees who have learned to view personal success as upward movement, career blockages can be very unsettling (Foot and Venne 1990). Generally, work organizations have not adapted to changing demographic pressures, maintaining pyramidal structures and recruitment policies that are no longer appropriate and focusing on downsizing for the past decade (Foot and Stoffman 1998: Chapter 4).

Figure 2.1 documents the workforce aging process by profiling the *age distribution* of labour force participants in 1976 and 2004. Note the decline in the relative proportion of teenagers (15 to 19 years) and young adults (20 to 24 years) over the 28-year period. Compare this with the increased size of the 35-to-44- and

FIGURE 2.1 *Age Distribution of the Labour Force by Gender, Canada, 1976 and 2004*

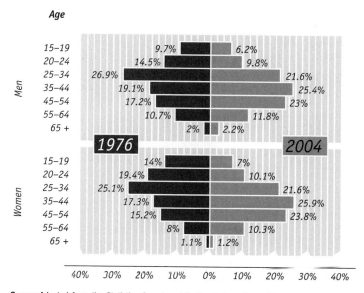

Source: Adapted from the Statistics Canada publication *Labour Force Historical Review*, 2004, Catalogue 71F0004XCB, February 18, 2005.

45-to-54-year-old cohorts over this time. While we have seen a marked trend toward "early" retirement, due to labour market restructuring and downsizing in the 1980s and 1990s, the proportion of 55- to 64-year-olds in the workforce has increased slightly as growing numbers of boomers reach "near retirement" age and move into this age category. Indeed, in 2002 the "near retirement rate" (the percentage of Canadians within 10 years of median retirement age) rose to 20 percent, up from just 11 percent in 1987 (Statistics Canada 2004d).

In 1921, only 5 percent of Canadians were 65 years of age and older. Over the next 50 years, this figure grew slowly to 8 percent. Then the aging of the Canadian population began to accelerate, and by 2000 almost 13 percent of Canadians were senior citizens. Statistics Canada predicts that, by 2026, seniors will account for 21.4 percent of the total population.[5] The relative size of the working-age population that funds government old-age security programs through payroll contributions will shrink considerably during this period. The earliest baby boomers are now in their late 50s, and are seriously planning for or starting to pursue their retirement options. Hence, pensions have become a prominent public concern, as has the issue of mandatory retirement and expected skill shortages. Will the Canada/Quebec Pension Plan (C/QPP), one of Canada's main *social safety nets,* be able to support all the retiring baby boomers, or will today's youth have to pay more to cover the added costs? Or, as an alternative solution, will Canadians have to wait until they are a few years older than 65 to begin collecting a government pension? Will employer-sponsored pension plans make up the difference? Probably not. Access to such plans has never been widespread and actually declined in the 1990s. In 2001, for example, just 31 percent of employees reported participating in an employer-sponsored pension plan and only 18 percent in a group Registered Retirement Savings Plan (RRSP).[6]

It would seem reasonable to expect that workforce aging would open up opportunities for younger workers, who are far outnumbered by middle-aged baby boomers (Easterlin 1980). For example, in 2001 roughly one-quarter of G.P.s, specialist physicians, and professors were 55 years or older. Carpenters, contractors, and other trades also have much older age profiles, sparking concerns over future skill shortages (Statistics Canada 2003a). Who will fill these jobs, especially in blue-collar areas, when there is growing emphasis on gaining postsecondary education? Consider as well the relatively small size of the 15-to-19-year-old and 20-to-24-year-old cohorts in today's labour force

(Figure 2.1). In another 10 years, when the first wave of baby boomers has entered retirement, there is a distinct possibility of labour shortages. Even so, we need to recognize that the generation in between continues to have difficulty finding satisfactory jobs.

Greater Workforce Diversity

Immigration is an important determinant of Canada's population mosaic. For centuries, Canada has been a country of immigrants. In the post–World War II era, Canadian governments have attempted to tailor immigration policy to labour market trends, partly in anticipation of labour shortages in key areas. Immigration levels have been kept low when unemployment was high, but have been allowed to rise when economic expansion required extra workers. In fact, there have been times (around 1910, for example, when the Western provinces were being settled) when immigration levels were considerably higher than they are today. But, along with economic factors, immigration policies have also been influenced by demographic trends. For well over three decades now, birthrates have been below replacement level, so low that our population would have declined unless supplemented by immigration.

Immigration is thus once again taking on increasing importance. In 2001, the proportion of immigrants in Canada reached its highest level in 70 years, with 18.4 percent of the population having been born outside the country. This increase reflects federal government quotas that are currently set around 220,000 to 245,000 immigrants per year (less than 1 percent of the total population), levels that are expected to rise further still (Citizenship and Immigration Canada 2005). Despite concerns by some that immigrants are taking jobs away from native-born Canadians, research shows that this is rarely the case. Immigrants frequently create their own jobs, or are willing to take jobs that others do not wish to have, and are carefully selected to match shortages of workers in specific occupational categories. In fact, with the impending shortage of labour force participants in coming years, there is much interest by federal and provincial governments in ensuring immigrants' skills and abilities are recognized and properly utilized.[7]

Until quite recently, there was far less *visible* diversity among immigrants to Canada. For example, in 1981, two-thirds of all immigrants living in Canada (including those who had been here for many decades) were born in

Europe. Today the source countries have dramatically changed. Between 1991 and 2001, 58 percent of immigrants came from Asia, with top countries including China, India, Pakistan, and the Philippines. Just 20 percent of immigrants came from Europe. Why? Canadian immigration policies no longer favour European immigrants but, more importantly, the demand to immigrate to Canada has declined in Europe, while it has increased dramatically elsewhere. Consequently, more immigrants are members of visible minority groups with distinctive cultural backgrounds.

According to the 2001 census, 13 percent of Canadians identified themselves as a member of a *visible minority* group (excluding Aboriginal Canadians), up from 5 percent in 1981. In Toronto and Vancouver, visible minorities made up over one-third (37%) of the city population. Chinese Canadians made up the largest visible minority group in Canada in 2001 (3.5% of all Canadians), followed by South Asians (3.1%), Blacks (2.2%), and Filipinos (1.0%). Given current immigration patterns, the proportion of visible minorities in Canada is expected to increase to 20 percent by 2017. Moreover, nearly 1 million (3.3%) of the nation's population is of Aboriginal origin. This figure will also increase in the decades ahead, given the younger age profile of Aboriginal Canadians (a median age of 24.7 years compared to 37.7 years for the non-Aboriginal population, according to the 2001 census). Immigrants are also younger, on average, so both Aboriginal Canadians and immigrants will make up a growing share of the workforce as older Canadians retire.[8] As discussed in later chapters, these demographic trends have prompted governments and employers to develop employment equity policies to facilitate greater workplace cultural diversity and reduce the potential for labour market discrimination.

A Better-Educated Workforce

On the whole, Canadians are becoming increasingly well educated, a trend that has also fundamentally changed the character of the labour force. Figure 2.2 shows the dramatic change in *educational attainment* of the labour force between 1975 and 2004. In 1975, 66 percent of females and 68 percent of males in the labour force had only a high-school education or less. Very few (only 7% and 10%, respectively) had a university degree. In the 30 years since, large numbers of both men and women obtained postsecondary credentials. By 2004, one in five Canadian labour force participants had a university degree.

FIGURE 2.2 *Educational Attainment of the Labour Force by Gender, Canada, 1975 and 2004*

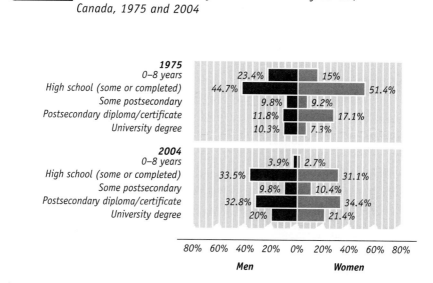

Source: Adapted from the Statistics Canada publications *Labour Force Historical Review,* 2004, Catalogue 71F0004XCB, February 18, 2005; and *Labour Force Historical Review,* 2001, Catalogue 71F0004XCB, February 22, 2002.

This rising educational attainment has been the backdrop for a debate on the role of education in a rapidly changing and increasingly global and technological economic environment. In fact, many analysts argue that a well-educated workforce is a nation's key resource in today's global marketplace. Business and government have argued that additional education and training, and an overhaul of the education system that makes it more job-relevant, will help Canada become more competitive internationally.[9] The latest variant of this perspective is found in the concepts of *lifelong learning* and *learning organizations.*[10] Essentially, these refer to continuous education, both formal and informal, in educational institutions and at work and home, throughout one's adult life. Educators, parents, and students have responded by raising concerns that closer links between education and the economy might mean too much business influence over what is taught at all levels of the educational system (Davis 2000; Turk 2000).

By international standards, Canadians are very well educated in terms of postsecondary credentials and enrollment levels.[11] But there is more to developing a nation's human resources than the acquisition of formal education. No doubt there are educational reforms that could lead to the graduation of more skilled, flexible, and enterprising labour force participants. Basic literacy and numeracy, along with other essential skills, are lacking in small pockets of the workforce.[12] But work-related training is not only the responsibility of the education system. Canada has a mediocre record of employer-sponsored workplace training. According to the most recent Adult Education and Training Survey (AETS) in 2003, only 30.1 percent of adults aged 25 to 64 years participated in employer-supported training (Peters 2004). While an improvement from 24.3 percent in 1997, Canada's training record remains average at best, when compared to the United States and western European countries (Kapsalis 1998). Moreover, it is well-educated workers already in good jobs who typically receive training, further accentuating labour market inequalities (see Chapter 3).[13]

The assumption underlying the "education and training" solution to economic growth is that our workforce is not sufficiently educated and trained, yet the above discussion suggests there is more to the problem. For one thing, while unemployment is higher among the less educated, a considerable number of Canada's unemployed are well educated. In addition, many Canadian workers are in jobs that require little education or training. In 1994 almost one in four (23%) of employed Canadians (aged 15 to 64) reported that they were overqualified for their job. Reflecting a different concept—that of *underemployment*—the 2004 Canadian Learning and Work Survey found that 34 percent of respondents felt underemployed.[14] This underemployment highlights the need to go beyond the supply side of the education-job equation to examine how job content and skill requirements could be upgraded—in short, how the demand for educated labour could be improved. Clearly, there are deeper problems in the structure of our economy as well, and concerted efforts by government, the private sector, organized labour, and professional associations are needed to address them. As two American sociologists concluded about the United States: "The problem is a shortage of good jobs to a greater extent than it is inadequate training and development."[15]

LABOUR FORCE PARTICIPATION TRENDS

Now that we have a better demographic understanding of the workforce, we can examine key labour market and employment trends that define where, how, and for whom Canadians work. *Labour force participation* (LFP) is the main indicator of a population's economic activity, at least from the perspective of paid employment. Whether relying on census or monthly Labour Force Survey data, calculations of labour force size or participation rates are based on the number of individuals 15 years of age or older who are working for pay (including self-employed individuals working on their own or employing others) and those who are looking for work. Hence, the *unemployed* (those out of work but who have actively looked for work in the past four weeks) are counted as part of the labour force. However, individuals performing unpaid household and child-care work in their home are not included in official labour force calculations, even though their labour makes essential economic and social contributions.

Using this official LFP definition to look at work patterns at the beginning of the last century, only 53 percent of Canadians (15 years of age and older) were participating in the labour force in 1901. The rate increased to over 57 percent by 1911 but did not go much higher until the 1970s (see Figure 2.3). By 1981 almost two-thirds (65%) of the eligible population was in a paid job or seeking one. Participation rates peaked in 1989 at 67 percent, and then fell a bit during the recession of the early 1990s. They have hovered between 65 and 67 percent since then. The 2004 LFP rate was 67.6 percent, representing a total of 17,183,400 labour force participants.

FIGURE 2.3 *Labour Force Participation Rates by Gender, Canada, 1946–2004*

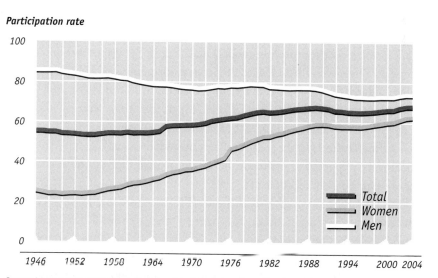

Participation rate

Source: Adapted from the Statistics Canada publications *Labour Force Historical Review*, 2004, Catalogue 71F0004XCB, February 18, 2005; and *Historical Statistics of Canada*, Catalogue 11-516, 1983, Series D205-222.

Gender Differences in Labour Force Participation

Breaking down the LFP rate by gender, we find striking differences. In 1901 only 16 percent of women aged 15 or older were in the paid labour force. However, this rate increased with each 10-year census. After World War II, the size of these increases was substantial (Figure 2.3). Between 1960 and 1975, female LFP jumped from 28 to 41 percent, and the following decade saw it rise to 55 percent. The landmark year was 1980, the first in which a majority of Canadian women were part of the labour force (which, as officially defined, leaves out unpaid family and household work).

The female LFP rate rose to almost 59 percent in 1990 and 1991, then dropped slightly for a number of years before rising further, to 62.1 percent in 2004. Thus, over the course of nearly a century, female labour force participation in Canada almost quadrupled, from 16 percent to 62 percent. As we will see in Chapter 4, the impact of this long-term trend on workplaces, families, and society as a whole has been immense. During the 1960s and 1970s, much of the growth in female LFP was due to larger numbers of women returning to paid employment after their children were in school or had left home. In contrast, most of the female employment growth in the 1980s and 1990s occurred among mothers who were working for pay and raising young children at the same time. In 2003, 65.7 percent of mothers with children under 6 had a paid job of some kind, compared to just 31.5 percent in 1976 (Statistics Canada 2004c: 20).

While female participation rates have been climbing, male rates declined over most of the past century, although not as steeply. Between 1901 and 1931, the male LFP rate remained close to 90 percent, dropping in the decades before and after World War II to percentages in the mid-80s. In the 1970s and 1980s, the male rates fluctuated between 76 and 78 percent. More recently, the male LFP rate dropped quickly from 77 percent in 1989 to just over 72 percent in 1996, staying at this level until 2000. Economic recovery and growth since the late 1990s have sparked a 4 percent increase in female LFP but have had little effect on male participation rates.

Thus, the long-term growth in total LFP rates in Canada has been the product of substantial increases for women and somewhat less dramatic declines for men. The latter change reflects two trends: men are living longer (with more men living past the conventional retirement age, the proportion of all men out of the labour force is increasing) and more men are retiring early (that is, before the standard retirement age of 65). The causes of growing female labour force participation, which are more complex, are discussed in Chapter 4. A similar convergence of female and male rates has occurred in other industrialized countries (the United States, Britain, Germany, France, Italy, and Japan, for example) but, over the past few decades, Canada has experienced the largest jump in female labour force participation.

It is important to note that these labour force participation rates are annual averages. Given the seasonal nature of some types of work in Canada (jobs in agriculture, fishing, forestry, construction, and tourism, for example), higher levels of labour force participation are typically recorded in the spring, summer, and autumn months. Furthermore, taking a full 12-month period, we observe higher proportions of Canadians reporting labour force participation at some point during the year. Thus, annual averages do not reveal the extent to which Canadians move in and out of different labour market statuses. For example, in 2001, 20 percent of employees experienced a change of status, either changing employers or becoming unemployed. While *job stability* (average length of time in a job) has increased in Canada in the 1990s, relative to the United States, the Canadian labour market still reflects a considerable amount of individual mobility.[16]

In addition, there are a number of groups that continue to have much lower than average rates of labour force activity. Many Aboriginal Canadians are economically marginalized, facing huge barriers to paid employment. For example, the 2001 census shows the employment rate of the Aboriginal population was 37.7 percent for those living on reserves, compared to 62.4 for the Canadian population. Those living in non-reserve areas had higher employment rates of 54.5 percent, up from 47.0 percent in 1996. Aboriginal Canadians also have much lower incomes, on average, and are much more likely to be living below the poverty line. In 2001, full-time workers living off reserve earned $34,700 a year, compared to $43,500 for the non-Aboriginal population. Those living on reserve earned just $28,355.[17] Similarly, research shows that disabled Canadians (who comprise 14% of the Canadian population) are much less likely to be in the paid labour force. In 2001, for example, just 43.7 percent of disabled Canadians were employed, compared to 78.4 percent of Canadians without a disability. While there is no significant difference in unemployment rates, over half of disabled Canadians (51.6%) were not in the labour force, compared to just 16.5% of Canadians without a disability.[18]

Labour Force Participation among Youth

With the exception of a small decline coinciding with the recession of the early 1980s, the labour force participation of Canadian youths (aged 15 to 24, females and males combined) rose more or less steadily from an annual average of 64 percent in 1976 to 71 percent in 1989. Figure 2.4 shows that the youth LFP rate then declined rapidly in the 1990s, dropping to 62 percent by 1996. Since then, it has risen again, to 67 percent in 2004. Over this period of time, the female and male youth LFP rates converged considerably, although male rates remained slightly higher (67.9%) than female rates (66.1%) in 2004.

Over the past two or three decades, teenage (ages 15 to 19) LFP rates have always been considerably lower than LFP rates for young adults (ages 20 to 24), since a larger proportion of the latter group have left the education system. In 2004, female and male rates for teenagers were very close (55.4% for girls and 53.7% for boys) but, among young adults, male LFP rates (81.5%) were considerably higher than female rates (76.4%) (CANSIM 282-0002).

FIGURE 2.4 *Youth Labour Force Participation Rates by Gender, Canada, 1976–2004*

Source: Adapted from the Statistics Canada publication *Labour Force Historical Review*, 2004, Catalogue 71F0004XCB, February 18, 2005.

Why did youth LFP rates decline in the 1990s? The recession that lasted till mid-decade sharply reduced the number of "good" entry-level jobs for youths. In fact, during this period youth wages declined more than for any other age group in the labour force. In turn, a larger proportion of youths decided to continue their education to improve their employment prospects. For example, between 1989 and 1998, the percentage of Canadian youths (aged 15 to 24) who were in school and not working rose from 29 to 40 percent. Furthermore, the size of the 20-to-24-year-old cohort declined, compared to the 15-to-19-year-old cohort. Since young adults have historically had higher LFP rates than teenagers, the overall youth LFP rate declined.[19] And, after dropping so dramatically, why have youth LFP rates increased in recent years? A stronger economy, reflected in lower unemployment rates, is certainly part of the answer. In addition, rapidly rising post-secondary tuition costs may have led some Canadian youths to choose work over school.

Labour Force Participation among Older Canadians

Over the past few decades labour force participation among older men has declined markedly. Between 1976 and 2004, the LFP rate for men aged 55 to 64 fell from 77 percent to 64.4 percent. As we would expect, it is the older members of this cohort who are most likely to leave the labour force. For example, in 2004, the LFP rate for men aged 55 to 59 was 75.6 percent, more than 20 points higher than for those aged 60 to 64 (53.2%). With somewhat better benefits provided to the elderly by government, compared with a generation or two earlier, and with a larger number of older men who have employer-sponsored pensions as well as other personal savings (RRSPs, for example), income security among older men has increased. Consequently, men's average retirement age fell from 65 years in 1976 to 61 years in 1997, rising back up to 63.3 years in 2003.[20]

Many factors influence retirement patterns. According to the 2002 General Social Survey, one-third of recent retirees stopped work for health reasons. Another one-third retired for a range of reasons and had no wish to continue working. But the remaining one-third of retirees would have continued working under different circumstances—for example, if they could work fewer days or hours, take more vacation time, or work part-time

(Morissette, Schellenberg, and Silver 2004). Those who retired "involuntarily" were far more likely to have retired due to health or downsizing, to have low income, and to report low life satisfaction. They were also more likely to lack pension benefits from a former employer and to return to work for financial reasons (Schellenberg and Silver 2004). In fact, in 2001, nearly 14 percent of men aged 65 years and over were still in the labour force, working in a range of jobs—most commonly farming, retail sales, personal services, and specialist professions (e.g., judges, doctors, senior managers). Higher education and self-employment status increased the odds of working as a senior (Duchesne 2004). These patterns challenge the traditional three-stage model of the work life—education, employment, then retirement. Instead, researchers are now using a *life-course perspective* that views individuals' roles as more fluid and examines the choices and constraints underlying transitions across roles.[21]

Despite the small trend toward re-employment among "retired" men, the fact remains that older men who lose their jobs have more difficulty finding work, since many employers are reluctant to hire and retrain older workers. In addition, older men are more likely to be employed in industries where employment is declining (manufacturing and the resource-based industries, for example). Hence, the chances of being re-hired by a former employer are lower than average (Tindale 1991). As a rule, most unemployed older men decide to retire. Some are offered early retirement packages by employers attempting to downsize their workforce, while others decide that "retired" would be an easier label to live with than "unemployed." Thus, while a shortage of jobs may have pushed some men over the age of 55 out of the labour force, an adequate pension encouraged others to retire voluntarily.

We see different work and retirement patterns among older women. In 2003 the median retirement age for women was 60.4 years, 3 years earlier than men. But LFP rates have still increased for older Canadian women because of their historically lower levels of participation in previous decades. In 2004, for example, the LFP rate for women aged 55 to 59 was 60.1 percent, up from 45 percent in 1989. For women aged 60 to 64, the LFP rate increased from 22 percent in 1989 to 34.5 percent in 2004. Financial reasons may be a key motivator (Marshall 2000). Women earn less on average than men and receive fewer benefits, including government and employer pensions (see Chapter 4). Hence, compared with men of the same age, fewer older employed women have the financial resources needed to retire. Furthermore, the growing trend toward wives becoming the sole wage earner in families is most evident in families with husbands aged 55 or older (Crompton and Geran 1995: 28). In some of these families, women may have continued to work after their husbands became unemployed or retired.

Unpaid Work

As stated in the introduction to this book, our emphasis is on paid work since this is the primary type of work for the majority of adults in an industrialized capitalist economy. Nevertheless, if we define work as activity that provides a product or service used or valued by others in society, it is obvious that paid work constitutes only a portion of all work done in Canada. The largest and most serious omission is *unpaid household and child/elder-care work.* In 1996 the census included questions about unpaid work for the first time. In 2001, the census found that 9 out of 10 Canadians had done at least some unpaid housework in the week prior to the census. 38 percent reported unpaid childcare and 17 percent reported unpaid eldercare. These figures are virtually identical as in 1996. As we might expect, women, particularly mothers of young children, do much more unpaid work. But, as we have noted earlier in this chapter, many of these young mothers are also members of the paid labour force. Labour market analysts have estimated that, if given a dollar value, unpaid work would add $235 billion, or another one-third, to Canada's gross domestic product (GDP).[22]

Also omitted from labour force participation rates and employment statistics are those Aboriginal Canadians (and a few non-Aboriginal Canadians), typically resident in the northern parts of the country, who continue to participate in traditional hunting and gathering work. Such groups form only a very small part of Canada's population. Nevertheless, their *subsistence work* clearly provides useful and valued goods and services for their families and communities.

In addition, *volunteer work* is overlooked if we rely only on official labour force statistics. According to the 1997 National Survey on Giving, Volunteering and Participation (NSGVP), 31 percent of Canadians (7.5 million individuals) contributed time and energy to volunteer activities in the previous year. In 2000, however, the NSGVP found a slight decline in volunteer activity. Just 27 percent of Canadians reported volunteering, though they contributed more time on average (162 hours per year) than those in 1997 (149 hours per year) (Hall, McKeown, and Roberts 2001). Though not directly comparable, another survey—the 2003 Survey of Non-Profit and Voluntary Organizations (NSNVO)—estimates that Canadians contributed more than 2 billion hours of volunteer time, worth the equivalent of 1 million full-time jobs.[23]

Volunteer workers are found in a wide variety of clubs and associations, as well as in religious and political organizations. However, a very large number work in the publicly funded service industries, assisting in the provision of health, social, recreational, educational, environmental, and other types of services. In recent years, as government departments at all levels have attempted to cut costs and reduce payrolls, volunteers have also been replacing paid employees. This trend will no doubt continue, so volunteer work is likely to become an even larger part of the total amount of unpaid work performed by Canadians. Recognizing this, the Voluntary Sector Initiative was launched in 2000 by the Government of Canada and the voluntary sector to help strengthen and improve volunteer contributions in Canada. A new survey—the Canadian Survey on Giving, Volunteering and Participating (CGSVP)—will now be done every three years to document the contributions of this important sector.

Official labour force participation statistics also omit some Canadians working in the *hidden economy*, or, as it is sometimes called, the underground or irregular economy. Such work includes illegal activities like selling drugs, prostitution, and gambling, as well as legal work done for cash or in exchange for other services. Mechanics working out of their backyard garages and not reporting this income and farmers assisting their neighbours in return for similar help at another time would be examples of this non-criminal group, some of whom would be represented in labour force surveys through their regular jobs. Others, if surveyed, might report themselves unemployed or out of the labour force in order to avoid detection by tax authorities.

There is little research on work in the hidden economy, since there is no direct way of systematically studying behaviour that people do not wish to have publicly documented. But the research available suggests that participation in the hidden economy is relatively common. During the mid-1980s, an assessment of concealed employment (legal but unreported work) in Western industrialized countries concluded that up to one in five adults might be involved at some point in a given year, although, for most of these individuals, this would reflect relatively few hours of work. Canadian researchers estimate that about 15 percent or more of total economic activity in Canada was not being reported to tax authorities in the mid-1990s.[24] The introduction of the Goods and

Services Tax (GST) several years earlier may have prompted more independent self-employed workers to begin asking for "under-the-table" payment.

In short, official statistics clearly underestimate the number of working Canadians and the value of their work. Various types of informal work (household and child/elder-care work, volunteer work, work in subsistence economies, and work in the hidden economy) are overlooked. This reflects how society values individuals' unpaid activities. Indeed, there is growing discomfort with conventional approaches to measuring economic activity. As articulated by three American economists, "the GDP [the nation's economic output] ignores the contribution of the social realm—that is, the economic role of households and communities. This is where much of the nation's most important work gets done, from caring for children and older people to volunteer work in its many forms. It is the nation's social glue."[25] Furthermore, it appears that informal work may be increasing. However, we could not even begin to provide an adequate overview of both paid employment and the various types of informal work in one volume. Consequently, with some exceptions, the rest of this book focuses on paid employment (including self-employment in the formal economy), which continues to represent the largest share of work performed by Canadians.

INDUSTRIAL CHANGES: THE EMERGENCE OF THE SERVICE ECONOMY

The work done by labour force participants can be categorized in several ways. We begin below by discussing industrial shifts and then examine occupational changes in the following section. *Industry* classifications direct our attention to the major type of economic activity occurring within the workplace. In the broadest sense, we can distinguish the *primary sector,* including agriculture, mining, forestry, and other resource extraction industries, from the *secondary sector* (manufacturing and construction), where goods are produced from the raw materials supplied by the primary sector, and the *tertiary sector*, where services rather than products are provided. A *service,* simply defined, is the exchange of a commodity that has no tangible form. These general industrial categories can be further subdivided into more specific industries. The service sector, for example, includes among others the finance, education, retail trade, government (public administration), and health service industries.

Service-Sector Growth

In 1891, the primary industries accounted for 49 percent of the Canadian labour force, the secondary sector for 20 percent, and the service sector for the remaining 31 percent (Matthews 1985: 36). In the following half-century, the primary sector lost its dominance, while the manufacturing industries expanded. At the same time, the service industries were also increasing in size. By 1951 almost half (47 percent) of all employed Canadians were working in the service industries, while the secondary sector accounted for almost one-third (31 percent) of employment (Picot 1987: 11). The relative size of the primary and secondary sectors have been declining steadily ever since. By 2004, 74 percent of employed Canadians held service-sector jobs, 21 percent were working in the secondary sector, and only 5 percent were employed in the primary industries (Bowlby 2000). Agriculture, a dominant industry a century ago, now accounts for just 2 percent of all employment in Canada. In contrast, the portion of workers in the service sector has more than doubled over this time. This sector has also come to represent a much larger share of the total value of goods and services produced in the Canadian economy.[26]

These statistics clearly demonstrate that we are now living in a service-dominated economy. A similar transition from a pre-industrial to an industrial to a service economy occurred in other capitalist countries such as the United States, the United Kingdom, Japan, Sweden, and Germany. But manufacturing never played as great a role in Canada's development as in that of other countries. Consequently, while three out of four workers are now employed in the service sector in both Canada and the United States, the service industries still account for a somewhat smaller share of the employed labour force in these other countries (OECD 2003).

The rise of the service sector can be attributed to a combination of factors. Productivity gains due to new technologies and organizational forms in manufacturing, but also in the resource industries, have meant that fewer people could produce much more. In primary industries, the growing size and declining number of farms in Canada is perhaps the best indicator. Toward the end of the 19th and into the early 20th century, factory mechanization greatly enhanced workers' productivity, despite a reduction in hours worked. In the last few decades, automated production systems (robotics and information technology) have once again accelerated this trend. Thus, the expansion of production in the primary and secondary sectors has been accompanied by a relative decline in the need for employees.

Industrial expansion and productivity gains over the years have also led to higher incomes and increased amounts of leisure time. These factors, in turn, have fuelled the demand for a wider range of services, particularly in the last few decades when the recreation, accommodation, and food service industries have been growing rapidly. In addition, the expansion of the role of the state as a provider of educational and social services and a funder of health services contributed significantly to the growth of the service sector. In the 1990s, governments' vigorous pursuit of deficit reduction resulted in reduced program funding, organizational downsizing, and the privatization of work previously done by civil servants, thus slowing the trend toward public-sector employment. However, both provincial and federal government departments have increased their hiring in the early years of the 21st century, in part because of the large number of baby-boomer employees approaching retirement.

Employment Diversity within the Service Sector

The wide range of specific industries within the broad service sector can be usefully categorized into six broad groups: distributive services (transportation, communication, and wholesale trade); business services (finance, insurance, real estate, and other services to business); the education, health, and welfare sector; public administration; retail trade; and other consumer services. Distributive services differ from the others by being the final link in the process whereby raw materials are extracted, transformed, and then delivered to the ultimate consumer. Business services also provide support to the primary and secondary industries (the goods-producing sector), but in a less tangible way. Like public administration, the education, health, and welfare sector contains primarily non-commercial services provided by the state, while retail trade and other consumer services (for example, food and beverage, accommodation, and the tourism industries) are commercial services aimed directly at consumers.[27]

It is useful to further combine the six service categories into an *upper tier* (distributive, business, education, health and welfare, and public administration) and a *lower tier* (retail trade and other consumer services) since, as we will demonstrate later, many more good jobs are located in the former. Table 2.1 uses these concepts to profile the industrial distribution of Canada's employed

TABLE 2.1 *Industry by Gender, Employed Population, Canada, 2004*

	Total	Women	Men
Industry	%	%	%
Goods-producing sector (total)	25.0	12.6	35.9
Agriculture	2.0	1.3	2.7
Natural resource based	1.7	0.6	2.8
Manufacturing	14.5	8.8	19.4
Utilities	0.9	0.4	1.1
Construction	5.9	1.5	9.9
Service-producing sector (total)	75.0	87.4	64.1
Upper-tier services (total)	52.2	60.4	45.0
Wholesale trade/transportation	8.7	4.9	12.0
Finance/insurance/real estate/leasing	5.9	7.5	4.7
Professional/scientific/technical services	6.3	5.7	7.0
Business/building/other support services	4.0	4.0	4.0
Educational services	6.5	9.1	4.2
Health care/social assistance	11.0	19.0	3.7
Information/culture/recreation	4.6	4.8	4.4
Public administration	5.2	5.4	5.0
Lower-tier services (total)	22.8	27.0	19.1
Retail trade	12.1	14.0	10.4
Accommodation/food services	6.3	8.1	4.7
Other consumer services	4.4	4.9	4.0
Total percent	100.0	100.0	100.0
Total (in 000s)	15,949	7,470	8,479

Source: Adapted from the Statistics Canada publication *Labour Force Historical Review*, 2004, Catalogue 71F0004XCB, February 18, 2005.

(other tables in this chapter also include only employed members of the labour force). In 2004, just one-quarter (25%) of employed Canadians were working in the goods-producing sector while, as already noted, 75 percent were employed in the service industries. Just over half (52.2%) of the employed were located in the upper-tier services, while not quite one-quarter (22.8%) had jobs in the lower-tier service industries.[28]

We have already observed that similar proportions of workers are employed in the service and goods-producing sectors in Canada and the United States. A more detailed comparison reveals only one major difference— the retail trade sector is considerably larger in the United States (about 17% of total employment, compared to 12% in Canada). However, we find a very different industrial distribution in Mexico, the other major North American economy. There, 20 percent of the employed labour force is located in the primary industries, compared to only 5 percent in Canada and 3 percent in the United States. Similar proportions are employed in manufacturing in all three countries, but the business services sector is only about one-fifth as large (in relative terms) in Mexico as in Canada and the United States.[29]

Gender and Age Differences in Service-Sector Employment

Table 2.1 also reveals distinct gender differences in industrial locations. In 2004, almost 36 percent of employed Canadian men were working in the goods-producing sector, compared with less than 13 percent of employed women. Women were more likely to be employed in the lower-tier services (27%, compared with 19.1% of men). Within the upper-tier services, women were much more heavily concentrated in the education, health, and welfare industries than were men. Men were considerably more likely to be employed in the distributive services (transportation, communication, and wholesale trade).

Age plays an important part in further stratifying the employed labour force across these industrial categories. Younger workers (aged 15 to 24) are much less likely to be employed in the upper-tier services. Instead, many young workers, particularly women, hold jobs in the lower-tier retail trade and consumer service industries.[30] Many of these young workers are students in part-time jobs. Thus, the expansion of the lower-tier service industries (for instance, shopping malls, fast-food restaurants, and tourism) since the 1970s has relied heavily on the recruitment of student workers, creating a distinct student-worker segment of the labour force. However, older female workers are also much more likely than older male workers to be employed in the retail trade and other consumer service industries.

OCCUPATIONAL CHANGES

While the industrial classification system is based on what is being produced, we also can categorize workers according to their *occupation*. Occupational distinctions are determined by the work an individual typically performs, the actual tasks that she or he completes. Thus, secretaries, managers, office cleaners, and accountants (occupational titles) work in mining companies, automobile factories, and government bureaucracies—that is, in the primary, secondary, and tertiary (or service) sectors of the economy. Other occupations are typically found only within specific industrial sectors. For example, teachers, nurses, and retail clerks are occupational groups largely unique to the service sector. So the two classification systems parallel each other to an extent but also overlap considerably.

For many years, Statistics Canada used the Standard Occupational Classification (SOC) system to categorize occupations. But a basic limitation of the SOC system was that it did not adequately capture the skill content of jobs. From our earlier discussions of education and training, it is clear that developing human resources—the skills of the workforce—has become a public policy priority. Consequently, in the 1990s, Human Resources Development Canada (HRDC) teamed up with Statistics Canada to develop a new occupational typology, the National Occupational Classification (NOC).

Table 2.2 displays the structure of the NOC with selected examples of occupations in each category. Essentially, the NOC identifies two dimensions of *skill*, providing information on skill level and generic skill type for each of the 25,000 detailed occupations in the Canadian labour market. Four basic skill levels are used, plus a separate level for management occupations, and each is directly related to the education and training required. For example, occupations in skill level A are mainly the professions and require a university education; occupations in skill level D require no formal education and only short demonstrations or some on-the-job training. The nine skill types indicate the kinds of tasks performed. Note that the highest skill groups are concentrated mainly in what we earlier referred to as "upper-tier services," while skill-level-D occupations are in lower-tier services and goods production.

Blue-Collar and White-Collar Occupations

Many but not all of the people employed in the primary and secondary industries would be classified as blue-collar workers, while most of those working in the service industries would be identified as white-collar workers. The term *blue-collar* has traditionally been used to distinguish occupations with potentially dirty working conditions (for example, farming, mining, truck driving, construction, and factory work) from *white-collar* (clerical, sales, managerial, and professional) occupations. The explanation, of course, is that for the former jobs wearing a white collar would be inappropriate, since it would not stay clean. In the past, white-collar occupations were viewed as having "higher status." But with the expansion of the service sector, especially lower-tier

TABLE 2.2 Classifying Occupations by Skill Type and Level—The National Occupational Classification*

	Occupational Skill Type								
Skill Level	Business, finance & administration	Natural & applied science	Health	Social sciences, education, government services, religion	Arts, culture, recreation, sport	Sales and service	Trades, transportation & equipment operators	Primary industries	Occupations unique to processing, manufacturing & utilities
Management	Senior managers, legislators, managers								
Skill Level A	auditors, accountants, human resource professionals, investment professionals	engineers, architects, systems analysts, computer programmers, physical and life science professionals	physicians, dentists, chiropractors, pharmacists, registered nurses, therapy professionals	judges, lawyers, professors, teachers, psychologists, social workers, clergy, probation officers, policy and program officers, researchers and consultants	librarians, writers, translators, public relations professionals, creative and performing artists				

(continued)

TABLE 2.2 Classifying Occupations by Skill Type and Level—The National Occupational Classification* (cont.)

Skill Level B								
clerical supervisors, administrative occupations, secretaries	technical occupations in physical and life sciences, engineering, architecture, electronics	medical and health care technologists and technicians	paralegals; social service, religious, and education occupations not classified above	technical occupations in libraries, museums, galleries; photographers, graphic artists, technical occupations in motion pictures, broadcasting, performing arts; announcers, athletes, and coaches	sales and service supervisors, technical sales specialists, insurance and real estate sales, chefs and butchers, police officers and firefighters	supervisors, machinists, electrical trades and telecommunications occupations, plumbers, carpenters, other construction trades, mechanics, train crew, tailors, jewellers	supervisors, underground miners, logging machine operators, fishing vessel operators, fishermen/women; contractors, operators, and supervisors in agriculture	supervisors, central control and processing operators

Skill Level C	clerical occupations, office equipment operators, library clerks, mail and message distribution occupations	assisting occupations in health services	retail sales clerks, travel and accommodation occupations, tour and recreational guides, food and beverage service occupations, childcare and home support workers	motor vehicle and transit drivers, heavy equipment operators, installers, repairers, and servicers	logging and forestry workers, mine service workers, oil and gas drilling operators, agriculture workers, fishing and trapping workers	machine operators and related workers, assemblers
Skill Level D			cashiers, food counter attendants, security guards, cleaners	trades helpers, labourers	labourers	labourers

* This table gives examples of occupational groups (based on 3-digit Standard Occupation Classification). For a complete list see *Job Futures 1996. Volume One: Occupational Outlooks* (Ottawa: Minister of Supply and Services Canada, 1996), xv–xviii, 455–57.

Source: "National Occupational Classification Matrix," http://www23.hrdc-drhc.gc.ca/2001/e/generic/matrix.pdf, National Occupational Classification—2001, Human Resources and Skills Development Canada. Reproduced with the permission of the Minister of Public Works and Government Services Canada, 2005.

services where low-paying jobs are quite common, the white-collar occupational category has come to include the majority of the labour force, including many people in less desirable jobs.

As primary-sector industries became less important over the course of the last century, we saw a huge decline in the size of some occupational groups and a proportional increase in others. A few examples can make the point. The 1911 census reported that 34 percent of the labour force worked in agricultural occupations, compared with only 4 percent in clerical occupations, and a similarly small proportion in professional occupations. By 1951, agricultural occupations had declined to 16 percent of the labour force, compared with 11 percent in clerical occupations and 7 percent in professions (O'Neill 1991: 10). The 1996 census revealed less than 5 percent of the labour force in occupations unique to primary industry, 9 percent in management, 19 percent in business, finance, and administrative occupations, and almost 20 percent in natural science, social science, government, and other white-collar occupations.[31] Thus, over the course of the past century, white-collar jobs have come to dominate the Canadian labour market.

Gender and Occupational Location

Many of the new white-collar positions have been filled by women, while the remaining blue-collar jobs in the primary and secondary sectors are still typically held by men. In fact, given the heavy concentration of women in clerical, sales, and service occupations, the term *pink collar* has been used to describe these occupational categories.

Using a revised version of the SOC to obtain comparability with earlier data, Table 2.3 presents a gender breakdown of Canadian occupational distributions for 1987 and 2004.[32] Note that the total employed labour force increased from 12.3 million to 15.9 million workers. Hence, even though some occupational categories may have declined a bit in relative (percentage) size, they might still have increased in absolute size between 1987 and 2004. Considering first the total distributions (female and male combined), we note some small changes. Managerial and professional administrative occupations have remained relatively stable (from 11% to 12% of all employed). At the same time, the other administrative and clerical category (office jobs without major planning and supervisory roles) decreased marginally, from 17.2 to

TABLE 2.3 *Occupation by Gender, Employed Population, Canada, 1987 and 2004*

	1987			2004		
	Women	Men	Total	Women	Men	Total
	%	%	%	%	%	%
Management occupations	6.0	10.6	8.6	7.0	10.8	9.0
Business, finance & administration						
Professionals	1.9	2.3	2.1	3.2	2.7	2.9
Administrative & clerical	29.7	8.0	17.2	24.3	7.1	15.2
Natural & applied science	2.3	7.0	5.0	3.0	9.7	6.6
Health occupations	9.2	1.8	5.0	10.1	2.1	5.9
Social science, education & government	8.2	4.7	6.2	11.4	4.7	7.8
Culture, recreation & sport	2.7	2.1	2.4	3.3	2.6	2.9
Sales & services						
Retail sales	9.7	3.0	5.9	9.3	3.8	6.4
Other sales & services	20.3	16.4	17.5	19.9	15.8	17.7
Trades, transportation & related	2.1	28.9	17.4	2.2	26.1	14.9
Occupations unique to primary industries	2.3	7.2	5.1	1.4	5.2	3.5
Occupations unique to manufacturing	5.8	9.1	7.7	4.8	9.3	7.2
Total percent	100.0	100.0	100.0	100.0	100.0	100.0
Total (in 000s)	5,310	7,024	12,334	7,470	8,479	15,949

Source: Adapted from the Statistics Canada publication *Labour Force Historical Review*, 2004, Catalogue 71F0004XCB, February 18, 2005.

15.2 percent of all employed. The remaining white-collar occupational categories (both professional, and sales and service) changed only to a limited degree, if at all. However, in the blue-collar occupations, we see relative declines in primary occupations (from 5.1% to 3.5%) and in manufacturing occupations (from 7.7% to 7.2%). The decline in blue-collar occupations is a result of the massive restructuring in the goods-producing sector outlined in Chapter 1, while the smaller decline in clerical occupations reflects a complex set of related changes, from advancing information technology to downsizing to the upgrading of some of these jobs to administrative categories.

Chapter 4 focuses on the unequal employment experiences and rewards of women and men, so we will comment only briefly at this point on the gender differences in Table 2.3. The 2004 data show a more even distribution of men than women across the occupational structure. Women are more heavily concentrated than men in "pink collar" occupations, with 24.3 percent employed in administrative and clerical occupations, 9.3 percent in retail sales, and 19.9 percent in other sales and service occupations, totalling 53.5 percent. In contrast, less than 27 percent of men are employed in these three occupational groups. Women are also relatively overrepresented in health (10.1%) and social science, education, and government (11.4%) jobs, which were cut in the 1990s but have expanded in the last few years.

However, it is important to note the changes over time in the gender composition of the occupational structure. In 1987, only 7.9 percent of employed women were in managerial or professional administrative occupations. Seventeen years later, 10.2 percent of women were working in these higher-status white-collar occupations. Table 2.3 also shows a small increase over time in the proportion of women in other professional positions, and a corresponding decline in the percentage of women in clerical jobs.

To conclude, there have been some shifts in the gender composition of the occupational structure, although female–male differences have not disappeared. But we must remember that these are broad occupational categories. Health occupations, for example, include doctors, nurses, orderlies, nursing assistants, and other support workers. Women are still more likely to be in the lower-status occupations within these broad categories, and are still typically supervised by men. Hence, it is necessary to look in more detail at the specific occupations within which women and men typically are employed, as we do in Chapter 4.

SELF-EMPLOYMENT TRENDS

As in other industrialized capitalist economies, self-employment has dropped sharply over the past century. In 1946, for example, 33 percent of employed Canadians were self-employed. By 1981 this figure had fallen to 10 percent (Riddell 1985: 9). This shift toward paid labour can be explained to a considerable extent by changes in the agricultural sector. Early in the 20th century, about three-quarters of those employed in agriculture were owner–managers, and most of the rest were unpaid family workers. But over the past 65 years, the number of farms in Canada has suffered a huge decline, from a high of 733,000 in 1941 to only 224,670 in 2001. At the same time, the remaining farms have been getting larger, and the proportion of paid workers in agriculture has been slowly increasing. By 1991, the 448,000 Canadians employed in agriculture consisted of paid workers (32%), unpaid family workers (9%), and self-employed individuals (59%). Thus, while wage labour has become more common in agriculture, a major decline in the number of family farms has also meant fewer self-employed individuals and unpaid family workers.[33]

Over the decades, self-employment in the secondary and service sectors also declined slowly. However, a reversal of this trend appeared about 20 years ago. In 2004, 2.4 million (15%) of all employed Canadians (including those in agriculture) were self-employed, up from 10 percent in 1981 (but down slightly from 17% in 1999).[34] A similar increase was observed in most other industrialized capitalist economies, including that of the United States, during the 1980s, but the trend has continued longer in Canada. Thus, while self-employment in Canada continued to rise in the 1990s, the self-employment rate in the United States remained around 10 percent. In Canada a decade ago, the number of *self-employed employers* (who hire others to work for them) and *own-account self-employed* (without employees) were roughly similar, but the latter group has now become larger. More than half of the growth in Canadian self-employment in the 1980s involved employers setting up new businesses. In contrast, in the 1990s, 90 percent of the growth involved the own-account self-employed. But while slightly more than 1 in 10 Canadian workers are now own-account self-employed, this figure is still very low compared to many less industrialized countries. For example, in 1999, 25 percent of the employed in Mexico were in this category.[35]

Self-employment increases with age. According to the 2001 census, the self-employed are about 8 years older on average than employees (46 versus 38 years). Indeed, while nearly one-quarter of the self-employed are 55 years and over, this is true for just 10 percent of employees. Self-employment is still more prevalent among men than among women. Thus, in 2004, 12.4 percent of employed women were self-employed, along with 19.5 percent of employed men. But the female self-employment rate has been rising steadily since the 1980s, and the rate of increase has been higher for women, especially as employers. By 2004, women made up almost 40 percent of all own-account self-employed in Canada, and over one-quarter of all employers. Some women have moved into self-employment to better balance work and family, while others seek greater challenge and opportunity (Hughes 2005, 1999; Miniti, Arenius, and Langowitz 2004).

The recessions in Canada at the beginning of the 1980s and the 1990s may have contributed indirectly to the rise of self-employment, particularly own-account self-employment. Indeed, in 2000, about one-fifth of women and men reported being forced into self-employment by job loss or lack of other opportunities (HRDC and Statistics Canada 2002). But a direct relationship between economic downturns and growth in self-employment is weak at best (Arai 1997; Hughes 2005; Manser and Picot 1999: 43). While many employers in both the private and public sectors reduced the size of their labour force during these recessions, leading to higher rates of unemployment, it was not necessarily those who lost their jobs who then set up their own businesses. The majority of the self-employed typically come directly from the ranks of wage-earners. Furthermore, managerial, professional, and technical workers are more likely to become self-employed than are clerical and manual workers (the groups most affected by unemployment). The self-employed are more highly educated than the rest of the labour force, suggesting that at least some of the growth in self-employment may have been a response by more advantaged workers to perceived job insecurity.

The self-employed are dispersed across a wide range of industries and occupations, with considerable variation in income and other rewards. According to the Survey of Self-Employment in Canada, a special survey conducted by Statistics Canada in 2000, 42 percent are involved in managerial or professional work, one-quarter are in service occupations, 21 percent work in blue-collar jobs, and the remainder (11.7%) work in primary industries such as agriculture (HRDC and Statistics Canada 2002). Given this wide dispersal of the self-employed across occupations and throughout the goods-producing and upper- and lower-tier services, it is not surprising that the income distribution of Canada's self-employed is bimodal. In 2002, 57 percent of the self-employed earned less than $20,000, compared to only 27 percent of paid workers. At the same time, 6 percent of the self-employed were among Canada's highest earners, reporting incomes in excess of $100,000, compared to only 1 percent of paid workers (Statistics Canada 2004c: 70).

THE CANADIAN LABOUR MARKET: REGIONAL VARIATIONS

So far, we have treated Canada as a single economic entity, putting aside important regional differences. Anyone who has driven across Canada will have a strong sense of its regional diversity. The fishing boats and lumber mills of the West Coast are left behind as one begins to head east toward the Prairies, replaced by oil wells and grain elevators. The flat landscape disappears a few hours after leaving Winnipeg, and one is faced by the rocks, forests, and water of the Canadian Shield. The smokestacks of Sudbury and other mining communities are reminders of the natural resource base of that region, but they eventually give way to the old grey barns of Ontario, symbols of an agricultural economy older than the one observed several days earlier. But this is unlikely to be the only memory of the trip through Ontario and into Quebec. The huge Highway 401 pushing its way past miles of warehouses and suburban factories will also leave a strong impression as it takes one through the industrial heartland of the country. Then, repeating the pattern observed in the West (but in reverse, without the mountains, and on a smaller scale), the farming economies of Quebec and the Atlantic provinces begin to merge with a forest-based economy. Eventually, one returns to a region where fisheries are once again important. In short, regional diversity means economic diversity.

This description obviously over-generalizes. There are high technology firms in the Fraser Valley, large factories in Calgary, and oil wells off the East Coast. And in the northern parts of Canada, energy development and mining compete uneasily with traditional Aboriginal hunting, fishing, and trapping. Still, it is essential to take into account the regional distribution of industries when considering work opportunities available to Canadians. Some regions are much more economically advantaged than others, as we point out below in our discussion of unemployment trends. Labour force participation rates also vary significantly across provinces. For example, in 2004, Alberta's labour force participation rate of 73.6 percent was much higher than the 59.3 percent rate in Newfoundland.

A brief comparison of industrial patterns of employment in 2004 highlights these regional variations (Table 2.4). For example, 7.2 percent of the employed labour force in Manitoba and Saskatchewan and 3.7 percent of that in Alberta had jobs in agriculture, compared with less than 2 percent in the other regions. Large concentrations of workers in the resource-based industries reflect the presence of the fishing industry on the East Coast and in British Columbia; forest-based industries in the Atlantic provinces, Alberta, and British Columbia; and the oil and gas industry in Alberta. Meanwhile, 17 percent of workers in Quebec and Ontario had jobs in manufacturing, compared with only 10 percent or fewer in the other regions. It is also interesting to note that Alberta and Ontario, the two provinces taking the lead in scaling down government over the past decade, have the lowest employment in publicly-funded industries (education/health/welfare and public administration).

TABLE 2.4 *Industry by Region, Employment, Canada, 2004*

Industry	Total %	Atlantic Provinces %	Quebec %	Ontario %	Manitoba/ Saskatchewan %	Alberta %	British Columbia %
Goods-producing sector							
Agriculture	2.0	1.6	1.4	1.2	7.2	3.7	1.8
Natural resource-based	1.8	4.1	0.9	0.5	2.3	6.3	1.7
Manufacturing	14.5	10.1	17.1	17.5	9.3	8.2	10.1
Utilities	0.8	0.8	0.9	1.0	1.1	0.9	0.4
Construction	5.9	5.9	4.5	5.8	4.9	9.1	7.0
Service sector							
Upper-tier services							
Wholesale trade	3.6	3.0	3.9	3.8	3.4	3.5	3.3
Business services	16.3	13.8	14.5	18.0	12.7	15.7	17.4
Education/health/ welfare	17.3	20.5	18.5	16.3	20.7	15.6	17.2
Public administration	5.2	6.5	6.0	5.0	5.9	4.0	4.7
Lower-tier services							
Retail trade	12.2	13.5	12.8	11.5	12.0	11.5	12.3
Other consumer services	20.4	20.2	19.5	19.4	20.5	21.5	24.1
Total percent	100.0	100.0	100.0	100.0	100.0	100.0	100.0
Total (in 000s)	15,949	1,073	3,685	6,316	1,055	1,757	2,059

Source: Statistics Canada 2005. *Labour Force Historical Review,* 2004. Catalogue 71F0004XCB.

Canada is not unique in having some regions in which primary industries are most important and others in which manufacturing is concentrated. But its economic history is marked by a reliance on exports of raw materials such as furs, fish, timber, wheat, coal, natural gas, and oil. This was Canada's colonial role in the British Empire, but even as an independent industrialized nation it has continued to provide natural resources to the global economy. The *staple theory of economic growth,* developed by Harold Innis and others, documents the economic, political, and social consequences of this dependence on the export of unprocessed staple products. The theory argues that overcommitment to extraction and export of a single or few resources makes a nation or region vulnerable in world markets. The relative absence of manufacturing industries means that large portions of the workforce remain employed in lower-skill primary- and tertiary-sector jobs. In addition, a weak manufacturing sector does not encourage significant quantities of research and development and does not generate spin-off industrial activity. These factors create a "staples trap" in which there are few economic development alternatives (Watkins 1991).

Indicative of Canada's natural resource base is the number of single-industry communities scattered across the country. These towns and cities, often situated in relatively isolated areas, exist only because the extraction of some natural resource requires a resident labour force. While some of these communities have attempted to diversify their economies, few have been successful. Instead, their economic activity remains dominated by the primary resource extraction industry (oil in Fort McMurray, Alberta, for example, or nickel in Sudbury, Ontario). If the market for the staple declines or the resource is depleted, the economic base of the community will crumble. Mine shutdowns in Uranium City, Saskatchewan, Schefferville, Quebec, and Elliot Lake, Ontario, in the 1980s are examples, as are the population declines in the 1990s in Labrador City, Newfoundland, and Kirkland Lake, Ontario.[36] By contrast, a stronger world market for oil has made it economically feasible to invest in oil-sands production, creating another economic boom in Fort McMurray and other northern Alberta communities.

Over the decades, the prospects of better jobs and higher incomes have led many thousands of people to "go down the road" to the industrial cities of central Canada. But many have also been attracted to the resource towns of the hinterland for the same reasons. Such communities have historically provided work for unemployed or underemployed migrants from other regions of the country. However, the life cycle of resource towns, as they move through fairly predictable stages of development, is one in which work opportunities for residents and migrants do not remain constant. Moreover, since mining, forestry, the railway, and other blue-collar industries traditionally have been male occupational preserves, women have had trouble finding satisfactory employment. Many single-industry towns are located in areas with sizable Aboriginal populations, yet these groups seldom benefit from the employment opportunities generated by the towns.

UNEMPLOYMENT TRENDS

Counting the Unemployed

Labour Force Survey estimates indicate that, during 2004, there were 17,183,400 Canadians aged 15 and older in the labour force, with 15,949,700 of them employed (including self-employed). The unemployed, numbering

1,233,700, made up the difference. To put this in perspective, the number of unemployed people in 2004 was larger than the total population (children included) of Nova Scotia and Prince Edward Island combined. The official *unemployment rate* is calculated by dividing the number of individuals out of work and actively looking for work (the unemployed) by the total number of labour force participants (including the unemployed). Thus, in 2004, the national unemployment rate was 7.2 percent, down from 7.6 percent a year earlier and a considerably higher jobless rate (between 9 and 11 percent) in the earlier part of the 1990s.

Such calculations reveal the percentage of labour force participants who are unemployed at a particular point in time (an annual average provides the average of 12 monthly estimates). But over the course of a year, many people find jobs while many others quit or lose them. Consequently, if we were to count the number of people who had been unemployed at some point during a year, this alternative unemployment rate would be higher. For example, in 1997 when the annual average unemployment rate was high (9.1%), over the course of the entire year, 17 percent of individuals and 28 percent of all families had experienced unemployment (Sussman 2000: 11). Moreover, while long term unemployment (defined as 52 or more weeks without a job) has fallen significantly since the 1990s, 10 percent of the unemployed in 2003 had not worked in at least a year (Dubé 2004).

The official definition of unemployment, whether we calculate rates for a single point in time or over the full year, identifies a state of being without paid work (including self-employment). It excludes students who do not wish to work while studying, individuals performing unpaid work in the home, the disabled who are not seeking work, and the retired, all of whom are considered to be outside the labour force. It also excludes potential labour force participants who, believing there is no work available in their community or region, have not actively sought employment in the previous four weeks. In the wake of the 1981–82 recession, the Labour Force Survey estimated the size of the *discouraged workers* group at 197,000 (March 1983). The number of discouraged workers declined as the economy recovered, and did not rise as high again during the recession that began in 1990, peaking at 99,000 in March 1992. This lower number of discouraged workers reflects other basic labour market changes already discussed: declining youth labour force participation, rising educational enrollments, and earlier retirement. In other words, rather than waiting for new jobs to materialize, a larger proportion of jobless Canadians may have stayed in or returned to school, or retired early, during the 1992–93 recession. In 2000, when the unemployment rate (6.8%) was at its lowest rate since 1975, the Labour Force Survey identified only 48,000 discouraged workers.[37]

Canadian Unemployment Rates over Time

Canadian unemployment rates reached their highest point last century (around 20 percent) during the Depression of the 1930s (Brown 1987). But these hard times were quickly replaced by labour shortages during World War II and in the immediate postwar years. The average national unemployment rate was only 2 percent during the 1940s. Since the end of World War II, unemployment has slowly climbed higher, responding to successive business cycles of recession and expansion. Unemployment rates peaked at over 11 percent during recessions at the beginning of the 1980s (11.9% in 1983) and the 1990s (11.4% in 1993).

In the early 2000s, the national jobless rate has fallen to around 7 percent, standing at 7.2 percent in 2004. Still, over the past half-century, we see a long-term upward trend (Figure 2.5). The 10-year average unemployment rate was

FIGURE 2.5 *Unemployment Rate by Age and Gender, Canada, 1946–2004*

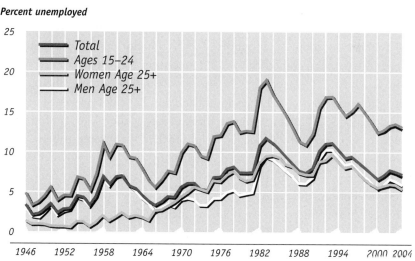

Percent unemployed

Legend:
- Total
- Ages 15–24
- Women Age 25+
- Men Age 25+

Source. Adapted from the Statistics Canada publications *Labour Force Historical Review*, Catalogue 71F0004XCB, February 18, 2005; and *Canadian Economic Observer*, Catalogue 71-001, "The Labour Force Survey: 50 Years Old," March 1996, vol. 09, no. 03.

4 percent in the 1950s, 5 percent in the 1960s, 6.7 percent in the 1970s, and 9.5 percent in the 1980s.[38] The 10-year average for the 1990s remained at the same level. Thus, some analysts argue that without a more proactive job creation strategy on the part of government, and more private-sector investments that generate demand for human capital, Canada's unemployment is unlikely to drop to the levels of the economically buoyant 1960s.

As unemployment rates have slowly climbed over the past decades, the duration of unemployment has also increased. In 1976, the average spell of unemployment was only about 14 weeks. This rose to 21 weeks by 1983, falling and then rising again to 26 weeks in 1994. In 1998, as the rate of unemployment had started to decline, the average duration of unemployment was still over 20 weeks. More than 1 in 5 (22%) unemployed Canadians had unsuccessfully tried to find work for 6 months or longer, and 10 percent of the unemployed had been seeking work for a full year or longer (CANSIM 282-0048). Older workers laid off during the recessions of the early 1980s and 1990s were most likely to experience prolonged periods of unemployment. In 2004, the average duration of unemployment fell to 15.5 weeks, a rate not seen since the late 1970s.

Regional Variations in Unemployment

National rates of unemployment conceal considerable variation across regions of the country. In 2004, Newfoundland and Labrador's unemployment rate averaged 15.6 percent, compared with 9.8 percent in New Brunswick, 8.5 percent in Quebec, 6.8 percent in Ontario, and 4.6 percent in Alberta. The Atlantic provinces, particularly Newfoundland and Labrador, have had higher-than-average rates of unemployment for decades, while the manufacturing provinces of Ontario and Quebec have typically had lower rates. Alberta, with its oil-based economy, has recorded particularly low unemployment rates when high international oil prices have fuelled the provincial economy. But with this exception, the regions most dependent on a few natural resources have typically experienced the most severe unemployment.

Unemployment also tends to be concentrated within specific occupational groups and industrial sectors. Seasonal work like fishing, logging, and construction carries a high risk of unemployment and helps account for some of the higher levels of unemployment in the regions of the country most dependent on natural resources industries. Manufacturing jobs are also prone to unemployment, since economic downturns often lead to layoffs and plant shutdowns. On the other hand, professional and technical occupations tend to be somewhat more protected from unemployment, though this changed somewhat in the downturn of the 1990s.

Aboriginal people continue to experience exceptionally high rates of unemployment. When combined with lower than average rates of labour force participation, it is clear that this group faces serious disadvantages in the labour market. According to the 2001 census, for example, the unemployment rate for the Aboriginal population was 19.1 percent, almost three times the rate for the non-Aboriginal population (7.1%).[39] Unemployment rates were far higher for those living on reserves (27.6%) than off reserves (16.5%). Isolated Native communities in regions such as northern Manitoba, Saskatchewan, Alberta, and the Territories offer few employment opportunities. In addition, low levels of education, limited work experience other than in short-term, low-skill jobs, minimal access to information about jobs and employers, and discrimination in hiring have all contributed to high rates of Aboriginal unemployment. In urban census metropolitan areas unemployment rates are lower (13.9%) but still high relative to the rest of the population. In the six Canadian cities with the largest Aboriginal populations (Winnipeg, Vancouver, Edmonton, Calgary, Toronto, and Saskatoon) unemployment rates are nearly three to four times higher for Aboriginal people, ranging from Winnipeg's 14.3 percent (versus 4.9% for non-Aboriginal people) to Saskatoon's 22.3 percent (versus 5.8%).

Gender and Unemployment

A comparison of female and male unemployment rates in the postwar years reveals some interesting shifts in relative position (see Figure 2.5). Looking at adult workers (25 years and older), during the 1950s and early 1960s, male unemployment rates were generally about twice as high as those for females. The two rates converged in the mid-1960s, and by the 1970s, female rates were typically about 2 percent higher than male rates. This trend continued until the recession of the early 1980s when, for a second time, the two rates came together. For the rest of that decade, female rates of joblessness were again higher than male rates, but the difference was always less than 1 percent.

During the 1990–92 recession and its aftermath, men were slightly more likely to be unemployed, but since the mid 1990s there has been a third convergence. In 2004, for adult (25 years and older) labour force participants, the female and male unemployment rates were 5.8 and 6.1 percent, respectively.

We would need a much more detailed analysis than is possible here to explain these shifts in female and male unemployment rates. Nevertheless, a large part of the explanation would focus on women's locations in the labour market. Female labour force participation was much lower in the 1950s and 1960s, and women were employed in a limited number of traditionally female occupations. Unemployment then, as now, was considerably higher in the blue-collar occupations, in which few women were found. Since that time, women have made their way into the rapidly expanding service sector alongside men. But many women continue to be employed in lower-level, less secure positions. Thus, during the past few decades, women have been somewhat more vulnerable to unemployment. However, the recessions at the beginning of the 1980s and 1990s led to widespread layoffs in blue-collar industries (manufacturing, for example). Since men continue to be overrepresented in these industries, their unemployment rate rose rapidly and passed the female rate when the economy weakened.

Youth Unemployment

Young Canadians (aged 15 to 24), particularly those with the least education (high-school dropouts, for example) have experienced high rates of unemployment for several decades (Lowe and Krahn 1999). Youth unemployment fluctuated between 9 and 14 percent in the 1970s (Figure 2.5). The 1981–82 recession pushed the rate to about 20 percent, but it had dropped to around 11 percent by 1989. The recession that began in 1990 reversed this decline; rates in the 18-percent range were reached in 1993 and 1994. By 2000, as the economy strengthened and the total unemployment rate dropped to 6.8 percent, the youth unemployment rate also declined, to 12.6 percent. Since then it has risen to 13.4 percent in 2004. Even though youth unemployment is affected by the same economic forces that determine the risk of joblessness for adults, young workers have fared much worse for several decades—the youth unemployment rate has not been below 10 percent since 1975. And even though young people typically have shorter spells of unemployment, because more of their previous employment has been part-time and part-year, they are less eligible for Employment Insurance (Betcherman and Leckie 1997).

A closer look at Figure 2.5 suggests that the gap between the youth and adult unemployment rates was largest in the early 1980s. Why? The shifting age distribution of the population provides a partial explanation. Canada experienced an exceptionally large baby boom during the 1950s and 1960s. The last of the children born in that era of high birthrates were making their way into the labour force during the early 1980s, just as a severe recession was reducing employment opportunities. Subsequent cohorts of school leavers became smaller. By way of example, in 1982, youths (aged 15 to 24) made up 24 percent of the total Canadian population, compared to only 14 percent in 2000. Thus, since the early 1980s, when youth unemployment rates peaked, smaller cohorts of youths have meant less competition for jobs typically held by young workers (Foot and Li 1986; Gunderson, Sharpe, and Wald 2000).

Throughout the past several decades, teenage (ages 15 to 19) unemployment rates have been higher than jobless rates among young adults (ages 20 to 24). Compared with teenagers, young adults are typically better educated and have more work experience. In 2004, for example, the unemployment rate for 15-to 19-year-olds was 18.1 percent, compared with 10.3 percent for 20- to 24-year-olds. While female and male adult (25 years and older) unemployment rates have again converged (see Figure 2.5), we still see a gender difference in joblessness among youth. For teenagers, the female unemployment rate in 2004 was 16.7 percent, compared to 19.4 percent for males. Among young adults, the female unemployment rate (8.3%) was also lower than the male rate (12%). These differences reflect, in part, the higher proportion of young males who seek employment in blue-collar industries where unemployment rates remain higher than in service industries (where a larger proportion of young women look for work).

Causes of Unemployment

We have already hinted at some of the causes of rising unemployment over the past several decades. Obviously, there is a demographic factor to consider. The sizable increase in the birthrate in the years following World War II led to rapid growth in the labour force several decades later. In addition, the increased proportion of women entering the labour force generated greater demand for jobs. But people working for pay are also people with money to spend, and the growth in labour force participation itself led to substantial job creation. Thus, econometric analyses conclude that these demographic shifts have had very little impact on the overall increase in unemployment in the past two decades (Gera 1991: 3).

Another basic type of explanation focuses on the characteristics of the jobless themselves. Such arguments are typically supported by anecdotal evidence and little else. One version suggests that many of the unemployed could be working if only they would accept the less attractive jobs that are available. But in many communities, the number of jobless far exceeds the total number of available jobs (Marsh, McAuley, and Penlington 1990). Furthermore, many of the jobs that may be available (and perhaps are even hard to fill) are part-time positions in the lower-tier services with pay rates so low that it would be impossible to support oneself, let alone a family. For many of the unemployed, accepting such work would be economically irrational, since it would force them to try to survive on very little income and discontinue actively searching for a better job.

A different version of the "blame the unemployed" explanation argues that they have insufficient skills and so are unable to compete for the jobs available in the new knowledge-based economy. However, recent studies have concluded that there is little evidence of a "skills gap" in Canada that might explain high rates of unemployment (Osberg and Lin 2000; Gingras and Roy 2000). In fact, as we argued at the beginning of this chapter, there is at least as much evidence of widespread underemployment in Canada as there is evidence of a skills shortage (Krahn and Lowe 1999; Livingstone 1999). Yet another version of this type of explanation of high unemployment is the presumed laziness of the unemployed, in combination with the generosity of the government (Swanson 2001). But it is hard to believe that more than a few Canadians would prefer the low level of government social assistance to a higher and more secure earned income. Furthermore, stricter eligibility requirements were introduced in 1996 when the old Unemployment Insurance program was replaced by the Employment Insurance program. By the late 1990s, only 36 percent of unemployed individuals received EI support, down from 74 percent in 1989 (Jackson et al. 2000: 153). Research on Canadians' work values (reviewed in Chapter 8) shows that few would choose the economic hardship and the stigma of unemployment over a regular job. Thus, as Betcherman (2000) concludes, we cannot blame our long-term high rates of unemployment on Employment Insurance policies, labour laws, or tax policies.

Economists often distinguish between *cyclical unemployment,* which rises during recessions and then declines as the economy recovers, and *frictional unemployment.* The latter results from the ongoing movement of workers in and out of jobs as they seek to match their skills and interests with the jobs offered by employers. Some unemployment, therefore, is normal, even in the strongest economy, since a perfect match between all jobs and workers is never possible. However, research suggests that periods of cyclical unemployment over the past few decades have had the permanent effect on the Canadian labour market of increasing the "natural rate of unemployment" to such a degree that it may be unlikely to go back down. Thus, economists recognize that Canada has developed a serious problem of *structural unemployment.* More deep-rooted and permanent than cyclical or frictional unemployment, this problem requires new approaches to economic policy.[40]

Some of the emerging industrial and employment trends discussed in the previous chapter have contributed to the high and persistent level of unemployment in Canada. With industrial restructuring, many corporations have shifted their activities to countries and regions where labour costs are lower, and where government legislation regarding labour relations, worker safety, and environmental protection is less developed. Automation and information technologies are replacing workers in some industries. In both the private and public sectors, many employers have responded to financial problems by downsizing, cutting full-time jobs, and relying increasingly on temporary workers (a trend discussed later in this chapter). In addition, parts of Canada's staple-based economy have been hit hard by recent global economic shifts. Some of our basic resource industries (wheat and lumber, for example) have lately encountered unstable world markets. Others, such as the East and West Coast fishing industries, are suffering because of declining resource stocks.

As noted in Chapter 1, the impact of government policies on unemployment trends is equally important and must also be scrutinized. High unemployment in previous decades has been linked to the federal government's pursuit of low inflation via high interest rates (Fortin 1996). With a market-oriented approach to economic development that included entry into a continental free trade agreement, the federal government may also have made it easier for large corporations to close Canadian factories and lay off workers.

International Comparisons

In 2004, Canada's unemployment rate of 7.2 percent seemed high when compared to Norway (4.4%), Japan (4.7%), Ireland (4.5%), Sweden (6.4%), the United Kingdom (4.6%) and the United States (5.5%). However, jobless rates were much higher in Germany (9.5%), France (9.7%), Greece (10.5%), and Spain (10.8%) (OECD 2005: Statistical Annex, Table A).[41]

Given that these countries are struggling with economic problems similar to those of Canada, why do some have high unemployment while others have nearly full employment? In terms of labour market policies, Canada lies between the *laissez-faire,* or free-market approach, of the United States and the more interventionist European approach. In countries like Norway, Sweden, and Japan, employers, organized labour, and governments frequently work together toward a common goal of full employment. Labour market programs are typically proactive. For example, in Sweden, 90 percent of government funds spent on labour market programs goes toward training, job creation, and assistance for workers moving to new jobs. The same approach has historically meant low unemployment in Germany, but reunification in the 1990s led to rapidly increasing jobless rates as the communist economy of East Germany was merged with the capitalist economy of West Germany. In Canada, a collective effort has been missing, as has a well-developed industrial strategy that might counter some of the global economic forces negatively affecting our economy. The government has relied on the private sector to create jobs and to train and retrain workers, even though it has not had a particularly good record in this regard.[42]

Of most interest, perhaps, are comparisons of Canada's unemployment trends with those in other North American countries. Between 1984 and 1995, Mexico's jobless rate stayed below 5 percent, lower than the U.S. rate (between 6% and 8%), and considerably lower than Canada's rate (between 10% and 12%). By 1995, Canada's unemployment rate was just under 10 percent, while the Mexican rate had risen to nearly meet U.S. rates of 6 percent. In the late 1990s unemployment rates and duration fell in all three countries, remaining lowest in Mexico.[43] Low rates in Mexico are, in large part, a function of the much larger informal economy and self-employed sector. Individuals seeking work are more likely to find jobs, albeit low-paying and insecure jobs, in the informal sector, or to support themselves with small-scale entrepreneurial activities (Martin 2000).

The difference in unemployment levels between the United States and Canada has a different source. Looking back over time, we find that unemployment rates were very similar in the two countries in the 1950s, 1960s, and 1970s. However, beginning in 1981, Canada's jobless rate began to outstrip the U.S. rate, and the gap widened further in the 1990s. To some extent, the growing gap can be attributed to the more severe recessions in Canada in the early 1980s and 1990s, and the stronger U.S. economy in the 1990s. However, there is also evidence that Canada's unemployed remain in the labour market searching for work longer than their American counterparts (Riddell and Sharpe 1998). As a result, larger numbers of the jobless in the United States are not officially counted as unemployed. In addition, the United States is a society that imprisons many more of its citizens. In 1993, 13 percent of young males (25% of young Black males) were either in prison or on parole or probation. While jobless, many of these young men would not be counted as unemployed.[44]

Canada has typically relied on the marketplace to reduce unemployment, while instituting Employment Insurance and social assistance programs to help deal with the social problems created by persistently high levels of unemployment. The present public policy challenge is how to create more employment opportunities for unemployed job seekers. While some other countries have devised more effective approaches to dealing with unemployment, we cannot simply import solutions. Canada has a unique economic history and industrial structure, a complicated political system where federal and provincial responsibilities are separate (and frequently disputed), and its own unique business and labour institutions. In short, we will have to fashion our own solutions. Nevertheless, the examples provided by other countries demonstrate that long-term high levels of unemployment need not be seen as inevitable, and that employers, government, and organized labour will all have to be active participants in any attempts to solve our structural unemployment problem.

HOURS OF WORK AND ALTERNATIVE WORK ARRANGEMENTS

In 1870, the standard number of hours in a workweek (the number beyond which overtime would typically be paid to full-time workers) in the Canadian manufacturing sector was 64. This was subsequently reduced because of trade union pressures and the introduction of new manufacturing technologies that could produce more goods in a shorter time. By 1901, 59 hours per week was standard in manufacturing. Twenty years later, 50 hours per week was typical, but this average changed little until after World War II. Between 1946 and 1949, the manufacturing standard dropped quickly from 49 hours per week to 44 hours. It then fell to 40 hours by 1957, fluctuated slightly around this point for the next two decades, and dropped to 39 hours per week by 1976 (Sunter and Morissette 1994). Since then we have seen a small further decline. In 2004, the average time (usual hours) worked per week for the total employed labour force was 36.5 hours. Thus, on average, Canadians spend considerably less time at work than they did a century ago.

A closer look at the decline of about two hours of work per week over the past two decades reveals a very important change in Canadian work patterns. Specifically, we have witnessed a distinct polarization in hours worked—more people have been working long hours and more people have been working part-time.[45] In 2004, only 59 percent of the Canadian employed actually worked between 35 and 40 hours per week in their main job, down from 66 percent in 1976. One in four (25%) worked less than 35 hours and 16 percent worked more than 40 hours per week. Thus, while the average number of hours worked has declined slightly, the much bigger story is the increase in both the proportion of part-time workers (less than 30 hours per week) and the proportion of Canadians working long hours.

We will return to the topic of part-time work in a moment. Meanwhile, a few observations about long working hours are necessary. For some workers, particularly self-employed professionals, longer hours may provide higher incomes. For others, the added work may simply be an attempt to avoid a decline in standard of living when real incomes stagnate. But longer hours also cut into leisure and family time, and they have been shown to have detrimental effects on health and work–life balance (Bunting 2004; Hochschild 1997; Kemeny 2002; OECD 2004; Shields 2000). Between 2000 and 2003 Statistics Canada noted a slight decline in average hours due in part to an aging workforce, longer parental leave, and a desire to better accommodate work–family balance (Galarneau, Maynard, and Lee 2005).

Shift work has long been common, but not always welcome, in some industries (e.g., health, consumer services), but there does seem to be some interest in alternative work schedules. *Teleworking* (working at home or in a remote site, often using computer technology), *flextime* (choosing the time to start and stop work), and *job sharing* have been receiving more attention in the past decade.

FIGURE 2.6 *Employees' Work Schedules and Arrangements, Canada*

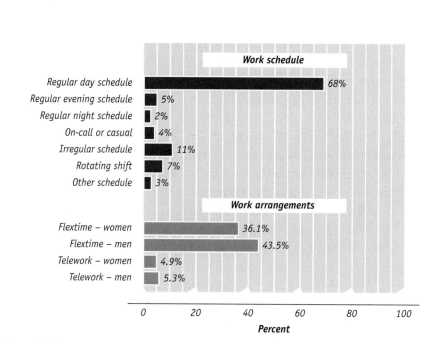

Source: EKOS Research Associates and the Graham Lowe Group Inc. (2005) *Rethinking Work: Understanding the Canadian Workforce and Workplace. 2004 Canadian Worker Survey Report;* Comfort et al. (2003) *Part-Time Work and Family-Friendly Practices in Canadian Workplaces.* Ottawa: Statistics Canada and Human Resources Development Canada. Cat. no. 71-584-MIE No. 6.

These work options can help parents balance childcare with their jobs. But they also have drawbacks, given that such arrangements often provide fewer benefits and can deprive workers of satisfying interactions with coworkers.[46]

But how common are flexible work arrangements in Canada? According to the 2004 Canadian Worker Survey, most workers had a regular day schedule (Figure 2.6). Just under 5 percent worked a regular evening schedule and 2 percent worked at night. Another 11 percent worked an irregular schedule, 7 percent a rotating shift, and 4 percent were on-call or casual. Additional Statistics Canada data from 1999 indicates that 36 percent of women and 43 percent of men had access to some type of flextime arrangement (where stop and start times vary). Telework is very uncommon. Historical comparisons are not available, but it seems that widespread flexibility in work arrangements is still a long way off in most workplaces.

Although most employed Canadians do not appear to want to work less (presumably because of the loss of income), *shorter workweeks* do offer a potential solution both to problems of overwork and of unemployment (Hayden 1999). A 1994 federal government advisory group, which included labour and industry representatives, recommended ways to redistribute work hours.[47] Canadian organizations such as the Work-Well Network and B.C.'s provincial "Work Less Party" actively promote reductions in working hours. Speaking on behalf of the Network, Bruce O'Hara (1993: 3) concludes that "working harder is the problem, not the solution" to Canada's economic difficulties.

Shorter workweeks have recently been negotiated in several western European countries, in part to reduce unemployment by sharing work among a larger number of workers. For example, in the mid-1990s, German autoworkers negotiated a new deal with Volkswagen in order to preserve jobs, achieving a reduced workweek with little pay loss. More significant, and contentious, has been France's legislated introduction in 2000 of a 35-hour workweek aimed at lowering unemployment. Though new legislation in 2005 now allows overtime and higher maximum hours for some workers, the 35-hour week remains the standard for many public- and private-sector workers in France. Longer vacations also help spread work within a society— as well as contributing to the quality of life. Unlike a number of European countries where vacations of four weeks or longer are required by law, many Canadian workers continue to have only two weeks per year of paid vacation (in addition to statutory holidays). For the majority of Canadian employees, vacation entitlements might rise to three or four weeks per year only after many years in the job (Kumar, Arrowsmith, and Coates 1991: 388–89).

NONSTANDARD WORK ARRANGEMENTS
Varieties of Nonstandard Work

Most employed Canadians still have a full-time, year-round, permanent, paid job. However, alternatives to this standard type of employment arrangement are increasing slowly, not just in Canada, but in all industrialized societies.[48] *Part-time work* is the most common type of nonstandard work, but the number of *multiple-job holders* has also been increasing slowly. The *own-account self-employed* category has been growing, as we have already noted. And there are also indications that *temporary* (contract or "contingent") work has become more widespread. These four types of nonstandard work combined (not counting overlaps) accounted for 28 percent of all employed working-age (15 to 64 years) Canadians in 1989, rising to one-third by 2002 (Krahn 1995; Vosko, Zukewich, and Cranford 2003: 43).

Nonstandard work may be mandated by employers or initiated by individual workers. Many employers, in both the private and public sectors, have responded to the economic difficulties of the past decades by replacing full-time with part-time workers (and sometimes part-year workers), and by eliminating permanent positions. The latter have sometimes been replaced with temporary (limited-term contract) positions or workers hired from a temporary help agency (a form of subcontracting). These employment strategies allow employers greater flexibility in responding to uneven demands for goods and services, and they clearly reduce labour costs (probably a more important factor, in many cases). This "flexible firm" model is discussed in Chapter 5. At the same time, a slowly increasing number of labour market participants have chosen to set up their own business (the own-account self-employed) or to take on a second job.

Some workers choose alternative forms of employment (part-time work or self-employment, for example) because of personal preference. But for many others, such choices are a response to a difficult labour market. Workers may create their own jobs because few paid jobs are available; accept temporary or part-time work only when permanent, full-time jobs are scarce; or take on a second job because their first job pays poorly. Since nonstandard jobs typically pay less, provide fewer benefits, are less likely to be covered by labour legislation, and have less employment security (Kalleberg, Reskin, and Hudson 2000; Vosko 2005), an increase in nonstandard employment also means an increase in the precariousness of employment (and income) for many Canadian workers. Furthermore, evidence is beginning to appear that a job history of nonstandard work carries a long-term cost—controlling for other factors, previous part-time work has a negative effect on current wages (Ferber and Waldfogel 1998).

We have already commented on the small increase in own-account self-employment in the past decades. Just over 10 percent of employed working-age Canadians are in this category, which includes farmers, doctors, lawyers, and business consultants, as well as the small entrepreneurs we typically associate with self-employment. Multiple job holding increased from 2 percent in 1977 to roughly 5 percent in 2001.[49] People take on a second job for a variety of reasons, including topping up an inadequate income, paying off debts, and saving for the future. Economic hard times undoubtedly have contributed to the growth in multiple job holding. However, since the majority are supplementing a full-time job, and since about one-third have a professional or managerial first job, we should be cautious about assuming that all Canadians with more than one job are in a precarious financial or employment situation. The same applies to the own-account self-employed (Kalleberg, Reskin, and Hudson 2000).

Nonstandard jobs are much more common in some sectors of the economy. In terms of industry variations, agriculture has the highest rate of nonstandard work, 65 percent in 1994 (Krahn 1995). A large proportion of these nonstandard workers are the own-account self-employed. Construction also has a high rate of nonstandard work (45%), reflecting extensive self-employment and temporary contract work. But it is the lower-tier service industries—retail and other consumer services—that are the main source of nonstandard employment, given their relative size in the economy. Thus, in 1994, more than 40 percent of Canadians working in retail or other consumer services held a nonstandard job. There is also considerable nonstandard work in the upper-tier services, particularly in the education/health/welfare sector (39%) where part-time and temporary work has become very common.

A worker's demographic characteristics also influence her or his likelihood of being employed in a nonstandard job. In 1994, for example, 40 percent of employed women were in nonstandard jobs, compared to 27 percent of men. Young workers were most likely to be in such positions (64% of females aged 15 to 24, and 52% of males in this age cohort). Older workers (55 to 64 years) also have somewhat higher rates of nonstandard employment (41% for women and 30% for men).

Part-Time Work

Half a century ago, in 1953, less than 4 percent of employed Canadians held part-time jobs. But during the 1960s and 1970s, part-time work became more common and today it is the most prevalent type of nonstandard work. Until quite recently, full-time work was defined by Statistics Canada as working 30 or more hours per week in total. But the rise in multiple job holding forced a rethinking of this definition, since some people holding several part-time jobs (totalling more than 30 hours) were being counted as full-time workers. Since 1996, part-time workers have been defined as those who work less than 30 hours per week in their main job.

FIGURE 2.7 *Part-Time Employment by Age and Gender, Canada, 1976–2004*

Part-time employment rate

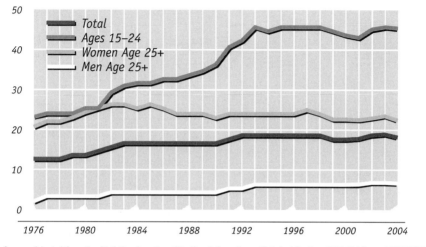

Legend:
- Total
- Ages 15–24
- Women Age 25+
- Men Age 25+

Source: Adapted from the Statistics Canada publication *Labour Force Historical Review*, 2004, Catalogue 71F0004XCB, February 18, 2005.

Figure 2.7 displays part-time rates for the past several decades, calculated using the new definition. Back in 1976, the national part-time rate for all employed Canadians (aged 15 and older) was 13 percent. The recession in the early 1980s pushed the part-time rate much higher, to 17 percent by 1983. The recession at the beginning of the 1990s led to another increase in part-time employment rates, to 19 percent in 1992. The part-time rate remained at this level until 1999 when it dropped slightly. Thus, by 2004, a total of nearly 3 million Canadians, 18 percent of all employed, were working part-time.

Part-time rates for adult women and men (aged 25 and older) are also plotted in Figure 2.7. It is apparent that few employed men aged 25 and older are working part-time, although the rate for this group climbed from 2 percent to 6.3 percent in 2004. As for adult women, their much higher part-time rates have fluctuated between 22 and 26 percent over the 30-year period. In 2004, 22.3 percent of adult women worked part-time. Overall, women comprise 69 percent of all part-time workers (aged 15 and older). The most prominent trend observed in Figure 2.7 is the increase in part-time employment among youth (ages 15 to 24). In 1976, the youth part-time employment rate was 21 percent. By 1983, the rate had soared to 31 percent, and it reached

41 percent by 1991. The youth part-time rate continued to rise, to 46 percent by 1995, before dropping marginally to 44 percent by 2004. Thus, at the beginning of the 21st century, almost half of the employed 15-to-24-year-olds in Canada, almost 1.1 million in total, were working part-time. While Figure 2.7 does not break down the youth part-time rate by gender, we should note that the male rate was 37.4 percent in 2004, compared to a part-time rate of 52.2 percent for 15-to-24-year-old females.

Some people choose part-time work because it allows them to balance work and family responsibilities, to continue their education while still holding a job, or simply to have more leisure time. Others, *involuntary part-time workers,* are forced to accept part-time jobs because they cannot find one that is full-time. The monthly Labour Force Survey asks part-time workers why they are working part-time, providing a range of answers. Figure 2.8 displays the reasons given by Canadian part-time workers in 1975 and 2004. While these data are

FIGURE 2.8 *Reasons for Part-Time Work, Canada, 1975 and 2004*

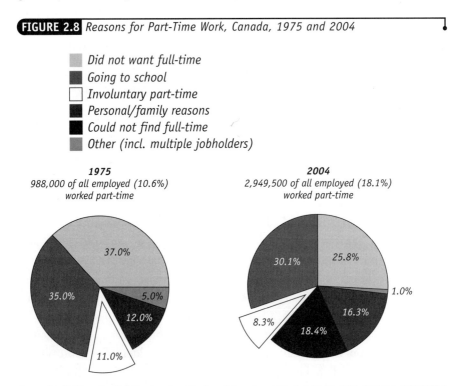

Did not want full-time
Going to school
Involuntary part-time
Personal/family reasons
Could not find full-time
Other (incl. multiple jobholders)

1975
*988,000 of all employed (10.6%)
worked part-time*

37.0%
35.0%
5.0%
12.0%
11.0%

2004
*2,949,500 of all employed (18.1%)
worked part-time*

30.1%
25.8%
1.0%
16.3%
8.3%
18.4%

Source: Adapted from the Statistics Canada publications *Labour Force Historical Review,* 2004, Catalogue 71F0004XCB, February 18, 2005; and *Labour Force Historical Review,* 2001, Catalogue 71F0004XCB, February 22, 2002.

not strictly comparable because of changes in how "reasons for part-time work" are measured, they give some sense of change over time. The most important difference is that in 2004 "involuntary part-time" is defined as being unable to find full-time work while actively looking for work. Another group, labelled "could not find part-time work," (not measured in 1975) are unable to find full-time work but have not actively looked in the past month.

In 1975, 37 percent of less than 1 million part-time workers stated that they did not want a full-time job. In 2004, this reason was given by a smaller proportion (25.8%) of a much larger cohort of part-time workers. While involuntary part-time work appears to have declined between 1975 (11% of part-timers) and 2004 (8.3%), we need to keep in mind that there was another 18.4 percent of part-timers in 2004 that reported not being able to find full-time work, though they had not actively looked in the past month. This long-term increase in those unable to find full-time work reminds us that there is more to the rise in part-time rates than simply changing worker preferences. Much of the increase reflects the actions of employers seeking ways to cut labour costs. That said, we also know that the involuntary part-time rate does decline somewhat when the economy improves (Marshall 2001a). In 1995, for example, when the unemployment rate was still 9.4 percent, 31 percent of the 2.5 million part-time workers could not find a full-time job.

Figure 2.8 reveals that almost one-third (30.1%) of part-time workers choose this type of work arrangement because they are attending school. For a majority of youth, part-time jobs fit well alongside attending high school, college, or university. The Labour Force Survey also shows that many adult women working part-time, especially those with children, are likely to cite personal or family reasons, or simply state that they do not want full-time work. But had these women been asked whether they would still prefer a part-time job if they had access to adequate and affordable childcare, it is likely that at least some would have expressed interest in full-time employment. In other words, traditional assumptions about female child-care roles, together with a shortage of affordable, quality childcare may create an underestimate of the real level of involuntary part-time employment among adult women (Duffy and Pupo 1992).

Part-time rates vary considerably across industries and occupations. In 2004, the part-time rate in the goods-producing sector was still very low (only 6%). Thus, most of the part-time jobs are in the service industries, in both the lower- and upper-tier services. For example, the part-time rate in the retail trade industry in 2004 was almost 34 percent. High rates are also found in education (26%) and health care (25%). However, many of the part-time jobs in the upper-tier services (nurses and teachers, for example) are better paid than the part-time jobs (often held by students) in the retail trade and other consumer services. As for occupational differences, only 6 percent of managers were working part-time in 2004, compared to 18 percent of clerical workers, 45 percent of retail sales clerks, and 50 percent of child-care and home-care workers. As we will see in the next chapter, there is a strong correlation between occupational status, income, and the prevalence of nonstandard work arrangements, including part-time work.

As in Canada, part-time employment rates have been rising in most other Western industrialized countries. Canada's current part-time rate is not unusual. For example, in 2003, about one in four workers were holding part-time jobs in Japan, the United Kingdom, and Australia, along with about 35 percent in the Netherlands. In contrast, Sweden, the United States, and Italy reported part-time rates lower than Canada's 19 percent rate in 2003 (OECD 2005: Statistical Annex, Table E).[50]

Temporary Employment and Part-Year Work

Some forms of temporary employment, like seasonal employment in the tourism or construction industries, have been part of the Canadian labour market for many decades (Marshall 1999; Vosko 2000). However, both private- and public-sector employers have begun to hire more employees on a limited-term (perhaps six months or a year) contract basis, rather than offering permanent positions as was the custom a decade or two ago. In addition, temporary help agencies have been hiring more workers to be sent out on assignment to employers seeking to fill a temporary employment need. But, in terms of detailed statistical trends, we know considerably less about the growth in *temporary employment,* since Statistics Canada only began monitoring this phenomenon on a regular basis in the late 1990s. Defining temporary jobs as those with a specific end date, the 2004 Labour Force Survey shows 12.8 percent of Canadians (1.7 million) in such positions. About one-quarter (23%) of all temporary workers were in seasonal jobs, and slightly more (27%) were in casual jobs (very short-term, with no specific contract). Most others (48%) held positions with a specific, contracted term. Of note, temporary jobs grew by 31 percent between 1997 and 2002, compared to a 12 percent growth rate for permanent jobs (CANSIM 282-0080; Statistics Canada 2004a).[51]

Adult (aged 25 and older) women and men are about equally likely to be in a temporary job (nearly 10% and 8%, respectively). Once more, as in part-time work, it is young workers who are heavily overrepresented in temporary jobs. In 2003, nearly 30 percent of 15-to-24-year-old employed Canadians were in a job with a specific end date. Young student workers are more likely to be employed in the lower-tier consumer services, where low-paying temporary jobs are more common. Also, in the upper-tier services (the education and health and welfare industries, for example), it is much easier to offer a new (young) employee a contract position than to move someone already in a permanent position to a contractually limited job. Quite often, contract employees work alongside permanent employees, performing similar tasks and for comparable pay, but without the guarantee of continued employment beyond six months or a year. While some people, often those with highly marketable skills, find such arrangements quite satisfactory, others are frustrated and worried by their employment and income insecurity.

FUTURE TRENDS

The 20th century recorded profound changes in how, where, and for whom Canadians work. The late 19th and early 20th centuries were times of rapid industrialization and workplace rationalization. The years that followed saw further growth in white-collar occupations, expansion of the service sector, and a decline in self-employment. By the 1960s, the Canadian work world had been largely transformed. But, as our examination of more recent trends has shown, the Canadian labour market was once again going through a significant restructuring at the end of the 20th century.

The service sector has expanded enormously, producing both good and bad jobs, with many more of the latter in the lower-tier service industries. Meanwhile, manufacturing industries have continued to contribute less to employment growth in Canada. We have seen a substantial increase in part-time employment, while other forms of nonstandard work (temporary or contract work, for example) have also become more prevalent. Unemployment rates declined in the late 1990s and early 2000s but are still high enough to cause concern, and could well rise again in the next several years. Compared to the mid-1990s, when the general public was highly concerned about unemployment, anxiety about employment and income security has abated somewhat. Nevertheless, some groups in the labour market (young workers, for example) have not really experienced the same economic recovery. Moreover, concern over the *quantity* of jobs is increasingly shifting to worries over job *quality* and the growing incidence of low paying, vulnerable work in Canada (Chaykowski 2005; Lowe 2000; Saunders 2005; Vosko 2005).

What lies ahead? By simply posing this question, we enter the debates and controversies surrounding the future of work. There is a burgeoning literature on the future of work, offering a bewildering range of contradictory scenarios. Students of the sociology of work and industry should be encouraged to critically appraise these writings for themselves. From our perspective, the evidence presented in this chapter does not support the extreme positions taken by many futurists. The trends we have documented suggest neither a "workerless world" as technology displaces humans (Rifkin 1995) nor a "de-jobbed world" in which work is repackaged into a variety of more flexible and rewarding entrepreneurial forms for a majority of Canadians (Florida 2002).[52]

We can be fairly certain that several of the trends we have documented will continue. Demographic factors are a given, especially workforce aging and the increasing diversity of the labour force. In addition, we can expect a bit more expansion in the service sector's share of total employment. Nonstandard work is still slowly increasing. The big unknown is unemployment: will jobless rates stay around 7 percent, as they were in 2004, or will they once again rise? Is an unemployment rate of, say, 5 percent—which we currently see in Alberta— even a remote possibility nationally? How would public policy have to mesh with employers' strategies to create adequate numbers of well-paying and satisfying jobs (Lowe 2000; Lowe and Schellenberg 2001)?

Examining Canadian work trends most likely to affect the future, Gordon Betcherman and Graham Lowe (1997) identify a cluster of interacting forces that are creating considerable economic anxiety for Canadians. Foremost among these are persistently high levels of unemployment and underemployment, especially among young people; the uncertainties created by spreading nonstandard work; and the creation of a more polarized labour market, resulting in the economic marginalization of poorly educated or unskilled individuals. By throwing more of the risks associated with economic change onto individuals and families, and by widening the gulf between haves and have-nots, these trends are eroding our sense of social cohesion.

Some of these trends—higher levels of structural unemployment, declining employment in manufacturing, and increasing nonstandard employment—are linked to global economic restructuring. An unstable world market for some of Canada's staple products has reduced the prospects of regions most dependent on primary industries. Increased international competition has prompted some Canadian-based firms to rationalize their operations, shut down or relocate, replace workers with automation, or resort to low-cost nonstandard employment. Canada's economy, then, is not insulated from global economic forces. The Canadian government's laissez-faire labour market policies have, in some ways, made working Canadians more vulnerable to the negative effects of these economic changes.

DISCUSSION QUESTIONS

1. What are some of the key demographic changes occurring in Canada? How will these trends affect employers and workers in the next 10 to 15 years?
2. What, if anything, can governments do to help respond to demographic changes occurring in the next 10 to 15 years?
3. How, and why, does labour force participation vary across Canada?
4. Work in Canada is very strongly shaped by regional location. Discuss how region influences key aspects of the work experience, such as unemployment, participation, occupation, and so on.
5. Discuss the concept of nonstandard employment and outline the trends in such types of jobs in Canada. What are some of the implications of these trends for workers, employers, and society in general?
6. What are the most important occupations and industries in the Canadian economy? How has the occupational and industrial structure changed over time, and why?
7. How does age shape one's work experience? How do the work experiences of "young" and "old" Canadians differ?
8. What are some of the key differences in labour market trends for women and men? How do we explain these?
9. Some observers of the Canadian labour market worry that, without additional training for workers, Canada will not be competitive in the global economy. Others worry about the extent of "underemployment" among Canadian workers. In your opinion, what should we worry about? Why?

10. Based on your readings in Chapters 1 and 2, what in your view are the most important challenges facing Canadian workers in the next 10 years?

NOTES

1. Excluded are residents of Nunavut and the Northwest Territories, people living on Indian reserves, full-time members of the armed forces, prison inmates, and others living in institutions such as hospitals or nursing homes (Statistics Canada 2005b). For additional information about this and other Statistics Canada data sources, visit the Statistics Canada website at http://www.statcan.ca.

2. See Statistics Canada (1995) for an overview of changes made to the Labour Force Survey (LFS) in the mid-1990s to reflect changing patterns of work in Canada.

3. Unless otherwise noted, the 2004 data and time series data (1976–2004) in this chapter come from annual averages compiled by Statistics Canada in the *Labour Force Historical Review 2004* (Cat. no. 71F0004XCB). As estimates from the LFS have been adjusted to reflect population counts based on the 2001 census, the figures presented here are not comparable to data presented in earlier editions of this text (Statistics Canada 2005d). In the few instances where data has not been available from this source, we have retrieved data from Statistics Canada's online database, CANSIM, located at http://www.statcan.ca/english/ads/cansimII/index.htm. We reference this data by noting the source as well as specific table and series numbers (e.g., CANSIM 282-0002).

4. Foot and Stoffman (1998: 8). These authors provide a fascinating account of how demographic trends, particularly the baby boom, influence all aspects of society. However, in our opinion, they underestimate the effects of the profound economic, political, ideological, and organizational changes that have also occurred in the past several decades. For additional discussion of demography and the labour market, see Sunter (2001).

5. Statistics Canada (2001). This projection is based on population estimates as of July 1, 2000. Projections take into account trends in fertility, mortality, international migration (immigration and emigration), non-permanent residents (NPR), and internal migration. Various scenarios by age and sex are provided to 2026 for Canada, provinces, and territories; and to 2051 for Canada.

6. See Morissette and Zhang (2004) and Statistics Canada (2004d). In the past several years there have been many useful discussions of Canadians' participation in private and public pension plans in Statistics Canada's quarterly publication *Perspectives on Labour and Income (*see in particular the Spring 2004 issue). The *Monthly Labor Review* provides similar articles on retirement and pension trends in the United States. See, for example, Purcell (2000) and Dohm (2000).

7. Immigration and visible minority statistics in this section are taken from Statistics Canada's (2003b) *Canada's Ethnocultural Portrait: The Changing Mosaic;* (2005c) *Population Projections of Visible Minority Groups, Canada, Provinces and Territories;* and *Ethnocultural Portrait of Canada: Highlight Tables,* 2001 Census – Cat. no. 97F0024XIE2001006. See Boyd and Vickers (2000) on Canadian immigration trends and policies over the past century. See Fullerton (1997) for similar trends in visible minority population growth in the United States, where the Black, Hispanic, and Asian populations are growing most rapidly.

8. Aboriginal statistics are based on Statistics Canada's (2003c) *Aboriginal Peoples of Canada: A Demographic Profile* and (2005a) *Aboriginal Conditions in Census Metropolitan Areas, 1981–2001.*

9. See Reich (1991); Drucker (1993); OECD, HRDC, and Statistics Canada (1998); and Laroche, Merette, and Ruggeri (1999) on the importance of education in today's global economy. The Economic Council of Canada (1992) and the Ontario Ministry of Education and Training (1996) are two examples of many calls for educational reform. In its discussion of Canada's potential role in the "knowledge economy," the Expert Panel on Skills (2000) presents a broader, more balanced set of observations and recommendations.

10. On learning organizations, see the discussion in Chapter 5; also see Lowe (2000). For a broader perspective on lifelong learning that emphasizes informal learning, visit the website for the Work and Lifelong Learning (WALL) Research Network run by David Livingstone and colleagues at http://www.wallnetwork.ca.

11. Zhao, Drew, and Murray (2000: 33). Also see Clark (2000), who provides a long-term overview of the growth of the formal education system in Canada, and Riddell and Sweetman (2000), who detail current educational participation and attainment trends.

12. Gilbert and Frank (1998). Also see Statistics Canada and HRDC (1996); OECD, HRDC, and Statistics Canada (1998); and Krahn and Lowe (1999) on workplace literacy concerns.

13. On workplace training, also see Betcherman, McMullen, and Davidman (1998), Statistics Canada and HRDC (1998: Chapter 3), Hum and Simpson (1999a), and Lowe (2000).

14. Figures for 1994 from Kelly, Howatson-Leo, and Clark (1997); 2004 figures from Livingstone (2005). The 1994 and 2004 figures are not comparable (due to wording) and do not necessarily demonstrate an increase in overqualification/underemployment. For additional discussions of underemployment, see Livingstone (1999); Krahn and Lowe (1999); Lowe (2000); Frenette (2001).

15. Bellin and Miller (1990: 187). Also see Green and Ashton (1992), Noble (1994), Krahn (1997), Krahn and Lowe (1999), Gingras and Roy (2000), and Smith (2001) on the debate over presumed skill shortages and their effects on competitiveness.

16. See Statistics Canada, *The Daily* (16 October 2002). For useful discussions on job stability, see Picot and Heisz (2000) and Heisz (2002).

17. Employment rates and income statistics come from the Census "Topic-based Tabulations" available on the Statistics Canada website at http://www.statcan.ca (Cat. no. 97F0011XCB-2001001 and 97F0011XCB-200100146). Drost (1996) and Maxim et al. (2000) present useful analyses of the labour market activity and economic conditions of Aboriginal Canadians based on earlier census data.

18. See Human Resources Development Canada (2003) *Disability in Canada: A 2001 Profile* for valuable statistics from the Participation and Activity Limitation Survey (PALS). Also see Hum and Simpson (1996). Hale, Hayghe, and McNell (1998) and Kruse (1998) provide information on labour market activity of disabled people in the United States.

19. See Morissette (1997), Sunter and Bowlby (1998), Marquardt (1998), Clark (1999), Sunter (2001), and Labour Market Ministers (2000) on labour market difficulties faced by Canadian youths in the 1990s. See also Picot (1998) on declining youth wages, and Bowlby (2000) on school-to-work transition trends, including information on the number of youths attending school and not working.

20. LFP rates for older workers come from Statistics Canada (2004d), Sunter (2001:31), and CANSIM Table 282-0002. Morissette and Zhang (2004) discuss participation in pension plans. Schellenberg and Silver (2004) discuss retirement preferences and experiences. See also Centre for Studies of Aging (1996) and Lindsay (1999) for more general discussions of aging in Canada, Purcell (2000) for similar comments on the U.S. experience, and Gendell (1998) on retirement trends in Germany, Japan, Sweden, and the United States.

21. Marshall and Mueller (2002) use a life-course perspective to discuss the aging workforce and social policy. Myles and Quadagno (2005) address population aging and retirement issues in an variety of countries.

22. Data on unpaid work comes from Statistics Canada (2003a) *The Changing Profile of Canada's Labour Force* and Census "Topic-based Tabulations" available on the Statistics Canada website (Canada's Workforce: Unpaid Work). Luxton and Vosko (1998) document efforts to have unpaid work measured on the census.

23. For more information about the voluntary sector, visit the website of the Voluntary Sector Initiative (http://www.vsi-isbc.ca). Data for 1997 and 2000 come from Hall, McKeown, and Roberts (2001). Data for 2003 come from Statistics Canada, *The Daily* (20 September 2004). Several articles in *Perspectives on Labour and Income* examine the volunteer activity of seniors and youths (F. E. Jones 1999, 2000) and employers' contributions (Luffman 2003). For a more theoretical analysis of volunteer work, based on trends in the United States, see Wilson (2000).

24. Methods of studying work in the irregular economy include unofficial surveys (assuming one can gain the confidence of respondents), examination of law enforcement records (in an attempt to estimate amounts of illegal economic activity), and assessments of the amount of cash in circulation in an economy (compared with taxation reports of the total amount of reported income). See Mirus, Smith, and Karoleff (1994) and OECD (2004: Chapter 5) for estimates of the size of the irregular economy in Canada. Felt and Sinclair (1992) present an interesting study of legal unpaid work in rural Newfoundland.

25. Cobb, Halstead, and Rowe (1995: 67). See Brown and Stanford (2000) and Osberg and Sharpe (1998) on quality-of-life indicators that can supplement standard economic measures (e.g., GDP) of societal progress.

26. Crompton and Vickers (2000) provide a century-long overview of labour force trends in Canada.

27. See Krahn (1992: 17–18) for further discussion of this classification system, which is quite similar to the industry typology developed by the Economic Council of Canada (1990) in its studies of the service sector. The Economic Council distinguished *dynamic services* (distributive and business services) from *non-market services* (education, health and welfare, public administration) and *traditional services* (retail trade and personal services).

28. Note that comparisons of the percentages in Table 2.1 to similar industry distributions in previous editions of this textbook are inappropriate, because of changes to Statistics Canada's industry classification system in 1999.

29. See Commission for Labor Cooperation (2003) for labour market profiles for Canada, the United States, and Mexico.

30. Betcherman and Leckie (1997). Also see Marquardt (1998).

31. Statistics Canada, *The Daily* (17 March 1998). As some occupations (e.g., agricultural occupations) have declined in relative size, Statistics Canada has begun to combine them with others (e.g., occupations unique to primary industry). Alternatively, as some categories have grown, they have been subdivided and re-named in Statistics Canada data presentations.

32. Starting in 1999, Statistics Canada began to use a revised Standard Occupational Classification System (SOC91), rather than the SOC80, for presentations of Labour Force Survey and census data. The biggest impact of the change was on the management category, with some occupations formerly coded in this category now being placed in other categories. Statistics Canada has only revised its databases back to 1987, so Table 2.3 begins with that year.

33. Statistics Canada, *The Daily* (4 June 1992; 26 April 1999; 18 April 2005); also see Bowlby (2002), Winson (1996), and Winson and Leach (2002). Ilg (1995) documents the same trend in the United States.

34. Statistics Canada (2004c) and Bowlby (2001: 14). In the 1990s, self-employment grew rapidly, accounting for 58 percent of employment growth from 1989 to 1998 (Picot and Heisz 2000). Rates declined in 2000 for the first time since 1986 (Bowlby 2001) but have since rebounded.

35. The Commission for Labor Cooperation (2003: 118) provides comparisons between Canada, Mexico, and the United States for own-account self-employed but notes the difficulties in comparing due to definitional

differences. See also HRDC and Statistics Canada (2002) for Canadian data on self-employment. The Global Entrepreneurship Monitor (GEM) project produces annual reports on international trends at http://www .gemconsortium.org.

36. See Lucas (1971), Clemenson (1992), and Angus and Griffin (1996) on single-industry communities; Statistics Canada, *The Daily* (28 April 1992; 15 April 1997) for census data on population decline in resource-dependent communities; Palmer and Sinclair (1997) on the collapse of the Newfoundland fishing industry; and Finnie (1999) on patterns of inter-provincial migration in Canada.

37. Akyeampong (1992) discusses the number of discouraged workers during the recessions in the 1980s and 1990s. See *Perspectives on Labour and Income* (Spring 2001, p. 64) for the 2000 estimate of discouraged workers.

38. For discussions of unemployment trends and policy responses, see MacLean and Osberg (1996), Betcherman and Lowe (1997), Battle and Torjman (1999), and a *Canadian Public Policy* special supplement (July 2000) on "Structural Aspects of Unemployment in Canada."

39. Unemployment rates come from the Census "Topic-based Tabulations" on the Statistics Canada website at http://www.statcan.ca (97F0011XCB2001045). See also Krahn, Wolowyk, and Yacyshyn (1998), Drost (1996), Grenon (1997), de Silva (1999), and Armstrong (1999) on the labour market position of Aboriginal Canadians.

40. See Osberg, Wein, and Grude (1995), Betcherman and Lowe (1997), Battle and Torjman (1999), and the July 2000 special issue of *Canadian Public Policy.*

41. OECD data has been adjusted to ensure comparability as far as possible using guidelines from the International Labour Office. See Sorrentino (1995) and Gitter and Scheuer (1998) on difficulties in comparing unemployment rates cross-nationally.

42. For discussions of the role of government in job creation, see Drache and Gertler (1991: 19–22), Tobin (1996), Battle and Torjman (1999), Roy and Wong (2000), and Riddell and St-Hilaire (2000).

43. See Commission for Labor Cooperation (2003: 97–99) for a discussion of unemployment trends in Canada, the United States, and Mexico.

44. Riddell and Sharpe (1998) provide a useful overview of research and debate about the Canada–U.S. unemployment rate gap. Their article introduces a

special issue of *Canadian Public Policy* (July 1998 Supplement) devoted to this topic. Sherman and Judkins (1995: 53) comment on the impact of high rates of imprisonment on unemployment figures.

45. Hall (1999) discusses the polarization of hours worked in Canada. See Rones, Ilg, and Gardner (1997) for American trends.

46. See Hamblin (1995), Nadwodny (1996), and Mirchandani (1999) on the benefits and costs of teleworking; Akyeampong (1993) on flextime work arrangements, and Marshall (1997) and Huberman and Lanoie (2000) on job sharing.

47. Human Resources Development Canada (1994); also see Yalnizyan, Ide, and Cordell (1994).

48. For discussions of nonstandard, "contingent," or "precarious" work, see Vosko, Zukewich, and Cranford (2003) and Lowe and Schellenberg (2001). Regarding such work in the United States, see Kalleberg, Reskin, and Hudson (2000) and Carré et al. (2000).

49. Sussman (1998) reports a 5.2 percent rate of multiple job holding in 1997; Marshall (2002) reports a 5 percent rate in 2001. Based on data from a smaller sample survey in 1994, Krahn (1995) provided an estimate of 7 percent. All sources, however, show a systematic increase in "moonlighting." For discussions of U.S. trends and implications, see Stinson (1997) and Kimmel and Powell (1999).

50. See also the Commission for Labor Cooperation (2003) for a discussion of nonstandard work in Canada, the United States, and Mexico, and Felstead, Krahn, and Powell (1999) for a comparison of nonstandard work trends in Canada and the United Kingdom.

51. Lowe and Schellenberg (2001: 12) suggest that, because some individuals working for temporary help agencies report themselves to be in permanent jobs, Labour Force Survey estimates of temporary employment may be too low. See Vosko (2000) for an analysis of temporary work in its historical context; see Henson (1996) for a personal account of temporary employment. Marshall (1999) analyzes trends in seasonal employment, and Grenon and Chun (1997) compare the quality of permanent and temporary jobs.

52. See Betcherman and Lowe (1997) and Lowe (2000) for counterarguments.

3

LABOUR MARKETS AND JOBS: OPPORTUNITIES AND INEQUALITY

INTRODUCTION

Why do high-school teachers earn more than retail sales clerks, or engineers more than construction workers? Presumably, these income differences are a result of teachers and engineers investing in extra years of education that, in turn, lead to more skilled and responsible jobs. So why do dentists make more money than child-care workers? While dentists spent more years training for their profession, caring for and teaching young children is also a very complex and responsible task. Why are the children of middle-class parents much more likely to go to university, compared with the children of less affluent Canadians? Why are women, the disabled, Aboriginal Canadians, and members of visible minority groups overrepresented in less rewarding jobs?

Such questions about variations in educational outcomes, career patterns, job security, and other job rewards are central to the sociological study of labour markets. We can define a *labour market* as the arena in which employers seek to purchase labour from potential employees who themselves are seeking jobs suitable to their education, experience, and preferences. In the labour market, workers exchange their skills, knowledge, and loyalty in return for pay, status, career opportunities, and other job rewards.

A number of other institutions and organizations support or interact with the operation of the labour market. Among their other functions, schools and families prepare individuals for entry (or re-entry) into the labour market. Government legislation affects how labour markets operate—minimum wage laws and legislation governing the activities of trade unions are examples. Government agencies may also assist the unemployed with financial support or job-training programs. Unions and professional associations are active in the labour market, looking after the interests of their members by bargaining for additional job rewards and, sometimes, by limiting access to better jobs. Organizations representing employers also try to influence labour market operations, sometimes lobbying governments to change laws regarding unions or to maintain a low minimum wage, or encouraging schools to include more employment-related subject matter in their teaching.

Labour economists and sociologists study many of these institutions and how they influence labour market operations.[1] Of particular interest to sociologists are the distributive aspects of the labour market. Specifically, does the labour market provide opportunities for hard-working individuals to improve their social position and quality of life, or does it reinforce patterns of inequality in society? Perhaps it does both?

The previous chapter discussed Canada's changing industrial and occupational structures, patterns of employment and unemployment, and growth in self-employment and other forms of nonstandard work. It indicated that some jobs are much more rewarding than others and that some people are at much greater risk of unemployment or underemployment. This chapter continues this discussion. It begins by reviewing some of the criteria by which individuals evaluate the desirability of jobs. We then ask several basic questions: Who has the good jobs? How did they get them? And how do they manage to keep them? Our attempt to answer these questions is built around a comparison of two alternative approaches to explaining labour market outcomes in societies like ours: the human capital model and the labour market segmentation perspective.

According to *human capital theory*, jobs requiring more effort, training, and skill typically receive greater rewards. This theory assumes that labour market participants compete openly for the best jobs, and that the most qualified people end up in the jobs requiring their particular skills. The outcome should be an efficient and productive economy and a fair allocation of job rewards. But in reviewing research on differences in job rewards, educational and occupational choice, status attainment, and labour market segmentation, we observe that the labour market does not always operate in this manner. In fact, there is evidence that the open competition assumptions of human capital theory need to be seriously challenged, and that a segmented labour market often allows the perpetuation of social inequalities.

Thus, the study of labour markets is not only about who gets better or worse jobs; it also addresses broader questions about social stratification and class structure.[2] Consequently, this chapter also discusses class inequalities within Canadian society and the growing polarization of living standards and life chances.

GOOD JOBS AND BAD JOBS

We have already begun to assess the quality of different jobs in our discussion of nonstandard employment relationships. Compared with being unemployed, a part-time or temporary job may be preferable. For most labour force participants, though, a full-time, full-year, permanent job would be much more desirable because of the income security it provides. Thus, recognizing that some people do prefer nonstandard jobs for educational, personal, or family reasons, and allowing that some nonstandard jobs (business consultants working on a contractual basis, for example) might pay very well, we would still classify only a minority of nonstandard jobs as "good jobs."

Clearly, the criteria for deciding whether a particular job is good or bad are not universal. Individuals compare the rewards a job provides against their own needs, preferences, and ambitions and against the personal costs of working in such a job. Since most workers are concerned about maintaining or improving their standard of living and quality of life, material or *extrinsic job rewards* are very important. How much does it pay? What kinds of benefits come with the job? Is it dangerous? Is it full-time and permanent?

In the following section, we examine variations in pay and other employment benefits in the Canadian labour market. We also present data on work-related injuries and deaths, our rationale being that a job with few risks to personal health and safety is, in a very crucial sense, a better job. We conclude with a look at occupational status (the prestige ranking of a particular job), even though it is not really an extrinsic job reward in the same sense as pay and benefits. However, occupational status is also not as subjectively (individually) determined as are the more *intrinsic job rewards* (the chance to be creative, to work independently, to develop friendships in a job, and so on) discussed in Chapter 8. While power relationships within Canadian workplaces are obviously also central to discussions of good and bad jobs, we provide a fuller discussion of such relationships in Chapter 6.

Income Differences

Statistics Canada uses the census to collect income information, along with a number of other annual surveys including the Labour Force Survey. We draw on several of these data sources to highlight some of the most important factors influencing income differences in the Canadian labour market.

Considering only *paid employees* (that is, excluding the self-employed), incomes in the service industries are typically lower than in the goods-producing industries. For example, in 2004, the annual average weekly wage (excluding overtime) was $1,015 in the natural resource-based industries and $768 in manufacturing. In contrast, the average weekly wage was only $502 in trade (wholesale and retail combined) and much less ($322) in accommodation and food services, both areas with substantial part-time employment. But the large and growing service sector contains both lower-tier and upper-tier services, with many better-paying jobs in the latter. Hence, in 2004, we also observed high average weekly wages in professional, technical, and scientific services ($887), educational services ($798), and public administration ($911). In short, there is more variation in employees' incomes within the service sector than within the goods-producing sector.

Comparing incomes of several major occupational groups, we find that (employed) managers reported average weekly wages of $1,181 in 2004, while Canadians in natural and applied science occupations earned an average of $1,014 per week. Individuals in health occupations, a broad category that includes both well-paid doctors and poorly paid orderlies and nursing assistants, reported an average weekly wage of $734 in 2004. But these earnings were still much higher than the weekly wages of Canadians in sales and service occupations ($406). While the Labour Force Survey cannot provide reliable income estimates for detailed occupational categories, we can use 2001 census information for this purpose. So, by way of example, we find that Canadian dentists earned an average of $108,034 in 2000, while lawyers reported average yearly earnings of $94,731. Cashiers, many employed part-time, earned only a fraction of this ($10,051), as did hotel clerks ($15,937) and hairstylists and barbers ($17,390).[3]

Higher earnings in some goods-producing and upper-tier services, and in managerial and professional occupations, can be traced, in part, to the presence of unions and professional associations that have bargained for higher incomes and full-time jobs. It is also apparent that workers with specific professional skills (for example, teachers, doctors, and engineers) and more formal education are generally paid much more than those with less training. Thus, some of the industrial and occupational differences in earnings are due to supply and demand factors in a labour market that rewards educational investments. Also important are differences in the bargaining power of the various groups participating in the labour market, a subject to which we return in Chapters 6 and 7.

These occupational earning patterns hide large gender differences. Thus, according to the 2001 census, among people working full-time and full-year, women's earnings were 71 percent of men's in 2000. More specifically, female dentists earned an average of $82,254, two-thirds (64%) of the earnings of their male counterparts ($129,104). Among senior managers, men earned an average of $114,519 compared to $69,993 for women. Female university professors reported 2000 earnings of $63,746—80 percent of the earnings of male professors ($79,993). Among the lower-paid occupations, we find male grocery clerks reporting annual 2000 earnings of $25,905, considerably higher than the earnings of their female counterparts ($20,691). Similar gender differences are observed in all occupational groupings but, as these specific examples demonstrate, the *female–male earnings ratio* does vary considerably.

These examples from the 2000 census compare women and men in full-time, full-year jobs. However, even among full-time workers, men average more hours per week (43.1 hours in 1997) than do women in full-time jobs (39 hours). This apparently small difference can still add up to five more weeks of extra paid work per year. When this difference in the annual volume of paid work is adjusted in the calculation of an *average hourly wage*, Labour Force Survey data for full-time, full-year employees reveal women earning about 83 percent of what men earned in 2004.[4]

The *gender wage gap* has been decreasing for several decades as women's earnings have been rising slowly while the earnings of men, on average, have stalled. Part of the reason for the long-term trend in rising female incomes is that more women have invested in higher education and, as a result, have gained access to better-paying jobs. However, if the explanation were this simple, we would expect female incomes to be higher than male incomes, since the proportion of female labour force participants with university degrees is now higher than the proportion for men (see Chapter 2). But women, on average, have less labour force experience and tend to choose fields of postsecondary study that lead to somewhat lower-paying jobs (Drolet 2002a). In addition, as we discuss in Chapter 4, gender-biased hiring and promotion practices can also affect the female–male earnings ratio.

By restricting our discussion to broad occupational categories, we also overlook the extreme ends of the *income distribution* in the Canadian labour market. At the top of the earnings hierarchy, chief executive officers (CEOs) of Canada's largest firms typically earn huge incomes (based on salaries, performance bonuses, and options to purchase stocks in their company at below-market value). In 2004, the CEOs of the top 160 companies listed on the Toronto Stock Exchange received a median compensation package of $1.8 million (Church 2005). The fact that the average compensation was much higher ($5.5 million) indicates that a few CEOs were exceptionally well paid (and that it was a good year to "cash in" stock options). Robert Gratton of Power Financial Corporation led the pack, with $173.2 million (including $169 million from stock options). Gerry Schwartz of Onex Corporation and Frank Stronach of Magna International were probably envious, taking home only $76.4 million and $52.5 million, respectively. It would take the average Canadian worker, with annual earnings of about $32,000 in 2000, almost 2,400 years to accumulate as much money as Schwartz and over 5,400 years to earn what Gratton received in one year.

At the other end of the wealth scale are numerous workers with very low incomes. Legislated minimum wages in Canada have always been very low, and in the past several decades they have not kept up with inflation. Thus, in 2005, if a worker earning a minimum wage worked 40 hours a week for 52 weeks in the year, she or he would earn $13,000 annually in Newfoundland (the province with the lowest minimum wage), $14,560 in Alberta, $15,496 in Ontario, $15,808 in Quebec, and $16,640 in British Columbia (with the highest minimum wage in Canada).[5] However, virtually no minimum wage workers would be employed year-round for 40 hours a week, so these estimates of annual income are too high for this group of workers.

Many, but certainly not all, of these lowest-paid workers are students working in the lower-tier services (retail trade and consumer services), frequently part-time. Some older, full-time workers also have minimum wage jobs, but many more are hired at pay rates only a few dollars above the minimum wage. As a second income in a household, such salesclerk, waitress, cashier, and service station attendant jobs might help pay some bills. However, if this were the only income, the household, especially if it contained children, would likely be living well below the official *low-income cutoff*, more commonly called the *poverty line*.[6]

Hourly wage rates in blue-collar occupations, such as construction or manufacturing, are typically at least twice as high as the minimum wage. Yet even with a $14-per-hour wage, and working full-time and year-round, these earnings would still make it extremely difficult to raise a family in most major Canadian urban centres. Furthermore, many of these blue-collar jobs are seasonal and are subject to frequent layoffs. Consequently, when we look closely at the characteristics of Canadian families living below the poverty line, we find that low wages, insufficient work (part-time or part-year), and periodic unemployment are usually the problem. Some of the *working poor* lack education or marketable skills, but many simply cannot find well-paid and secure employment.[7]

Other Employment Benefits

Additional *employment benefits*—a form of indirect pay and increased income security—are another important dimension of the quality of jobs. Canadian employers are legally required to contribute to Employment Insurance (EI), the Canada/Quebec Pension Plan, and Workers' Compensation. Many, particularly large firms and public-sector organizations, also spend large amounts on additional benefits including paid vacation; maternity/paternity leave; sick leave; medical, dental, disability and life insurance; and private pension plans. Over the past four decades (1953 to 1998), the costs of non-wage benefits have doubled in Canada, from about 15 percent of total labour costs to more than one-third. Given the amount employers spend on such "non-wage forms of compensation," it is probably inappropriate to call them "fringe benefits" as has been the custom (Budd 2004: 597).[8]

A national survey in 2000 revealed that between 50 percent and 60 percent of Canadian workers had an employer-sponsored pension plan, health insurance, a dental plan, and paid sick leave (Figure 3.1). Somewhat fewer (46%) reported that they had access to paid maternity or parental leave.[9] But almost one-third (30%) received none of these five benefits (a small proportion were able to access some benefits through a spouse's plan). In contrast, 29 percent received all five benefits (Lowe and Schellenberg 2001). In other words, the distribution of fringe benefits is highly polarized. As shown in Figure 3.1, full-time employees are more than twice as likely as part-time workers to receive each of these benefits, and permanent employees about

FIGURE 3.1 *Job Benefits by Labour Market Status, Canada, 2000*

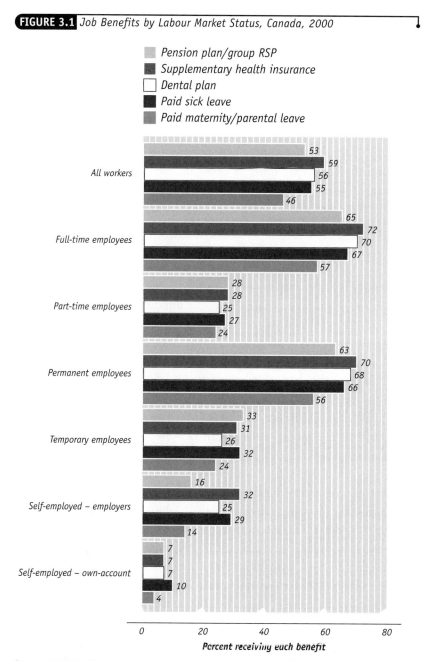

Pension plan/group RSP
Supplementary health insurance
Dental plan
Paid sick leave
Paid maternity/parental leave

All workers
53
59
56
55
46

Full-time employees
65
72
70
67
57

Part-time employees
28
28
25
27
24

Permanent employees
63
70
68
66
56

Temporary employees
33
31
26
32
24

Self-employed – employers
16
32
25
29
14

Self-employed – own-account
7
7
7
10
4

0   20   40   60   80

Percent receiving each benefit

Source: CPRN/EKOS, *Changing Employment Relationships Survey* (Lowe and Schellenberg 2001). Reprinted by permission from Canadian Policy Research Networks.

twice as likely as temporary workers. The self-employed, particularly the own-account self-employed, typically do not receive such benefits, since it is expensive (and sometimes quite complicated) to "purchase" health and dental plans for oneself. These types of benefits are also more frequently provided in larger firms (where pay is also typically higher) and much more frequently in unionized workplaces.[10] Furthermore, workers in larger workplaces and in unionized settings are more likely to receive formal job-related training (Jackson 2005: 47–50) which, in time, can translate into better jobs and higher income. It is apparent, then, that pay differences within the Canadian labour market are accentuated by the uneven distribution of fringe benefits. And, as the costs to employers of providing such benefits continue to increase, we may see relatively fewer workers receiving them (Waldie 2005).

Risks to Personal Health and Safety

We can extend this discussion of job quality by examining the personal risks of different kinds of jobs. For example, every year hundreds of Canadian workers die from work-related injuries or illnesses. To be precise, between 1993 and 2003, a total of 9,050 Canadian workers—an average of more than two every day—died as a result of their job. While work-related deaths may receive less attention, the fact remains that the chances of being killed on the job are still greater than those of being killed on the road by a drunk driver. From the mid-1980s until the mid-1990s, *industrial fatality rates* declined, to 5.14 deaths per 100,000 workers in 1996 (a total of 703 industrial fatalities). However, the number of fatalities then began to rise, to 787 in 1999 and 963 in 2003. This most recent fatality count translates into a rate of 6.12 deaths per 100,000 employed Canadians.[11]

Industrial fatality rates are generally much lower in the service industries than in the goods-producing sector (Marshall 1996). For example, over the period 1988–93, the manufacturing sector saw an average of 8 deaths per 100,000 workers per year. The construction and transportation industries had much higher fatality rates, with about two dozen work-related deaths per year per 100,000 workers. But these rates were dwarfed by those found in the resource-extraction industries: between 1988 and 1993, the annual fatality rates were 63 (per 100,000 workers) in mining, 82 in logging and forestry, and 113 in fishing and trapping. The 26 miners killed in Nova Scotia's Westray mine on May 9, 1992, are a tragic example of the high risks involved in working in Canada's resource industries.

Certain occupations within these industries carry extremely high risks (Marshall 1996). The most dangerous are those involving cutting, handling, and loading of materials in the mining industry (281 deaths per 100,000 workers between 1988 and 1993) and working with insulation in the construction industry (246 deaths). However, since relatively small numbers of people work in these and other high-risk jobs (including fishing and trapping), the three leading causes of work-related death are exposure to harmful substances (20% of all industrial fatalities between 1988 and 1993), accidents involving transportation vehicles (19%), and being struck by an object. Since the high-risk jobs are typically filled by men, 96 percent of the workers who died because of work-related reasons between 1988 and 1993 were men.

The incidence of *work-related injury and illness* rose in Canada during the 1980s. In 1982, Workers' Compensation Boards and Commissions across the country compensated 479,558 individuals for work-related injuries and illnesses resulting in time loss or permanent disability. This number increased to 620,979 in 1989, but then began to decline steeply, to 410,464 in 1995. While the rate of decline then slowed, the number continued to drop, to 345,854 in 2003. Thus, while the rate of industrial fatalities started increasing again in the late 1990s, we have seen the opposite trend for workplace injuries and illnesses.[12] Even so, Canadian workers are aware of and worry about health and safety risks at work. A 2002 national survey revealed that 32 percent of employed Canadians felt that their "health or safety [was] at risk because of [their] work (Brisbois 2003: 38). The compensation costs for workplace injuries are huge—in 1998, they totalled $4.65 billion (Human Resources Development Canada 2000b).

As with work-related fatalities, rates of *time-loss work injuries* are much higher in the goods-producing industries, particularly in forestry, manufacturing, and construction (Figure 3.2). However, the distributive services (transportation and wholesale trade) also have high injury rates. Time-loss work injury rates for men are more than twice as high as for women, while young men (aged 15 to 29) have particularly high rates (43.3 injuries per 1,000 workers in 1996). The explanation for these gender and age differences is found in the different industrial sectors in which women and men tend to be employed. Young men are most likely to be working in the most dangerous occupations in the goods-producing sector and in the distributive services.

Not all work-related injuries and illnesses get reported and compensated, and the effects of some do not become apparent for many years. Hence, it is

FIGURE 3.2 *Time-Loss Work Injuries by Industry,* * *Gender, and Age, Canada, 1996*

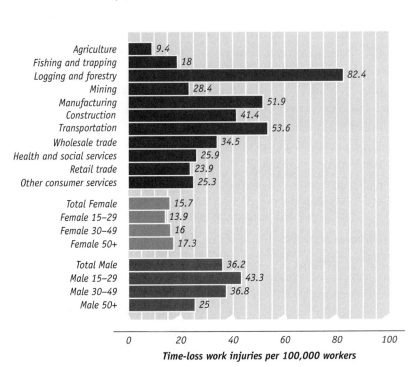

Time-loss work injuries per 100,000 workers

*Data for business services, education, and public administration not provided by source.

Source: Adapted from the Statistics Canada publication *Statistical Report on the Health of Canadians,* Catalogue 82-570, Table 61, September 16, 1999. (This publication is the result of a collaborative effort by Health Canada, Statistics Canada, and the Canadian Institute for Health Information.)

useful to also look at research on *workplace stressors* that might be harming workers' health. In a 2000 national survey, for example, 1 in 3 (34%) Canadian workers aged 15 and older indicated that their job was stressful because of too many demands or too many hours of work. About 1 in 8 also experienced excess worry or stress because of risks of accident or injury (13%), concerns about job loss or layoff (13%), and poor interpersonal relationships in the workplace (15%). Such stressors are associated with negative health outcomes. For example, the more physically demanding the job, the higher the rate of self-reported work injuries. And the higher the job strain, the more frequent the reports of poor physical and mental health.[13]

In addition, many workers are exposed to *workplace hazards* that can be harmful to their health. A 1991 national study revealed that 34 percent of employed respondents had been exposed to dust or fibres in the air at their job site in the previous 12 months. One in 4 (26%) reported exposure to loud noise, 22 percent had been exposed to poor quality air, and 18 percent said they had been exposed to dangerous chemicals or fumes. Blue-collar workers were much more likely to be exposed to dust or fibres, loud noise, and dangerous chemicals, while white-collar workers were more likely to report poor quality air in their work environment.[14]

There are a number of important conclusions to be drawn from this brief discussion of workplace health and safety in Canada. First, it is clear that work-related stress, injuries, illnesses, and fatalities constitute an extremely serious social problem. Hundreds of Canadians die in work-related injuries each year, and hundreds of thousands are injured or become ill. But even though work time lost to injuries and illnesses continues to far exceed the time lost due to strikes and work stoppages in Canada (15.4 million days versus 2.5 million days in 1998, for example), the latter is much more often perceived to be a problem.

Second, workplace health and safety problems constitute a huge cost for employers. In addition to the direct costs of compensation ($4.65 billion in 1998, as noted above), there are indirect costs resulting from production losses, damage to equipment, lower efficiency, decreased employee morale, and lost supervisory time. Although such indirect costs are notoriously hard to calculate precisely, estimates of total costs are typically about twice as high as the direct costs of compensation.[15] In other words, the costs of workplace health and safety problems in the late 1990s were probably approaching $10 billion annually. Third, industrial and occupational differences in injuries and fatalities demonstrate that the risks of injury, illness, and death are an additional important dimension of overall job quality. Some of the best-paid groups within the labour market experience many fewer risks to their personal health and safety. Workers with lower incomes and fewer benefits also tend to be at greater risk of ill health and injury because of their job (Thomason and Pozzebon 2002).

Occupational Status

We seldom find doctors, lawyers, scientists, or professors avoiding the question when they are asked, "What do you do?" But for clerks, janitors, parking-lot attendants, and many others, the same question might elicit an apologetic "I'm just a . . ." In short, there is considerable consensus in our society about which jobs have higher status. While less important than income and benefits (which directly determine one's standard of living) and job security (which ensures continuity of that standard), *occupational status* (or prestige) is something we must also consider when comparing different jobs. To a great extent, our self-image and the respect of others is determined by our occupational status, although there is probably also a tendency for individuals situated higher within a stratification system to take occupational status more seriously (Ollivier 2000).

Over the past half-century, the status of professional occupations has always been much higher than that of most sales/service and blue-collar occupations. However, with the rapid expansion of the lower-tier service industries, the proportion of Canadians employed in skilled trades and manufacturing occupations, and in better-paying unionized service-sector jobs, has declined (see Chapter 2). This may account for the somewhat unexpected finding in a recent study that the status of many of these trades occupations (e.g., welder, plumber, carpenter) as well as that of firefighters and police officers appears to have increased between 1975 and 2000.[16] At the same time, the relative status of many professions (e.g., lawyers, psychologists), as well as members of Parliament, appears to have declined.

Despite these shifts over time, there is still a strong relationship between occupational status and income. It may seem that both higher pay and higher status are a direct result of the greater skill and responsibility required by certain jobs. Generally, higher status jobs do require more education, cognitive ability, and skill (Hunter and Manley 1986). But it is also possible that some occupations have come to be seen as more prestigious because, over time, incomes in this line of work have risen. Higher incomes, in turn, might be the result of skill increases, but they may also reflect the ability of a powerful occupational group to limit entry into its field or to raise the prices for the services it provides. It is also clear that some occupations have traditionally had higher social status because they were viewed as men's work rather than women's work (see Chapter 4), although there is some evidence that gendered assumptions about occupational status are changing.[17] Thus, for a variety of reasons, jobs defined as "better" in terms of extrinsic rewards (e.g., pay, benefits, worker safety) typically also have higher status in society.

The links among income, education, and occupational status have led to widespread use of *socioeconomic status* (SES) scales to locate individuals or families within a social hierarchy or stratification system. In Canada, until fairly recently, the most commonly used SES measures were the *Blishen scores.* Bernard Blishen used information on occupational prestige rankings, along with census data on the average income and education of Canadians employed in various occupations, as well as the proportion of women employed in these occupations, to estimate SES scores for almost 500 different occupations (Blishen, Carroll, and Moore 1987). But the Blishen scores have become dated as new occupations (in the field of information technology, for example) have emerged and the relative status of other occupations has changed.

CANADA'S CLASS STRUCTURE

The question of who has better and worse jobs in the Canadian labour market is really a question about social inequality. Therefore, answering the question requires us to examine the class structure of our society. The term *social class* is used in a variety of ways by social scientists. Some treat it as equivalent to socioeconomic status, discussed above, referring to an individual's position in a fluid social hierarchy determined by occupation, education, and income. Others, including Robert Reich (1991) writing about "symbolic analysts" and Richard Florida (2002) describing the growth of a "creative class," use the concept to distinguish between occupational groupings that have become more prominent and dominant in contemporary society (see Chapter 2). However, social class has a much more precise sociological meaning. It refers, at one level, to a particular position within a stratified social structure and, on another level, to the power relationships among groups directly engaged in the production process. These "relations of ruling"[18] have an impact on individuals' life chances (their living standard and the opportunities to improve it), and, sometimes, on the way they perceive the social world around them and react to it.

Marx on Social Class

Class relationships were central to the social and economic theories of Karl Marx. In his analysis of capitalism, Marx focused primarily on the relationship between the class of *capitalists,* who owned the means of production, and the *proletariat.* The latter class owned no production-related property, and so had little choice but to exchange labour in return for a wage. Marx also discussed the middle class of small business owners (the *petite bourgeoisie*) and predicted that, in time, it would largely disappear; small businesses would be swallowed up by bigger competitors, and self-employment in agriculture would gradually be replaced by wage labour. The result would be an even more polarized society. Conflict between an increasingly impoverished class of workers and the capitalist class would eventually lead to the emergence of a new type of egalitarian society.

As documented in Chapter 2, over the past century Canada has moved in the direction of less self-employment, as have other industrialized capitalist societies. However, we have also observed an increase in self-employment in the past two decades. In addition, expansion of the service industries, and a growing number of skilled (and well-paid) workers in white-collar occupations, force us to rethink Marx's model of social class relationships. What, then, are the characteristics of Canada's class structure?

Postindustrial Class Structure of Canadian Society

Marx's original classification system cannot accurately map the complexities of social class in Canadian society today. For example, while an individual whose store has three employees is classified as an employer, her or his responsibilities, power, and job rewards are hardly comparable to the president of a corporation employing thousands of workers, wielding a great deal of power, and taking home a pay package of several million dollars a year. Similarly, while both a senior manager in a large corporation and a sales clerk in a retail store may be paid workers, their work experiences and rewards are very different. A variety of attempts have been made to incorporate such important distinctions into theories of social class.

A widely used approach was developed by Erik Olin Wright (Wright et al. 1982). In his classification of labour force participants, Wright took into account ownership, the employment of others, the supervision of others, and control over one's own work. He distinguished between large employers and small employers, two groups that have legal ownership and also employ others, and the *petite bourgeoisie,* who own their businesses (or farms) but do not have others working for them (the own-account self-employed). Large employers (the group most closely resembling Marx's class of capitalists) are in a different category from small employers with only a handful of employees, since the latter would typically be directly involved in the production process (working alongside employees).

The large class of paid employees was separated by Wright into managers, who, while not owning the enterprise, would be involved in decision making, along with (or in the absence of) the legal owners; supervisors, who are not involved in planning and decision making but who nevertheless exercise authority over others; and workers, who have no ownership rights, decision-making power, or authority over others. One final class grouping, *semi-autonomous workers* (for instance, social workers, university professors, and other salaried professionals), are identified by the relatively greater control they retain over their own work.[19]

Census and Labour Force Survey data do not contain information on planning, decision making, and the exercise of authority over others, so these sources cannot be used to profile the contemporary Canadian class structure. However, a series of Ontario-wide surveys have been using Wright's approach to class analysis to profile the class composition of that province since the early 1980s (Livingstone 1999: 158). In 1996, this research vehicle described the Ontario employed labour force as comprising the following: corporate capitalists (1%); small employers (8%); the own-account self-employed or *petite bourgeoisie* (14%); managers (8%); supervisors (4%); professional and semi-professional employees (19%); and service and industrial workers (46%). The small sample size in this study (about 600 respondents) means that these estimates are not precise, and the single province focus does not allow us to generalize to the rest of the country. Even so, this study clearly demonstrates that the large class of paid workers contains several distinct types with varying amounts of decision-making authority and, as we have discussed earlier in this chapter, large differences in income, benefit packages, and occupational status. Even within the "worker" category, those employed in unionized blue-collar settings, for example, would be much better paid and receive more benefits compared to workers in less secure jobs in the lower-tier services. Consequently, there is merit in continuing our discussion of labour market outcomes by examining other approaches to the subject.

Furthermore, the class-based approach does not directly address questions about how a society allocates better or worse jobs to individuals and groups, how individual workers might improve their employment situation, or how some groups of workers might manage to exclude others from access to better jobs. Such questions about *labour market processes* are central to the human capital and labour market segmentation perspectives discussed in the remainder of this chapter.

THE HUMAN CAPITAL MODEL OF LABOUR MARKET PROCESSES

We have argued that, on a variety of dimensions, some jobs are better than others, while recognizing that some jobs also require more skills and training. Ideally, jobs with specific requirements would be filled by individuals most suited for these positions. If being suited for a particular job meant that one must first obtain an advanced education, then it would seem only reasonable that, having done so, one would be rewarded with a better job. Stripped to its essentials, these are the basic premises of *human capital theory,* a major economic explanation of how the labour market operates.

This theoretical perspective assumes that a job's rewards are determined by its economic contribution to society. It also predicts that more dangerous and unhealthy jobs should be paid more, since workers would have to be compensated for these greater risks. The model assumes that labour market participants are all competing for jobs in a single, open labour market. Information about available jobs is widely circulated. All potential employees with the necessary qualifications have equal access to job openings. When it comes to choosing whom to hire, employers make rational decisions, based on an assessment of an individual's ability.

In order to "get ahead," it makes sense for job seekers to get more education and training. This means delaying entry into the labour market and forgoing immediate earnings. Yet this is not a permanent loss because by obtaining more education one is investing in *human capital,* which can later be "cashed in" for a better job. In short, the human capital model emphasizes the *supply side* of labour markets, and largely overlooks the behaviour and characteristics of employers and work organizations (the *demand side*). It also ignores questions about class structure and unequal power relationships within the labour market. The human capital perspective on labour markets is premised on a *consensus* view of society, in contrast to the assumptions of *conflict* underlying class-based and labour market segmentation approaches.[20]

The basic logic of human capital theory is compelling, given the evidence that better-educated individuals generally are less likely to be unemployed and more likely to hold well-paying, high-status jobs. But it is also difficult to ignore the contrary evidence. As we have seen, the most dangerous and unhealthy jobs are not necessarily the best-paid jobs. There are also many examples of well-trained and highly motivated people working in poorly paid, low-skill jobs. For example, although most recent university graduates have managed to find reasonable jobs, compared to graduates a decade or two ago, a larger minority have had difficulty finding full-time, permanent work that matches their training.[21] The problem is clearly not one of insufficient education or effort.[22] Instead, much of this youth underemployment can be traced directly to organizational downsizing and large-scale cutbacks in the hiring of entry-level workers by employers in both the public and private sectors. Similarly, even though poorly paid Canadian workers were better educated, on average, in the 1990s than in the 1980s, they were no more likely to make their way out of this disadvantaged position in the 1990s than they had been a decade earlier (Morissette and Zhang 2005).

To make the same point, while many of the wealthiest members of our society are no more educated than the rest of us, they frequently appear to have had a head start in the career race. We continue to see the powerful influence of huge firms largely controlled by individuals who inherited their money (Yakabuski 2001). Thus, the social class of one's family can strongly influence later labour market outcomes. In addition, effort and competence do not always correlate neatly with labour market rewards. For example, a 1999 study showed that up to a third of CEOs of money-losing companies listed on the Toronto Stock Exchange still received performance bonuses. In some cases, these bonuses involved hundreds of thousands of dollars, on top of already huge salaries, while company losses for the year were in the millions of dollars (Church 2001).

Thus, the relationships among initiative and effort, education and training, and risk, on the one hand, and occupational attainment and income on the other, are not nearly as consistent as the human capital model would suggest. In fact, research shows that some groups are systematically less likely than others to have benefited from their investments in human capital, and that reasonable returns on education and training are obtained only in certain industrial sectors or only from some types of employers. Furthermore, a great deal of evidence shows that some groups in society are more likely to have access to higher education. To understand who gets the good jobs, we must look beyond the labour market to families, schools, and other institutions that shape labour market outcomes.

Social Structure and Occupational Choice

The human capital model attempts to explain how people are sorted into different occupational positions by focusing on the characteristics of individual workers. People whose skills and abilities are more valued by society, and who have invested more in education and training, will be leading candidates for the better jobs, according to the model. Another assumption it makes is that individuals choose among work options, eventually settling in the occupational niche that best suits them. But to what extent do individuals actually have a choice from among the wide range of occupations? Does everyone start the "career race" from the same position, or are some groups disadvantaged at the outset? Does chance also play a role in matching individuals and jobs?

We probably all know a few people who accidentally ended up in their present job, with little planning, or who landed a great position by being in the right place at the right time. We probably also know people who, as children, decided they wanted to be a teacher or a doctor and then carefully pursued the educational route to these goals. Such behaviour and outcomes are consistent with human capital predictions, although the theory does not ask why these individuals had such high occupational goals.

In fact, we can easily imagine how socioeconomic origins might influence career patterns. It would be hard to picture a member of the wealthy Bronfman or McCain families aspiring to be a bus driver. Similarly, many people who make their living as farmers would probably cite growing up on a farm as a major career influence, while growing up in a working-class family in a single-industry community would likely channel a youth into one of the local mills or mines. It is also clear that many women in today's labour force were constrained in their career choices by society's attitudes about appropriate gender roles. Until quite recently, it was typically assumed that women could work as teachers and nurses, for example, but not in traditionally male occupations (see Chapter 4). Thus, while aptitudes and investments in education do play an important role, it is also clear that for many working Canadians, family circumstances, social class background, and community of origin, along with personal attributes such as gender, race, and ethnicity, are important determinants of educational and occupational choices and outcomes.

Equality of Educational Opportunity

One of the core values underlying our education system is that of *equality of opportunity*. This is the belief that gender, ethnicity, family background, region of residence, or other individual characteristics should not be an impediment to obtaining a good education and, through this, access to good jobs and a decent standard of living. As we have already seen, basic gender differences in educational attainment have disappeared (although, as Chapter 4 discusses, there are substantial gender differences in types of educational choices). But research continues to show large differences in educational attainment depending on one's family background.

Analysis of data collected in a 1994 national survey reveals that among Canadian adults aged 26 and over, a total of 16 percent had acquired a university degree. But among those from families where one or both parents had a degree, 59 percent had completed university, compared to only 12 percent of those from families where neither parent had a degree. Thus, among previous generations of potential university students, being a member of a family with a tradition of university attendance increased the chances of completing university by almost five times.[23]

How do such patterns of *intergenerational transfer of advantage* develop? Over the past several decades, a number of different studies have highlighted how middle- and upper-class parents have higher expectations of their children, serve as role models for postsecondary educational participation and successful careers, and are more likely to financially support their children's postsecondary activities. For example, an Ontario study of the "class of 1973" tracked a large number of high-school graduates for several decades and conclusively demonstrated that a more advantaged background leads to higher *educational and occupational aspirations,* greater educational attainment and, in time, higher status occupations and higher incomes (Anisef et al. 2000).

But the world of postsecondary education has changed dramatically since the 1970s. The proportion of Canadian youth going on to college or university has risen sharply with the opening of new postsecondary educational institutions, continued demand by employers for higher qualifications, and the introduction of new student finance systems.[24] Have social class differences in educational attainment been reduced? Research reveals that the efforts to improve access to higher education largely eliminated gender differences in educational attainment and reduced class differences somewhat (Wanner 1999). Even so, children from more affluent and better-educated families continue to be significantly overrepresented among university students (Wotherspoon 1995; de Broucker and Lavallée 1998; Krahn 2004). And, as a comparison of Canadian *school–work transition studies* from the

1970s, 1980s, and 1990s demonstrated, children from more advantaged families continue to have higher occupational aspirations (Andres et al. 1999).

Findings from other studies of this type clearly reinforce this conclusion. In 1985, we began a longitudinal study of high-school graduates in three Canadian cities (Edmonton, Sudbury, and Toronto). As in the 1970s Ontario studies, we found much higher educational aspirations among young people from higher-SES families. For example, 64 percent of high-school graduates from families where at least one parent had a supervisory, managerial, or professional occupation planned to obtain a university degree, compared with only 38 percent of those from lower-SES families. Four years later in 1989, we observed that 43 percent of sample members from higher-SES families had attended university during at least three of the four intervening years, compared with only 21 percent of those from lower-SES families.

A more detailed analysis of the Edmonton sample, and of a similar 1989–93 study in Vancouver, revealed that respondents from families where at least one parent had a university degree were much more likely to graduate from university themselves. In Edmonton, the odds of acquiring a university degree were about three times higher for respondents from families with a university-attendance tradition, compared to those whose parents were not university-educated. In Vancouver, the difference was not as large. Even so, 48 percent of sample members with a university-educated parent obtained a degree within four years of leaving high school, compared to only 25 percent of those whose parents did not have a degree (Andres and Krahn 1999).[25]

Economic Advantage and Cultural Capital

The fact that children of university-educated parents are more likely to graduate from university themselves is not simply a function of their higher educational and occupational aspirations. More highly educated parents also have higher incomes, on average, and more money means access to more and better postsecondary education (Davies and Guppy 1997). The rapid expansion of Canada's postsecondary system several decades ago increased access to higher education, and the introduction of new student finance systems allowed more children of working-class parents to take advantage of the new universities, community colleges, and technical schools. However, it is very likely some of these gains were lost over the past decade, an era when postsecondary tuition fees rose rapidly, as did student loan debts (Allen and Vaillancourt 2004).[26] Although the gap in university participation rates between low and high income families changed little between 1993 and 2001, tuition rates have continued to rise steeply since then and we may yet see the participation gap widen. Furthermore, higher tuition fees may also begin to affect postsecondary dropout rates and are already creating additional stresses for low-income families trying to juggle their other financial responsibilities with the growing costs of educating their children.[27]

So, parents' money continues to matter. But the process whereby advantaged backgrounds translate into educational success and better jobs begins long before students reach the end of their high-school studies and start to pay university and college tuition fees. Statistics Canada's National Survey of Children and Youth reveals the disturbing fact that children in Canada's poorest families are three times more likely than children of the wealthy to be in remedial education classes at school. In contrast, the most advantaged children are twice as likely to be in classes for "gifted" students (Mitchell 1997). A long list of studies has also demonstrated that, even though the high-school dropout rate has been declining over the past few decades, teenagers from less affluent families are still much more likely to leave school without a diploma.[28]

Children in Canada's poorest families frequently go to school hungry. Parents with low incomes may have to work long hours to cover basic living costs, and will have less time and money to invest in their children's education. The schools in poorer neighbourhoods may also have fewer resources. These are among the more obvious explanations of the relationship between economic disadvantage and less successful educational outcomes. In his analysis of social stratification systems, Pierre Bourdieu (1986) introduces the concept of *cultural capital* to further explain why middle-class youth perform better in the education system. Schools encourage and reward the language, beliefs, behaviour, and competencies of the more powerful groups in society. Middle-class youth bring more of this cultural capital to school with them, and so have a distinct advantage. They are more likely to speak like their teachers, to be comfortable in a verbal and symbolic environment, to know something about the subjects being taught, to have additional skills (music training, for example), and to have access to learning resources in their home.[29]

A 1994 national survey of Canadian adults provides some examples of this process. Considering only those respondents with children aged 6 to 18,

78 percent of those with a university degree reported that their children read every day for pleasure, and 70 percent stated that their children had a specified daily time for reading. Among parents without a university degree, only 49 percent said that their children read daily, while 41 percent reported a specified daily time when their children read at home.[30]

Thus, schools are not neutral institutions. Instead, they are part of the process whereby structural inequalities and power differences within society are reproduced, within and across generations, because the culture of the more powerful classes is embedded within them. Obviously, not all middle-class youths do well in school, and not all children from less affluent families do poorly. But the probability of doing well in school, graduating from high school, successfully completing postsecondary studies, and choosing an educational path that leads to a higher-status, better-paying job is much greater for middle-class youth.

Differences by Gender and Region

Several decades ago, Canadian studies showed that young women had lower educational and occupational aspirations than young men. More recent studies show equivalent aspiration levels among young women and men (Andres et al. 1999), although female and male teenagers still report quite different types of specific career goals. As female labour force participation has increased and gender role attitudes have changed, the proportion of young women going on to higher education has risen to match the male rates. However, young women are still more likely to enroll in traditional "female" areas of study such as education, nursing, and the arts and humanities. As the next chapter documents, gender differences in educational choice continue to shape labour market outcomes for women and men.

The Ontario longitudinal study from the 1970s also examined the effects of students' region of residence on educational and occupational aspirations. The researchers noted that rural youths were less likely to plan on and participate in higher education, and speculated that the more "limited horizons" of rural youth might be due to their underexposure to attitudes favouring higher education (Anisef et al. 2000: 143). In addition, the absence of local institutions of higher learning forces rural youth to leave their home communities to go to college or university, making the transition more difficult. More recent studies in

Nova Scotia and British Columbia (Looker and Dwyer 1998; Andres and Looker 2001) have reinforced such conclusions about problems faced by rural youth wishing to participate in the postsecondary system. The construction of community colleges in smaller urban centres in some provinces has had a beneficial impact on this pattern. For example, a longitudinal study of 1996 Alberta high-school graduates showed that, in 2003 at age 25, those who had completed high school in larger cities were only marginally more likely to have obtained a postsecondary credential, compared to young people from smaller communities and rural areas. However, the study also showed that youth from large cities were significantly more likely to have received a university degree.[31] Thus, even though the specific patterns are shifting, rural youth continue to acquire fewer and less valuable postsecondary credentials than young people in large urban centres, limiting their occupational choices and their future earning potential.

Occupational Mobility Research

The term *social mobility* generally describes how individuals or groups move from one position within a social hierarchy to another. For example, we might see that the status of an immigrant group has risen over several generations in Canada. Or we might document the misfortunes of middle-class individuals and families who lost their jobs and became downwardly mobile because of industrial restructuring.[32] Because social standing is to a great extent influenced by occupational position, most social mobility research has examined occupational mobility.

We can usefully distinguish between *intergenerational* and *intragenerational* mobility.[33] The former involves comparisons between an individual's occupational status and that of someone in a previous generation, most frequently his or her parent (or grandparent). Because of the gendered nature of occupational structures, it is usually most informative to compare the occupations of daughters and mothers, and sons and fathers. Intragenerational mobility involves comparisons between an individual's present and previous occupations, for example, between an individual's first full-time job after completing formal education and her job in mid-career.

There is also an important difference between *structural mobility* and *circulatory (or exchange) mobility.* Chapters 1 and 2 documented massive changes in industrial capitalist societies over the past century: a huge decline in agriculture and a parallel expansion of the urban workforce, the rise of large bureaucracies, and extensive growth in the service industries. These changes precipitated enormous growth in white-collar occupations. Because of this occupational shift, we would expect to find much larger proportions of labour force participants in middle-level occupations today than a generation or two ago.

Using data from surveys asking respondents about their occupation and the occupations of their parents, researchers employ sophisticated statistical techniques to determine how much structural mobility has been occurring within a society. Such research addresses a potentially more interesting question: how much intergenerational mobility do we observe independent of the effects of changing occupational structures—in other words, how open is a given society to occupational mobility?

A basic democratic value, and a core principle of human capital theory, is that the most talented people end up in the more responsible and rewarding jobs. If this were the case, we would observe a great deal of circulatory mobility across generations, rather than a closed society in which occupational positions are typically inherited. Theoretically, circulatory intergenerational mobility occurs when those with more talent, skill, training, and motivation are able to move into the higher occupational positions. Merit, not social origin, determines occupational position. But no society has ever realized the goal of becoming a true *meritocracy*, raising the question: To what degree are occupational position and social status inherited in a given society?

An equally interesting question is whether some societies are more open than others to occupational mobility. Early mobility studies examined intergenerational mobility across three basic occupational categories—agricultural, manual, and non-manual work. Seymour Martin Lipset and Reinhard Bendix (1959) compared the United States with a number of European countries. Many might have predicted that occupational inheritance would be more pronounced in European countries, with their entrenched class structures, and less significant in the United States, with its stronger democratic values. However, few important cross-national differences were observed. Studies using more recent data (from the 1970s and 1980s) and a larger number of occupational categories suggest few differences between mobility patterns and trends in Canada and the United States. However, they do indicate that Canada may have a somewhat more open stratification system than Sweden, the United Kingdom, France, the Netherlands, and Australia (Wanner 2004: 144).

To examine shifts over time in Canadian mobility patterns, researchers have compared results from three national surveys on occupational mobility completed in 1973, 1986, and 1994 (Wanner 2004). The Canadian occupational structure became somewhat more open for both women and men, but in different ways. The steep decline in agricultural employment since the middle of the 20th century meant that many men moved out of the agricultural occupations held by their fathers. For women, the major shift was away from the housework that had been the main female occupation for their mothers' generation. For both sexes, expansion of postsecondary educational opportunities, leading to higher status occupations, played an important role (Wanner 2004). However, most of this opening of the stratification system occurred between 1973 and 1986. Little changed in terms of mobility opportunities in the following decade, with the exception of continued slow movement of women into managerial positions.

Overall, these Canadian mobility studies find only a limited amount of direct occupational inheritance across generations and conclude that the occupational stratification system became somewhat more open during the 1970s and 1980s. Yet those at or near the top of the occupational hierarchy are still more likely to pass their advantages on to their children: "Canada is still a stratified society characterized by a considerable amount of inheritance of privilege" (Wanner 2004: 144).

Status Attainment Research

Social mobility and *status attainment* studies address similar questions, but the latter focus more explicitly on intragenerational mobility and on the role of education in determining occupational outcomes. Using the same data analyzed in the mobility studies described above, Canadian researchers find that the most important determinant of an individual's current job is his or her first job, that is, the level at which the person enters the labour market. In turn, the status of that first job is heavily influenced by the amount of education obtained.[34]

Taken alone, these findings lend support to the human capital model—greater investments in education allow one to enter the labour market at a higher level, and, consequently, to move up to even higher status positions. However, the intergenerational component of the status attainment model also yields clear evidence of status inheritance by way of education. Canadians with more education (and, hence, who have entered the labour market at higher levels and have advanced further) tend to come from families with well-educated parents in high-status occupations (Wanner 1999).

While these are not new insights, they do reinforce our earlier conclusions. First, education strongly influences occupational outcomes. Second, parents' occupational status is often transmitted to their children via differences in the amount of children's educational attainment. Status attainment studies do not explain how social origin affects labour market opportunities, but we have already mentioned some of the processes. Obviously, wealthier parents can afford more (and better) education for their children, and can provide the cultural capital needed to succeed in school. They can have a strong influence on aspirations. Being better situated in the labour market, they may also be able to provide more information about how to find good jobs or to put their children in touch with useful contacts.

On average, Canadian women today are more likely than men to acquire postsecondary credentials, and recent status attainment studies have shown that the effects of gender on occupational status have largely disappeared (Wanner 1999). Even so, Canadian women are still less able to "cash in" these credentials for jobs with higher pay and more career potential. As we will see in Chapter 4, this is partly due to the different types of education acquired by women and men. If women typically become nurses and men become engineers, then perhaps the higher status (and pay) of men's jobs makes sense. But this line of reasoning simply raises other questions: Why do traditionally female jobs have lower status and less pay? Are the contributions made by individuals in these occupations really less valuable to society?

A large part of the female–male difference in career mobility is a function of the gender-segregated nature of the labour market; women are more likely to be employed in industries and occupations where promotion opportunities are limited. Some of the gender difference in mobility also can be traced to women's interrupted careers due to child-rearing responsibilities. Of course, the human capital model would predict that less career experience (as a result of interruptions) would translate into less upward mobility. But to fully understand the careers of women and men, we must also examine the societal values and institutions that put the biggest share of responsibility for child rearing on women (see Chapter 4).

Recent Canadian studies have shown that the effects of language (French or English) on educational and occupational status attainment have basically disappeared (Wanner 1999). However, we continue to find that, despite their higher education on average, foreign-born Canadians experience less upward movement in the course of their careers. Limited knowledge of the Canadian labour market, language difficulties, licensing requirements that keep some immigrants from practising in their area of training, and discrimination in hiring and promotion all can create substantial career barriers. (Krahn et al. 2000; Reitz 2001).

Comparisons of status attainment findings from the 1970s, 1980s, and 1990s also highlight that, even though education continues to matter in terms of occupational attainment, the value of a first university degree has declined over the past several decades (Wanner 2000, 2004). Similar Canadian research comparing the labour market entry experiences of different age cohorts and different generations suggests that, despite their greater investments in higher education, young Canadians today are facing a more difficult labour market than did their counterparts in earlier generations (Baer 2004).

Summing up, we have found some support for the human capital model by identifying the central role education plays in determining one's occupational status and income. However, this must be weighed against evidence of considerable status inheritance, frequently passed on through the better education obtained by children in more well-to-do families. The fact that access to good education is not all that equal suggests that, to some degree, the issue of who gets the good jobs is class-based rather than an outcome of merit-based labour market processes.

By further examining the status attainment process, we observed that the Canadian labour market is not equally open to all labour force participants. Some groups of workers—men, individuals born in Canada, and members of the baby-boom generation, for example—are better able to cash in their education and training for higher-status and higher-income jobs. So there may be more than one labour market in operation, a possibility not considered by the human capital model. And factors other than skill, training, and effort may determine access to preferred jobs.

LABOUR MARKET SEGMENTATION

Labour market segmentation researchers begin with the proposition that there is not a single, open labour market operating in our society. Instead, better and worse jobs tend to be found in different settings and are usually obtained in distinctly different ways. Certain types of labour force participants (women, visible minorities, and youth, for example) are concentrated in the poorer jobs.

Segmentation theories also highlight the slim chances of moving out of the *secondary labour market* into jobs in the *primary labour market*. This is a key proposition, since the human capital model does not deny that some jobs are better than others; it simply maintains that those individuals who have the most ability and initiative and who have made the largest investment in education and training will be more likely found in the highly skilled and rewarding jobs. However, the segmentation perspective emphasizes the barriers that limit access to the primary labour market for many qualified individuals, as well as the ability of primary labour market participants to maintain their more advantaged position.[35]

There are actually several varieties of labour market segmentation research. All share these basic propositions, but they differ in their explanations of the origins of segmentation and in their breadth of analysis. The broader dual economy approach has evolved out of the work of Marxist scholars tracing the growth of dual economic sectors in industrial capitalist societies. The narrower approach has grown out of studies of labour market dynamics within private-sector firms and government bureaucracies (internal labour markets), and out of examinations of tactics used by various occupational and professional groups to restrict entry into their field of work (labour market shelters).

Dual Economies

The *dual economy* model assumes that an earlier era of *competitive capitalism* was succeeded in the 20th century by an age of *monopoly capitalism*.[36] Thus, a few large and powerful firms came to dominate automobile manufacturing, the oil and mining industries, the transportation sector (airlines and railways), and the computer industry, for example. Similarly, the finance sector came to be controlled by a handful of large banks, investment firms, and insurance companies. These dominant firms exert considerable control over suppliers and markets and are also able to manipulate their political environment. Examples of the power of such firms include instances in which automobile manufacturers have been able to limit foreign imports, mining and oil companies have influenced government environmental policies, and a handful of giant computer companies have come to dominate the IT industry.

Although global economic restructuring during the last decade has led to more intense competition among some of these large *core-sector* firms, they still operate in an economic environment very different from the *periphery sector*, where we find numerous smaller firms. Because they have less control over their environment and generally face more intense competition, such businesses have a much greater chance of failure. Many lower-tier service-sector firms (for example, small retail shops and restaurants) would be found here, as would some smaller firms in the upper-tier services (for example, small office supply firms in the business services) and in the goods-producing sector (for example, small manufacturing companies).

Such enterprises are typically less profitable since they cannot control their markets and suppliers, they have low capital investments, and they are often less technologically advanced and generally more labour-intensive. Compared with the core-sector firms, these enterprises typically do not require the same level of skill, education, and commitment from employees. Indeed, replaceable unskilled or semiskilled employees are often ideal.

This segmentation model basically proposes that the core sector contains a primary labour market with better jobs, while the periphery sector contains secondary labour markets. Capital-intensive core enterprises, by definition, require fewer workers to equal or exceed the productivity of firms that are more labour-intensive, although they frequently require workers who are more highly trained and better educated than those hired in the periphery sector. The large and bureaucratic nature of these enterprises means that there are reasonably good opportunities for upward mobility. The workers in these types of enterprises tend to be well paid and to have good benefit packages, are more likely to receive training, and may have greater job security.[37] Segmentation writers usually place government employees in the primary labour market because of the above-average job rewards that they typically receive.

Why would core-sector employers be willing to pay more than the going rates in the secondary labour market? Higher profit margins make it easier to do so, but this observation does not really answer the question. Part of the answer lies in the collective strength of labour unions and professional associations, which have been much more active in the primary sector. Labour market segmentation theorists also argue that it would be too costly for core-sector firms not to pay well. These enterprises have sizable capital investments and thus wish to avoid costly shutdowns due to labour disputes. For government employers, the cost of long labour disputes over pay and benefits would be public annoyance over the loss of essential services (e.g., health care, education). For both private- and public-sector employers, high labour turnover resulting from low wages would lead to the expense of training many new employees. Hence, it is simply good business practice to offer job security (negotiated with a union, if necessary), provide generous wage and benefit packages, and attempt to improve working conditions.

In the peripheral sector, employers are much less likely to provide high wages, extensive benefits, and long-term job security. We may also find more unhealthy working environments in this sector. Here, smaller profit margins, more intense competition, difficulties in passing along increased labour costs to customers, and greater vulnerability to economic cycles keep wage levels down. In addition, labour turnover is less of a problem for periphery-sector employers; lower skill requirements and little on-the-job training make workers easily replaceable. Consequently, low-paying jobs are common and labour turnover is high, making union organizing more difficult. The term *job ghettos* has frequently been used to describe work in such labour markets. Job insecurity, higher chances of unemployment, and irregular career earning patterns are a fact of life in secondary labour markets.

Barbara Ehrenreich (2001) describes how difficult it is to make a living when employed in the secondary labour market. When researching her book *Nickel and Dimed: On (Not) Getting By in America,* she worked for periods of time as a waitress, an assistant in a nursing home, an employee of a "cleaning maid" company, and a salesperson at Wal-Mart. Her coworkers, who impressed her with how hard they worked, were often women, immigrants, members of visible minority groups, and students. The jobs they held were characterized by low pay, absent benefits, limited training, physically demanding work, limited authority, lack of respect (both from employers and members of the public), and lack of protection from exploitation and harassment. The difficult working conditions encountered by her coworkers in the secondary labour market were matched by the equally difficult times they faced finding affordable housing and adequate transportation to get to work. While the hard work and stress often led to health problems, her coworkers simply could not afford to be sick, not only because they had no health benefits but also because a week without a paycheque could mean a missed rent payment.[38]

Internal Labour Markets

There are typically fewer chances for career mobility within firms in the periphery sector. A car-wash attendant, a cleaning maid, or a mechanic in a small automobile repair shop would have few promotion prospects simply because the workplace would not have a bureaucratic hierarchy with well-defined career ladders. But many employees of IBM, Shell Canada, the Royal Bank, and other large private-sector firms do have such mobility opportunities, as do university professors, government employees, and many health care workers. In contrast to the often dead-end jobs within secondary labour markets, most large corporations and public institutions have a well-developed internal training and promotion system, or what may be called an *internal labour market*. From the perspective of the work organization, such internal labour markets help retain skilled and valuable employees by providing career incentives; they also transmit important skills and knowledge among employees. From the perspective of employees, internal labour markets mean additional job security and career opportunities.

Researchers who study internal labour markets have usually sidestepped larger questions about the changing nature of capitalism, concentrating instead on specific features of these self-contained labour markets. Focus has been on *ports of entry*, specifically, the limited number of entry-level jobs that are typically the only way into such an internal labour market; *mobility chains*, or career ladders, through which employees make their way during their career within the organization; and training systems and seniority rules, which govern movement through the ranks.[39] Essentially, these concepts elaborate Max Weber's theory of bureaucracy while also emphasizing the career advantages held by workers in such labour markets.

There is some obvious overlap between studies of internal labour markets and the broader dual economy perspective. Early segmentation researchers tended to assume that all core-sector firms contain well-developed internal labour markets open to all employees. However, there are important exceptions. For example, many major corporations have long career ladders, but these are open to only some of their employees. Clerical workers (usually women) are often restricted from moving into higher ranks (Diprete and Soule 1988; Hachen 1988). Some large corporations, particularly in the fast-food industry, are built around small local franchises. These workplaces typically have all the characteristics of a secondary labour market. In addition, corporate and government downsizing and the increased use of temporary and other nonstandard workers (Vosko 2000; Vosko, Zukewich, and Cranford 2003) mean that the number of advantaged workers in internal labour markets has declined dramatically.

Labour Market Shelters: Unions and Professions

Most internal labour markets have been put in place by employers. In contrast, unions and professional associations have tried to improve job and income security for their members by setting up *labour market shelters,* sometimes within specific work organizations but more often spanning a large number of similar types of workplaces. Some occupational groups have restricted access to certain types of work since government legislation requires that such tasks be completed only by certified trades (for example, electricians). Public safety is the rationale for this legislation, but it also serves to protect the jobs and (relatively high) incomes of members of these often-unionized trades.

Industrial unions, such as the Canadian Auto Workers, bargain collectively with employers to determine the seniority rules, promotion procedures, and pay rates within manufacturing establishments (see Chapter 7). Such contracts shelter these industrial workers from the risk of job loss or pay cuts for a specified period of time. Negotiated staffing restrictions also restrict nonmembers from access to the better jobs by giving laid-off members priority if new jobs open up.

Professional associations using tactics different from those of unions also serve as labour market shelters for their members. *Professions* have been distinguished from occupations in a variety of ways, but certain key features appear in most definitions. Professionals such as doctors, lawyers, engineers, and psychologists work with specialized knowledge acquired through extensive formal education in professional schools. They enjoy a high level of work autonomy and are frequently self-regulating, policing themselves through their own professional associations. Professionals also exercise considerable power over their clients and other lower-status occupational groups, and typically emphasize the altruistic nature of their work.

For example, doctors spend many years training in medical schools where they acquire knowledge specific to their profession. They go on to determine their own working conditions, whether they are employed in large health care institutions or work as self-employed professionals. In fact, in the past decades, many doctors have increased their work autonomy by setting up professional corporations. Through their own professional associations, doctors can censure colleagues who have acted unprofessionally in the course of their work. Doctors have a great deal of power over their patients and those in assisting occupations, such as nurses and medical laboratory technicians. Finally, members of this profession emphasize their code of ethics, by which they are committed to preserving human life above all other objectives.

In short, professions are far more powerful than occupations.[40] Randall Collins (1990) uses the term *market closure* to describe the ability of professionals to shape the labour market to their advantage, rather than simply responding to it as do most occupational groups. Like the craft guilds of an earlier era, professional groups in contemporary Western society can restrict others from doing their type of work by controlling entry into their profession. Professional associations have close relationships with their (typically university-based) professional schools and can strongly influence both course content and entrance quotas. Many professional groups have actively lobbied for and obtained legislation that both requires practitioners of their profession to be officially licensed and prohibits anyone else from performing certain tasks. Over time, however, some occupational groups have gained more autonomy and power, while others may have experienced a process of de-professionalization.[41]

Barriers to Primary Labour Market Entry

Returning to our original focus on structural barriers to labour market movement, we can now see how internal labour markets and labour market shelters not only provide more work rewards to some groups of labour market participants but also contribute to their ability to maintain labour market closure. Generally, there are only a limited number of positions within the primary labour market, and entry is often restricted to those with the proper credentials—a degree, union membership, or professional certification. Often, superior credentials (an MBA from a prestigious university, for example) will further improve one's chances of getting into the system and moving up through it. And, as we argued earlier, a more advantaged family background typically gives one a significant head start (Davies and Guppy 1997).

Specific credentials are not needed for all jobs in the primary labour market, but information about vacancies is not always widely distributed. In some cases, it may be passed only through small and informal networks. Since a large portion of all jobs are obtained through personal referrals, it is clear that access to information is extremely important. Without contacts within the primary labour market, qualified applicants might never be aware of job openings (Petersen, Saporta, and Seidel 2000).

Barriers to mobility out of the secondary labour market may also be more subtle. For some individuals, a history of employment in marginal jobs may itself act as a barrier. In some regions of Canada, the prevalence of seasonal work means that many people are forced into an annual cycle where income is obtained from part-year jobs, employment insurance, and welfare assistance. A work record showing frequent layoffs or job changes may simply be reflecting the nature of the local labour market. But it could also be interpreted as an indication of unstable work habits. Similarly, a long sequence of temporary jobs, a common experience for many well-educated and highly motivated young Canadians, could nevertheless handicap them because employers might assume the problem originates with the job applicant rather than the labour market.[42]

Geography itself can limit one's work opportunities. A highly motivated and skilled worker living in Newfoundland, where the cod fishery has been decimated (Palmer and Sinclair 1997), or in Uranium City after the mine was closed, is clearly at a disadvantage compared with residents of large cities in Ontario or Alberta where unemployment rates have historically been much lower. Our discussion in Chapter 2 of employment opportunities in single-industry communities made the same point. There are simply fewer good jobs available outside of the wealthier regions of Canada.

Some years ago, a research team in Atlantic Canada spent considerable time examining that region's employment opportunities. They argued that the high level of poverty and unemployment in the region cannot simply be explained with reference to inadequate education, training, or effort on the part of local residents. A "casualty" model of poverty, which would hold that most of the poor are incapable of working, or of working in good jobs, does not fit the facts. Rather, an explanation that focuses on the nature of the local labour market is much more useful. In short, many Atlantic Canadians have spent their working lives in a *marginal work world* where low pay, few benefits, limited career opportunities, and little job security are a way of life.[43]

Historical Patterns of Racial and Ethnic Discrimination

Employer *discrimination* in hiring and promotion can also block movement into better jobs, thus contributing to the overrepresentation of specific disadvantaged groups in secondary labour markets. Before we discuss the forms such discrimination might now take, it is instructive to look back in Canadian history at some examples of officially sanctioned racial and ethnic discrimination.

After Canada was colonized, members of various Aboriginal nations frequently participated in the pre-industrial labour market, acting as guides, fighting for the French against the English (and vice versa), and providing the furs sought by the trading companies for their European markets.[44] But as central Canada began to industrialize, and later as the West was opened for settlement, the traditional economies of the Aboriginal nations in these regions were largely destroyed. Usually the presence of Aboriginal groups was seen as an impediment to economic development. The most common solution was to place them on reserves, out of the way.

Nevertheless, some Aboriginal workers were employed in the resource industries (forestry, mining, and commercial fishing and canning, for example), in railway construction, and in agriculture, particularly in Western Canada. Not all of this paid employment was completely voluntary. For example, in the 1950s and 1960s, by cutting off social assistance benefits, the provincial and federal governments forced many Aboriginal people to migrate to southern Alberta to work in the sugar-beet industry (Laliberte and Satzewich 1999). But the decline in demand for large numbers of unskilled and semiskilled workers in Canada's resource and transportation industries also meant declining wage-labour opportunities for Aboriginal people. Even when jobs were available, racist attitudes and labour market discrimination were rampant (Knight 1978). This should not be surprising in a society that clearly viewed members of First Nations as second-class citizens, not even allowing them to vote until 1960.

As Canada industrialized in the late 19th century, manufacturing enterprises began to take root in Quebec and Ontario. However, patterns of ethnic stratification based on English ownership of key industries meant that French-Canadian workers were typically underrepresented in the higher-status, better-paying positions. For example, in 1938 in the Quebec industrial town of "Cantonville," French Canadians provided most of the factory labour while

English-speaking outsiders filled the management and technical positions (Hughes 1943: 46). French-Canadian men frequently also worked as labourers in the natural resource sector and in railway construction where discrimination on the part of English-Canadian employers and supervisors was common.

During the 1880s, many Chinese men were recruited to work on railway construction in Western Canada and in the resource-based industries. They were generally paid about one-half to two-thirds of what others received for the same job and encountered a great deal of hostility from the white population (Anderson 1991: 35). Labour unions lobbied for legislation restricting Chinese workers from entering specific (better-paid) trades. Chinese workers were forced into marginal jobs in a racially split British Columbia labour market. In addition, the British Columbia government passed legislation that severely restricted the civil rights of Chinese residents of the province. For its part, the federal government required Chinese immigrants to pay a head tax on entering Canada, while other immigrants were not required to do so, and, later, it passed the Chinese Immigration Act which, for several decades, virtually stopped all Chinese immigrants from entering the country (Li 1982; Anderson 1991; Chui, Tran, and Flanders 2005).

Perhaps the discrimination faced by Chinese workers was the result of white workers' fears that they would lose their jobs to non-whites who were willing to work for lower wages. However, studies of this era suggest that racist attitudes—fears about the impact of too many Asian immigrants on a white, Anglophone society—were frequently the underlying source of publicly expressed concerns about job loss and wage cuts. White workers seldom tried to bring Chinese workers into their unions (a tactic that would have reduced employers' ability to pay lower wages to non-whites), even though the latter frequently showed interest in collectively opposing the actions of employers.[45]

Peter Li (1982) has described how, in the first half of the 20th century (1910 to 1947), Chinese Canadians living in the Prairie provinces encountered similar forms of racial discrimination. Barred from entry into many trades and professions, Chinese workers were mostly restricted to employment in *ethnic business enclaves,* where, by supporting one another and working extremely hard, they could make a living. Chinese restaurants and laundries, for example, provided employment for many members of the Prairie Chinese community.

Similar experiences of racial discrimination were encountered by Black Canadians in pre- and early-industrial Canada. Some members of the Canadian

Black community were descendents of slaves brought to Nova Scotia by Loyalists leaving the United States (Milan and Tran 2004). Others were former American slaves who had left the United States, often via the "Underground Railroad" (a network of sympathetic individuals and groups in the United States and Canada who assisted slaves fleeing north), and had settled in the area west of Toronto to the Detroit River. During the mid-19th century, white Canadians frequently lobbied to have Blacks sent back to their former owners in the United States (Pentland 1981: 1–6). By the 1860s, officially sanctioned discrimination against Blacks had largely disappeared, but labour market discrimination continued. For example, many Blacks worked for railway companies but were restricted to a limited number of lower-paying jobs, such as sleeping-car porter. White workers were found in the better-paying, less menial jobs such as sleeping-car conductor or dining-car steward. The system was supported by societal values that took for granted that Blacks should work in menial jobs. It wasn't until 1964 that the arrangements (embedded in the contracts between railway companies and their unions) that legitimized this *racially split labour market* were completely abolished (Calliste 1987).

Disadvantaged Groups in Today's Labour Market

Today, labour laws and human rights legislation make it more difficult for employers to discriminate against specific types of labour market participants. In fact, a series of recent studies have demonstrated that French–English differences in educational attainment and income have disappeared.[46] Nevertheless, some groups are still severely disadvantaged in the labour market. Some of this inequality of employment opportunity results from unequal access to higher education and specialized training. Some of it is a product of regional differences in employment opportunities. But there is also evidence of continued patterns of discrimination in hiring and promotion in the Canadian labour market, although it is typically not as blatant and institutionalized as in the past.

By 2001, visible minorities (not including Aboriginal people) made up 13.4 percent of the Canadian population (just under 4 million people), up from 4.7 percent (1.1 million people) 20 years earlier. Seventy percent of these non-white Canadians were immigrants (born outside of Canada). Compared to the rest of the adult population, recent immigrants (those who arrived in the previous decade) are much better educated. In 2001, 45 percent of recent male immigrants (aged 15 and older) had a university degree, as did 37 percent of recent female working-age immigrants, compared to only 23 percent of the remainder of the working-age (female and male) Canadian population.[47] Recent immigrants are also more likely to be in the labour force, because of their higher level of education and their somewhat younger age profile. Since Canadian immigration policies favour better-educated immigrants, these higher education and labour force participation patterns are not surprising.

Nevertheless, recent immigrants report considerably lower incomes. For example, in 2001, male immigrants who had been in Canada for one year were earning 63 percent of the amount earned by their male Canadian-born counterparts. For those who had been in the country for 10 years, the comparable figure was 80 percent (similar patterns were observed for female immigrants). These statistics suggest that, in time, immigrants slowly catch up with Canadian-born workers. However, a three-decade comparison of the experiences of different cohorts of immigrants tells a somewhat different story. Specifically, the 1991 census showed that one-year male immigrants were making 63 percent of the amount earned by Canadian-born men (the same percentage observed in the 2001 census). In contrast, 10-year immigrants were earning 90 percent (compared to 80 percent in the 2001 census). Going back another decade, in 1981 one-year male immigrants were earning 72 percent of what Canadian-born men were earning, while 10-year male immigrants had completely caught up. Similar over-time changes were seen among female immigrants.

In short, over the course of the past two decades, the average education of immigrants to Canada has been rising but it appears to be taking longer for immigrants to catch up to their Canadian-born counterparts in terms of income.[48] This is clearly a serious social and economic problem, not only for immigrants who are having difficulty "cashing in" their investments in human capital (i.e., their educational attainment and prior work experience),[49] but also for society as a whole since the under-utilization of recent immigrants' skills and training comes at a significant cost to the larger economy.

It is important to note the "changing colour of poverty in Canada" (Kazemipur and Halli 2001). The majority of recent immigrants to Canada have been members of visible minority groups, so when discussing the difficult labour market experiences of recent immigrants we really are talking primarily about newcomers from South and East Asia, the Middle East and, to a smaller extent, Africa, Central America, and South America. Related to this, evidence is beginning to accumulate that, with a few exceptions, Canadian-born visible minority group members earn as much as their white counterparts with similar amounts of education and training.[50] In other words, it is not visible minority status *per se* that is associated with labour market disadvantage, but the combination of visible minority and immigrant status. Thus we find much higher than average proportions of non-white recent immigrants underemployed in the secondary labour market, earning less than they could and should. Taxi drivers, janitors, domestic workers, parking-lot attendants, security guards, and workers in ethnic restaurants are obvious examples.

A substantial part of the problem lies in the non-recognition or undervaluing of non-Canadian educational credentials. Canadian employers and professional licensing organizations (e.g., medical or engineering associations) often assume that university degrees or technical diplomas acquired outside of Canada are inferior, and use such assumptions to avoid hiring (or licensing) well-educated immigrants who were enticed to come to Canada with promises of finding good jobs in their areas of training. For example, a 1999 study of recent refugees who had settled in Alberta showed that, even though 39 percent had held management and professional positions in their home country, only 11 percent had found similar employment in Canada (Krahn et al. 2000). For many of the former professionals interviewed in this study, the problem lay in Canadian employers and educational institutions being unwilling to recognize their foreign credentials.[51]

Along with the undervaluing of their non-Canadian credentials, some recent immigrants may be held back by English or French language deficiencies and by having fewer personal contacts within the primary labour market.[52] In addition, racial discrimination continues to handicap visible minority immigrants, as well as some Canadian-born visible minority groups, seeking to improve their employment situation. Statistics Canada's 2003 Ethnic Diversity Survey revealed that one in five visible minority labour force participants had experienced discrimination or unfair treatment in the previous five years based on their ethnicity, language, accent, or religion, compared to only 5 percent of non-visible minority Canadians.[53]

Such treatment might involve *systemic discrimination* that is subtle and hidden beneath the everyday practices within a workplace bureaucracy, as in situations where employers routinely pass over qualified visible minority candidates when making hiring or promotion decisions (Beck, Reitz, and Weiner 2002). Or it may be very deliberate and blatant. A recent Toronto study of Filipino and West Indian women recruited to work in Canada as nannies and nurses provides some telling examples. The owner of a nanny agency explained her hiring and referral practices as follows: "If you're from Jamaica, I won't even interview you. I know this is discrimination but I don't have any time for this . . . Jamaican girls are just dumb. They are not qualified to be childcare workers." Black nurses told the researchers how their ward assignments typically required much heavier physical labour, involved considerably less skilled work, and offered few supervisory opportunities. The long-term impact of such employment experiences would be less upward career mobility.[54]

In 2001, Canada had about 1.1 million Aboriginal citizens, 3.4 percent of the total population. The Aboriginal population is growing much faster than the rest of the population, and is also considerably younger. Hence, a much higher proportion of Aboriginal people are of working age or will soon move into this age category. For example, by 2017 the number of young Aboriginal adults (aged 20 to 29) is expected to increase by over 40 percent, compared to a projected increase of only 9 percent for non-Aboriginal young adults.[55] At the same time, Aboriginal people have lower labour force participation rates (61% in 2001, compared to 67% for non-Aboriginal people), much higher unemployment rates (19% in 2001, versus 7% for non-Aboriginal people), and are more than twice as likely to be living in low-income households.[56] Why?

On average, Aboriginal Canadians are not as well educated as other Canadians (Davies and Guppy 1998), although the proportions with high-school and postsecondary diplomas have been rising slowly. Nevertheless, in 2001, only 6 percent of Aboriginal Canadians aged 25 to 64 had a university degree, compared to 20 percent of all Canadians in this age range.[57] Access to good schooling has been a serious problem for generations, as has racism and discrimination within the education system (Schissel and Wotherspoon 2003). But lack of education does not fully explain the marginal labour market position of Aboriginal Canadians.

Almost one-third of Canada's Aboriginal population still lives on the reserves (O'Donnell and Tait 2004) where their great-grandparents were forced to settle when white settlers took over their ancestral lands, destroying their livelihood and making them dependent on a far-from-charitable government (Shewell 2004). Employment opportunities are severely limited in these frequently very isolated communities. Large industrial megaprojects in Northern Canada (mines, mills, pipelines, dams, and so on) have typically offered only limited employment to residents of local Aboriginal communities. At the same time, these projects have often damaged traditional hunting, trapping, and fishing economies and further undermined traditional culture (Waldram 1987; Stabler and Howe 1990; Niezen 1993).

The employment situation of urban Aboriginal Canadians is slowly improving, but there is still a huge gap separating them from non-Aboriginal urban residents.[58] The former may have a patchy employment record, which can seriously restrict entry into the primary labour market. They may also often encounter prejudice among employers, which, along with a lack of contacts and limited job-search resources (such as transportation), can further handicap them in their search for satisfactory employment. Thus, an explanation for the disadvantaged labour market position of Aboriginal Canadians must examine the many other factors beyond the acquisition of human capital that contribute to the segmenting of the Canadian labour market and to the generation of urban poverty.

Disabled Canadians are another group characterized by low labour force participation and high unemployment rates, fewer hours of work, and lower incomes. As with the Aboriginal population, disabled people who do manage to obtain employment are frequently found working in low-skill, poorly paid jobs in the secondary labour market. A 2001 Statistics Canada study estimated the number of working-age (15 to 64) disabled Canadians at just under 2 million. The labour force participation rate for this segment of the population was very low at 49 percent, while the unemployment rate was higher than average at 10.6 percent.[59] While some severely disabled Canadians might be incapable of paid employment of any kind, it is important to remember that only about one-third of working-age disabled Canadians are severely disabled. Many of the people who are mildly and moderately disabled would be eager to find jobs, or better jobs, if they could.

However, huge barriers keep disabled people from obtaining satisfactory employment. Many employers, and even some of those involved in assisting the disabled, assume that all those with visual, hearing, cognitive, and movement handicaps are equally impaired, though many of these individuals could cope well in a variety of jobs, with limited assistance. Some employers are uncomfortable with or fearful of disabled people. While a significant number of employers say they would hire disabled workers, few actually do so. In addition, many disabled people also face serious transportation problems and are held back by buildings designed without their needs in mind.[60] While there has been a little progress in improving their labour market position in recent years, disabled people continue to be significantly overrepresented in the secondary labour market. In fact, many find work only in special institutional settings where pay rates are low and career opportunities are nonexistent.

Women have faced prejudice and discrimination in the Canadian labour market for decades. Employment and human rights legislation have eliminated much of the systemic discrimination against women in recent decades, but more subtle forms of discriminatory hiring and promotion behaviour remain. For example, male employers or managers may sometimes decide that a female applicant, no matter how qualified, does not have the personality supposedly needed for a particular high-level position, or the career commitment that the corporation thinks it requires. We will address such labour force difficulties and their effects in more detail in the next chapter.

This discussion of discrimination has focused so far on the experiences of visible minorities (many of whom are immigrants), Aboriginal Canadians, disabled people, and women, the four groups of labour force participants designated for special attention in the federal government's efforts to promote *employment equity*. But other groups that can also encounter discrimination are not necessarily protected by legislation.

Age discrimination occurs when older workers, laid off in a plant shutdown and applying for another core-sector job, are rejected in favour of younger applicants because of the belief that older workers are harder to retrain and will have difficulty adapting to new technologies.[61] On the other hand, youth attempting to find their first permanent job also face age discrimination. Finding employment in the part-time student labour market within the lower-tier services is generally much less difficult than getting an interview for a full-time career position in a major corporation or a government department, even if one has the necessary educational qualifications. "Lack of experience" is a term heard all too frequently by unsuccessful young job applicants.

Discrimination on the basis of sexual orientation may be prohibited by human rights legislation, but this does not ensure that all employers act accordingly. Gay and lesbian Canadians continue to face a less-than-welcoming climate in many workplaces (Mitchell 1999), while same-sex couples continue to struggle to obtain the employment benefits received by opposite-sex couples (Carter 1998).[62] These examples, and the range of other research reviewed above, demonstrate that employer discrimination continues to be one of the factors accounting for persistent patterns of labour market segmentation in Canada.

NEW PATTERNS OF LABOUR MARKET SEGMENTATION

In a sweeping historical analysis of labour market segmentation trends in the United States, David Gordon and his colleagues separated the last two centuries into three major epochs (Gordon, Edwards, and Reich 1982). During the period of *initial proletarianization* from about 1820 to 1890, a large, relatively homogeneous industrial working class was emerging. While modern forms of capitalist management were still in their infancy, craftworkers were beginning to lose control over the labour process. However, this trend did not develop fully until the era of *homogenization of labour,* from the end of the 19th century until the start of World War II. This period saw extensive mechanization and deskilling of labour, the growth of large workplaces, and the rise of a new managerial class within American capitalism. The period following World War II to the early 1980s (when these authors wrote their book) was one of *segmentation of labour,* in which distinct primary and secondary labour markets emerged.

Canada began to industrialize later and has always remained more of a resource-based economy. Thus, the same three stages do not precisely describe our economic history, although there are some similarities (see Chapter 1). However, the emphasis on change in this historical overview encourages us to ask: Have further changes in labour market segmentation been occurring as we moved into 21st century? The answer is yes. There is considerable evidence of fundamental realignments in the industrial and occupational structures, and in the nature of employment relationships, not only in Canada, but also in the United States and other Western industrialized countries.

Labour Market Polarization

The goods-producing industries have declined significantly over the past several decades. Factory closures due to increased overseas competition in the 1980s were followed by the movement of manufacturing bases to low-wage countries such as Mexico in the 1990s and, more recently, China (see Chapter 1). The widespread introduction of microelectronic and robotic technologies allowed other manufacturing sectors to maintain production levels with fewer employees. A similar situation developed in Canada's resource industries, where unstable world markets, in combination with technological innovations that reduced the need for labour, led employers to substantially reduce their workforces.[63] During the 1990s, downsizing and hiring cutbacks were extensive in many large corporations in the business and distributive service industries, and by the turn of the century, many white-collar jobs were being outsourced to countries like India (Friedman 2005). As for the public sector, downsizing became the norm in the 1990s as governments (both federal and provincial) cut deficits by laying off employees and privatizing some of the services they had previously provided directly to the public.[64]

Although a shrinking workforce has been observed in both the goods-producing and service sectors, there has still been employment growth over the past two decades. Most of it has been in the service industries. While new high-skill, well-paying jobs have been created, leading some observers to proclaim the emergence of a "knowledge economy" (Drucker 1993; Florida 2002), there has also been a significant increase in the proportion of jobs in the secondary labour market. Many of the new jobs created have been part-time or temporary, paying less and offering fewer benefits and career opportunities than the full-time, permanent jobs that have disappeared. In other work settings, traditional employment relationships have been altered when workers retired—temporary positions and contract work have replaced the previously permanent positions. Thus, nonstandard jobs with less security have been increasing.[65]

Industrial restructuring and the growth in nonstandard work have led to *labour market polarization*. Compared to the 1970s and 1980s, the 1990s saw more Canadians unemployed and relatively fewer people employed in the well-paying, full-time, permanent jobs that used to be taken for granted. Put another way, the primary labour market shrank during the 1990s, leaving fewer workers with access to its good jobs, as defined by income, benefits, and job security. The economic recovery at the end of the last century and the beginning of the 21st century slowed, but did not reverse, this polarization trend as relatively fewer of the new jobs were "good jobs."

Within the broadly defined secondary labour market, two fairly distinct new segments appear to have emerged. The first is a *student labour market,* consisting mainly of part-time jobs, most of them in the lower-tier consumer services and retail trade industries (Marquardt 1998; Lowe and Krahn 1999). Many young Canadians participate voluntarily in this labour market, using their low-wage part-time jobs to earn discretionary income or to pay for their education. But this part-time workforce also contains non-students, women who cannot work full-time because of their family responsibilities, and others who simply cannot find a full-time job (see Chapter 2). The new *temporary/contract labour market* exists in both the upper- and lower-tier service industries (Lowe and Schellenberg 2001; Vosko, Zukewich, and Cranford 2003). As employers in both the private and public sectors have come to rely more heavily on temporary workers for both low- and high-skill jobs, a larger proportion of well-educated young adults are finding themselves spending a number of years in such jobs before they are able to obtain more secure employment. While many of these temporary and contract jobs require higher levels of skill and more training, they nevertheless offer less job security, fewer benefits, fewer training and career opportunities (Hoque and Kirkpatrick 2003), and less income (Galarneau 2005).

Rising Income Inequality

One of the results of labour market polarization has been an increase in individual-level income inequality in Canada. While real earnings (taking inflation into account) rose systematically in Canada in the three or four decades following World War II, this trend stopped in the 1980s. For many workers, real incomes have declined since then. A smaller proportion have seen their real incomes increasing. Compared to women's incomes, inequality in men's incomes has been more pronounced (Picot 1998).[66]

Even more significant has been the growing income difference across age groups. Despite their high levels of education, on average, the relative earnings of young Canadians have been dropping over the past two decades. For example, the 2001 national census showed that, for full-time, full-year employed Canadians, men under 40 years of age were earning only marginally more (in relative terms) than their counterparts a decade earlier (in 1991), and considerably less than this age cohort in 1981. For women, the dividing line was age 30. In contrast, women older than 30 and men older than 40 were doing relatively better in 2001 compared to these age cohorts in earlier decades. In other words, as Statistics Canada concluded based on these analyses, "[t]here is a clear generational divide in the labour market."[67] Thus, it is apparent that industrial restructuring and new forms of labour market segmentation (i.e., an increase in nonstandard jobs) have seriously disadvantaged younger workers, particularly young men.

At the household level, however, there is less evidence of an over-time increase in earnings inequality. While the strong growth in real family incomes that characterized Canadian society in the 1950s, 1960s, and 1970s had stopped by the 1980s, the rising proportion of employed women and dual-earner families since then has meant that family income inequality has not really increased over the past several decades (Rashid 1999).[68] But we have seen a significant change in the composition of low-income households. In the 1970s, the elderly made up a sizable portion of Canada's poor. Since then, improvements in income security for senior citizens have reduced poverty among the elderly (Myles 2000). However, the combined labour market trends noted above point to an increase in problems of poverty among working-age Canadians. The *working poor* have become a larger group in the Canadian labour market. Younger households (Morissette 2002b), particularly female-headed single-parent households (Rashid 1999), have been particularly disadvantaged.

Some observers have used the term *declining middle class* to describe these trends toward greater inequality. Others have commented on the emergence of an *underclass* of citizens, individuals largely marginalized from the labour market. But neither concept has been precisely defined. Some writers use the term "declining middle class" to refer to a declining proportion of well-paying blue-collar and mid-level white-collar jobs. Others emphasize growing income inequality without necessarily distinguishing between individual and family income.[69] Similarly, when proposing that an underclass has emerged, some users of the concept are referring to a growing number of long-term unemployed. Others include people unable to work and dependent on declining levels of social assistance in the category, while still others reserve the term to describe the homeless and absolutely destitute.[70]

Because of this conceptual confusion and the need for more research to test some of the assumptions underlying these concepts, we prefer to use the broader term "labour market polarization" to refer to the interrelated trends we have described—industrial and occupational realignment, rising levels of non-standard work, and growing individual income inequality. Nevertheless, we strongly emphasize that social inequality in Canada has increased as a result of these labour market polarizing trends.

POLICY RESPONSES TO LABOUR MARKET BARRIERS AND GROWING INEQUALITY

What might be done to create better jobs and to counter the trend toward greater labour market polarization? Responses based on a human capital perspective would primarily emphasize access to education as the most important public policy goal. In fact, this was a central concern of Canadian governments throughout the second half of the 20th century, particularly during the 1960s and 1970s when a wide array of new postsecondary educational institutions were opened and student loan systems were introduced.

These policy efforts did have an effect. The average educational attainment of the Canadian population rose dramatically, and some of the systemic barriers to equal access to education were reduced. Even so, as we have seen in this chapter, research continues to show that not all groups in Canadian society have equal educational opportunities. Consequently, continued policy responses targeting this goal are needed. In fact, with rising postsecondary tuition costs, we may well see a reversal of the slow trend toward greater equality of educational opportunity that characterized the latter decades of the 20th century.

Over the past several decades, government education policies shifted from building postsecondary institutions to encouraging young Canadians to stay in school and promoting skill upgrading among older labour force participants, particularly the unemployed. At the same time, more emphasis was placed on policies promoting formal training and skill upgrading in the workplace.[71] Concerns about reducing labour market inequalities via improved access to postsecondary education gave way to worries about Canada's competitive position in the global economy. Underemployed youth, older workers affected by industrial restructuring, and the marginal labour market position of other groups such as immigrants and Aboriginal Canadians came to be seen not so much as social problems of inequality, but as economic problems. Unless we invested even more heavily in and took more advantage of current stocks of human capital, it was argued, Canada would fall behind in the global economy (Laroche, Merette, and Ruggeri 1999; Expert Panel on Skills 2000).

Clearly, greater investments in education and training are important (Lowe 2000). Indeed, some groups of Canadian labour force participants continue to be educationally disadvantaged. Some could benefit from improvements in their literacy skills. The Canadian record on workplace training is not very good, and there are many barriers to receiving needed training (Sussman 2002). Thus, with global competitiveness as the goal, greater investments in human capital are part of the longer-term solution. But we also need to look at other structural factors, including the currently popular and misguided (in our opinion) belief that governments have little or no role to play in stimulating the economy,[72] a limited emphasis on research and development by Canadian firms, and trade agreements that have been more beneficial to some sectors (and regions) than others.

If reducing labour market inequalities is our goal, human capital investments are only part of the solution. We know that additional education and training would assist some Canadians in improving their labour market position. At the same time, we are also aware that one in four working Canadians consider themselves to be overqualified for their jobs, including many with university degrees. Research on literacy requirements in the workplace document large proportions of workers who are seldom required to use their literacy skills in their jobs.[73] Thus, *underemployment* is a serious problem in the Canadian workplace (Lowe 2000). So too are the very low incomes of Canada's working poor. Legislation that would significantly increase the minimum wage in all provinces would make a difference, as would legislation protecting the rights of unions to organize low-income workers (see Chapter 7).

While the human capital explanation of labour market processes focuses our attention on education as the key policy variable, the labour market segmentation perspective encourages us to identify the barriers that keep large numbers of qualified individuals out of better jobs. Obviously, a shortage of good jobs is part of the problem. So too is labour legislation that still largely overlooks the rights and needs of the working poor, and of nonstandard workers, a majority of whom are women (Fudge and Vosko 2001). We also continue to need strong legislation that discourages discrimination against disadvantaged groups in the Canadian labour market. Equity programs that help disadvantaged groups to catch up with mainstream labour market participants are also part of a policy package that could help reduce labour market inequality. Improved school-to-work transition programs are needed to assist young people in their search for rewarding jobs and careers, given that young Canadians have been particularly hurt by labour market polarization trends.[74] Finally, we believe that unions have a larger role to play in the Canadian labour market. If more Canadian workers were covered by union contracts, some of the inequality within the labour market could be reduced. If unions were involved on a more equal basis in decisions about national, regional, and firm-level industrial strategies, we might also move further toward the goal of global competitiveness. We will return to many of these topics in the following chapters.

CONCLUSION

After profiling good jobs and bad jobs in terms of income, benefits, risks to personal health, and occupational status, we began our analysis of these patterns by discussing Canada's class structure and the power relationships that underlie it. To answer questions about who gets the better jobs, we shifted our attention to the human capital and segmentation explanations of labour market processes. While the human capital model highlights the central role of education in determining occupational outcomes, it fails to account for the many examples of qualified and highly motivated individuals working in unrewarding jobs, and ignores intergenerational transfers of advantage. In contrast, the segmentation approach recognizes inequalities in labour market outcomes and provides a better account of how such power differences are created and maintained. In turn, studies of status attainment demonstrate both the importance of educational attainment and the extent to which status and power are inherited, frequently through unequal access to higher education.

While not directly applying the concepts of class analysis, we address the issues examined in class analyses in our descriptions of structured patterns of labour market inequality, intergenerational transfers of advantage, labour market polarization, and the intersecting effects of gender, race, ethnicity, and immigration status. Questions about who gets the good jobs, and why, are essentially questions about social inequality and the power relationships that underlie it. In the following chapters on gender, management strategies, conflict in the workplace, and industrial relations we will revisit this central theme.

DISCUSSION QUESTIONS

1. Based on material discussed in Chapters 2 and 3, what are the key factors that distinguish "good jobs" from "bad jobs" in the Canadian labour market?

2. Do the concept of "social class" and the sociological theories focusing on class structure, class inequality, and class conflict still have relevance for our understanding of the Canadian labour market and society in general? Why or why not?

3. Compared to several decades ago, does education play a larger or smaller role in determining labour market outcomes? Why?

4. Many people would argue that, in Canadian society today, everyone has more or less the same chance to get ahead in life. Do you agree or disagree? Why?

5. Based on material discussed in Chapters 2 and 3, how effectively do human capital and labour market segmentation theories explain present and past patterns of social inequality in Canada?

6. In your opinion, which groups of Canadian workers are really most disadvantaged? Have these patterns of labour market inequality changed over time?

7. Should the state (i.e., provincial and federal governments) get involved in trying to reduce labour market inequalities? If no, why not? If yes, what would be some of the most effective approaches?

NOTES

1. See van den Berg and Smucker (1997) on the neoclassical economic approach and more sociological analyses of labour markets.

2. See Clement and Myles (1994), Tilly (1998), and Spilerman (2000) on labour markets, social inequality, and class analysis.

3. Labour Force Survey estimates of weekly wages for 2004, by industry and occupation, appear in the Spring 2005 issue of *Perspectives on Labour and Income* (p. 67). Census data on 2000 earnings were retrieved from Statistics Canada's website at http://www12.statcan.ca/english/census01/home/index.cfm.

4. *Perspectives on Labour and Income* (Spring 2005: 67). This method of calculating the female–male earnings ratio differs from the method used by Statistics Canada until the late 1990s: it uses a different data source (the Labour Force Survey rather than the Survey of Consumer Finances); adjusts for average hours worked; is based on hourly pay rather than

annual earnings; excludes overtime pay; and excludes the self-employed. See Galarneau and Earl (1999) and Drolet (2002b) for useful discussions of these and other approaches to measuring the gender wage gap.

5. For additional information on federal and provincial labour legislation, including minimum wage rates, see the Human Resources and Skills Development Canada (HRSDC) website at http://www.hrsdc.gc.ca/en/home.shtml.

6. Statistics Canada's "low-income cutoff" (LICO) is the measure most often used to index poverty in Canada (National Council of Welfare 2005). A family (of a given size) is considered to be in the low-income category if, compared to the average Canadian family of that size, it spends at least 20 percent more of its total income on food, clothing, and shelter. LICOs are calculated separately for communities of different sizes since the cost of living is higher in larger communities.

7. See Morissette (2002a), Chung (2004), and Sussman and Tabi (2004) for additional statistics on Canada's working poor. For a more qualitative treatment of this subject, check out the CBC website at http://www.cbc.ca/paidtobepoor/. Winson and Leach (2002) and Ehrenreich (2001) provide excellent book-length discussions of the problems faced by the working poor in Canada and the United States, respectively.

8. Marshall (2003) notes the growing costs to Canadian employers of fringe benefit packages. See Lettau and Buchmueller (1999) on benefit costs in the United States. In a discussion of soaring health care insurance costs in the United States, Waldie (2005) reports that General Motors spends more on health care for its employees than it spends on steel for automobile production.

9. Different surveys, using somewhat different language, lead to marginally different estimates of the proportion of Canadian workers receiving specific benefits (Akyeampong 2002; Marshall 2003).

10. Grenon and Chun (1997: 26) compare benefits received by permanent and temporary workers, while Zeytinoglu and Cooke (2005) document the disadvantaged position of nonstandard workers. Drolet and Morissette (2002) describe better benefit packages among "knowledge workers" and Akyeampong (2002) does the same for union members. See Budd (2004) on similar patterns in the United States.

11. Marshall (1996) provides work-related fatality data from 1976 to 1993. More recent data are from the Canadian Centre for Occupational Health and Safety (CCOHS) website at http://www.ccohs.ca/.

12. Over-time data on work injuries and illnesses are from the CCOHS website at http://www.ccohs.ca/. See Conway and Svenson (1998) on a similar decline in occupational injury and illness rates in the United States, which they attribute to legislative reforms brought on by increased compensation payments to injured workers.

13. The 2000 survey results are from Statistics Canada's General Social Survey (Williams 2003). Additional observations about the link between work stress and ill health are based on analyses of the 1994–95 Population Health Survey (Statistics Canada 2000a); also see Wilkins and Beaudet (1998) and Shields (2003).

14. Grayson (1994) discusses self-reported work hazards using data from the 1991 General Social Survey.

15. See Human Resources Development Canada (2000b) for an assessment of the costs of occupational injuries and fatalities.

16. Goyder (2005) reports these results from a single-city (in Ontario) study first conducted in 1975 and then replicated in 2000; also see Goyder, Guppy, and Thompson (2003).

17. Goyder, Guppy, and Thompson (2003) report that, in 1975, higher occupational prestige was assigned to an occupation if research participants were told to imagine a man employed in this job compared to a woman. By 2000, this gender difference had disappeared and, in fact, slightly higher status was now assigned to "people-oriented" occupations (e.g., lawyer, psychologist) if women were employed in them. However, while there appears to have been some convergence in gender differences in occupational prestige, Goyder, Guppy, and Thompson (2003: 433) note that there are still large differences in the labour market position of, and rewards received by, women and men (also see Chapter 4).

18. Clement and Myles (1994); also see Livingstone and Mangan (1996: Chapter 1) and Grabb (2002) on differing theoretical treatments of social class.

19. See Wright (1997) for his more recent theorizing about social class and social mobility, including the intersection of gender and class.

20. Becker (1975: first published in 1964) is credited for developing human capital theory, although its basic premises originate in neoclassical economics.

As we describe it here, the model is closely linked to the functionalist theory of stratification (Davis and Moore 1945). See Laroche, Merette, and Ruggeri (1999) on the concept and measurement of human capital.

21. See Morissette (1997, 2002b), Marquardt (1998), Picot (1998), Livingstone (1999), Lowe (2000), Labour Market Ministers (2000), Frenette (2001), and Crompton (2002) on problems of underemployment among Canadian youth.

22. Statistics Canada (2004b) documents the continued high levels of investment by Canadians in postsecondary education. Jackson (2005: 47) summarizes results from various training surveys and notes that young labour force participants are more likely than their older counterparts to have received formal job-related training.

23. Unpublished analysis (by H. Krahn and G. S. Lowe) of data from the 1994 Canadian component of the International Adult Literacy Survey (Statistics Canada and Human Resources Development Canada 1996). De Broucker and Lavallée (1998) conducted a similar analysis of the same data set. Also see Statistics Canada, *The Daily* (18 January 2005) and Knighton and Mirza (2002) for similar findings from Statistics Canada's School Leavers Survey and Survey of Labour and Income Dynamics (SLID), respectively.

24. Clark (2000) presents a long-term overview of the growth of the formal education system in Canada. Also see Statistics Canada (2004b).

25. See http://www.arts.ualberta.ca/transition/ for additional publications based on the Edmonton study. Similar research demonstrating how family background influences educational aspirations and outcomes has been conducted in Nova Scotia (Looker 1994, 1997; Looker and Dwyer 1998; Looker and Thiessen 1999), Newfoundland (Sharpe and White 1993), and British Columbia (Andres 1993; Andres and Looker 2001).

26. In 1995, Canadian universities received 24 percent of their total revenues from student fees, up from 13 percent in 1980 (Little 1997: 11). Since 1995, university tuition fees have continued to rise (Statistics Canada, *The Daily*, 2 September 2004) as has average student debt (Statistics Canada, *The Daily*, 26 April 2004). For a wide range of excellent studies of postsecondary financing issues, see the Canada Millennium Scholarship Foundation website at http://www.millenniumscholarships.ca/.

27. See Statistics Canada, *The Daily* (16 February 2005) on the low–high income gap in university participation rates in 1993 and 2001. Hemingway

and McMullen (2004) discuss the need to conduct research on the impact of higher postsecondary costs on low-income families.

28. See Corak (1998) for a collection of empirical studies examining the impact of family background on physical and mental health, educational attainment, and income. Gilbert et al. (1993) and Tanner, Krahn, and Hartnagel (1995) review Canadian research on high-school dropouts. Alexander, Entwisle, and Horsey (1998) and Duncan et al. (1998) present similar information for the United States.

29. Also see Esping-Anderson (2004: 308–9). Looker (1994, 1997) and Davies (2004) offer Canadian analyses of the impacts of cultural capital on educational outcomes; Marjoribanks (2002) focuses on similar patterns in Australia in his study of "family and school capital."

30. Unpublished analysis (by H. Krahn and G. S. Lowe) of data from the 1994 International Adult Literacy Survey.

31. Unpublished analysis (by H. Krahn) of data from the 2003 Follow-up Survey of 1996 Alberta High School Graduates. Statistics Canada (*The Daily*, 18 January 2005) reports similar findings from the 1991 and 1995 national School Leavers Survey—rural youths were only a bit less likely to participate in the postsecondary system, but significantly less likely to attend university.

32. See Swift (1995) and Winson and Leach (2002) on the downward mobility experiences of Canadian families in the 1990s. Newman (1989), Dudley (1994), and Ehrenreich (2001) tell similar U.S. stories about the impact of economic restructuring on mobility opportunities. McBrier and Wilson (2004) document the greater downward mobility of African-Americans during this era.

33. Wanner (2004) provides a useful discussion of social mobility definitions and the important stratification questions addressed by social mobility research. See Birdsall and Graham (2000) for recent U.S. contributions to this research tradition. Harrington and Boardman (1997) study people who have done exceptionally well in American society despite their disadvantaged family background.

34. Most status attainment studies are modelled on Blau and Duncan's (1967) original study of American men. Creese, Guppy, and Meissner (1991: Chapter 7) present Canadian status attainment results based on 1986 national survey data. Wanner (2004) compares Canadian findings from

1973, 1986, and 1994 national surveys. See Hauser et al. (2000) for recent U.S. status attainment analyses.

35. See Ashton (1986: Chapter 3), Rubery (1988), and Kalleberg (1988) on labour market segmentation theory and research.

36. Edwards (1979), Hodson and Kaufman (1982), and Gordon, Edwards, and Reich (1982) are examples of this perspective on labour market segmentation.

37. See Akyeampong (1997: 52) and Lowe and Schellenberg (2001) for Canadian data on better jobs in large, capital-intensive firms, and Kalleberg and Van Buren (1996) for U.S. data.

38. See Aguiar (2001) for a detailed analysis of low-paid, low-skill work in the contract building-cleaning industry in Toronto.

39. Smith (1997) and Osterman and Burton (2005) provide overviews of research on internal labour markets. Examples of specific studies of internal labour markets include Ospina (1996), Walsh (1997), Barnett, Baron, and Stuart (2000), and Petersen, Saporta, and Seidel (2000).

40. See Freidson (1986), Leicht and Fennell (2001), Rossides (1998) and Evetts (2003) on the sociology of professions.

41. Cant and Sharma (1995) discuss the "professionalization" of practitioners of non-traditional medicine, Wallace (1995a, 1995b) writes about variations in professional commitment among lawyers, and Randle (1996) analyzes the "de-professionalization" of scientists. Tracy Adams (1998, 2000, 2003) documents the interplay of gender and professionalization trends in the fields of dentistry and dental hygiene.

42. See Burchell and Rubery (1990) and Ferber and Waldfogel (1998) on the liabilities of irregular work histories.

43. See Apostle, Clairmont, and Osberg (1985) on the *Marginal Work World* research project, and Apostle and Barrett (1992), Osberg, Wein, and Grude (1995), and Palmer and Sinclair (1997) for more recent labour market research conducted in the same region.

44. See Jenness (1977: 250–51) and Patterson (1972); Pentland (1981: 1–3) reports that some Aboriginal people were also kept as slaves in Quebec, but this practice had largely disappeared by the early 1800s.

45. Baureiss (1987); Creese (1988–1989); Muszynski (1996); Anderson (1991).

46. Shapiro and Stelcner (1997); Lian and Matthews (1998); Hum and Simpson (1999b); Wanner (1999).

47. Statistics Canada, *The Daily* (21 January 2003; 11 March 2003).

48. Statistics Canada, *The Daily* (11 March 2003). Chui and Zietsma (2003) and Li (2003) use both a different data source (the *Longitudinal Immigration Database*) and different methods (longitudinal analysis of individual records) to compare the earnings of immigrants in the 1980s and 1990s. Li (2003) is more optimistic about the "catch up capacity" of recent immigrants (i.e., how long it takes for the incomes of equivalently-educated immigrants to match those of their Canadian-born counterparts). Hum and Simpson (2002) use yet another data source (the Survey of Labour and Income Dynamics) to conclude that immigrants arriving in the 1990s were not doing as well as those who had come to Canada in earlier decades.

49. Reitz (2001) used 1996 census data to estimate that if immigrants received full value for their education and work experience, they would be earning about 20 percent ($15 billion) more.

50. Hum and Simpson (1999b). See also Bauder (2001), Tran (2004), Palameta (2004), Galarneau and Morissette (2004), and Maximova and Krahn (2005).

51. Also see Li (2001), Palameta (2004), and Galarneau and Morissette (2004).

52. Lamba (2003) demonstrates how more extensive social networks can assist refugees in their search for good jobs. Creese and Kambere (2003) describe how African accents of English-speaking African immigrant women negatively affect their employment and housing opportunities.

53. Palameta (2004: 37). For further discussion of racial discrimination in the Canadian labour market, see Reitz (1988), Das Gupta (1996), Satzewich (1998), Kunz, Milan, and Schetagne (2000), and Milan and Tran (2004). See Petersen, Saporta, and Seidel (2000), Baldi and McBrier (1997), and Deitch et al. (2003) on racial discrimination in the United States.

54. Stasiulis and Bakan (2003: 78, 128–36). Also see Milan and Tran (2004) on the disadvantaged labour market situation of Blacks in Canada. They report that 32 percent of Black Canadians surveyed in the 2003 Ethnic Diversity Survey reported experiences of racial discrimination in the previous five years, compared to 20 percent of all visible minority group members and only 5 percent of non-visible minorities (p. 7).

55. Statistics Canada, *The Daily* (28 June 2005).

56. See http://www12.statcan.ca/english/census01/home/index.cfm. For additional discussion of the labour market situation of Aboriginal

Canadians, see de Silva (1999), Armstrong (1999), Maxim et al. (2001), and White, Maxim, and Gyimah (2003).

57. Statistics Canada, *The Daily* (11 March 2003).

58. Statistics Canada, *The Daily* (23 June 2005).

59. Statistics Canada (2003d). Also see Shain (1995), Hum and Simpson (1996), and Federal, Provincial and Territorial Ministers Responsible for Social Services (2000: 32) on the labour market situation of Canadians with a disability, and Prince (2004) on the inadequacy of Canadian public policy regarding disabled people. Hale, Hayghe, and McNeil (1998) and Kruse (1998) provide U.S. data on the employment situation of disabled people.

60. Rioux (1985) and McKay (1991) discuss employment barriers faced by disabled people; see Wilgosh and Skaret (1987) on employers' attitudes toward hiring disabled people.

61. See Chiu et al. (2001) for a recent comparative study (United Kingdom versus Hong Kong) of discriminatory attitudes toward older workers.

62. See Busby and Middlemiss (2001) on the limited protection against discrimination on the basis of sexual orientation in the United Kingdom.

63. Clement's (1981) study of technological change at INCO documented the beginning of this trend; see Osberg, Wein, and Grude (1995) and Bob Russell (1999) for more recent examples.

64. On downsizing and layoffs in the Canadian private and public sectors, see Swift (1995), Osberg, Wein, and Grude (1995), Sharpe (1999), Gunderson and Riddell (2000), Dubé (2004), Morissette (2004), and Harrison (2005).

65. See Krahn (1995), Vosko (2000), Lowe and Schellenberg (2001), Vosko, Zukewich, and Cranford (2003), and Galarneau (2005) on trends in non-standard work.

66. See Picot (1998), Rashid (1999), Morissette, Zhang, and Drolet (2002), and Chawla (2004) on inequality trends in Canada, Bernstein and Mishel (1997) on U.S. trends, and Wolfson and Murphy (1998) for Canadian–U.S. comparisons.

67. Statistics Canada, *The Daily* (11 March 2003). For more on the declining labour market position of young workers, see Morissette (1997), Betcherman and Leckie (1997), Kapsalis, Morissette, and Picot (1999), Picot (1998), Marquardt (1998), and Labour Market Ministers (2000).

68. While family income inequality has not increased substantially, there has been a noticeable rise in wealth inequality in Canada since the mid-1980s (Morissette, Zhang, and Drolet 2002), in large part because of rising family debt levels (wealth is defined as total assets minus debt).

69. See Picot (1998) and Wolfson and Murphy (1998) on definitions of income inequality and polarization.

70. In some media accounts, the term "underclass" is used to describe poor families who, *presumably* having become dependent on social assistance, have passed such values of "welfare dependency" along to their children. For discussions of the term "underclass" and/or criticisms of the notion of "welfare dependence," see Wilson (1997), Schwartz (1999), Schecter and Paquet (1999), and Buckingham (1999).

71. On work-related training, see Myles (1996), Leadbeater and Suschnigg (1997), Betcherman, McMullen, and Davidman (1998), Statistics Canada and Human Resources Development Canada (1998; Chapter 3), Hum and Simpson (1999a), Sussman (2002), and Jackson (2005, Chapter 3). Muhlhausen (2005) describes the limited effectiveness of U.S. training programs for increasing the income of participants.

72. See Osberg and Fortin (1996), Battle and Torjman (1999), and Riddell and St-Hilaire (2000) on the importance of government involvement in Canada's economy and labour market; see Bluestone and Harrison (2000) for similar arguments about the United States.

73. See Livingstone (1999), Lowe (2000), Frenette (2001), and Crompton (2002) on self-reported underemployment in Canada; on literacy under-utilization, see Krahn (1997).

74. See Pauly, Kopp, and Haimson (1995) and Stern et al. (1995) on high-school-based programs, and Gitter and Scheuer (1997) and Lehmann (2000) on youth apprenticeship programs.

4

WOMEN'S EMPLOYMENT

INTRODUCTION

In the past few decades women's work has changed enormously. In 1970, just over a third of adult women in Canada worked outside the home for pay. By 2004, this figure had jumped to 62.1 percent.[1] Traditional barriers to female employment have slowly been eroding. Feminism challenged the conventional wisdom about women's capabilities. Rising education levels and job opportunities made more women career-minded, and the economic necessity of being self-supporting or contributing to the family income left some women little choice but to find a paying job.

This chapter examines these transformations in women's employment patterns. We will review the dramatic increases in female labour force participation rates during this century, showing how the labour market has become segmented into men's and women's jobs. We will also discuss two seemingly contradictory trends: on the one hand, significant gains in women's employment opportunities and rewards, and on the other, the persistence of major work-related gender inequities. A central theme is that women's work has been undervalued and poorly rewarded throughout history. Consequently, gender remains a key determinant of inequality in our society. The chapter ends optimistically, however, by reviewing two significant policy initiatives—employment equity and pay equity—that have started to redress this long-standing problem.

To better grasp the causes and consequences of gender inequity in the work world, we shall address the following questions: What forces have either pushed or pulled women out of the home and into paid employment in recent decades? Why are women concentrated in a limited range of jobs at the bottom of the occupational ladder? How did these jobs come to be labelled female, and, conversely, how have the most challenging and rewarding jobs come to be defined as male? What barriers still prevent equality of opportunity and rewards for women in the labour market? And finally, but perhaps most importantly, how can greater equality for women be achieved in the workplace?

A word of clarification before we proceed. To avoid confusion, we should define how we use the terms *sex* and *gender*. Basically, sex is the biological distinction between men and women; gender is socially constructed in the sense that it refers to how a particular society defines masculine and feminine roles. Thus, one can observe the sex segregation of occupations. But to explain this it is necessary to look at how jobs are *gendered*, that is, how they take on societal images of appropriate male or female behaviour. We tend to prefer the term gender, given that male–female differences in employment are almost solely a product of socially created gender roles and ideologies.

WOMEN'S ECONOMIC ROLE IN HISTORICAL CONTEXT

Although history has for the most part been written from the perspective of men, even a quick glance back to the past reveals that women have always performed a vital, if somewhat unacknowledged, economic role. Native women, for example, were indispensable to the fur trade, the major industry during much of Canada's colonial period (Van Kirk 1980). White male traders relied on Native women to act as interpreters, prepare food, clean pelts for market, and teach them wilderness survival skills. Little wonder that fur traders sought out Native women as wives. When an agrarian economy began to develop in Upper Canada (Ontario) during the 19th century, the family was the basic production unit, in which women played a key role. Men worked the fields, while women looked after all domestic work associated with child rearing, tending the livestock and garden, making clothes, and preparing food.[2] Similarly, in the West in the early 1900s, women contributed significantly to agricultural development by working on their family farms and by making the farm home "a haven of safety and healthfulness."[3]

Industrialization and Women's Work

In Chapter 1, we noted that one consequence of the rise of large-scale factory production was a growing separation between men's and women's work. Men were drawn into the industrial wage-labour market; women were increasingly confined to the domestic sphere of the household. Marjorie Cohen's feminist analysis of economic development in 19th-century Ontario reveals that, prior to wide-scale industrialization, an integration of family and household existed within the emerging market economy (Cohen 1988). Traditional theories of industrialization, Cohen points out, tend to ignore the contribution of households to the economy. Ontario's early economy was primarily based on two staple exports, wheat and timber, which were subject to unstable international markets. Consequently, women's household labour had to fulfill two functions: generate family income by producing agricultural goods to sell in the local consumer market, and perform the domestic chores necessary for the family's survival. Cohen's research underlines the importance of examining how the public and private spheres of market and household have been intertwined in diverse ways during all phases of economic development, with women always performing pivotal roles, albeit quite different from those performed by men.

The absence of a wage-labour market and the necessity of contributing to the household economy meant that few women were employed outside the home prior to the rise of industrial capitalism. Even in late-19th-century Canada, only a fraction of women were engaged in paid employment. In 1891, for example, 11.4 percent of girls and women over the age of 10 were employed, accounting for 12.6 percent of the entire labour force (Lowe 1987: 47). But as factories sprang up in the late 19th century, they began to re-define women's economic role. Employers in some light industries, such as textiles, recruited women as cheap unskilled or semiskilled labourers who, according to prevailing stereotypes, would be less likely to unionize and more tolerant of boring tasks.

By focusing only on paid employment, we risk ignoring the work activities of the vast majority of women during this era. The *unpaid domestic labour* of women—raising the future generation of workers and feeding, clothing, and caring for the present generation of workers—was an essential function within capitalism. Out of these competing pressures on women emerged a gendered division of labour that persists today. As we document below, this now takes the form of the *double day* (or, in equally graphic terms, the *second shift*), whereby most married women spend their days in paying jobs, yet still assume most of the responsibilities of childcare and domestic chores when they get home.

Early-20th-century attitudes about women's economic roles distinguished between single and married women. Expanding manufacturing and service industries had an almost insatiable demand for both blue- and white-collar workers. The employment of young, single women prior to marriage came to be tolerated in domestic, clerical, sales, and some factory jobs. Once married, women were expected to retreat into the matrimonial home. Of course, some wished to remain in the labour force, and others were forced to stay through economic necessity. In these cases, married women laboured at the margins of the economy in domestic and other menial jobs that usually had been abandoned by single women.[4]

Industrialization accentuated age and gender divisions in the economy. Examining the work patterns of working-class women in Montreal, Canada's first large industrial city, Bettina Bradbury documents how age and sex determined who was drawn into wage-labour. Women made up about 35 percent of the city's industrial workforce during the 1870s. In certain industries, such as domestic work and the sewing and dressmaking trades, four out of five workers were women, and the vast majority of them were single. Given the scarcity of wage-labour for wives because of strong sanctions against their employment, such women could make a greater contribution to the family economy by being *household managers*. In this role, wives stretched the wages of male family members and single daughters as far as possible, occasionally supplementing this by taking in boarders or turning to neighbours or charities for help (Bradbury 1993).

The Family Wage Ideology

Powerful social values justified this division of labour. Especially influential in perpetuating women's subordinate role as unpaid family workers was the ideology of the *family wage*. As working-class men began organizing unions to achieve better wages and working conditions, one of the labour movement's demands was that wages should be high enough to allow a male breadwinner to support a wife and children. The labour movement's successes in this regard had the effect of drastically reducing women's presence, and the cheap labour they provided, in the workplace (a policy that was typical of unions until a serious male labour shortage arose in World War I). Middle-class reformers also lobbied for restrictions on female industrial employment due to its presumed harmful personal and social effects. In response, employers limited their hiring mainly to single women, further reinforcing this ideology. Despite its sexist tone, the family wage ideology may have benefited those women who were dependent on a husband by raising the standard of living in working-class families. The price, of course, was the restriction of women's labour market opportunities to areas in which they would not compete directly with men—hence the endurance of the term "male breadwinner."[5]

However, the family wage ideology tells only part of the story of women's lives during early industrialization. Joy Parr's case studies of two Ontario industrial towns, Paris and Hanover, between 1880 and 1950, caution us against thinking in binary oppositions—masculine/feminine, market/non-market, public/private, waged/non-waged (Parr 1990: 8). Parr shows how the specific combinations of community context, gender roles, household forms, and industrial development patterns interacted to fashion different gender ideologies. Her study shows the variations in outcome within the broad contours of gendered work patterns.

On the surface, both Paris and Hanover appeared to be small, thriving manufacturing communities. But the knit-goods industry based in Paris relied on a largely female workforce, while in Hanover's large furniture factory, the workers were almost exclusively male. Consequently, different gender identities based on women's employment patterns arose to maintain each town's labour force. In turn, these were reinforced by attitudes and behaviour in households and the community. Certainly Paris was the exception at the time in Ontario. Nonetheless, its flow of daily life during the 80 years covered by Parr's study forces us to reconsider a model of industrialization based on one dominant mode of production in which males are the breadwinners.

This brief historical sketch has identified a number of prominent themes. First, although their widespread participation in the paid labour force is a recent development, women have always made essential economic contributions. Second, women's entry into paid employment occurred in ways that reproduced their subordinate position in society relative to that of men, although the specific forms this took varied across time and place. Third, the changing interconnections among households, families, and the wage-labour market are crucial to understanding women's roles in the continuing evolution of 21st-century capitalism.

FEMALE LABOUR FORCE PARTICIPATION PATTERNS

Few changes in Canadian society since World War II have had as far-reaching consequences as evolving female employment patterns. Virtually all industrial nations have experienced rising *female labour force participation rates* since the end of World War II. This trend has been especially rapid in Canada.

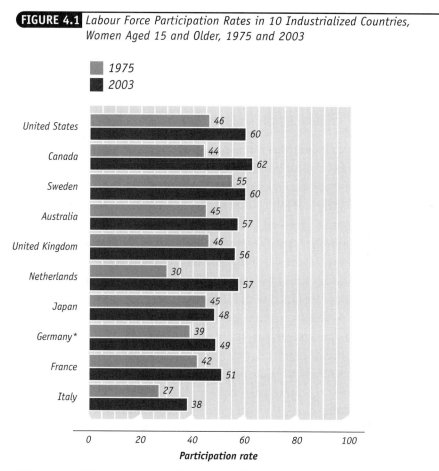

FIGURE 4.1 *Labour Force Participation Rates in 10 Industrialized Countries, Women Aged 15 and Older, 1975 and 2003*

■ 1975
■ 2003

Country	1975	2003
United States	46	60
Canada	44	62
Sweden	55	60
Australia	45	57
United Kingdom	46	56
Netherlands	30	57
Japan	45	48
Germany*	39	49
France	42	51
Italy	27	38

Participation rate

*West Germany in 1975; the unified East and West Germany in 2003.

Source: U.S. Department of Labor, Bureau of Labor Statistics, 2005. Comparative Civilian Labor Force Statistics, Ten Countries 1960–2004.

Figure 4.1 provides a comparison of labour force participation rates among adult women (aged 15 and older) for 10 major industrial nations between 1975 and 2003. Note that rising female employment is a general cross-national trend. Canada and the United States experienced larger increases than all but the Netherlands. The tremendous expansion of white-collar service-sector jobs, coupled with rising educational levels and a declining birthrate, drew millions of Canadian women into employment at an accelerated rate. By 1980, Canada's female labour force participation rate surpassed all 10 countries in

Figure 4.1 except the United States and Sweden. The latter is an interesting case, given that it experienced the least change of all 10 countries over the 1975–2003 period, largely because women's employment outside the home has been actively encouraged by state policies for decades.

Influences on Women's Employment

What social and economic factors account for this remarkable increase in Canada's female labour force participation? Clearly there was no single cause. The huge postwar baby-boom generation was completing its education and flooding into an expanding job market by the late 1960s. Young women were becoming much better educated, which raised their occupational aspirations and made them more competitive with men in the job market. Traditional stereotypes of women's work no longer fit reality. The massive expansion of white-collar service-sector jobs boosted the demand for female labour, and the growth of feminism contributed to more liberal social values regarding women's work roles. As we have seen, service-sector growth also fuelled a trend to part-time jobs, which are convenient for women with family responsibilities. Shrinking family size allowed married women to pursue employment more readily. Moreover, rising separation and divorce rates forced a growing number of women to find their own source of income. And in many families, the reality (or threat) of declining living standards made a second income essential.

The interaction of this multitude of supply and demand factors underlay the sharp rise in female employment.[6] Interpreting the overall impact of these changes on women's lives, Charles Jones, Lorna Marsden, and Lorne Tepperman (1990: 58–59) point out that as more women have entered paid employment, their work patterns have not come to mirror those of men. Rather, largely because women are heavily concentrated in the secondary labour market and often have to juggle family responsibilities with a job, their lives have become increasingly individualized. Compared with their mothers' generation, younger women today have adopted new ways of combining work, education, and family. Hence, since the 1950s, there is now much greater variety, fluidity, and idiosyncrasy in women's roles. For Jones, Marsden, and Tepperman, the *individualization process* is defined by these elements: (1) variety, through greater opportunities for employment and education; (2) fluidity, in terms of increased movement among these roles and domestic/household roles; and

(3) idiosyncrasy, in the sense that it is now exceedingly difficult to predict whether, when, and where a woman will be working for pay.

How women actually make choices and tradeoffs between paid employment, marriage, and child rearing has been the subject of much debate. For some researchers, such as Catherine Hakim (2002), it is largely women's "preferences" that explain the growing diversity in female employment. Downplaying the roles of segregation or discrimination, Hakim argues that women choose one of three patterns—"home-centered" (where family is the priority), "career centered" (where work is the priority), or "adaptive" (where non-career jobs are blended with family responsibilities). But while work preferences play a role, there are many other factors that shape women's paid work. Access to childcare, household resources, and the availability of suitable employment are just some of the considerations that filter into women's decisions. Thus, as McRae (2003) argues, understanding women's employment "depends as much on understanding the constraints that differentially affect women as it does on understanding their personal preferences."

Indeed, there is no question that many factors have a bearing on the likelihood of a woman participating in the labour force. Figure 4.2 identifies important variations in 2004 participation rates by region, age, marital status, family circumstances, and educational level. Provincially, women in Alberta have the highest participation rate (67.3%), followed by those in Prince Edward Island (64.2%), Manitoba (63.4%) and Ontario (63%). Nova Scotia and Newfoundland have the lowest rates (58.9% and 54.3%, respectively). Certainly local and regional job opportunities have a direct bearing on female participation rates, as do regional differences in age structure and educational attainment. For example, Alberta has a relatively strong service sector and a young and well-educated workforce—hence its high rate of labour force participation for both women and men. In contrast, Newfoundland has limited employment opportunities for women (and men) due to chronic economic underdevelopment and the decline of the fisheries. Consequently, much work occurs outside the sphere of paid employment in a thriving informal economy and through self-provisioning (Sinclair and Felt 1992).

Personal characteristics also influence work patterns. Among youths (aged 15 to 24) single women are less likely than their married (or cohabiting) counterparts to work for pay—because more of them are full-time students. But what's changed in the past several decades is a convergence of participation rates

FIGURE 4.2 *Female Labour Force Participation Rates by Selected Characteristics, Canada, 1996 and 2004*

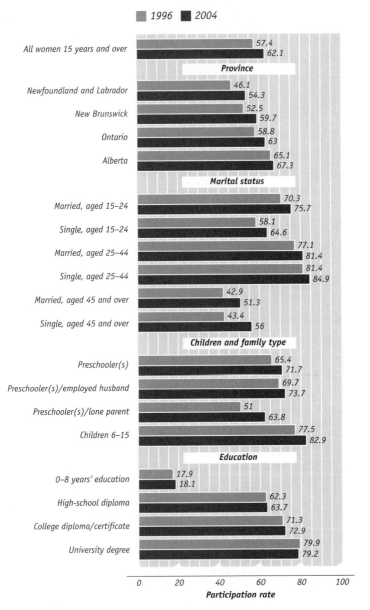

■ 1996 ■ 2004

	1996	2004
All women 15 years and over	57.4	62.1
Province		
Newfoundland and Labrador	46.1	54.3
New Brunswick	52.5	59.7
Ontario	58.8	63
Alberta	65.1	67.3
Marital status		
Married, aged 15–24	70.3	75.7
Single, aged 15–24	58.1	64.6
Married, aged 25–44	77.1	81.4
Single, aged 25–44	81.4	84.9
Married, aged 45 and over	42.9	51.3
Single, aged 45 and over	43.4	56
Children and family type		
Preschooler(s)	65.4	71.7
Preschooler(s)/employed husband	69.7	73.7
Preschooler(s)/lone parent	51	63.8
Children 6–15	77.5	82.9
Education		
0–8 years' education	17.9	18.1
High-school diploma	62.3	63.7
College diploma/certificate	71.3	72.9
University degree	79.9	79.2

Participation rate

Source: Adapted from the Statistics Canada publication *Labour Force Historical Review,* 2004, Catalogue 71F0004XCB, February 18, 2005.

for married and single women over age 25. Indeed, 82.3 percent of women between the ages of 25 and 44 are in the labour force. By comparison, just under half of women aged 45 or older are in the labour force. Women with higher levels of educational attainment are more active in the labour force—79 percent of women with university degrees compared to 64 percent who have a high-school diploma. Family situation also matters. However, because a high proportion of mothers who have children under the age of 15 are relatively young and well educated, their labour force participation rates are above average. For example, 72 percent of mothers with preschool children participate in the labour force.

Financial necessity is a major factor, often the most important one, in many women's decisions to seek employment. Generally, wives in low-income families or women who are lone parents may be compelled to work to meet, or help meet, basic expenses. In fact, the most common dual-earner family combination in Canada is a blue-collar husband and a wife in a clerical, sales, or service job (Chawla 1992: 24). The participation rate for single mothers with preschool children at home was 64 percent in 2004, while the rate for mothers of preschoolers who had an employed husband (or partner) was 74 percent. For many single mothers, finding or keeping a job may be difficult, given such factors as lower educational attainment and difficulties arranging adequate and affordable childcare. Low-paid, unstable work is a major reason for the high incidence of poverty in this group.[7] In 2002, 35 percent of female-headed families had incomes below Statistics Canada's low-income cutoff (LICO), compared to just 5 percent for two-parent families with children (Statistics Canada 2002). Growing concern about child poverty in Canada is clearly linked to the difficulties lone-parent females face in the labour market (Canadian Council on Social Development 2001). In contrast, better-educated, urban, middle-class women who are married and whose spouse or partner is a professional or a manager have greater opportunities and the luxury of choice with regard to employment (Lowe and Krahn 1985: 4). Among families with the highest incomes in Canada, it is very likely that the female partner is a high-earning manager or professional.[8]

WORK AND FAMILY

Traditional gender roles and values have defined women's work as household tasks and child rearing. This enduring male–female division of labour underlies much of the gender inequality we see today. For example, domestic responsibilities continue to limit women's availability for paid work, keeping many married women financially dependent on their husbands. Some experts argue that women's inequality is the result of two systems of domination: capitalism and patriarchy. Capitalism incorporated earlier patriarchal social arrangements. Broadly speaking, *patriarchy* refers to male domination over women, but more specifically it describes forms of family organization in which fathers and husbands hold the power.[9] Remnants of patriarchy still reinforce stereotypes of women as cheap, expendable labour. Furthermore, women's traditional family roles of wife and mother often restrict their employment opportunities and, for those who are employed, create a double day of paid and unpaid work.

Changing Family Forms

In the 1950s, many women saw the home as their major priority. Though some women needed to work for pay (Sangster 1995), the most typical pattern was to work before having children, stay home to raise them, and return to the labour force in middle age after the children had grown up.[10] Because most wives in the 1950s and 1960s responded to the demands of child rearing by leaving the labour force, employers assumed that women must have a weaker attachment to paid work than men. A 1952 national study found that the major reason married women under the age of 40 left their employment was for child-raising responsibilities (Boyd 1985). Similar family constraints on employment for women still exist. However, in the 1980s and 1990s, there was a pervasive national trend for mothers with children at home to seek paid jobs, with this group accounting for most of the increase in female participation. As a result, many more women are juggling paid work with family roles.

Today, far fewer women leave the labour force, and those who do leave for a shorter time. Consequently, the dual-earner family is now the norm in Canada. Since the 1960s, the traditional family in which the husband was the sole earner has been eclipsed by dual-earner families, which make up 62 percent of all husband–wife families. At the same time, the number of lone-parent families is on the rise. According to the 2001 Canadian census, 25 percent of families with children present are headed by lone parents, more than 81 percent of whom are women.[11]

The Domestic Division of Labour

Given the dramatic changes in women's paid work, has there been a similar shift in the gender division of household chores and caregiving? Evidence gleaned from *time budget studies,* which ask people to record in detail how they spend their time, leads to the conclusion that, despite some change, many working women put in what Arlie Hochschild calls a "second shift" in the home.[12] After working seven or eight hours on the job, some wives return home to cook, clean, shop, and look after children—the domestic chores their mothers and grandmothers did as full-time housewives. This echoes the findings of a 1971 study by Martin Meissner and his colleagues, which found that taking a job outside the home meant a decrease of 13.5 hours per week in leisure time for a sample of British Columbia wives. Yet their husbands experienced no loss of leisure time because they had not shared the housework to begin with.[13] The researchers concluded that time pressures and added workloads create a "double burden" for many employed wives.

Though recent investigations show some change, women still continue to shoulder this double burden. Statistics Canada's 2001 census documented unpaid work trends for all Canadians. As Figure 4.3 shows, women devote much more time on average to housework. Men are nearly twice as likely to spend just a few hours a week, whereas nearly half of women spend 15 hours or more. While the General Social Survey (GSS) will soon gather new data on unpaid work trends, we know from past GSS surveys that in dual-earner families where both partners worked full-time, more than half of the women had sole responsibility for housework (meal preparation and cleanup, cleaning, and laundry). Tasks were shared equally in just 1 in 10 such families.[14] Yet there are important variations in the domestic division of labour. In addition to age and education—younger and university-educated women tend to have less traditional roles—recent Canadian studies have identified single parenthood, gender ideologies, and women's relative earnings as influencing time spent on household tasks and how these are divided.[15] Studies also indicate more egalitarian patterns in same-sex households (Dunne 1996; Sullivan 1996).

Population aging also has implications for women in that they often provide care to elderly and dependent parents and relatives. *Eldercare* is a pressing issue amid predictions of a "caregiving crunch," as fewer women are available to meet such demands because of their own jobs. According to the 2002

FIGURE 4.3 *Hours of Unpaid Housework, Women and Men, Canada, 2001*

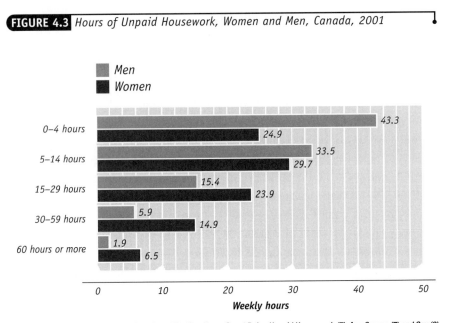

Source: Adapted from the Statistics Canada publication *Hours Spent Doing Unpaid Housework (7), Age Groups (7) and Sex (3) for Population 15 Years and Over, for Canada, Provinces, Territories, Census Metropolitan Areas and Census Agglomerations, 1996 and 2001 Censuses—20% Sample Data,* Catalogue 97F0013XIE2001001, February 11, 2003.

General Social Survey (GSS), roughly 1 in 5 Canadians 45 years and over provide care to a senior, most typically a parent. Most of these caregivers also work in a paid job.[16] The heaviest burden falls to the "sandwich generation," adults who combine childcare with caring for their elderly parents. In 2002, about 27 percent of families with children under 25 years of age were also providing eldercare to one or more seniors (Williams 2004). On average, women devoted more time—about 29 hours a month, compared to 12.5 hours for men. "Sandwiched" workers report higher stress, and are more likely to have had to reduce hours, reduce income, or turn down promotions to cope with their caregiving responsibilities. Recognizing the growing demands of eldercare, the federal government introduced *Compassionate Care* benefits in January 2004, which provide six weeks of paid leave to care for a gravely ill family member. Like parental leave benefits, however, many individuals are ineligible unless they have worked 600 hours in the preceding 52 weeks.

Womens' household work varies, depending on their social class position and their partners' occupation. Working-class women have a higher probability of facing the combined stresses of poverty and menial employment. Women with managerial and professional partners may be expected to comply with the traditions of a two-person career. It has been commonly assumed in large corporations that male managers climbing the career ladder will have the unpaid services of a wife. Success in the corporate world demands unwavering dedication to one's career, often resulting in very low family involvement. Wives are expected to put on dinner parties, accompany their husbands to company social functions, make travel arrangements, and plan household moves. They are what Rosabeth Moss Kanter (1977) calls "unpaid servants of the corporation." These antiquated vestiges of male corporate culture are eroding as more women pursue their own careers. However, women attempting to establish themselves in management still face disadvantages, given lingering expectations about two-person careers. Unfortunately, their husbands are seldom willing to take on extra childcare and housework.

Balancing Work and Family

"Work has changed. Women have changed," writes Arlie Hochschild. "But most workplaces have remained inflexible in the face of the family demands of their workers and at home, most men have yet to really adapt to the changes in women. This strain between the change in women and the absence of change in much else leads me to speak of a 'stalled revolution'" (Hochschild 1989: 12). One consequence of this stalled revolution is the rise of *job–family conflict*—or what is also called work–life conflict. Drawing on two large-scale surveys in 1991 and 2001, Duxbury and Higgins (2001) found rising levels of *role overload* (an excess of demands), *work to family conflict* (where work interferes with family life), and *family to work conflict* (where family responsibilities interfere with work). For example, in 1991, 47 percent of Canadians reported high overload. By 2001, this had jumped to 59 percent. As Figure 4.4 shows, mothers, in particular, were very likely to feel overloaded, with nearly three-quarters reporting overload from too many demands on their time and energy.

Juggling work and family can affect quality of life. Figure 4.4 shows a striking increase in job stress from 1991 to 2001. While stress increased for mothers, it increased for fathers too. This suggests that while child-care and family responsibilities often fall to women, men are increasingly contributing to the changing configuration of work, gender, and family by re-defining their

FIGURE 4.4 *High Role Overload and High Job Stress by Gender and Parental Status, Canada, 1991 and 2001*

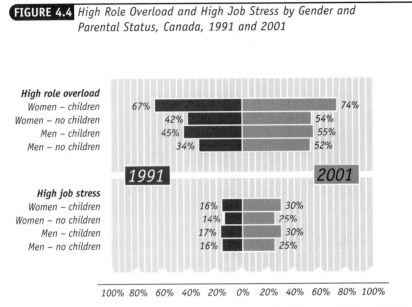

Source: Duxbury and Higgins (2001) *Work-Life Balance in the New Millennium. Where Are We? Where Do We Need to Go?* Ottawa: Canadian Policy Research Networks. Reprinted by permission from Canadian Policy Research Networks.

roles.[17] Male gender roles are changing slowly; for example, there were "stay-at-home-dads" in 6 percent of all families with a parent at home in 1997, up from 1 percent in 1976 (Marshall 1998: 11). It will be important for future research to monitor male attitudes and behaviour on work–family balance issues.

The pressures created by work–family imbalance have costs for employers too. For example, Duxbury and Higgins (2001) found in their study that high role overload and work–family conflict was related to higher absenteeism, job dissatisfaction, lack of commitment to the organization, and intention to quit. Indeed, absenteeism is a significant source of lost productivity for employers. According to Statistics Canada, female rates of absenteeism have risen steadily since the late 1970s, mainly due to family or personal obligations as growing numbers of mothers with preschool children seek employment (Statistics Canada 2000b). In 2004, employed women lost an average of 10.5 days, compared to 8.0 for men, due to illness, disability, or personal or family responsibilities.

While some employers have moved to create family-friendly work environments, firms in the most dynamic sectors of the economy face a range of pressures that make work–life balance difficult. A study of knowledge workers in the seven largest Ottawa-area high tech workplaces found that work cultures in these firms were defined by heavy workloads and long hours. This culture is not sustainable, however. It may suit younger, single workers who have no dependent care responsibilities, but lack of attention to work–life balance issues will cause increasingly serious problems—notably high quitting rates, especially among key female knowledge workers—as the high tech workforce moves into its 30s, marries, and starts raising children (Duxbury, Dyke, and Lam 2000).

Looking at factors that can help employees better integrate their job, family, and personal life, Duxbury and Higgins (2001) note the importance of more flexible work hours and schedules, improved personal and family-related leave, the ability to refuse overtime, and the presence of supportive managers. Slowly, in part due to studies such as these, heightened awareness of the economic and individual costs of job–family conflict is challenging employers to reconsider some long-standing policies. Yet many managers seem rigidly locked into a nine-to-five day and five-day-week schedule; and paid maternity or paternity leave (in addition to the period covered by EI) or elder-care support are provided by only a minority of employers.[18]

Of all types of support, quality childcare is perhaps the most pressing need. According to the 2004 *Childcare and Early Education Report,* child-care arrangements present an ongoing problem for many parents—although preferences for different forms of childcare vary widely among working parents (Beaujot 1997; Statistics Canada 2004a). Despite the federal government's recent pledge to commit $5 billion to childcare, the number of licensed childcare spaces by no means filled the need. In 2004, there were 745,254 regulated child-care spaces across Canada, up from 373,741 in 1992. This represents a near doubling of available spaces—a very positive change, given the historical heel dragging that has accompanied the promise of a national daycare program (Timpson 2001). But even with this increase, regulated child-care spaces still only accommodate 15 percent of children up to age 12 in Canada (Friendly and Beach 2005). Many children of working parents are looked after informally by sitters, neighbours, nannies, or relatives (Statistics Canada 2004a: 55) So far, Quebec is the only jurisdiction in Canada to make a concerted effort to address these issues, having introduced in 1997 a comprehensive program of universal subsidized daycare, early childhood education, support for parental leaves, and improved family allowances.[19]

All in all, what we have just documented surely amounts to a "stalled revolution." Over a decade ago, the Conference Board predicted that more employers would respond to these needs as the drain on productivity became more visible and employee pressure for such policies mounted (Paris 1989). Unfortunately for many working parents, there are few tangible signs of this happening. And for some employees who do have flexible work arrangements, recent research suggests that these may not meet expectations in terms of reducing the stresses of balancing job and family. After examining a variety of alternative work arrangements, Janet Fast and Judith Frederick (1996) concluded that self-employment, flex-place (working some hours at home), and shift work had no effect on workers' perceived time stress. Both women and men in part-time jobs are less time-stressed than full-timers, although far more women than men choose part-time work for family-related reasons. Flextime (employees choosing when their work day starts and stops) helps to reduce time pressures, but it is less available to women than to men. Compressed workweeks (fewer days per week with longer hours each day) seem to increase women's level of time stress, likely because they interfere with family routines. Job sharing, which now is available to only 8 percent of part-time workers, was not examined in this study, but it may have the potential to contribute to reduced time pressure and better integration of work and family (Marshall 1997).

However, the barriers to achieving work–family balance are numerous and complex. Arlie Hochschild's research suggests that even in organizations that provide family-friendly policies and programs, employees may experience difficulties finding enough time for their families. One of the largest barriers is the fact that society places great value, and confers rewards, on people who are successful in their careers. The result, especially among professionals and managers, is that employees spend more, not less, time at work. As Hochschild (1997: 198–99) suggests, "in a cultural contest between work and home, working parents are voting with their feet, and the workplace is winning."

GENDER SEGREGATION IN THE LABOUR MARKET

At the heart of gender inequality in the work world is the structuring of the labour market into male and female segments. *Occupational gender segregation* refers to the concentration of men and women in different occupations. A potent combination of gender-role socialization, education, and labour market mechanisms continue to channel women into a limited number of occupations in which mainly other females are employed.

Female Job Ghettos

Female *job ghettos* typically offer little economic security and little opportunity for advancement; furthermore, the work is often unpleasant, boring, and sometimes physically taxing. Women in job ghettos lack ready access to the more challenging and lucrative occupations dominated by men. These male segments of the labour market operate as *shelters,* conferring advantages on workers within them through entrance restrictions.[20] The concepts of ghettos and shelters emphasize the unequal rewards and opportunities built into the job market on the basis of a worker's sex. It is especially important to recognize that job opportunities determine an individual's living standard, future prospects, and overall quality of life—in Max Weber's words, her or his *life chances.*

One of the fundamental mechanisms underlying segmentation is the *gender labelling* of jobs. Employers do not always make hiring decisions on strictly rational grounds, despite what economics textbooks would have us believe. If all hiring decisions were totally rational, women would have been recruited much earlier in the industrialization process and in far greater numbers, given their cost advantage as cheap labour. As already mentioned, men typically opposed the employment of women in their occupations for fear of having their wages undercut. Furthermore, traditional values narrowly defined female roles as child rearing and homemaking. Women, therefore, were relegated to the less rewarding jobs that men did not want. Because these occupations came to be labelled "female," future employers would likely only seek women for them, and, regardless of the skills demanded by the job, pay and status would remain low.

Dominant social values about femininity and masculinity have long been used to define job requirements. For instance, by the late 19th century, teaching, social work, nursing, and domestic work were socially acceptable for women. Society could justify this on the ideological grounds that these occupations—caring for the sick, the old, and the unfortunate, transmitting culture to children, and performing domestic chores—demanded essentially female traits. Exclusive male rights to the better jobs and higher incomes thus went unchallenged, and the role of women as homemaker and wife was preserved. Once a job was labelled male or female, it was difficult for workers of the opposite sex to gain entry (Lowe 1987).

Trends in Labour Market Gender Segregation

In Chapter 2, we outlined gender differences in the occupational distribution of the labour force. Figure 4.5 explores this further, summarizing employment distribution and concentration in 2004. The left-hand bars in Figure 4.5 identify the percentage of employees in each occupation who are women. Health occupations, such as nursing, have very high concentrations of women (over 80%). So do occupations in business/finance/administration (71.1%) which include administrators, secretaries, and various professionals such as accountants and financial planners. Social sciences and government services occupations (which include teachers, civil servants, social workers, and lawyers) are just over 68 percent female, and sales/service occupations (for example, jobs in restaurants, bars, hotels, tourism, hairdressing, and child-care facilities, and domestics and building cleaners) are around 57 percent female. All these could

FIGURE 4.5 *Employment Concentration and Occupational Distribution of Women, Canada, 2004*

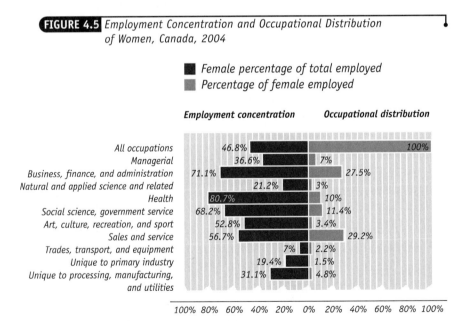

Source: Adapted from the Statistics Canada publication *Labour Force Historical Review*, 2004, Catalogue 71F0004XCB, February 18, 2005.

be labelled job ghettos in the sense that the majority of employees are women, although the pay and security provided in some (particularly nursing, teaching, and other professions) are relatively good. By contrast, women form a small minority of workers in natural and applied sciences (e.g., engineers, computer programmers), primary occupations (e.g., forestry, mining, fishing, farming), and trades, transportation, and construction.

The right-hand series of bars in the figure shows the distribution of the female labour force across all occupations. Sales and services jobs employed well over one in four women workers in 2004. So too did occupations in business/finance/administration, which are largely comprised of clerical and secretarial workers. Taken together, these two occupational categories account for over half of the female labour force. This attests to women's continuing over-representation in lower-status occupations.

For a more complete understanding of where men and women are located in the labour market, we need to investigate the industrial distribution of employment. You will recall from Chapter 2 that Canada's service-based economy is generating polarization between good and bad jobs. How then are men and women distributed within the upper and lower tiers of the service sector? Referring back to Table 2.1 will help to answer this question. Over one-third of men work in the goods sector. Potentially this has negative implications for the male labour force, given the massive restructuring occurring in manufacturing. Women's jobs are far more likely to be found in service industries. Overall, 87 percent of employed women in 2004 were in the service sector, with over one-quarter employed in its lower tier (retail trade and consumer services). In comparison, 19 percent of men worked in the lower-tier services. Interestingly, a higher proportion of women than men also worked in upper-tier services (60% and 45%, respectively). The main explanation for this is the concentration of women in the education/health/welfare sector. But in overall terms, our discussion of service-sector jobs in Chapter 2 suggests that relatively more women than men are in nonstandard, low-quality jobs.

Indeed, there is a growing body of research on the gendered nature of nonstandard work. The very notion of the "standard job" can be considered a masculine norm of employment—the husband as family breadwinner, working full-time and in a more or less permanent job with the same employer. These standard job assumptions underlay the postwar employment contract, which has changed dramatically with the spread of a range of nonstandard employment relationships (Vosko 2000, 2005). A major reason women seek self-employment is to gain the independence and flexibility they feel they need to balance work with their family and personal life. But there are costs, including considerably lower wages than their counterparts who are employees and lack of benefits and career supports, such as access to training or business networks (Hughes 2005; Arai 2000). Linked to gender, there also is a racial dimension to nonstandard work (Zeytinoglu and Muteshi 2000). Visible minority women, often recent immigrants, are employed in some of the most economically marginal forms of work, in family-run businesses, domestic labour, or the garment industry (Sassen 2002). Perhaps the most feminized form of nonstandard work is temporary employment. As Leah Vosko (2000) argues, the expanding temporary help industry is based on traditional stereotypes of "women's work," even though the flexible and impermanent employment relationships it offers are seen as a hallmark of the "new economy."

How much has gender-based job segregation diminished over the course of this century? Table 4.1 shows that in 1901, 71 percent of all employed women were concentrated in five occupations. Within this small number of socially acceptable women's jobs, that of domestic servant employed the greatest portion of working women at 36 percent. Next in importance were those of seamstress and school teacher, each employing roughly 13 percent of the female labour force. These three occupations fit our description of a job ghetto, given that three-quarters or more of all workers in the job were female. In the early 20th century, outright exclusion was the prominent form of gender-based labour market segregation, with professions such as law and medicine barring women, while others, such as dentistry, allowed women to enter but under conditions that ensured their marginalization (Adams 1998).

This was not the case, however, for the two other main female occupations. Clerical work at the turn of the century was still a man's job, although by the 1940s the gender balance had shifted toward women. Interestingly, clerical work was one of the few traditionally male jobs to undergo this *feminization* process. Behind this change was the rapid expansion of office work accompanied by a more fragmented and routinized division of labour. As a result, a layer of new positions emerged at the bottom of office hierarchies, opening office doors to women (Lowe 1987). Finally, the fact that farmers and stock raisers appear as one of the five prominent female occupations may seem peculiar. Considering that Canada was an agricultural economy in 1901, it is

TABLE 4.1 *The Five Leading Female Occupations in Canada, 1901, 1951, 2001*

Occupation	Number of Employed Women	Percentage of Total Female Employment in Occupation	Females as Percentage of Employment in Occupation
1901			
1. Domestic servants	84,984	35.7	87
2. Seamstresses	32,145	13.5	100
3. School teachers	30,870	13.0	78
4. Office clerks	12,569	5.3	78
5. Farmers and stock raisers	8,495	3.6	2
All five occupations		71.1	
1951*			
1. Stenographers and typists	133,485	11.5	96.4
2. Office clerks	118,025	10.1	42.7
3. Sales clerks	95,443	8.2	55.1
4. Hotel, café, and private household workers n.e.s.†	88,775	7.6	89.1
5. School teachers	74,319	6.4	72.5
All five occupations		43.8	
2001			
1. Retail salespersons	355,465	4.8	60.1
2. Secretaries	265,375	3.6	97.9
3. Cashiers	237,560	3.2	85.9
4. General office clerks	233,870	3.2	87.4
5. Registered nurses	218,540	2.9	94.2
All five occupations		17.7	

*1951 census does not include the Yukon or the NWT.

†n.e.s.: Not elsewhere specified. Excludes a few persons seeking work who have never been employed.

Source: Adapted from the Statistics Canada publications *Canada's Workforce: Paid Work, 2001 Census,* Catalogue 97F0012XCB2001017, February 11, 2003; *Women in Canada: A statistical report,* 2nd Edition, Catalogue 89-503E, 1990; *Labour Force: Occupations and Industries, Catalogue 98-1951, Ninth Census of Canada, 1951,* Volume IV; and *Occupational trends in Canada, 1891–1931,* Catalogue 98-193, 1939.

understandable that women would form a small part of the paid agricultural workforce.

The 20th-century march of industrialization saw a decline of some female jobs, such as those of domestic servant and seamstress, and the rise of new employment opportunities in booming service industries. By 1951, the two leading female occupations were in the clerical area. Along with sales clerks, hotel, café, and domestic workers, and teachers, they made up 44 percent of the female workforce just after World War II.

As we have seen, female job ghettos still exist. Office and sales jobs have topped the list since World War II. However, women are now entering a broader range of occupations and professions. For example, according to the 2001 census, the five main female-dominated job titles accounted for only 18 percent of the entire female labour force. This is a positive sign of occupational diversification, but how far have women moved into non-traditional (in other words, male-dominated) occupations? Karen Hughes's analysis of the 484 detailed occupations in the 1971 census discovered that 86 percent of women worked in traditionally female occupations.[21] Over the next 15 years, the proportion of women in these traditional occupations declined to 79 percent, then changed little between 1986 and 1991. This raises questions about the degree to which women can move out of gender-typed occupations over their working lives—the idea that "female" jobs are a way station through which some women pass on their way to better jobs that are not gender-typed (Chan 2000).

Yet, at the same time, women were making gains in what can be called *non-traditional jobs,* those in which men have predominated. By 1991, women had achieved equal representation with men in optometry and financial management, and nearly equal representation in sales management and government administration (Hughes 1995, 2000b).

Additional insights can be gleaned about the shifts taking place in female employment by focusing on professional occupations. Because professions are organized around specific bodies of knowledge and expertise, usually acquired through a university degree, women's access to these jobs should improve as their level of education rises. The proportion of all university degrees granted to women has been rising steadily. Women's share of undergraduate enrollment jumped from 43 to 53 percent between 1972–73 and 1992–93, and to 56 percent in 2001–02 (see Figure 4.6). There were also large increases at the master's and doctoral levels. All major program areas saw rising female enrollments. Math, science, and engineering fields are the only disciplines in which women do not make up more than half of all students; in fact, women are significantly underrepresented in these areas.

Consistent with the direct connection between educational and occupational attainment discussed in Chapter 3, women accounted for two-thirds of the growth in professional occupations between 1981 and 1986 (Marshall 1989). Especially strong gains were made in male-dominated professions—a trend that has continued through the 1990s and 2000s. For example, in 2004,

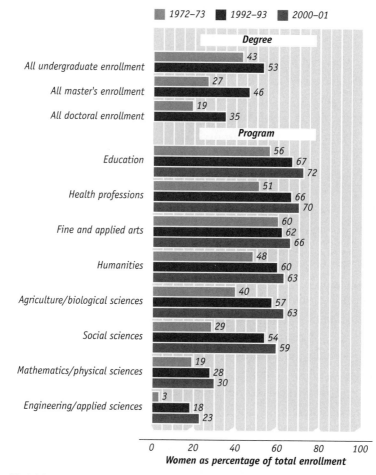

FIGURE 4.6 *Women as a Proportion of Total Full-Time University Enrollment, by Degree and Program, Canada, 1972–73, 1992–93, and 2000–01*

■ *1972–73* ■ *1992–93* ■ *2000–01*

Degree

All undergraduate enrollment — 43, 53

All master's enrollment — 27, 46

All doctoral enrollment — 19, 35

Program

Education — 56, 67, 72

Health professions — 51, 66, 70

Fine and applied arts — 60, 62, 66

Humanities — 48, 60, 63

Agriculture/biological sciences — 40, 57, 63

Social sciences — 29, 54, 59

Mathematics/physical sciences — 19, 28, 30

Engineering/applied sciences — 3, 18, 23

Women as percentage of total enrollment (0, 20, 40, 60, 80, 100)

Source: Adapted from the Statistics Canada publications *Women in Canada: A gender-based statistical report, 2000,* Catalogue 89-503, page 94, September 14, 2000; and *The Daily,* Catalogue 11–001, University enrolment by field of study, Monday, March 31, 2003.

60 percent of new graduating physicians were women (Schmidt 2005). Pharmacy is also an interesting case, having become 50 percent female by the mid 1980s. Research examining Ontario pharmacists sheds light on this feminization process. Despite gender parity, pharmacy remains segregated horizontally and vertically. Women pharmacists are more likely to work in chain

stores or hospitals than in independent pharmacies, they are more likely to be employees than owners or franchisees, and their earnings are significantly lower (Tanner et al. 1999). However, women are more satisfied with their current job and careers than are men. Men, on the other hand, have shown no signs of retreating from pharmacy as more women entered.

Still, many traditional male professions present hurdles to female entry (Wirth 2004). Equally important, there has been little movement of men into female-dominated occupations. For instance, since 1987 women's share of nursing and other health-related professions (excluding medicine and dentistry) has remained steady at around 87 percent.[22] Thus, to the extent that occupational gender segregation is breaking down, it is due to women moving into male-dominated areas, not vice versa.

Gender Stratification within Occupations

Thus far, we have traced broad historical patterns of *horizontal* occupational gender segregation. Now we will examine a related obstacle women face, *vertical* segregation. This refers to how a gendered division of tasks, statuses, and responsibilities exists within specific occupations. As a rule, men tend to be in positions of greater authority in organizations and, consequently, usually receive better job rewards than women. Consider the teaching profession. Even though women made up 71 percent of all elementary and secondary school teachers and counsellors in 2001, they occupied only 46 percent of school administrative and principal positions. Approximately 53 percent of the women in teaching are in elementary school classrooms, compared to just 20 percent of men. Men are concentrated in the higher-status, better-paying jobs in universities, colleges, high schools, and educational administrations.[23]

The situation is much the same in other professions. Women have made significant inroads recently, particularly in law and medicine, although there is a tendency for them to opt for lower-status specialties such as family law or general medical practice.[24] This reinforces assumptions often made by the male establishment in these professions about female colleagues being best suited to tasks calling upon their "natural instincts" as wives and mothers.

In their probing investigation of the changes in legal careers in Canada, John Hagan and Fiona Kay portray a gender-stratified profession in which work environments are not family-friendly. Large numbers of women entered law in the 1970s and 1980s, when the profession was expanding and reorganizing into larger firms. According to Hagan and Kay,

> women were recruited into the profession during a period when they were needed to fill entry- and intermediate-level positions and were perceived to be compliant employees. . . . It is possible that . . . many partners in law firms were encouraged by the belief that as in the teaching profession of an earlier era, women would assume entry-level positions in the profession, work diligently for a number of years, and then abandon their early years of invested work to bear children and raise families, leaving partnership positions to men assumed to be more committed to their occupational careers. (1995: 182)

These assumptions were largely borne out in research findings. The changing professional climate of the 1980s restricted opportunities for both men and women to attain partnerships—a financial and management stake in the firm that results in much higher earnings. However, women lost out even more than men, mainly because of the conflicts and compromises engendered by the interplay of professional demands and expectations with family roles. Still, many women who continue to practise law full-time after having children have high work commitment and high job satisfaction, despite lower earnings than men. But as other researchers have noted, women are far less likely than men to have children once they become lawyers, and far more likely to leave the profession due to work–family conflict, low pay, and discrimination (Brockman 2001).

In addition to law, women have also made breakthroughs in the male domains of business, medicine, and dentistry. According to Table 4.2, women made up nearly 31 percent of general practitioners and 23 percent of dentists. They also made up 35 percent of managers at all levels (last figure not shown in table). However, a detailed analysis of the managerial category shows that women are far less likely to be at senior management levels. For example, among the full-time, full-year workers shown in Table 4.2, women accounted for just 21 percent of senior managers in financial, communications, and other business services, and only 12 percent of senior managers in

TABLE 4.2 *Average Employment Income of Full-Time, Full-Year Workers in the 16 Highest-Paying Occupations for Women, Canada, 2000*

	Income, Men	Income, Women	F–M Earnings Ratio	Total Number Employed	Number of Women Employed	Women as % of Total
Judges	146,008	131,663	90.2	1,825	440	24.1
Specialist physicians	160,833	98,383	61.2	12,480	3,845	30.8
General practitioners and family physicians	133,789	96,958	72.5	22,040	6,780	30.8
Senior managers—Financial, communications carriers, and other business services	141,829	90,622	63.9	40,920	8,810	21.5
Dentists	129,104	82,254	63.7	8,710	2,000	23.0
Lawyers and Quebec notaries	114,894	77,451	67.4	47,285	14,660	31.0
Senior managers—Goods production, utilities, transportation, and construction	120,914	75,267	62.2	44,625	5,170	11.6
Information systems and data processing managers	80,466	67,203	83.5	38,795	9,735	25.1
Senior managers—Trade, broadcasting, and other services, n.e.c.	108,527	67,161	61.9	37,685	6,700	17.8
School principals and administrators of elementary and secondary education	71,288	64,692	90.7	23,335	10,625	45.5
University professors	79,993	63,746	79.7	30,550	9,010	29.5
Optometrists	89,419	63,656	71.2	2,075	795	38.3
Engineering, science, and architecture managers	87,002	62,466	71.8	18,545	2,470	13.3
Physicists and astronomers	73,736	62,153	84.3	2,225	400	18.0
Mining engineers	73,297	62,066	84.7	1,985	180	9.1
Senior managers—Health, education, social, and community services and membership	83,595	61,217	73.2	14,170	7,005	49.4

Source: Adapted from the Statistics Canada publication *Number and Average Employment Income (2) in Constant (2000) Dollars, Sex (3), Work Activity (3) and Occupation—1991 Standard Occupational Classification (Historical) (707A) for Population 15 Years and Over with Employment Income, for Canada, Provinces, Territories and Census Metropolitan Areas, 1995 and 2000—20% Sample Data*, Catalogue 97F0019XIE2001003, March 11, 2003.

key goods-producing industries. Likewise, there are few female managers in male-dominated areas such natural sciences and engineering or in the mining and oil sector. Instead, women managers are concentrated in female enclaves such as health, education, and community services, where, as Table 4.2 shows, they make up almost half of managers.

Women still encounter a *glass ceiling*—subtle barriers to advancement that persist despite formal policies designed to eliminate them (Wirth 2004). Looking at the 458 employers covered by the federal Employment Equity Act (discussed in more detail below) in 2003, women made up 44 percent of the 620,000 employees. Yet close to two-thirds of these women worked in three occupational groups: clerical personnel; administrative and senior clerical personnel; and supervisors of clerical, sales, and service personnel. While women's share of managerial and professional occupations was up slightly from 2002, their representation in management jobs was inversely related to the authority and rewards of these positions. Moreover, even though employers covered by the Act have been actively recruiting more women into upper-level management, women held only about one in five of these senior positions in 2003.[25] At the executive and board level, it has been exceedingly difficult for women to gain entry in most corporations. Despite some appointments of women to the top corporate jobs, such as president, chief executive officer (CEO), executive vice-president, or chief operating officer, Canadian women occupied just 7.1 percent of top corporate jobs in 2004 (Catalyst 2004)—a trend in line with other industrialized countries (Burke and Mattis 2000; Hughes 2000a).

In sum, although the evidence of wider employment horizons for women bodes well for the future, progress toward full gender equality in the workplace is halting and uneven. As Figure 4.5 illustrates, achieving this goal would see women represented in each occupation proportional to their overall share of the labour force, which is now at 47 percent. This immediately raises the question (which we address shortly) of how best to reach this objective. But would women want full access to all male jobs, given that a good number are dirty, dangerous, physically exhausting, insecure, low-paying, or at risk of being restructured out of existence? The implication, of course, is that unless all jobs at the bottom of the occupational hierarchy are upgraded in terms of pay and working conditions, gender equality as defined above would only be a partial advance for women.

THE WAGE GAP

One of the most obvious consequences of labour market segmentation is the *gender wage gap* (or female–male earnings ratio, as we labelled it in Chapter 3). Figure 4.7 shows that women who work full-time for the entire year earned an average of 71 percent of what similarly employed men earned in 2003. This is an accurate way to compare male–female earnings differences, because

FIGURE 4.7 *Average Earnings for Full-Time/Full-Year Workers* by Gender, Canada, 1969–2003*

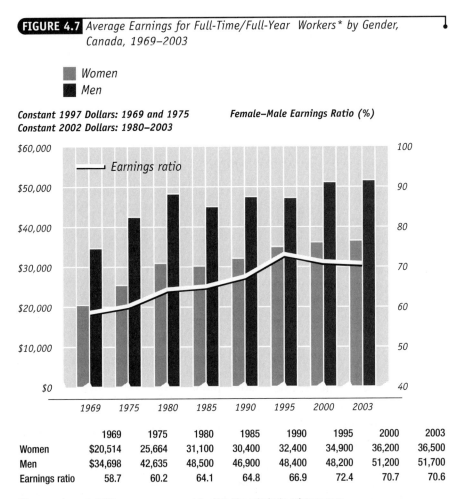

	1969	1975	1980	1985	1990	1995	2000	2003
Women	$20,514	25,664	31,100	30,400	32,400	34,900	36,200	36,500
Men	$34,698	42,635	48,500	46,900	48,400	48,200	51,200	51,700
Earnings ratio	58.7	60.2	64.1	64.8	66.9	72.4	70.7	70.6

*A person who worked 30 hours or more per week for 49 to 52 weeks in the reference year.

Source: Adapted from the Statistics Canada publications *Women in Canada: A gender-based statistical report,* 2000, Catalogue 89-503, page 155, September 14, 2000; and the Statistics Canada CANSIM database at http://cansim2.statcan.ca, Series V1542060 and V1542064.

part-time or part-year workers, who have lower earnings and are disproportionately female, are not included. Despite women entering some of the
higher-paying managerial and professional jobs, the wage gap narrowed by just
9 percentage points from 1980 to 1995, and has since plateaued at around 71
percent. This partly reflects the way in which employment growth trends often
counteract each other. During the 1970s, the number of women in the 20
highest-paying occupations swelled more than fourfold, compared with a
twofold increase for men (Boulet and Lavallée 1984: 19). But this trend was
offset by an expansion of female employment at the lower end of the pay scale.
While not used in our calculation of the wage gap, the big rise in the number
of part-time jobs is important because it has reduced the earning potential of
many women. This distinction between full- and part-time earnings reflects
the growing employment disparities among women and between the sexes in
the service economy.

Gender-Based Inequality of Earnings

Trends in earnings between 1969 and 2003 deserve brief comment. Figure 4.7
displays *real earnings,* which means that the effects of inflation have been eliminated by adjusting earnings. Looking at the 1980–2003 period (in 2002 constant dollars) illustrates the startling fact that male real earnings actually
decreased slightly between 1980 and 1995, while women's earnings nudged
upwards. Quite different labour market and organizational processes underlie
these divergent earnings profiles. While women were being recruited into
intermediate-level professions and junior and middle-level management, males
were losing ground as traditionally well-paying (often unionized) manual occupations became less common and downsizing pushed older professionals and
managers (mostly men) into early retirement. Within the labour market as a
whole, some of men's loss was women's gain. A rising proportion of employed
women, with incomes somewhat higher than the incomes of women a generation earlier, have kept family incomes from dropping as far as they otherwise
might have. Without the wife's income, many husband–wife families now in
the middle or working class would likely join the ranks of the poor.

Wages are a basic indicator of overall job quality. Recall from Chapter 3
that high wages also are part of a larger package of extrinsic and intrinsic job
rewards. Stated simply, in addition to paying relatively well, "good jobs" usually offer other advantages: a range of benefits, job security, advancement

FIGURE 4.8 *Job Benefits Received by Employees and Self-Employed, by Gender, Canada, 2000*

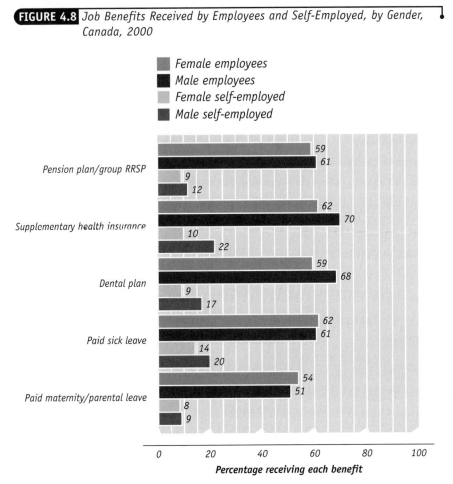

■ Female employees
■ Male employees
□ Female self-employed
■ Male self-employed

Percentage receiving each benefit

Source: CPRN/EKOS, *Changing Employment Relationships Survey,* microdata file (n = 2500). Reprinted by permission from Canadian Policy Research Networks.

opportunities, and interesting and challenging work. Turning to Figure 4.8, we find that female employees and self-employed individuals are less likely than men to receive benefits such as supplementary health insurance and a dental plan. Furthermore, 40 percent of female workers in 1994 were in nonstandard jobs, compared with 27 percent of men (Krahn 1995: 40). Men are also more likely to receive promotions. Jobs occupied by men tend to offer more decision-making authority, have higher skill requirements, and be less repetitive (Krahn 1992: 96). In short, the gender wage gap is part of a broadly based discrepancy between the quality of jobs performed by men and women.

Women's lower earnings also mean reduced living standards. In other words, women in Canada are more likely than men to be poor. Some 42 percent of lone-parent mothers with children under age 18 lived in poverty in 2001, compared with 9.5 percent of two-parent families with children. Along with female lone-parents, women over the age of 65 are among the groups in Canadian society most vulnerable to poverty. In 1998, 45.6 percent of all elderly unattached women lived in poverty, mainly due to inadequate pensions. On the positive side, improvements in income support programs for the elderly have helped to reduce this number from 71 percent in 1980.[26]

Shifting our focus from the bottom to the top of the income distribution, average male and female earnings in the 16 highest-paying occupations for women are displayed in Table 4.2. First observe that in several of these jobs the female–male earnings gap is greater than in other occupations. Specifically, the ratio is lower than or equal to 71 percent (see Figure 4.7) in about half of these jobs. The gap is narrowest for judges (90%), and widest among specialist physicians (61%). Thus, in the most well-paying occupations, as elsewhere, women have a greater chance of receiving a lower salary. Still, women who do succeed in entering Canada's top-paying occupations can expect to earn considerably more than the average female income. But we also need to note that, with one exception (senior managers in health, education, social, and community services) all of the highest-paid occupations are male-dominated. In fact, if we examine the list of Canada's 50 best-paid CEOs in 2004, we find that none are women.[27]

When interpreting such data, we must keep in mind that because of the quite recent entry of women into managerial and professional jobs, gender differences in age and experience will account for part of the earnings gap—but by no means all of it. Studies of women and men with comparable education, age, and work experience have found a wage gap (albeit a small one) immediately after graduation.[28] While differences in experience, education, field of study, occupation, and industry help explain part of the wage gap, roughly one half remains unexplained (Drolet 2002b).

Accounting for Women's Lower Earnings

Two forms of discrimination affect female earnings. The first is *wage discrimination,* whereby an employer pays a woman less than a man for performing the same job. The second stems from the gender-segregated structure of the labour market documented earlier. Occupational gender segregation results in lower female wages by channelling women into low-paying job ghettos which, in turn, provide few opportunities for mobility into more rewarding jobs.

To what extent does each form of discrimination contribute to the overall wage gap? To answer this question, one must look at an employee's education, training, and work experience, as well as at the occupation, industry, and geographic location of employment.[29] These factors may influence a worker's productivity and thus his or her earnings. Research by Morley Gunderson (1994), a labour economist, shows that when these productivity factors are taken into account (or are statistically controlled), the earnings ratio in Ontario increases from around 60 percent to between 75 and 85 percent. Differences in work experience, combined with the segmented structure of the labour market, emerge as the main determinants of gender differences in earnings. Applying this analysis to men and women employed in the same jobs within a single establishment, the wage gap narrows even further, as the earnings ratio rises to between 90 and 95 percent. In short, there is little direct pay discrimination by employers, at least in Ontario; rather, it is the division of the labour market into male and female jobs that creates pay inequality.

Human capital theory emphasizes that education is the great equalizer in the job market. Ideally, people with identical educational credentials should have the same amount of human capital and, therefore, be equally competitive in their earning power. But as Chapter 3 emphasized, human capital theory fails to explain why some individuals with equivalent education are much better paid. Nonetheless, education does matter. Ever since World War II, obtaining a good education has been a route to upward mobility for many young Canadians, particularly males from working-class families.

Does the same hold true for women? There is little doubt that women have become better educated. As noted previously (see Figure 4.6), women have entered a wider range of programs and increasing numbers go on to graduate studies. Dramatic gains have been made in male-dominated professional faculties, such as law, medicine, and business. But women continue to be concentrated in traditional areas: in 2001, one-fifth of all women enrolled in Canadian universities were pursuing education degrees (Statistics Canada 2003f: 34). Professional training in science and technology also remains heavily male-dominated. In 2001–02, 76.4 percent of students in Canadian engineering, architecture, and related programs were men, down slightly from

81 percent in 1992–93 (Statistics Canada 2004b). The challenge of attracting more women into engineering has been the focus of a study jointly sponsored by the federal government and the engineering profession. In its 1992 report, the Canadian Committee on Women in Engineering highlighted the complex social and cultural barriers that have obstructed women's way into engineering.[30] More recently the Canadian Coalition of Women in Engineering, Science, Trades and Technology (CCWEST) has launched a national initiative to improve the representation of women in these fields. Women are not only underrepresented at the university level, but also in the construction, electrical, and computing trades where they make up only a fraction of apprentices (Statistics Canada 2003b).

Given that more women than ever before are graduating with strong education credentials, a question worth considering is, have they been able to find the types of jobs for which they have been educated? Overall, a university education is a good predictor of higher occupational attainment for both women and men. However, this generalization requires careful qualification to account for the combined impact of gendered enrollment patterns, mentioned above, and occupational gender segregation. And, as also documented above, a well-educated woman cannot expect to earn as much as a comparably educated man. In short, human capital theory alone fails to explain the employment realities of Canadian female university graduates.

THEORETICAL PERSPECTIVES ON WORK-RELATED GENDER INEQUALITY

Our key question is, how has occupational gender segregation become so deeply entrenched? Once we can answer this, we will be better able to inform public policy discussions about strategies for achieving equality in the workplace. However, a comprehensive theoretical model capable of adequately explaining the origins, development, and perpetuation of work-related gender inequities has yet to be formulated. Feminist analysis has provided a necessary antidote to the long-standing male bias in much of the sociological research on work. Feldberg and Glenn characterized research in the 1970s in terms of *job* and *gender models* of the workplace. Studies of male workers focused on working conditions and organizational factors (the job model); explanations of women's employment discussed their personal characteristics and family roles (the gender model).[31] Since then, feminist analysts of women's work have probed the link between the private and public spheres of domestic work and paid employment, illuminating how the formal economy depends on unpaid caring work (largely done by women), while at the same time rewarding the "model" worker who is free of caregiving responsibilities. While gender remains a central consideration, research is increasingly "intersectional," exploring how factors such as motherhood, race, ethnicity, age, and sexuality work to create differences and similarities in women's economic lives.[32]

Human Capital and Labour Market Segmentation Models

Two of the main competing theories of how labour markets operate are *human capital theory* and *labour market segmentation* (or *dual labour market*) *theory* (see Chapter 3). We have already criticized the human capital model for its inability to explain the persistence of gender segregation.[33] While the human capital model is concerned with factors influencing the characteristics of workers—the supply side of the labour market—the segmentation model examines how job requirements within organizations create a demand for particular kinds of workers. Labour market segmentation theory distinguishes between secondary and primary labour markets. Secondary labour markets are located in marginal, often uncompetitive industries that must struggle constantly to keep wages and operating costs down. In contrast, corporations and state bureaucracies in the primary sector offer relatively high wages, decent benefits, job security, and pleasant working conditions. They also have well-developed internal labour markets providing employees with good career opportunities and employers with a stable, committed workforce.

For the segmentation model to accurately explain why men and women hold different kinds of jobs, women would have to be concentrated in the secondary segment of the labour market. This is partly true, since, it will be remembered, relatively more women than men are employed in lower-tier service industries, which form a large part of the secondary sector. However, critics point out that many women are also employed in the primary sector as typists, cleaners, or food servers. In short, a general segmentation perspective cannot account for gender differences in employment within the same industry or establishment, or inequalities among women. Nor does it explain how gender segregation developed in the first place.[34]

Additional factors need to be considered to explain why gender is a major source of inequality in the labour market. Some researchers, for instance, link women's labour market position to their traditional roles within the family. Women's family roles have centred on activities such as raising children and caring for dependents; cooking, cleaning, and other services essential to enabling family members to hold down jobs; and contributing a secondary income to the family budget—roughly in that order. Many employed women experience a conflict when forced to choose between their job and their family. According to this view, women's loyalty usually goes to the latter, and employers consequently often view these women as lacking commitment to their work. Barriers to interesting and better-paying jobs consequently persist (Garnsey, Rubery, and Wilkinson 1985).

Gender-Role Socialization

But how are employers' stereotypes of women, women's expectations and restricted employment opportunities, and women's family roles linked to their employment patterns within the workplace? Are such patterns best explained with reference to women's early socialization, including awareness of family roles and feminine qualities, or as a response to their employment conditions? Clearly, both processes may be involved.

The socialization of girls and boys into traditional gender roles creates cultural norms and expectations that they will carry into the workplace as adults. For example, a survey conducted in the early 1980s of children aged 6 to 14 in Ontario, Quebec, and Saskatchewan documented the effects of *gender-role socialization*. While the girls recognized the expanding occupational horizons of women, at a personal level they still held very traditional aspirations. As the researchers conclude: "Many seem to be saying 'Yes, women can become doctors, but I expect to be a nurse'" (Labour Canada 1986: 55). Further discussions with these girls revealed that when they imagined themselves as adults, most saw women who were mothers with small children, supported by a husband.

Have these attitudes changed? Our survey of 1996 Albertan grade-12 students shows that while gender-role attitudes are becoming more egalitarian among youth—the vast majority of female and male respondents agree that a woman should have the same job opportunities as a man—more traditional work preferences continue to influence career choices. Nurse, social worker, and teacher were the popular career choices for females, compared with computer programmer/analyst, engineer, and auto mechanic for males.[35] As Jane Gaskell's research illuminates, the choices young working-class women make while they are in school, and upon leaving it, prepare them for domestic labour and employment in female job ghettos. Gaskell (1992: 134–35) explains that the young women whom she studied

> . . . end up recreating the world they are not happy with. They see what men are like; they see what jobs pay and how one gets trained for them; they see the lack of child-care options. Within the world as they experience and know it, they do their best. The young men take their gender privilege equally for granted. The result is the reproduction of gender divisions, not because they are desired, but because these young people don't believe the world can be otherwise.

Gender Inequality and the Organization of Work

Also indispensable to an understanding of female work behaviour is how socialization patterns are strengthened and reproduced within the workplace. Rosabeth Moss Kanter's research suggests that the main sources of gender inequality must be located within work organizations. Kanter's core argument is that "the job makes the person." She elaborates:

> Findings about typical behaviour of women in organizations that have been used to reflect either biologically based psychological attributes or characteristics developed through a long socialization to the female sex role turn out to reflect very reasonable and very universal responses to current organizational situations. (1977: 67)

In Kanter's view, then, men and women employed in similar jobs in an organization will react in similar ways to their job conditions. (We return to this subject in Chapter 8.) A good example is Donald Roy's (1959–60) study of male machine operators. These men escaped the drudgery and isolation of their work—"kept from 'going nuts,'" in Roy's words—by engaging in idle chatter, playing games, and fooling around. Female keypunch operators, file clerks, or assembly-line workers may seek similar relative satisfactions to cope with the numbing tedium of their work. In either case, management can use the coping behaviour as evidence that these workers are incapable of performing more demanding jobs. This creates a double bind for employees: the most effective ways of personally coping in jobs at the bottom of the organization are also indications to management that workers in these jobs deserve to be kept there.[36]

Kanter does not ignore possible gender differences in socialization or non-work roles. Rather, she underlines the pervasive influence of an individual's job content and organizational position on her or his attitudes and behaviour. This perspective is similar to what Pat and Hugh Armstrong (1994: Chapter 6) call the *materialist* explanation of the gendered division of labour. These researchers document the strong connection between women's self-perceptions and the kinds of work they do. The historical fact that men and women have performed different tasks in the home and in the labour force creates gender differences in work orientations. For example, women may be aware of their subordinate economic role, but they either rationalize it as an outcome of their domestic responsibilities, or else feel powerless to change it.

Gender is a potent controlling device within organizations. Male managers draw on social stereotypes of women as more oriented toward pleasing others, and therefore compliant, to devise paternalistic methods of supervision. Another dominant perception employers have is that women put their domestic responsibilities first, therefore devoting less effort and commitment to their employment. The assumption that women choose to allocate more effort to their domestic than to their employment roles is central to human capital theory's account of why women receive lower wages than men and have limited career opportunities. However, this myth was exploded by Denise and William Bielby (1988), who document that women do not allocate less physical and mental energy to work than men. In fact, women are more efficient workers, given that the majority put greater effort into their work than men with similar attributes. This finding leads the researchers to suggest that women, therefore, should be preferred by employers.

Kanter advances the analysis of gender divisions in organizations by showing how management is a social process that relies heavily on trust and conformity.[37] To reduce uncertainty in decision making—the essence of managerial activity—a premium is placed on recruiting individuals who are predictable. The best way to achieve this is to recruit only those people who have social characteristics identical to the existing group of managers. This social cloning reproduces male dominance in management and, moreover, creates an enormous barrier to women.

In sharp contrast, the role of the secretary is built on a combination of female stereotypes. Kanter uses the term *office wife* to describe the subordinate, paternalistic, and almost feudal relationship secretaries usually have with their male employers. The job becomes a trap through the dependency of managers on their secretaries and the personal loyalties that result. Many competent secretaries are not promoted because their good performance makes them indispensable to the boss. Unlike other positions in modern bureaucracies, there are no safeguards in place that would curb or restrict the exercise of managerial authority and the expectation that personal services, like fetching coffee, are part of the job. Because secretaries are often stuck in this role, their behavioural responses, such as self-effacement or timidity, may undermine advancement prospects.

Men who have more opportunities take advantage of them, developing behaviours, values, and work attitudes that help them do so. Once young male management trainees are identified as "fast trackers," the resulting halo effect creates the impression that they do not make mistakes and gives them the momentum to move up. Conversely, women in dead-end jobs quite rationally decide to give up, losing both work commitment and motivation. This signals to supervisors that they do not deserve promotions or raises.

Women who succeed in entering management face the problem of *tokenism*. Kanter argues that a goldfish bowl phenomenon, resulting from being an identifiable minority, leads women or members of visible minorities to work harder to prove themselves. In turn, this confirms the dominant group's impression that these individuals are different and, therefore, do not belong. Furthermore, tokens lack the support systems so essential for surviving in the middle and upper ranks of organizations. But the problem is not sex or race, per se; rather, it is one of being part of a group underrepresented in a particular environment. Other empirical research into tokenism concludes that it is limited to women in traditionally male occupations and, further, that by focusing on the imbalance in numbers as the causes of the problem, it deflects attention away from the more complex mechanisms of discrimination (Yoder 1991). However, Kanter's more general point still holds: the *structure of opportunities* in an organization—basically, who has access to which positions, resources, and rewards—tends to create self-fulfilling prophecies that only serve to reinforce the subordinate status of women.

Recent research pursues these issues raised by Kanter in several directions. New evidence suggests that *ascription*—allocating opportunities or positions, such as in hiring and promotion decisions, on the basis of ascribed characteristics such as a person's sex—operates through an organization's personnel practices. Ascription can be reduced or eliminated if formalized procedures replace informal networks to identify candidates for management jobs (Reskin and McBrier 2000). Other research uses gender as an explicit analytic category. For example, while acknowledging that women's work experiences are shaped by organizational structure and power, Rosemary Pringle's (1989: 162) study of secretaries adds the dimension of sexuality:

> Far from being marginal to the workplace, sexuality is everywhere. It is alluded to in dress and self-presentation, in jokes and gossip, looks and flirtations, secret affairs and dalliances, in fantasy, and in the range of coercive behaviours that we now call sexual harassment. Rather than being exceptional in its sexualization, the boss–secretary relation should be seen as an important nodal point for the organization of sexuality and pleasure.

By focusing on sexuality, researchers are able to advance our understanding of how employment relations and work structures contribute to defining gender, both for men and for women. This new area of *gender studies* challenges us to rethink how masculinity, femininity, and heterosexuality are socially constructed categories reinforced by existing work institutions.[38]

On a larger scale, *sexual harassment* is the most draconian use of male power over women in the workplace. In one of the most comprehensive surveys ever conducted on this topic, the 1993 national Violence Against Women Survey found that 23 percent of Canadian women aged 18 and over had experienced some form of workplace sexual harassment, the most common being inappropriate comments about their bodies or sex lives. Other Canadian and U.S. surveys suggest higher rates, since there is likely underreporting of "poisoned environment" harassment—the display of a demeaning attitude toward women or unwanted sexual attention.[39] Some researchers consider harassment to be one form of the rising trend of aggression and violence in the workplace (DiGiacomo 1999). Harassment, or the fear of it, may lead women to quit jobs, or avoid entering non-traditional occupations (Barling, Kelloway, and Frone 2005; Pringle 1989: 164). Of note is the fact that, in a service economy, it is not only coworkers or employers who harass, but customers as well (Hughes and Tadic 1998).

ACHIEVING WORKPLACE EQUALITY

We have catalogued the inequities women face in terms of work opportunities and rewards. Some of the evidence leads to optimism that the stumbling blocks to equal labour market opportunities and rewards are slowly being pushed aside. But the agenda of equality in employment can be achieved more effectively and sooner through bold public policy initiatives. This chapter will conclude by assessing two major policy thrusts: employment equity and pay equity. Unions, too, can play a decisive role in women's quest for greater workplace equality; their role will be examined in Chapter 7.

Employment Equity

The 1984 Royal Commission on Equality of Employment (chaired by Rosalie Abella, now a Supreme Court judge) defined *employment equity* as a strategy to eliminate the effects of discrimination and to fully open the competition for job opportunities to those who have been excluded historically. Four groups were identified as disadvantaged in terms of employment: women, visible minorities, Aboriginal people, and persons with disabilities. It is important to recognize that women do not constitute a homogeneous group. Indeed, we can speak of women who are also Aboriginal, or disabled, or members of a visible racial or ethnic minority as facing a double disadvantage. Focusing on women, the Abella Commission asserted that

> . . . equality in employment means first a revised approach to the role women play in the workforce. It means taking them seriously as workers and not assuming that their primary interests lie away from the workplace. At the same time, it means acknowledging the changing role of women in the care of the family by helping them and their male partners to function effectively both as labour force participants and as parents. And it means providing the education and training to permit women the chance to compete for the widest possible range of job options. In practice this means the active recruitment of women into the fullest range of employment opportunities, equal pay for work of equal value, fair consideration for promotions into more responsible positions, participation in corporate policy decision-making through corporate task forces and committees, accessible child care of adequate

quality, paid parental leaves for either parent, and pensions and benefits. (Canada 1984: 4)

The Abella Commission articulated the growing belief that all individuals, regardless of their personal characteristics, should be treated fairly in recruitment, hiring, promotions, training, dismissals, and any other employment decisions. The commission provided the rationale for the federal 1986 Employment Equity Act by arguing that *systemic discrimination* creates employment barriers that can only be dismantled through strong legislation. Systemic discrimination is the unintentional consequence of employment practices and policies that have a differential effect on specific groups. This form of discrimination is built into the system of employment, rather than being the conscious intent of individuals to discriminate. For example, minimum height and weight requirements (using white males as the norm) for entry into police or fire departments have the effect of excluding women and members of certain visible minorities, even though this was not the intent.

The 1986 Employment Equity Act covers federal government employees, 370 federally regulated employers and Crown corporations, and employers that have 100 or more employees and bid on government contracts worth more than $200,000. The thrust of the act is twofold. First, it requires these employers to identify and remove employment practices that act as artificial employment barriers to the four *designated groups* (women, visible minorities, Aboriginal people, and disabled people). Second, it establishes targets and timetables for achieving a more representative workforce that will reflect the proportion of qualified and eligible individuals from designated groups in the appropriate labour pool (not in the entire population). As the preamble to the act spells out, the ultimate goal is "to achieve equality in the work place so that no person shall be denied employment opportunities or benefits for reasons unrelated to ability" (Canada 1985).

Employment equity policy also recognizes a need for positive measures that will rectify historic imbalances in staff composition (for example, special training programs for Aboriginal people) and the reasonable accommodation of differences (such as changing the RCMP dress code to permit Sikhs to wear turbans and Aboriginal officers to have braids) to make workplaces more accessible and hospitable for a greater diversity of individuals. In this way, organizations are able to become more representative of the increasingly varied composition of Canadian society.

The Employment Equity Act requires employers to file annual reports showing their progress in recruiting and promoting members of the four designated groups. Generally, these reports demonstrate the slow pace of change. However, recent research does provide a measure of success, suggesting that increased access for designated groups to professional, supervisory, and upper management positions has contributed to reducing the wage gap between these groups and white males.[40] Yet there is wide-ranging criticism of employment equity, to which the Special Committee of the House of Commons that reviewed the act responded by recommending stricter monitoring and enforcement; broadening the act to include more workplaces; greater employer commitment to the equity goals, timetables, and plans; and the establishment of a comprehensive national employment equity strategy.[41] Overcoming male resistance to gender equality in organizations and backlash against what some men view as the threat of feminism in action will require these measures and more.[42]

Stronger legislation may indeed speed progress toward equality. But business motives may be a more powerful influence. As discussed in Chapter 2, many employers realize they will face growing labour shortages as baby boomers begin to retire in the near future. Employers are being confronted with the reality that women and members of the other three designated groups will make up a large majority of new labour force entrants in coming years. Employment equity seems least effective at eradicating racial discrimination, based on the growing number of racial discrimination cases coming before human rights tribunals (Jain and Al-Waqfi 2000). But in the current language of employers, "diversity" is becoming a major human resource management challenge, which means making workplaces welcoming for everyone, regardless of their gender or race or their disability status.

Surveying the progress made to date for women under the Employment Equity Act, Leck (2002) notes both positive and negative outcomes. On the plus side, evidence to date suggests that organizations with equity programs have generally improved their human resource practices, increased the presence and status of women, and narrowed the wage gap between women and men. On the downside, however, are growing complaints from employers about the administrative costs of equity programs, perceived declines in productivity, and evidence of a rising male "backlash." In Leck's view, these negative outcomes stem largely from a poor understanding of what equity programs seek to do. Even among university students educated in human resource management and equity programs, she notes, there is a widespread belief that programs involve quotas set by government (untrue), or are unnecessary because discrimination against women no longer occurs.

In order for equity programs to thrive, Leck suggests organizations improve practices in a number of areas. First, effective communication is critical, ensuring that employment equity is understood not just by supervisors and managers responsible for human resource issues, but by employees throughout the organization. Employee forums to discuss equity and diversity issues, diversity booths at company events, equity councils, and written articles in internal company newsletters are just some ways Leck believes organizations can better educate employees about equity and diversity issues. In addition, involving employees, unions, and other employee groups in formulating and reviewing practices is crucial for increasing acceptance, reducing backlash, and easing fears about some of the myths of equity programs. Well-designed and well-delivered training is also needed to ensure that human resource practices, such as performance appraisals, recruitment, and selection, are bias free and that diverse groups are made to feel welcome in organizations.

Pay Equity

A related policy initiative is *pay equity.* Pay equity, or *comparable worth* as it is called in the United States, focuses specifically on the most glaring indicator of gender inequality in the labour market: the wage gap. Pay equity is a proactive policy, requiring employers to assess the extent of pay discrimination and then to adjust wages so that women are fairly compensated. Frequently a topic of heated public debate, pay equity recognizes that occupational gender segregation underlies the wage gap. It attempts to provide a gender-neutral methodology for comparing predominantly female jobs with predominantly male jobs that are in different occupational classifications under the same employer. For example, a secretarial job (female) would be compared with the job of maintenance technician (male) in the same organization. A standardized evaluation system assigns points to these jobs on the basis of their skill level, effort, responsibility, and working conditions.

Underpinning the process is the recognition that women's work is valuable but has been under-rewarded in the past. The objective is to pay employees on the basis of their contribution to the employer. This will establish *equal pay for work of equal value*, a more far-reaching concept than *equal pay for the same work*, which only compares women and men in the same (or "substantially similar") jobs. However, developing and applying a truly gender-neutral system for evaluating jobs has proved exceedingly difficult because it challenges long-standing assumptions about the nature of skill (Steinberg 1990).

Pay equity legislation mainly covers the public sector, although in Ontario and Quebec, which have the most extensive legislation, parts of the private sector are also included.[43] Pay equity policies are explicit in their intent. As the Pay Equity Task Force notes in its 2004 report, pay equity begins from the premise that "differences in pay for comparable work which is based solely on differences in sex are discriminatory, and that steps should be taken to eliminate these differences" (Canada 2004). Opponents claim that the economy cannot afford the resulting wage increases, that some businesses may be driven into bankruptcy, or that it interferes with the operation of the labour market. But for millions of employed women, pay equity may provide long overdue recognition of their economic contributions to society.

Pay equity legislation has gone some distance toward creating fairer, gender-neutral compensation schemes. One example of the successful implementation of pay equity is found in the Manitoba provincial government. Five thousand women employees received wage adjustments averaging 15 percent at a cost of 3.3 percent of the provincial civil service payroll.[44] Ontario pay equity legislation has been heralded as the most comprehensive in North America. Yet loopholes, exclusions, and cumbersome procedures have muted its impact. For instance, casual workers and private-sector firms with fewer than 10 employees are exempted, as are women who are unable to find a suitable male job to use for comparison within their place of employment.[45] Under the scrutiny of feminist analysis, the technical issues and problems of pay equity become the barriers to eliminating the gender wage gap and re-defining the value of women's economic worth. So while pay equity policies cannot eliminate the female–male wage gap, they have contributed to narrowing it. Furthermore, by increasing earnings at the bottom of the female earnings distribution, they have benefited more than just middle-class women.[46]

Perhaps one of the most publicized pay equity cases has involved the federal government itself, which after 14 years was forced to settle with its employees (librarians, clerks, and others) in 1999. Following this episode, the federal government established a Pay Equity Task Force to review existing legislation. Reporting in 2004, the Task Force made over 100 recommendations for improving federal pay equity legislation. Central among these were moving away from complaint-based to proactive legislation; extending coverage to all employees (part-time, casual, seasonal, temporary); and extending pay equity to visible minorities, Aboriginal people, and people with a disability. In addition the Task Force recommended the creation of a federal Pay Equity Commission to administer the pay equity law and conduct investigations and audits, and a Pay Equity Tribunal with expertise in pay equity and human rights to adjudicate disputes. While the federal government has promised action on this issue, by 2005 no new legislation had been introduced (Canada 2004).

Another policy instrument for achieving gender equality in employment is section 15 of the Canadian Charter of Rights and Freedoms. This section of the Charter became law in April 1985. It established for the first time in our history a constitutional entitlement to full equality for women in law, as well as in the effects of law. This latter point is a crucial one, for regardless of the wording or intent of a law, if in practice it results in discrimination against women, the courts can rule it to be unconstitutional. This legal process may gradually come to have an influence on gender inequality in the workplace.

CONCLUSION

We are beginning to rectify some of the most glaring problems women confront in the work world. The employment equity and pay equity programs now operating in Canada are reform oriented, aiming to modify existing employment institutions, values, processes, and social relationships. But how effective will workplace reforms be in altering women's traditional non-work roles—especially within the family—and the supporting socialization processes and ideologies? The limitations of such reforms prompt some to argue for greater recognition of women's unpaid caregiving work and greater involvement of men in such activities. For them, equity can only be achieved when women and men begin to share this work, and employers begin to recognize that all workers have family responsibilities. Clearly any successful strategy must recognize that employment inequities are firmly embedded in the very structure and values of our society. So while we look for ways of reducing the effects of patriarchy in our families, schools, and other institutions, we also need to continue to focus directly on the organizational barriers that stand in the way of gender equality in the workplace.

DISCUSSION QUESTIONS

1. Discuss women's economic contributions in the early days of Canada's industrialization. To what extent did women's work differ from men's and why?
2. When you compare the work women have done historically and the work they do today, what has changed most dramatically? What has largely stayed the same?
3. How does women's labour force participation vary internationally? Nationally? What are some of the key factors that shape Canadian women's labour force participation?
4. Given that more women are now engaged in paid work, what types of changes have we seen in the "second shift" of domestic work, childcare, and eldercare?
5. What is work–family conflict and how does it affect working women and men? How might employers and governments assist in reducing work–family conflict?
6. What is gender segregation and why is it so important for understanding inequalities in work?
7. Over the past few decades, many more women have been moving into the professions of medicine, dentistry, and law. But fewer have been choosing engineering and science professions. Furthermore, relatively few men have been choosing traditionally "female" occupations and professions. Why might this be?
8. What are the recent trends in women's and men's earnings? How do we explain the wage gap?
9. Compare the processes and goals of employment equity and pay equity. How useful do you think these types of policies are for improving equity in the workplace?
10. Of the theories discussed in Chapter 4, which do you feel are more helpful for understanding gender inequalities at work?

NOTES

1. The female labour force participation rate is the number of women in the paid labour force (employed, or unemployed and actively looking for work) divided by the female population aged 15 years of age and older (14 and older, prior to 1975). This calculation undervalues women's economic contribution because it overlooks unpaid household work.

2. See Cohen (1988). Useful sources for women's history in Canada are Strong-Boag and Fellman (1997), and the labour studies journal *Labour/Le Travail.*

3. Danysk (1995: 150) examines the gendered construction of work roles in Western Canadian agriculture. The quotation is from a 1923 prairie farm publication. Rollings-Magnusson (2000) presents further useful information on the vital contribution of women to the development of the West.

4. For historical background on the transformation of women's work inside and outside the home during the rise of industrial capitalism, see Cohen (1988), Bradley (1989), and Armstrong and Armstrong (1994: Chapters 2–3).

5. On the family wage, see Humphries (1977), Land (1980), and Bradbury (1993: 80, and Chapters 3 and 5). For a related discussion of how the norm of the "male breadwinner" has recently changed in working-class steelworker families in Hamilton, Ontario, see Livingstone and Luxton (1996).

6. These factors are discussed in Armstrong and Armstrong (1994: Chapter 6) and Jones, Marsden, and Tepperman (1990: Chapter 12). For an overview of trends related to female employment see Statistics Canada (2000b, 2003e) and Cooke-Reynolds and Zukewich (2004).

7. On lone mothers in the workforce, see Crompton (1994). See Schellenberg and Ross (1997) for a detailed analysis of the working poor in Canada.

8. Rashid's (1994: 48) analysis of the top percentile of all families shows that wives in these families had incomes four times greater that the average for all wives, and that without this income 57 percent of these families would not reach the top income percentile. Statistics Canada's National Longitudinal Survey of Children and Youth revealed that, in 1994, 1 in 6 children under the age of 12 lived in lone-mother families, and that

these children face far greater disadvantages than those in two-parent families (*Canadian Social Trends* 1997: 7–9).

9. For differing perspectives on patriarchy, compare Johnson (1996), Walby (1990), Hartmann (1976), and Lerner (1986).

10. This post–World War II pattern is described in Ostry (1968). Also see Canada, Department of Labour (1958).

11. 2001 Census, Topic-based Tabulations – Income of Individuals, Families and Households (Cat. no. 97F0020XCB2001097); Families and Household Living Arrangements (Cat. no. 95F0312XCB2001004).

12. Hochschild (1989). See Harvey, Marshall, and Frederick (1991) for a Canadian time-budget study using the General Social Survey.

13. Meissner et al. (1975). For studies of the household division of labour, see Luxton (1980) and Pupo (1997).

14. Marshall (1993: 12). Data are from the 1990 General Social Survey (GSS). As a note of caution, there is some bias in how the GSS measures housework, because it is based on the perceptions of only one respondent per household (either male or female).

15. Coltrane (2000) and Doucet (2001) provide excellent reviews of research in this area. See also Davies and Carrier (1999); McFarlane, Beaujot, and Haddad (2000); and Looker and Thiessen (1999).

16. See Cranswick (2003) and Frederick and Fast (1999) for recent national trends.

17. For an overview on work–family literature, see Greenhaus and Parasuraman (1999). On the changing role of fatherhood, see Gerson (1993) and Coltrane (1996).

18. For a review of work and family policies in Canada, see Skrypnek and Fast (1996). For a detailed analysis of the U.S. situation, see Spain and Bianchi (1996).

19. For an analysis of family-related policies and programs in Canada and other countries see Jenson and Thompson (1999) and Mahon (2002).

20. See Feuchtwang (1982: 251) for a definition of *job ghetto*. Freedman (1976) discusses the concepts of labour market segments and shelters (see Chapter 3 on these concepts).

21. Hughes (1995, 2000b). Traditional occupations are those in which women constitute a greater proportion of employment than they do in the labour force as a whole.

22. Statistics Canada (2000b: 128). On the nursing profession, see Sedivy-Glasgow (1992) and Armstrong, Choiniere, and Day (1993). On men in non-traditional jobs, see Williams (1995, 1989).

23. Data derived from Statistics Canada 2001 Census, Topic-Based Tabulations, 97F0012XCB2001017.

24. On women in medicine in Canada, see Gorham (1994). Also see Brockman (2001) and Hagan and Kay (1995) for a detailed study of gender issues in Canadian lawyers' careers.

25. Data is from the 2004 *Annual Report—Employment Equity Act* produced by Human Resources and Skills Development Canada.

26. See National Council of Welfare (2004) *Poverty Profile 2001* at http://www.ncwcnbes.net.

27. "2004 Rankings: The ROB Survey on Compensation," *The Globe and Mail*, 4 May 2005. The total compensation packages of the top 50 ranged from $5.2 million to $173.1 million.

28. Using the National Graduate Survey, Davies, Mosher, and O'Grady (1996) found a rise in the female–male earnings ratio between 1978 and 1988 from 84 to 91 percent (or stated differently, a closing of the wage gap).

29. The literature on the wage gap is reviewed by Drolet (2002a). For a more sociological analysis, see Coverman (1988).

30. Canadian Committee on Women in Engineering (1992: 1). On the male culture of engineering see McIlwee and Robinson (1992: Chapter 6) and Miller (2004).

31. See Feldberg and Glenn (1979) and Acker (1990) for critical discussions of the male biases in the sociological study of work and organizations.

32. On feminist approaches to understanding gender and work see Armstrong and Armstrong (1994), Andrew et al. (2003), and Reskin and Padavic (2002).

33. On human capital explanations of sex differences in occupations and earnings, see Blau and Ferber (1986: Chapter 7).

34. For critical discussions of occupational gender segregation, see Siltanen (1994), Phillips and Phillips (1993: Chapter 4), and Armstrong and Armstrong (1990: Chapter 1). Hagan and Kay (1995) critically utilize human capital theory and a more general version of labour market segmentation theory in their study of gender inequalities in the legal profession. Developments in the measurement of occupational gender

segregation and the underlying factors are discussed in Rubery and Fagan (1995) and Blackburn, Jarman, and Siltanen (1993).

35. Unpublished data from the 1996 Alberta High School Graduate Survey. Virtually all females (97%) and 85 percent of males agreed or strongly agreed with the statement "a woman should have the same job opportunities as a man." Career choices and other study findings are presented in Lowe, Krahn, and Bowlby (1997).

36. For an interesting discussion of the significance of workplace conversations about household work, see Hessing (1991).

37. For a summary of related research, see Powell (1993).

38. The growing area of *men's studies* examines the social construction of masculinity in work and unemployment (Morgan 1997); on sexuality and power in organizational life, see Hearn et al. (1989). Coming at sexuality and work from another angle, Ross (2000) documents the history of erotic dancers—strippers—in North America.

39. See Johnson (1994: 11) for the 1993 survey, conducted by Statistics Canada. A survey reported by Crocker and Kalemba (1999) found, in contrast, that 56 percent of working women surveyed had experienced workplace sexual harassment in the year prior to the survey. See Sev'er (1999) and Bowes-Sperry and Tata (1999) for discussions of research evidence, theoretical interpretations, and measurement issues on this topic.

40. Leck (2002) discusses a number of studies documenting a narrowing of the wage gap.

41. Canada (1992). For discussions of employment equity see Abu-Laban and Gabriel (2002) and Bakan and Kobayashi (2000). On affirmative action in the United States, see Reskin (1998).

42. Resistance to equity policies by males and managers (often the same) and the organizational barriers to the elimination of discrimination are analyzed by Cockburn (1991). See Jo-Ann Wallace (1995) for an analysis of resistance to employment equity policy at the University of Alberta, and Creese (1999) for how gender, ethnicity, disability, and union politics combined to shape debates about employment equity at BC Hydro.

43. For an overview of pay equity legislation in federal, provincial, and territorial jurisdictions, see the final report of the Pay Equity Task Force (Canada 2004).

44. For discussions of the history and methodology of pay equity, see Armstrong and Cornish (1997) and the final report of the Pay Equity Task Force (Canada 2004). For a comparative perspective see Gunderson (1994).
45. Neale (1992). Under the former New Democratic government, Ontario extended pay equity into the private sector, but one evaluation of its impact suggests that pay equity awards have been small and that there are considerable difficulties administering and gaining compliance with the act (McDonald and Thornton 1998).
46. For critiques of pay equity and how it has been implemented, see Fudge and McDermott (1991) and Armstrong and Cornish (1997). Forrest (2000) discusses the large federal government pay equity settlement with female members of the Public Service Alliance of Canada in 1999. Simulations of the U.S. economy show that the wage gap would close 8 to 20 percent if comparable worth were nationally applied (Gunderson 1994: 111). Figart and Lapidus (1996) provide a positive evaluation, showing that such policies would generally contribute to reducing earnings inequality between the sexes and among women. However, other researchers suggest that the closing of the wage gap (at least in the United States) is in large part due to the increasing stagnation of white male earnings (Bernhardt, Morris, and Handcock 1995).

THE ORGANIZATION AND MANAGEMENT OF WORK

INTRODUCTION

We live in an organizational society. Attending school, buying weekly groceries, negotiating a student loan, volunteering at our community food bank, searching the Internet, or watching our favourite TV program all bring us into contact with formal organizations. Even more important, at least from our sociological perspective, are the organizations in which people work. In this chapter, we examine the organization and management of work, sampling selectively from an enormous and diverse literature on organizational analysis and management theory and practice. We consider different perceptions of work organizations. While a manager or executive may assess a particular work organization in terms of efficiency or profitability, other employees may use totally different criteria. Given these variations in the objective features and subjective experiences of work organizations, a number of interrelated questions arise.

First, what are the major factors that determine the structure—that is, the patterned regularities—of work organizations? Second, looking specifically at process, how are organizations transformed over time, and to what degree is organizational change a response to a changing external environment or to internal forces? Third, how much do the actions and beliefs of managers, employees, and other groups such as clients and customers influence the form, goals, and internal dynamics of an organization? Finally, our discussion raises a further question: how will the new organizational forms that emerged in the late 20th century continue to transform working life in the future?

Managers have played a key role in constructing modern organizations, mainly because they make critical decisions about goals, structure, personnel, and technology. We will therefore examine the various schools of management that have emerged over the past century. This chapter's focus on mainstream organization and management literature is reflected in the prominence given to the themes of *consensus* and *integration*. Then, in Chapter 6, we present a more critical perspective on organizations and management, focusing on the themes of power, control, resistance, and conflict. While some of the basic problems of organizing and managing work in a capitalist economy will be introduced here, a more systematic critique appears in the next chapter.

WHAT'S WRONG WITH BUREAUCRACY?

As we saw in Chapter 1, a major step along the road to industrialization was the division of craftwork into simpler components. Less skilled labourers could then perform each of these narrow tasks more cheaply. But once all the parts of a craftworker's job had been simplified and reassigned, coordinating and integrating these tasks became a problem. This helps to account for the role of managers, who, since the late 19th century, have become central actors in most work organizations. Questions about how best to integrate and coordinate the activities of large numbers of workers within a single enterprise gave rise to early theories of management.

Weber, Bureaucracy, and Capitalism

The organizational structure adopted by 19th-century businesses was the bureaucratic hierarchy. As we have already seen, Max Weber considered *bureaucracy* to be the organizational form best able to efficiently coordinate and integrate the multitude of specialized tasks conducted in a big factory or office. Without bureaucracy, he predicted, "capitalist production could not continue."[1] Although he described the general features of bureaucracy, Weber did not provide a practical guide for managers on how it should be organized. Nevertheless, his description has become the model for a highly mechanistic type of organization. Gareth Morgan defines bureaucratic organizations as those "that emphasize precision, speed, clarity, reliability, and efficiency achieved through the creation of a fixed division of tasks, hierarchical supervision, and detailed rules and regulations."[2]

No doubt bureaucracies were an improvement over the tradition-bound "seat of the pants" methods used in running most 19th-century businesses. Their predictability greatly increased the productivity of industrial capitalism. Furthermore, a bureaucracy is a system of authority. Its hierarchical structure, formal lines of authority, and impartial rules and regulations are designed to elicit cooperation and obedience from employees. One advantage of a hierarchy with an elaborate division of labour was that employers could gain employee commitment by offering the prospect of career advancement. This is the internal labour market described in Chapter 3. Given that a bureaucracy's specialized division of labour meets Adam Smith's conditions for increasing industrial productivity, noted in Chapter 1, it would seem that bureaucratic work organizations are ideal.

The Ills of Bureaucracy

All things considered, most sociologists would agree that bureaucracies have serious flaws. Organizational researchers have amassed considerable evidence showing that bureaucracies are often overly complex and, therefore, difficult to manage, resistant to change, and unable to cope with uncertainties. Working conditions in bureaucracies are frequently unsatisfying, as we will show in Chapter 8 (Daft 1995; Perrow 1986). A paradox of bureaucracy is that, far from achieving machine-like efficiency, it often unintentionally creates inefficiency, a problem Weber largely failed to see. Furthermore, sociologists have documented a dark side of organizations, where things can go seriously wrong (Vaughn 1999). Examples of such mistakes, mishaps, or misconduct are plentiful, and would include the *Challenger* space shuttle disaster, the collapse of Enron and WorldCom, the town of Walkerton's water contamination, and the fatal Westray mine explosion.

A number of classic sociological studies challenge the notions of bureaucratic rationality and efficiency. For example, Robert Merton described how employees who slavishly obey the rules could undermine the efficiency of a bureaucracy (Merton 1952). The rules become ends in themselves, rather than the means of achieving organizational goals. Officials acquire a *bureaucratic personality,* compulsively following procedural manuals to the last detail. Similarly, Peter Blau's fieldwork in American government agencies revealed how workers behaved according to their own unofficial rules.[3] In doing so, they were directly responding to organizational pressure to attain certain goals, such as handling a quota of clients or cases in a given period. Blau identified the tension between the official rules of the bureaucracy and workers' counter-rules. By striving to make their jobs easier, employees often erode bureaucratic efficiency.

Another example is Alvin Gouldner's (1954) classic study of an American gypsum plant. He argues that organizational life cannot be made as fully predictable as the bureaucratic model would suggest. Events such as promotions, layoffs, or dismissals are unpredictable. Moreover, rules reducing uncertainty for management may be a major source of discontent among employees. In fact, Gouldner argues, often the bureaucratization process is implemented not for reasons of greater efficiency, but as a result of power struggles between workers and managers. In short, there is nothing inevitable about bureaucracy; rather, its many faces reflect the shaky balance of power between workers and managers within an organization.

The notion of bureaucratic efficiency rests, in part, on the assumption that employees will readily submit to *managerial authority.* In Weber's view, workers accept the legitimacy of the existing authority structure, abide by the rules, and obey their bosses because they believe the basis for such authority is impartial and fair. Underlying capitalist bureaucracies is a *rational–legal value system.* However, the assumption that there is general acceptance of goals is contradicted by the realities of employee–employer relations, which are punctuated by conflict and resistance (see Chapter 6). People in positions of power set organizational goals.

These goals are, therefore, "rational" from management's perspective, but not necessarily from the perspective of workers. What is rational for them is what reflects their own interests, such as higher pay, a safer and more comfortable work environment, or more scope for making work-related decisions. These goals are often in conflict with those of management. Hence, conflicts over the distribution and use of decision-making power are normal.

Weber's model of bureaucracy neglects employee resistance to such authority. As we will also see below, resistance can be frequently detected in workers' informal group norms and codes of conduct that run counter to the official systems of production. Not surprisingly, a major concern of management has been to find ways of gaining worker cooperation. Managers have gone to great lengths to convince workers to accept organizational goals as their own personal goals, and to justify their decisions to those below them. The development of *management ideology* justifying superior rewards and the right to give orders is part of this process. Reinhard Bendix (1974: 13) explains: "All economic enterprises have in common a basic social relation between the employers who exercise authority and the workers who obey, and all ideologies of management have in common the effort to interpret the exercise of authority in favourable light."

Bureaucracy and Fordism

As an organizing principle, the bureaucratic hierarchy was integrated with technology to create the huge factories that formed the backbone of industrial capitalism. When Henry Ford introduced the moving assembly line in 1914, launching the era of mass-production manufacturing, he was building on the machine-like logic of bureaucratic organization. *Fordism*, based on assembly-line mass production, combined technology with bureaucracy to shape 20th-century work and economic growth.

Robert Reich (1991), in *The Work of Nations*, provides a critique of Fordism. Reich's treatise ranges over the sprawling terrain of the emerging global economy, which has shifted the locus of economic activity (as we documented in Chapter 1) from goods production to services. The key to prosperity used to be large volume production in giant factories by corporations that dominated world markets. In today's global marketplace the distinction between goods and services is blurred. Specialized expertise in finance, research and development, and marketing adds value to products and responds to constantly changing market needs. Furthermore, service and production networks tie together working units that are scattered across many countries.

Reich's portrayal of post–World War II corporate America is an indictment of Fordism and bureaucracy. It is also a recipe for how not to organize work in the new global economy:

> America's corporate bureaucracies were organized like military bureaucracies, for the efficient implementation of preconceived plans. It is perhaps no accident that the war veterans who manned the core American corporations of the 1950s accommodated so naturally to the militarylike hierarchies inside them. They were described in much the same terms as military hierarchies—featuring chains of command, spans of control, job classifications, divisions and division heads, and standard operating procedures to guide every decision. . . . As in the military, great emphasis was placed upon the maintenance of control—upon a superior's ability to inspire loyalty, discipline, and unquestioning obedience, and upon a subordinate's capacity to be so inspired. (Reich 1991: 51)

A crucial question, then, is whether alternatives to the Weberian model of bureaucracy have emerged. Consider three examples that reveal a shift away from bureaucracy. The first is the decline of internal labour markets in the 1990s, as large organizations downsized and increasingly relied on contingent workers who did not have access to internal careers. Research on managerial careers suggests that today fewer managers are likely to spend their entire career climbing the ladder of one organization than in the 1950s or 1960s— the heyday of the "organization man" (Osterman 1999). The second example revisits the Weberian emphasis on rational and impersonal norms of worker behaviour. Researchers have studied how some customer service organizations cultivate employees' use of emotions and informality. For example, Martin, Knopoff, and Beckman (1998) described how work relations in The Body Shop were defined by "bounded emotionality," encouraged by a corporate leader who emphasized the values of caring, sharing, and love. The third example is how work has become deinstitutionalized, moving outside the structure of a single large organization. One response to rapidly changing markets and globalization is the rise of networked forms of business organization, including franchises, joint ventures, outsourcing agreements and strategic alliances (Podolny and Page 1998; Hendry 1999). These decentralized and flexible forms of business organization raise anew the old questions about how coordination and control are achieved.

In short, the triumph of Western industrial capitalism was founded on what is now recognized as an outmoded model of organization and management. This is a leading theme in the new management literature we review below. However, Enron and other recent cases of corporate fraud have underscored the need for more ethical and accountable forms of governance—issues that gave rise to bureaucracy. These corporate scandals also raise questions about how otherwise upstanding individuals in their communities rationalize fraudulent behaviour and are socialized within the corporation to accept corrupt activities as normal (Anand, Ashforth, and Joshi 2004). As organizational researchers Samuel Culbert and Scott Schroeder (2003: 105) observe: "When it comes to the conduct of hierarchical relationships and how those with organizational authority are expected to direct and account, the 20th century concluded the way it began—with a steady stream of abuses, scandals, and exposés of hierarchy gone awry." So the debate on better ways to organize and manage work continues with renewed vigor.

THEORETICAL PERSPECTIVES ON ORGANIZATIONS

Organizational theory offers a range of perspectives from which to analyze the structure, functions, and dynamics of organizations. This section will touch on several major theoretical currents in this rich literature: contingency theory, organic models, sociotechnical systems, and strategic choice. Some of these ideas will resurface later in the chapter—for example, the sociotechnical model in our discussion of work redesign, and strategic choice in our discussion of managerial decision making. A newly emerging theoretical interpretation of the changes re-shaping organizational life comes from the application of postmodernist theories. As we will see, this critique of conventional organizational theory shares common elements with the much more practically oriented new management literature. We have already provided an overview of feminist perspectives on organizations in Chapter 4. We can expect that issues of gender, ethnicity, and race will receive increasing attention from researchers as workplaces continue to become more socially diverse (Reskin 2000; Elliot and

Smith 2004). The radical critique of organizations, itself a distinct perspective, will form part of our exploration of conflict, control, and power in Chapter 6.

Structures, Systems, Strategies

The weaknesses and gaps in the Weberian model of bureaucracy have led to alternative theories of organizations. Weber's thesis that a bureaucratic structure is most appropriate for any type of capitalist economic activity has been largely replaced by *contingency theory*. It proposes that organizational structures and processes are contingent upon the immediate problems posed by their environment. There is no single best way to organize or to manage work. A basic question guiding organizational research today is, What structures and strategies does a particular organization require to survive?[4]

The many possible answers to this question are typically based on an organic model, which views organizations as social systems adapting to internal and external changes. This view of organizations is grounded in the general sociological theory of *structural functionalism*. Hence, there is an emphasis on goal attainment, functions, structural adaptations, and a value consensus among organizational members. Conventional organization theory has been criticized for being too concerned with structure per se, and, as a result, losing sight of organizations' larger socioeconomic, political, and historical context. Other weaknesses include ignoring the attitudes and behaviours of employees and, as we will see in Chapter 6, glossing over the realities of workplace power and conflict.[5] Nonetheless, some useful insights about organizations can be found in this empirical literature.

For example, in the 1950s, Burns and Stalker studied synthetic fibre, electronics, and engineering industries, proposing a continuum of organizational forms from the *mechanistic* (or highly bureaucratic) to the *organic*. More open and flexible organizational structures and management styles were required, they concluded, in industries with rapidly changing technologies and market conditions (Burns and Stalker 1961). Joan Woodward's studies revealed a direct relationship between production technology, on the one hand, and structural forms and management styles on the other. She concluded that bureaucracy and tight management controls are more appropriate in industries with mass-production technology. When the workflow is varied, and production processes are non-routine and complex—as in craft or continuous process technologies—more flexible, organic structures function best (Woodward 1980).

Woodward's research attests to the importance of technology as a key variable in organizational analysis. The concept of a *sociotechnical system,* introduced by researchers at London's Tavistock Institute of Human Relations, takes this idea one step further. In their study of the mechanization of British coal mining, Trist and Bamforth documented how social relations in teams of miners were dependent on the technology used.[6] Replacing the traditional method of mining with mechanical conveyer belts reduced productivity by destroying the technical basis for the work teams and creating negative social–psychological consequences for workers. Consequently, optimizing the fit between social and technical aspects of production has become an important goal in work design. The information technology revolution continues to raise concerns about the impact of technological change on work organization, the quality of working life, workers' skills, and organizational performance (Liker, Haddad, and Karlin 1999; Burris 1998). Research on work in the information-based service industries reveals increasing complexity and diversity, especially among frontline workers dealing with customers. Indeed, the growing emphasis on consumers—such as in call centres, retail, or professional services—has made customer relations a new dimension of work organization, in addition to employee–manager and coworker relations (Korczynski et al. 2000).

There is a tendency in the literature on workplace structures to *reify* organizations, that is, to discuss them as if they had a life of their own, independent of the actions and decisions of their members. We should remember that organization theory is largely about decision making. Given the constraints of technology, markets, government regulations, labour negotiations, and so on, what are the most appropriate choices for managers to make? This issue is taken up by the *strategic choice perspective* on organizations.[7] In contrast to contingency theory's emphasis on environmental and other constraints imposed on decision making, the concept of strategic choice draws our attention to the ways in which dominant groups (coalitions) actually choose among various strategies for structuring an organization, manipulating its environment, and establishing work standards. To fully comprehend the nature of work organizations, we must examine the guiding theories, actual practices, and supporting ideologies of management.

Despite the persistence of these established approaches, organizational theory is evolving. As Chapter 4 documented, feminist researchers have effectively challenged the one-sided male perspective traditionally found in organizational literature, illuminating how workers' experiences, opportunities, and rewards in workplaces are shaped by gender.[8] The latest attempt to recast it comes from *postmodernism*. This emergent perspective on organizations touches on the major trends re-shaping the workplace, thereby contributing to the debate about the nature of postindustrial society. According to this critique, organization theory was the product of *modernity*—the age of the bureaucracy and the assembly line (Clegg 1990; McKinlay and Starkey 1998). The driving force was rationality and an increasing division of labour, and everything that flowed from these principles. However, in the postmodern age, the principles of uncertainty, flexibility, and complexity shape organizational life.

Change and Innovation

There is a vast management literature on strategic planning. While reminiscent of strategic choice theory, mentioned above, strategic planning has a more practical thrust, helping managers achieve specific goals. Still, it raises sociological questions about how managers are able to influence an organization's performance and shape its future. Certainly the new management literature assumes that conscious decisions and strong leadership form the basis of survival strategies. But as Henry Mintzberg (1994) argues, managers have not learned from the failures of strategic planning. The problem with strategic planning, according to Mintzberg, has been its formalization of the planning process, which stifles strategic thinking about creative alternatives and visions for the future of the organization. This is yet another example of how bureaucratic procedures can inhibit innovative behaviour in organizations.

A basic definition of strategy is "a plan for interacting with the competitive environment to achieve organizational goals" (Daft 1995: 45–49). This is vague, so to add some precision, Mintzberg and his colleagues (Mintzberg, Ahlstrand, and Lampel 1998) identify 10 distinct schools of strategy, each with its own definition of strategy formation in organizations, how this happens, and who the key actors are.

For example, one perspective views strategy formation as an emergent process, in which organizations learn from their strengths, continuously adapting them to the future. This describes many high tech firms that are scrambling to make order out of a chaotic environment and therefore cannot have a fixed master plan (Brown and Eisenhardt 1998). Another school focuses on power. From this perspective, some sociologists argue that the core issue is whether organizations really act according to strategic plans, or whether these plans are post hoc explanations by social scientists trying to make sense of organizational changes and power structures (Knights and Morgan 1990). Strategy becomes primarily a mechanism of power. To understand it, we must uncover how corporate strategies are rooted in power relations and serve to reproduce existing organizational hierarchies and inequities, but nonetheless are subject to ongoing negotiations with interest groups inside the organization. While it is difficult to know whether a particular strategy actually makes a difference, it could have the implicit effect of making managers and employees more goal-oriented and self-disciplined (Knights and Morgan 1990).

More broadly, an organization's strategy provides a description of how it deals with internal and external change. Increasingly, the focus in management literature is on how to measure the effectiveness of a strategy. This shifts the emphasis from the process of developing a strategy to the actions taken to implement it. This is especially critical in decentralized business organizations that operate in many locations around the world. For example, when a new management team at Mobile Oil's North American division rolled out a "customer-focused" strategy, they introduced a "balanced scorecard" to measure progress toward financial and non-financial goals (Kaplan and Norton 2001). Balanced scorecards and similar performance measures have been expanded to include a company's goals for the environment, the community in which it operates, and its employees.

Understanding Organizational Change

The difficulties of strategic planning reflect the larger challenges of organizational change. How do organizations change? Observing that organizational turmoil became the norm in the 1980s, Bob Hinings and Royston Greenwood (1988) argue that it is essential to recognize and manage change. Their research on local governments in Britain focuses on "strategic organizational design change," which is change that signals a fundamental shift in the orientation of the organization. Their study emphasizes that change and stability are two sides of the same coin. Theoretically, Hinings and Greenwood introduce the concept of a *design archetype* to describe an organization's structure, management system, and supporting beliefs and values. They also suggest that organizations can be classified according to *tracks*. The metaphor of a track captures the extent of change occurring, specifying whether an organization remains within one design archetype over time or moves between archetypes.

Hinings and Greenwood's theory is intended to account for both successful and unsuccessful attempts at organizational change. Because the late 20th century saw an increase in organizational decline, this process has become the subject of considerable research. Distinguishing between decline, on the one hand, and cutbacks and retrenchment on the other, William Weitzel and Ellen Jonsson (1989: 94) suggest that the root cause of decline is the failure of organizations "to anticipate, recognize, avoid, neutralize, or adapt to external or internal pressures that threaten the organization's long-term survival." Whether it is avoiding decline or pursuing growth and innovation, these issues reflect two strong themes in contemporary organizational and management literature: how change can be "managed," and the conditions under which change leads to innovation in products and services (Hage 1999; Tushman and Anderson 2004). One basic research finding is that change is disruptive for organizations. The challenge this presents for firms in volatile sectors, such as high tech, is that new organizational models designed to adapt to rapidly evolving markets create employee turnover because of how they alter employment relationships. This in turn can undermine future firm performance (Baron and Hannan 2001).

Technology has, historically, precipitated huge changes in the organization of work, as we saw in Chapter 1. To more fully understand technological change, it is useful to view it as a process of managerial decision making. This helps to demystify technology by showing how it, too, reflects organizational and human constraints and choices. John Child (1985) identifies four goals that managers have when they introduce new technology: reducing costs and boosting efficiency; increasing flexibility; improving quality; and achieving greater control over operations. These goals, which often overlap, are "evolving" and the end results may not be predictable. They also depend on the type of technology, its application, and the nature of the business. For example, customer service jobs in call centres span a wide continuum, from the skilled and rewarding in financial services, to the unskilled, routinized, and low-paid in telemarketing (Buchanan 2002; Kleemann and Matuschek 2002).

Various case studies underscore major organizational and human constraints on management's ability to plan and implement technological change. The complexity of new technology invariably requires accompanying organizational change and job redesign. Moreover, the full potential of automation cannot be realized without recognition of the human dimension. Only by involving workers in planning and implementing new technology, and giving them the responsibility to operate it, can a smooth transition be attained. For instance, often workers' detailed knowledge of old systems is indispensable to making new technology function (Hayes and Jaikumar 1988; Forester 1989; Wood 1989a). Shoshana Zuboff's (1988: 389) conclusion, drawn from her study of several high tech firms, aptly expresses the contingent and often unpredictable quality of technological change:

> Even where control or deskilling has been the intent of managerial choices with respect to new information technology, managers themselves are also captive to a wide range of impulses and pressures. Only rarely is there a grand design concocted by an elite group ruthlessly grinding its way toward the fulfillment of some special and secret plan. Instead, there is a concentration of forces and consequences, which in turn develop their own momentum.

THE ROLE OF MANAGERS

We have alluded to the key role managers have in organizing and regulating work, but exactly what do managers do? We know that within any hierarchical organization authority resides at the top. This gives rise to one of the most vexing problems confronting management: how to obtain employee compliance and prevent opposition to authority. In this respect, the workplace is a microcosm of the larger society. Maintaining orderly and harmonious social relations among people who are not equals has always been a problem for those in power. Another pressing concern involves motivating workers to achieve the quantity and quality of output considered necessary by management. Employers do not have complete control over the amount of effort expended by employees. Hence, conflict often erupts over different perceptions of the right amount of effort given a specific wage and work situation.

Management Ideology and Practice

Managers became a prominent new social group in the early 20th century. *Cost accounting* techniques for calculating how much each factor of production, including labour, contributes to profits was one way capitalists initially tackled the problems of running increasingly large and complex enterprises. Appointing trained managers, often factory engineers, was an equally important social innovation. As business historian Alfred D. Chandler Jr. suggests, the visible hand of the corporate manager replaced the "invisible hand" of market forces (Chandler 1977). Corporate boards of directors, representing the shareholders, delegated to managers the authority to operate the business profitably. Recall from Chapter 1 that this growing power of managers and the separation of corporate ownership from daily control functions has been labelled the *managerial revolution.* In striving to meet their broad mandate, managers undoubtedly have transformed the organization and control of work.

In considering the evolving role of managers, it is important to distinguish among *ideology, theory,* and *practice.*[9] Management ideologies, we noted earlier, are used to justify existing authority relations in organizations. Joseph Smucker's research shows how Canadian managerial ideologies have undergone several shifts since 1900. The self-perception of managers moved away from that of autonomous achieving individuals to one of organizational team workers, while managers' image of workers was transformed from recalcitrants to associates (Smucker 1980: 163). The economic disruptions of the late 20th century gave rise to an even more individualized ideology that encourages workers (including managers) to be ceaselessly flexible and adaptable. As one researcher points out, self-help business books with titles like *Who Moved My Cheese?* and *Swim with the Sharks without Being Eaten Alive* have generated a new Darwinian discourse in North American corporate culture (Brown 2003). The downside for employers is that encouraging employees to be lifelong learners, upgrading their skills and designing their own careers, serves to weaken employment relationships and make commitment and loyalty—which are becoming important in the 21st-century competitive labour market—more difficult to achieve.

Henry Mintzberg's distinction between "folklore" and fact is a useful way of deciphering what managers actually do. The words that supposedly guide managers are, he argues, often little more than vague goals: "If you ask managers what they do they will most likely tell you that they plan, organize, coordinate, and control. Then watch what they do. Don't be surprised if you can't relate what you see to those four words" (Mintzberg 1989: 9). This underscores the importance of looking closely at how management theories are actually applied and at the results—especially at a time when there are so many management fads. Indeed, management researchers are now seeking to understand the factors shaping the discourse of these new management approaches, such as Total Quality Management (discussed below). Often, the discourse leads to overly optimistic assessments of their impact and a gap between rhetoric and practice (Zbaracki 1998; Pollitt 1995). The resulting contradictions can undermine the credibility of corporate leaders in the eyes of employees.

Another interpretation of managers views their activities as directed at defining and pursuing common problems, goals, and solutions, interpreting their actions as responses to situational constraints (Godard 1994: Chapter 6). This view helps to demystify management, showing how a group possessing formal authority still faces the limits that institutions place on human agency. For example, today's managers and supervisors increasingly are expected to be "coaches" and "mentors" to their staff. However, when these new roles are added to managers' already heavy workloads, time constraints become barriers to follow-through. Seeing management as a socially-constructed practice avoids the trap of reifying organizations. This *social action perspective* on management also raises one of the thorniest problems in studying organizations: how to conceptualize individuals within organizations. As Charles Perrow (1986: 66) puts it, "How can you talk about organizations without talking about individuals?" In response, it should be said that much of organization theory does just that.

Managers as Decision Makers and Leaders

This brings us to the topic of *decision making,* which has received much attention in writings on management. Perrow shows how one of the classic statements about the role of managers, Chester Barnard's 1938 book, *The Functions of the Executive,* falls into the trap of presenting a "nonpersonal" model of organizations (Barnard 1938). For Barnard, even the decisions made by executives are not personal choices (that is, individually determined) because they are embedded in a coordinated system. The logic and rationality often attributed to managers, for him, are actually imposed in a disciplined way by the organizational environment. Barnard's influential contribution to subsequent thinking about organizations is basically flawed, argues Perrow, because it "glorifies the organization and minimizes the person" (Perrow 1986). Organizational policies take on a moral tone, and are assumed to be functional for everyone involved, from executives to workers and customers.

Research on decision making examines the role of the manager and whether she or he, as an individual, can make a difference in the organization (Hickson 1987). What seems to matter more than the individual manager's role, however, is the process by which decisions are made, especially how other organizational actors are involved in this process (Rodrigues and Hickson 1995). Media reports of larger-than-life corporate empire builders or turnaround artists suggest that successful business leaders are a rare breed. This image of managers as decisive visionaries—what leadership expert Warren Bennis (1999) terms the myth of the triumphant individual—is challenged by researchers who argue that leaders' roles are constrained by the specific context of their organizations and their wider environment. In fact, there are some who debate whether leadership skills that can truly change organizations can be taught or are innate characteristics (Micklethwait and Wooldridge 1996: Chapter 8; Yammarino, Dansereau, and Kennedy 2001). The underlying sociological issue is how leadership practices are embedded in workplace relationships, group dynamics, and the overall authority structure of an organization. So we need to look beyond the CEO and executive team to understand the conditions that enable people at all levels to take initiative in pursuit of an organization's goals.

A related question is the relationship between leadership and an organization's performance. Studies show that corporate executives do affect their firm's performance, but that this effect is overshadowed by other factors (for instance, industrial sector, markets, technology, and size) that differentiate firms (Thomas 1988). Now the issue of executive compensation also is viewed through the lenses of ethics (what is fair and reasonable) and accountability, to ensure that executives answer to shareholders, customers, and other "stakeholders" for their actions (Nichols and Subramaniam 2001). Certainly the most widely read argument that leaders directly affect firm performance is Jim Collins's (2001) book, *Good to Great*. As the title suggests, some companies make the leap to greatness and Collins traces this to what he calls "level 5 leadership": having a CEO "who blends extreme personal humility with intense professional will" (Collins 2001: 21) to do what's required to create an outstandingly successful company.

Collins challenges conventional thinking about ego-driven CEOs. So, too, do other perspectives on leadership that emphasize social and psychological attributes. Using the example of former U.S. president John F. Kennedy, one scholar suggests that leaders need to be able to "develop fresh approaches to long-standing problems and open issues to new options," accomplishing this mainly by rallying people around a compelling vision of future possibilities (Zaleznik 2004: 77). Such inspiration requires the kind of creativity typically found in artists, not managers, because the latter are bound by structures and processes of the organization. Other currently influential perspectives on leadership suggest that excellent leaders have a high "EQ"—emotional intelligence quotient. Based on research by Daniel Goleman (2004) in hundreds of global corporations, the concept of emotional intelligence goes beyond IQ and technical skills to emphasize the critical importance of "soft" skills. EQ is based on three self-management skills (self-awareness, self-regulation or control over feelings, and motivation) and two skills related to managing relationships with others (empathy and social skill). Goleman's research shows that emotional intelligence can, with time and perseverance, be learned. Yet whether it is called EQ or something else, leadership often involves very basic activities like listening and informal chatting, which become extraordinary because of the positions of the people who engage in these behaviours (Alvesson and Sveningsson 2003).

An overriding management objective is to create motivated and cooperative employees, so this is a key criterion for assessing a manager's effectiveness, as well as organizational success. On this latter point virtually all theories of management agree. But as we now will show, various management theories have advocated different methods for achieving organizational goals. During the past century, several distinct schools of management appeared. The two most influential were scientific management and the human relations approach.

SCIENTIFIC MANAGEMENT

Charlie Chaplin's classic movie *Modern Times* humorously depicts the impact of scientific management on working conditions. The comedian plays a harried factory worker whose job has been analyzed and redesigned by stopwatch-wielding efficiency experts. Every few seconds, Chaplin tightens a nut as another identical piece of equipment zips past him on the assembly line. Chaplin is little more than an automaton whose actions are programmed by the production system. Once out on the street, he continues repeating the motions on anything that fits his two wrenches.

Taylorism

Scientific management began in the United States as a set of production methods, tools, and organizational systems designed to increase the efficiency of factory production. The term itself was coined in 1911. But this new approach to factory management and organization, popularized by an engineer named Frederick W. Taylor, had been developed by the end of the previous century. Taylor and other "efficiency experts" extolled the virtues of scientific management, and the method soon spread across North America and, to a lesser extent, Britain and Europe. Taylor's theories were at the cutting edge of a "thrust for efficiency" that contributed to the rise of 20th-century industrial capitalism (Palmer 1975). These early management consultants advocated workplace reorganization, job redesign, and tighter administrative and employee controls, all in the name of efficiency and profits.

Taylor laid out the following steps for rationalizing the labour process: (1) shift the decision-making responsibility for doing a job from workers to management; (2) use scientific methods to determine the most efficient way of executing a job and redesign it accordingly; (3) provide a detailed description of how to perform each step in a job; (4) select the best worker to perform the job; (5) train workers to execute the job efficiently; and (6) closely monitor workers' performance. Taylor believed his management techniques benefited all parties involved. Yet their overriding effect was to give management tighter control over workers' activities by making all major work decisions (Bendix 1974; Haber 1964; Nelson 1980). Critics view Taylorism as the cornerstone of all 20th-century management (Braverman 1974; Littler 1982). As we will elaborate in Chapter 6, according to the *labour process perspective,* Taylorism degraded labour, minutely fragmenting tasks, reducing skill requirements, and eliminating workers' input about how their jobs should be done. While many have contested Braverman's claim about the pervasive influence of Taylorism, there can be little doubt that the basic principles of scientific management continue to influence the amount of control and discretion workers can exercise when performing their jobs.

Taylor was convinced that worker *soldiering,* or deliberate laziness, was the scourge of industry. He believed that workers consciously restricted production by keeping bosses ignorant of how fast a job could be done. His solution was to determine "scientifically" the one best way of performing a job through *time and motion studies* of each step. A base rate of pay was then tied to a production quota. If workers exceeded the quota, they received a pay bonus. Lazy workers unable to achieve the quota would be forced to quit because their base rate fell below a minimum level. Essentially, scientific management was founded on the assumption that workers were motivated by economic gain alone. Taylor preached that the scientific basis of a "fair day's wage" and the productivity gains through more efficient work methods would bring about a new era of industrial cooperation and harmony.

Taylor's view of human nature was coloured by his preoccupation with technical efficiency. In his mind, the ideal worker was more like a machine than a human being. Taylor was a leading ideologue for early 20th-century management, articulating its deep concerns about the "labour problem." Industrial cooperation would only replace class conflict, he predicted, once a complete mental revolution had taken hold of both management and labour. As Taylor stated in *Industrial Canada* (April 1913: 1224–25), a prominent Canadian business magazine just before World War I:

> The new outlook that comes to both sides under scientific management is that both sides very soon realize that if they stop pulling apart and both push together as hard as possible in the same direction, they can make that surplus [i.e., profits] so large that there is no occasion for any quarrel over its division. Labour gets an immense increase in wages, and still leaves a large share for capital.

Scientific management failed to provide a successful formula for labour–management cooperation. Taylor's package of managerial reforms was seldom adopted completely. Yet various aspects of scientific management were quickly introduced into Canadian factories and, by the 1920s, were used to overhaul large corporate and government offices.[10] Still, Taylor's rhetoric of scientific objectivity was an ideological justification for greater management control over labour. Furthermore, careful analysis of Taylor's original documentation shows inaccuracies in his published claims about efficiency gains at the Bethlehem Iron Company, where in 1899 he redesigned the work methods and equipment for shovelling heavy chunks of metal called "pig iron" (Wrege and Hodgetts 2000).

The Legacy of Scientific Management

Regardless, scientific management's emphasis on detailed job descriptions, planned workflows, and time and motion studies continues to influence job design and work organization. For instance, in the service sector, Burger King attempts to maximize food sales and minimize labour costs by emphasizing SOS—"speed of service" (Reiter 1991: 85). Complying with this motto requires highly standardized work procedures and strict management controls in all Burger King outlets. Time and motion studies dictate how long it takes to prepare the fries, burgers, and drinks, exactly how workers should do these tasks, and where they should be positioned depending on the design of the kitchen. Building cleaning is another type of service work that has been reorganized along scientific management lines. A study of the building cleaning industry in Toronto found a shift away from a semi-autonomous and multi-skilled "zone cleaning" approach to "mono-task" assignments (Aguiar 2001). The latter resulted in simplified and repetitive tasks assigned to "cleaning specialists" who did only mopping, dusting, restrooms, or other specific tasks. While these changes were intended to improve quality and customer service, software was used to analyze building cleaning needs and design efficient ways to do more cleaning with fewer workers. New technology, such as "backpack" vacuums, contributed to these efficiencies. Taylorism also is very much alive in manufacturing. One recent case study of an appliance factory in Wales documented the persistence of time study, piece-rate payment, and efficiency gains through reduced operator discretion, despite efforts by senior management to change these practices (O. Jones 2000).

History provides many examples of workers resisting scientific management. Skilled industrial workers besieged by managerial rationalizations in the early 20th century responded by striking (Heron and Palmer 1977; Kealey 1986). Autoworkers were one occupational group that experienced a barrage of scientific management, along with extensive technological change. Chapter 8 will comment further on how Fordism combined scientific management techniques with mass-production, assembly-line technology to create some of the most alienating and stressful working conditions in modern industrial society.[11]

In many ways, Taylorism represented the practical application of bureaucratic principles to a manufacturing setting. Henry Ford's moving assembly line used technology to develop the logic inherent in scientific management and bureaucracy. The resulting increased monotony and speed of production sparked a huge increase in employee turnover. Only by doubling wages was Ford able to induce workers to accept the new production methods. Worker opposition to Fordism is never far from the surface. When General Motors opened a Vega plant in the early 1970s at Lordstown, Ohio, the relentless pace of the line, coupled with more restrictive management, led to massive labour unrest and sabotage (Aronowitz 1973: Chapter 2). This iconic example of worker resistance has faded into history, however, and in the 1990s management at the Lordstown plant secured workers' consent to reduce the workforce from 12,000 to 3,000 through outsourcing, in the interests of keeping the factory competitive (Sallaz 2004). As described in Chapter 6, *lean production,* the latest Japanese-inspired management system in North American auto factories, has retained elements of Fordism.

In short, major human costs often accompany increased efficiency, productivity, and profits. Other schools of management have attempted to counteract the harshness that resulted from Taylorism and Fordism by developing more humane working conditions. As one organizational researcher aptly concludes, scientific management principles "make superb sense for organizing production when robots rather than human beings are the main productive force, when organizations can truly become machines" (Morgan 1997: 26).

THE HUMAN RELATIONS MOVEMENT

Bureaucracy, scientific management, and mass-production technologies transformed work in the 20th century. Many jobs became routinized and monotonous, stripped of opportunities for workers to use their minds or develop their skills and abilities. Employee dissatisfaction, often in the form of high turnover and absenteeism rates or industrial unrest, threatened to undermine the machine-like efficiency of the new industrial system. Taylor, Henry Ford, and a host of "efficiency experts" sought to redesign production systems so that control would be firmly in the hands of management. Technical efficiency was paramount.

Toward Normative Control: Workers as Human Beings

However, gaining the cooperation of workers within an increasingly bureaucratized, mechanized, and regimented labour process remained difficult. Some employers responded with programs, broadly known as *corporate welfare* or *industrial betterment,* which emphasized the need to treat workers as human beings. Popular in leading North American firms by the 1920s, corporate welfare programs tried to reduce the alienating effects of bureaucracy and worker dissatisfaction with routinized tasks. The goal was a loyal and productive workforce; the means were healthier work settings and improved job benefits. Recreation facilities, cafeterias, cleaner and more pleasant work environments, coherent personnel policies, medical care, and pensions are major examples of corporate welfare efforts.

Taylor and other efficiency experts were quick to dismiss these schemes as a waste of money. Yet many firms committed to scientific management also used corporate welfare measures to gain greater cooperation from staff (Jacoby 1985). More than anything, the corporate welfare movement showed that the principles of bureaucracy and scientific management failed to address the key ingredient in modern industry—human beings. Not until the *human relations school of management* began to systematically examine some of the same concerns in the 1930s did the scientific management model face a serious challenge.

Control within organizations depends upon rank-and-file employees complying with management directives. Such *compliance* can be achieved in three different ways: coercive management techniques that rely on penalties and harsh discipline; utilitarian methods by which employees are motivated by economic self-interest; and normative approaches that assume that workers equate their own interests with organizational goals, thus becoming motivated to work hard (Etzioni 1975). Scientific management combined coercive and utilitarian methods, with mixed results at best. The normative approach cultivates a community of interests throughout the organization. Workers are more likely to identify with management goals, pursuing them as their own. A precondition for this is a relationship of mutual trust between employees and managers (Heisig and Littek 1995). Trust means having positive expectations about another person's or group's conduct in a relationship, or put differently, believing that they are concerned about your best interests (Lewicki, McAllister, and Bies 1998; Lowe and Schellenberg 2001). In short, a normative approach to employee relations sharply contrasts with scientific management in its assumptions regarding human nature and motivation.

Theoretically, then, the big happy corporate family of the human relations school replaced the carrot of incentive wages and the stick of harsh discipline.

In practice, however, scientific management and human relations often operate side by side. And beneath the changing workplace structures is a layer of fragile social relationships that influence individuals' work experiences and organizational success.

The Hawthorne Studies

The human relations school of management originated in the *Hawthorne Studies*, a program of research conducted by Harvard Business School researchers between 1927 and 1932 at Western Electric's Hawthorne Works on the outskirts of Chicago.[12] Western Electric management was initially concerned with the effects of fatigue and monotony on production levels. These were central concerns of industrial psychologists at the time. Various studies examining the effect of rest pauses, hours of work, and lighting levels on productivity led researchers to unexpected findings.

In the Relay Assembly Test Room study, workers were placed in two separate rooms. Researchers then recorded production while varying light intensity for one group but not the other. To their surprise, productivity in both groups increased, regardless of lighting level. Only when light intensity was reduced to that of bright moonlight did productivity decline. Several variations on this study came up with the same puzzling findings. Searching for possible explanations, the researchers speculated that a fundamental change had occurred in the workplace. Involving workers in the study had the unintended effect of raising their morale. They now felt that management cared about them as individuals. Productivity, concluded the researchers, increased as a result. This is the famous *Hawthorne effect,* in which people's awareness of being part of a study may itself confound the results. A basic axiom of human relations management had been discovered: the humane treatment of employees and the creation of an *esprit de corps* improves their motivation to cooperate and be productive.

Later, interviews with employees revealed that work groups were governed by informal behavioural codes (Wilensky and Wilensky 1951). The Bank Wiring Observation Room study further probed work-group behaviour. For 7 months, 14 employees were observed as they wired telephone switching banks. Researchers documented how *informal group norms* replaced formal directives from management. The work team set its own production quotas, making sure that no member worked too quickly or too slowly. Infractions of the group's rules, such as reporting violations of company policy to management, were punished. Generally, strong pressures to conform to the group norms prevailed (Homans 1950: Chapter 3). This exposed a hidden side of the workplace, where workers consciously engage in practices to oppose or subvert established authority. After the Hawthorne studies, workplaces were viewed as social systems rather than in purely technical terms.

Cooperation and Conflict in Human Relations Theory

Human relations theory emphasizes how workers' attitudes, values, emotions, psychological needs, and interpersonal relationships shape their work behaviour. This approach to management seeks the best match between the worker, given her or his personal background and psychological makeup, and the job. Careful recruitment and effective training of employees, as well as good supervision and communications, are therefore essential.

The human relations perspective rejects the utilitarian assumption of scientific management, assuming that people want to cooperate. Yet Elton Mayo and other human relations theorists claimed that workers are unaware of their cooperative instincts, acting instead on the basis of irrational sentiments or beliefs. For Mayo, the survival of society depended on cooperation, so he advocated a new industrial order run by an administrative elite. The leadership of this elite would encourage the development of work environments that would bring out the cooperative instincts and productive potential of employees (Mayo 1945).

Mayo and other human relations advocates consider workers unable to act in their own best interests. Less authoritarian leadership, careful selection and training of personnel, an emphasis on human needs, and job satisfaction shaped the human relations approach. Yet these management tactics rest on contradictory views of the workplace (Burawoy 1979: 7). Human relations theory assumes that industrial harmony is healthy and that conflict is destructive, which leaves no place for unions. However, human relations theory also argues that workers must be closely regulated by management in the greater interests of cooperation and harmony. These inconsistencies have led some critics to label human relations theory as an elaborate justification for management's manipulation of workers (Rinehart 2006).

Many strands of management thinking make up the human relations tradition. One durable strand, which draws on psychologist Abraham Maslow's concept of a hierarchy of human needs, was developed by Douglas McGregor in the 1950s and 1960s. McGregor (1960) saw in bureaucracy and Taylorism a mechanistic approach to managing and motivating people that stifles human potential by relying mainly on economic incentives. He called this Theory X, and contrasted it with Theory Y, which views a manager's main task as creating an environment that aligns workers' personal needs for growth, development, and participation with an organization's goals. McGregor reminded managers that Theory X and Y are not strategies, but sets of assumptions about human nature. While an influential idea, the humanistic Theory Y did little to influence day-to-day management practice when it was published 40 years ago. Today, McGregor's ideas are once again very much in vogue, as shown in our review below of current approaches to human resource management (Heil, Bennis, and Stephens 2000). Most human resource managers want to treat workers as human beings. Yet the structures and cultures of most large organizations still pose major barriers to putting these ideas into practice. In this sense, McGregor's thinking remains futuristic.

ORGANIZATIONAL CULTURE

The Hawthorne studies document how workers construct their own culture in the workplace, replacing formal rules and norms with informal ones. Consequently, the potential for management–worker conflict is increased, since the official goals of management and the unofficial, personal goals of workers and work groups are not the same. This issue is examined in Chapter 6. Yet informal workplace relations and organizational culture also provide management a means to increase employee conformity and integration. This apparent paradox between workplace social relations and culture as means for employee resistance on the one hand, and as the basis for a value consensus on the other, is rooted in the two divergent theoretical traditions we first identified in Chapter 1. The former, which derives from Marx's view of society, emphasizes the dynamics of conflict. The latter draws on the work of French sociologist Émile Durkheim and his concerns about achieving social stability (Ray 1986; Ouchi 1981: Chapter 6).

The Informal Side of Organizations

By focusing only on the formal structure of organizations and the role of managers, we overlook vital features of life in work organizations. Less visible but equally important in the daily operations of an organization is its *informal* side, where employees reinterpret, resist, or adapt to work structures and management directives (Blau and Scott 1963). Thus, to get a complete picture of a work organization, we must examine it from the vantage point of employees' informal practices and social relations. The organizational charts found in corporate annual reports present management's image of how things ought to operate. Employees further down the hierarchy often see things differently, and act accordingly. Despite management's efforts to regiment work procedures, employees often respond in ways that do not fit these plans. Total control of people by bureaucratic structures, management discipline, and technology is difficult to imagine. Furthermore, there is an underlying tension between the cooperation necessary for any organization to function and the internal competition among individuals and groups over the distribution of power, rewards, and other scarce resources.

Since the Hawthorne studies, the *informal work group* has been under the microscope of social scientists. Donald Roy's (1952) participant observer study of a Chicago machine shop graphically portrays work-group dynamics and counter-systems of production. Roy worked as a machinist in a shop that ran on a *piecework payment system*. Productivity was regulated by management through bonus payments for output that was above established quotas. Machine operators constantly battled over the base wage with time-study personnel trained in scientific management. The machinists invented ingenious shortcuts to maximize their wages while minimizing effort. Their informal system of production involved what in shop-floor jargon was called "gold-bricking" on "stinker jobs." That is, on difficult jobs for which they could not possibly earn a bonus, workers relaxed and enjoyed some free time. On easy jobs, however, they could "make out" by exceeding the quota to receive a bonus. Any production beyond an unofficial quota was stored up in a "kitty" to be used to boost output figures on a slow day. Management rarely challenged the workers' system because production usually fluctuated within an acceptable range. Ironically, the workers had designed a more predictable and efficient production process.

Managers, too, participate in similar manipulations of informal work practices. With limited authority over workers, many middle- and lower-level managers and supervisors feel constrained in their roles. Consequently, they may not entirely follow the directives of top executives. Melville Dalton's classic study, *Men Who Manage* (1959), illustrates how managers bend rules and short-circuit bureaucracy in order to achieve their objectives. Especially fascinating is his analysis of how official and unofficial rewards often complement each other, giving management a wider repertoire of means to achieve a particular goal. In practice, unofficial systems guided actions in these bureaucracies. Obtaining a promotion depended less on merit than it did on social characteristics such as, for example, being a member of the Masonic Order or the local yacht club. Employees at all organizational levels regularly engaged in what, to the outside observer, was dishonest activity. The company's resources were dispensed as personal rewards in order to keep the bureaucracy running, to acknowledge someone's special services to the firm, or to solidify social relationships in order to get a job done. Workers borrowed tools and equipment; managers had home improvements done at company expense. According to Dalton, this sort of "freewheeling" frequently occurs just beneath the thin veneer of bureaucracy.

Defining Culture

Based on insights like these about informal group relations, the concept of *organizational culture* has become a prominent theme in literature on management and organizations (Trice and Beyer 1993; Schein 2004). Social-anthropological methods are useful in studying organizational culture. Viewed from the perspective of their members, organizations become mini-societies. The shared beliefs, customs, rituals, languages, and myths serve as "social glue," binding together the diverse elements of the organization. *Culture,* then, refers to a system of shared meanings about how organizational life ought to be conducted. It can express how things really get done at an informal level. Culture focuses our attention on the processes by which employees actually carry out and collectively interpret work activities (Morgan 1997: Chapter 5).

Shaping the *dominant culture* of an organization—that is, encouraging employees to identify strongly with the goals of the corporation—has become a management preoccupation. Viewed from this perspective, the norms and values of employee work groups become *countercultures* challenging the dominant organizational values. Corporations with strong internal cultures, as evident in Toyota's commitment to quality or The Body Shop's culture of social responsibility, send a clear signal to employees and customers. A strong corporate culture can create a self-selection process, whereby individuals are attracted to a firm because of its values and those who are already there either buy in or leave. Senior managers attempt to "brand" their organization by marketing their core values, partly in an effort to attract the kind of employees they want. Attempts to develop strong organizational cultures are a means of achieving what Amitai Etzioni (1975) calls normative control. Critics argue that this management approach is used to enforce conformity and prevent dissent. This may not serve the overall goals of the organization if it stifles the kind of healthy debate that generates solutions to problems (Ray 1986).

Managing Consensus and Change through Culture

Contributing to the boom in organizational culture studies in the 1980s was the belief among North American managers that strong collective values explained the growing success of Japanese firms. William Ouchi's (1981) research suggested that successful North American corporations that emulated this Japanese approach had developed clan-like internal cultures that made the firm's values the main guides for what employees think and do.[13] Encouraged by such studies, many employers attempted to create a strong corporate culture, a dominant system of beliefs and behaviour to which, ideally, all members of the organization would be committed. These values were reinforced by mythology that revered the founders of the organization and legendary past leaders, and by rituals such as annual awards dinners at which employees' achievements were recognized.

The popular management literature often features organizations whose strong culture underlies success. There is no doubt that Toyota's culture has directly contributed to its exceptional product quality, customer satisfaction, and growing market share. As Jeffrey Liker describes in his book, *The Toyota Way*, the auto manufacturer's Toyota Production System (TPS) is the most innovative approach to lean production (see Chapter 6 for details) anywhere. One of the TPS management principles is a culture on the plant floor that enables any worker to stop production to fix a problem, however small it may be. This requires every worker and every team to become thoroughly involved in troubleshooting and problem solving. That's how Toyota "builds in" quality. Liker explains:

> So here we have a paradox. Toyota management says it is OK to run less than 100% of the time, even when the line is capable of running full-time, yet Toyota is regularly ranked among the most productive plants in the auto industry. Why? Because Toyota learned long ago that solving quality problems at the source saves time and money downstream. By continually surfacing problems and fixing them as they occur, you eliminate waste, your productivity soars, and competitors who are running assembly lines flat-out and letting problems accumulate get left in the dust. (Liker 2004: 130)

There also is growing emphasis on the role of culture in organizational transformation. This recognizes that changes to structures and processes can only be effective if they are accompanied by cultural change. A good example of cultural-driven organizational change is Baptist Health Care in Pensacola, Florida (Stubblefield 2005). This not-for-profit health care organization employs 5,500 employees in 5 acute care hospitals, nursing homes, mental health facilities, and outpatient centres. In the mid-1990s, patient satisfaction ratings and employee morale hit a low and it faced aggressive competition from private-sector health care conglomerates. Executives decided that the solution was to reinvent the culture, "empowering Workers to become Owners and Winners"—what is referred to as a WOW culture in the workplace (Stubblefield 2005:xiv). Baptist Health Care attributes its current high levels of service excellence to the transformation of its culture and work environment, guided by three principles: employee satisfaction, patient satisfaction, and leadership development. Responsibility for renewing the culture, and sustaining these changes, was handed over to employee-led committees. There are teams on culture, communication, customer loyalty, employee loyalty, and physician loyalty. Teams use a variety of measures to create transparency and accountability for key goals. Regular surveys of employees, physicians, and patients inform continuous communication and action planning. Baptist Health Care is on *Fortune* magazine's list of "100 best companies to work for" in America and has won awards for the quality of its patient care.

Cultural Differences

An organization's culture reflects many other characteristics, including the larger social and historical context. Research casts doubt on the view that an employer can "manage" culture. Using employee attitude data collected from IBM employees in 64 countries, Geert Hofstede and his colleagues investigated national variations in organizational culture (Hofstede et al. 1990; Hofstede 1984). In particular, the researchers asked whether measurable cultural differences among organizations are a function of the distinctive features of these organizations, or are shaped by the larger environment in which the organization is situated.

Hofstede's (1984: 252) international study of IBM employees found that national cultures directly influence individuals' behaviour in work organizations, the values they hold, organizational design, and even the theories used to explain organization. Follow-up studies frequently confirm that employee values are not the product of their organization, but are far more likely to vary by nationality, age, or education. Rather than values being at the core of organizational culture, the researchers found shared perceptions of daily practices (what needs to be done and how to do it) to be central (Hofstede et al. 1990: 311). The practical implications of this research are twofold. First, employee values are more likely to enter the organization through the hiring process than through on-the-job socialization that attempts to inculcate the dominant culture. Second, the effectiveness of popular management techniques, organizational design, and work humanization efforts will vary according to national context.[14]

Other researchers, often sociologists or anthropologists, examine the meaning of the organization to participants, and how a diversity of understandings and values coexist or at times are in conflict (Rowlinson and Procter 1999; Harris and Ogbonna 1999). Postmodernist researchers have criticized the unitary, dominant model of organizational culture by highlighting the wide range of symbols and stories that have relevance for different groups in an organization. Saturn Corporation attempted to create illusions of empowerment among customers and employees through its advertising campaigns for the cars it produces. Researchers focus on Saturn's use of "storytelling" as a mechanism for creating a symbolic community that appeals to the needs of consumers and workers for a sense of connectedness (Mills, Boylstein, and Lorean 2001). However, given the diversity of social, cultural, and demographic backgrounds that individuals bring into organizations, it is important to understand how distinct groups experience these dominant cultural symbols (Chemers, Oskamp, and Costanzo 1995). Feminist researchers argue that gender is constructed in workplaces through the use of language and other symbols, a process shaped by the interaction of workforce characteristics, the nature of the industry, and the larger cultural context—none of which managers can really control (Poggio 2000).

In fact, most organizations have a variety of cultures—dominant and alternative—depending on the degree to which particular groups within them share similar work experiences. This cultural fragmentation can result in organizations saying one thing in their philosophy or mission statement, while their day-to-day operations reflect something quite different. As well, work groups can develop cultural practices that are essential for performing effectively but that may be undermined by senior management because of their distance from the group. For instance, an anthropologist who studied Xerox photocopier repair technicians discovered that these workers had devised an elaborate oral culture, based on the circulation of stories that contained vital information about how to fix these idiosyncratic machines. However, this oral culture was invisible to senior managers, who planned the technicians' work on the erroneous assumption that corporate standards of design excellence ensured predictable machine operations (Orr 1996). In sum, research from a variety of perspectives clearly shows that the idea of a unitary corporate culture, fashioned by management to gain employee commitment to goals they had no say in setting, greatly oversimplifies the nature of organizational life.

IN SEARCH OF NEW MANAGERIAL PARADIGMS

Even a glance at management literature over the last two decades leaves the impression that corporate North America is in constant upheaval. In many respects, the large bureaucracies that have come to dominate the economic landscape are like dinosaurs: cumbersome, slow to respond and adapt, and, according to some critics, a dying breed. Turbulent economic times have prompted even the most vocal champions of bureaucracy to search for better ways to manage organizations. Managers, consultants, and some business school academics have created a growth industry of popular management books. These seek a new *paradigm* that would be capable of inspiring managers to meet the huge challenges of what many observers perceive to be an increasingly global, customer-focused, knowledge-based, technology-dependent economy.[15]

Tom Peters and Bob Waterman Jr. set the tone in their enormously influential 1982 book, *In Search of Excellence,* in which they documented how rigid, inflexible organizations are less able to survive, let alone grow and profit, in today's rapidly changing economy (Peters and Waterman 1982). Despite the fact that many of Peters and Waterman's "excellent" firms later experienced problems (*Business Week,* 5 November 1984), the writers fuelled debates about the need to rethink old ways of organizing and managing derived from the legacies of Weber, Taylor, and Ford. In this respect, much of the *new management literature* corresponds with the sociological critique of bureaucracy.[16] It also raises issues at the core of organizational studies: how change occurs, the role of managers, issues of power, and the possibilities for gaining cooperation and forging a consensus among all organizational participants. A brief overview of several major contributions to the new management literature will elaborate these themes.

The Critique from Within

Tom Peters has become one of the best-known management gurus. In his book *Thriving on Chaos: Handbook for a Management Revolution,* he recants some of the ideas in *In Search of Excellence,* arguing that there is no such thing as an excellent company because a firm can always be run better and rapid change requires constant adaptation (Peters 1987). His model of what he calls a "winner" is based on five sets of characteristics: an obsession with responsiveness to customers; constant innovation in all areas; the full participation and empowerment of all people connected with the organization; leadership that loves change and promotes an inspiring vision; and non-bureaucratic control by simple support systems. Peters's obsession with embracing change sets a new goal for managers: create internal stability in an organization so that it can be constantly adapting to an ever-changing external environment.

Canadian management expert Henry Mintzberg's (1989: Chapter 11) view of fluid, flexible, innovative organizations also highlights how organizations constantly must adapt. Mintzberg uses the concept of *adhocracy* to define an innovative organization, one that has a fluid and decentralized operating structure. This contrasts with the inflexibility of what Mintzberg calls the "machine organization," or traditional bureaucracy, and even with the "entrepreneurial organization," which is too performance-oriented. The work of an adhocracy is performed by multidisciplinary teams. Hierarchical top-down control, rigid lines of authority, and narrow job functions are replaced by a *matrix structure,* which enables experts to move between functional units for specialized administrative tasks and multidisciplinary project teams. Canada's National Film Board is an operating adhocracy, with its artists, technical experts, and administrators forming and re-forming teams to create the documentaries and animated films for which the agency is famous. Peter F. Drucker (1993: 59), the dean of management gurus, argues that in a knowledge- and technology-based economy, successful organizations must be able to continuously change and renew themselves. Essentially, these writers are elaborating the concept of the "learning organization."

But continuous change is not easy. Rosabeth Moss Kanter evokes the image of elephants learning how to kick up their heels in her book *When Giants Learn to Dance* (Kanter 1989). She observes that change has become the new management orthodoxy and wonders if the resulting managerial innovations designed to harness change will infiltrate the old elephantine corporations most in need of reinvigorating. Bureaucracy is a hindrance to a cooperative, creative, and flexible workplace. But new, entrepreneurial firms also face problems finding the best way to manage growth and change. Kanter uses Kodak and Apple Computer as examples of old elephantine and new entrepreneurial corporations. Kodak was stagnating, while Apple was chaotic and without a shared vision of the future. These two contrasting firms launched themselves on similar new management strategies. They reorganized according to their strengths, established closer relationships with suppliers and customers, emphasized cooperation and the generation of new ideas, cut out management layers, and developed self-managing employee teams. The result, in Kanter's (1989) words, was "a triumph of process over structure."

Later, Kanter (1995) examined successful "cosmopolitan" firms that rely on ever-changing global business networks. Using the perspective of globalization, she identifies a basic problem: "as cosmopolitan businesses extend their market scope, tap new and better supply sources, and link up with global chains, they inevitably loosen local ties and relationships" (Kanter 1995: 145). This results in increased temporary work, outsourcing, and growing workloads and longer hours for core staff, with no guarantee of future security. Kanter advocates human resource policies that create new forms of security within the context of rapid change. She gives examples of global firms that have introduced a "new social contract" based on the concept of employability—ensuring that employees receive the kind of ongoing training, learning opportunities, and career counselling that will keep them employable.

So it is not surprising that revolution is a common metaphor in this new management literature. For example, in their book *Workplace 2000: The Revolution Reshaping American Business,* Boyett and Conn (1991) portray a plant employing 130 people, where, in a 6-year period, output rises 80 percent, absenteeism drops to 2 percent, and turnover drops to 1 percent. There are no time clocks, no supervisors, no job classifications, and no definite work assignments. Everyone is salaried, and productivity reaches that of plants employing a third more staff. The concepts guiding this remarkable turnaround include work teams, participative management, employee empowerment, flatter organization structures, flexibility, creativity, pay-for-knowledge, continuous learning, quality, and doing more with less.

This language also pervades management discourse in the 21st century. Indeed, globalization and information technology have put stronger emphasis on the need to break down hierarchies and centralized control. One way to achieve this is to rely on decentralized business networks, as described by Thomas Malone:

> NIKE, the athletic apparel manufacturer, outsources all its manufacturing to other companies. Hewlett-Packard surveys its own employees to see whether they think it should merge with Compaq. British Petroleum divides itself into ninety business units, each with an average of fewer than six hundred employees. One of the best computer operating systems in the world is developed not by a company but by a loose network of thousands of volunteers. (Malone 2004: 27)

Pressures to re-shape organizations also come from a growing emphasis on customers. Max Weber's ideal bureaucracy was a vertical structure, which critics claim can't meet today's needs for product and service quality. Frank Ostroff is typical of many management thinkers who advocate horizontal organizations. He defines the horizontal organization as one in which "the emphasis shifts from top down to focusing across at customers, from compliance with executive orders to meaningful participation in the production of customer satisfaction, quality, and team excellence"—all of which is essential in "today's radically different business world" (Ostroff 1999: 74).

With this brief sampling of key issues in this now-vast new management literature, we now will consider three additional themes—quality, downsizing, and learning—that have come to characterize the challenges that employees and managers face in putting "new paradigms" into practice.

The Quality Mission

Putting quality and customers first are two axioms of the new management. To this end, some organizations have embraced the *total quality management* or TQM (sometimes called *total quality improvement*) model. Skeptics may view TQM as just a catchy label applied to a particular combination of customer-service initiatives, often with the help of computers (Micklethwait and Wooldridge 1996: 26). However, TQM champions claim that the model reduces costs and increases productivity, customer satisfaction, quality of working life, and firm competitiveness.

TQM was North American management's response to the competitive edge Japanese firms had gained in the 1980s through an obsession with product quality. However, its roots lie in the U.S. government's efforts to set quality standards for suppliers to the Defense Department during World War II and the pioneering work of American statistician W. E. Demming on quality control systems, which Japanese firms readily adopted (Aguayo 1990). TQM advocates a customer-based emphasis on quality. "Customers" can be external consumers or citizens in the case of public-sector organizations, or other units or individuals within the organization ("internal customers"). TQM's emphasis on continuous improvement focuses on involving employees in identifying and quickly resolving problems, or anticipating future problems. Like other management schools, there is no unifying model of TQM. Some perspectives on TQM complement ideas on organizational learning, reviewed below, because quality depends on more learning and creativity among employees (Argyris 1999). In addition to quality control systems, typical elements of TQM include performance measures, improved communications and feedback systems, problem-solving work teams, employee involvement, and a culture of trust and cooperation (Clarke and Clegg 1998: 254–65).

Earlier we described the Toyota Production System, which consistently has achieved exemplary quality targets. Another rigorous approach to quality is called Six Sigma, which was pioneered by General Electric (GE), the giant manufacturing and financial services conglomerate that had $74 billion (U.S.) in revenues in 2004. Since the 1980s, GE developed a culture that looked at all business processes from the customers' perspective. This required reducing bureaucracy, breaking down rigid boundaries, and empowering workers to bring forward their ideas about what needed improving. As the GE website explains, Six Sigma "is not a secret society, a slogan or a cliché. Six Sigma is a highly disciplined process that helps us focus on developing and delivering near-perfect products and services" (http://www.ge.com/sixsigma/). Six Sigma is a statistical concept that reflects 3.4 defects per million opportunities for each product or service, which means striving for perfection. Like Toyota, at GE quality has to be the responsibility of each and every employee to achieve success. This is achieved, in part, by extensive measurement of all work processes guided by "Black Belt" team leaders who relentlessly seek opportunities to improve processes that will increase customer satisfaction and improve productivity (Devane 2004).

As with other management trends, it is difficult to sort out real change from the rhetoric. Overall, it does seem that if a quality-improvement strategy is part of a comprehensive set of high-performance work practices (discussed below), then there is a higher likelihood that there will be tangible movements toward less bureaucracy, greater customer responsiveness, organizational flexibility, and employee participation. For example, two of the winners of a national U.S. quality award, Xerox Corporation and Milliken & Company (a textile manufacturer), invest heavily in employee training and use teams extensively. However, systematic analysis is needed to determine how specific changes actually contribute to measurable improvements in quality. One example of this kind of analysis can be found in the health care sector, where in some hospitals patients are treated as "customers" and quality committees rethink processes and procedures in order to improve the quality of patient care and to cut costs. A study of the introduction of TQM at an Edmonton hospital found that much of it turned out to be counterproductive, because it was linked in employees' minds with downsizing, which was happening at the same time. Increased workloads, reduced competence resulting from multi-skilling, and the overall impact of doing more with less created "role ambiguity, conflicts and demoralization"—with potentially negative effects on health care costs and quality (Lam and Reshef 1999: 741).

Some analysts argue that the "quality movement" has stimulated thinking about how to change workplaces so that employees become the key factor in improved organizational performance (Pfeffer 1994: Chapter 9). But like other major changes, there is frequently resistance to TQM changes at various levels of the organization, largely because it may be perceived as disrupting the existing power structure. Middle managers balk at the idea of empowering employees, knowing that it is their own power that will be redistributed downward. Employees and unions may resist bringing the logic of the marketplace inside an organization—such as treating university students as "consumers" and measuring their "satisfaction" through course evaluations. Furthermore, employees no doubt have trouble with such vague goals as "continuous improvement," which suggests that their best efforts today may not be good enough tomorrow.

The Downside of Downsizing

Despite the people focus of human resource management in the last decade, the most common management strategy has been *downsizing*. Job cuts, accompanied by organizational restructuring, occurred in close to half of Canadian firms with 100 or more employees between 1991 and 1996, and one-third expected more downsizing to come (*Report on Business* April 1996: 16). In the public sector, deep budget cuts in the 1990s aimed at reducing government deficits resulted in more significant staff cuts than in the private sector (Lowe 2001). In the United States, the 100 largest firms on the *Fortune 500* list downsized an average of 15 times between the early 1980s and mid-1990s, eliminating 2.5 million workers from their payrolls (McKay 1996: 56). Downsizing has continued across the industrialized economies in the 21st century, more often as a result of mergers, acquisition, and large-scale reorganization than of cost cutting per se. In addition, the movement of jobs by multinational companies based in industrialized countries to developing economies like India and China—a process called "offshoring"—has added a new dimension to job loss through restructuring.

There is now considerable evidence that downsizing has inflicted damage on individual employees, communities, and on the downsized organizations

(Baumol, Blinder, and Wolff 2003). Economically, the short-term cost savings associated with downsizing often do not translate into improved profits. Expenses may actually increase over time as managers realize they need to re-hire workers or contract out work that otherwise would not get done. Organizationally, downsizing is often a quick fix that is not part of a comprehensive change strategy. When organizations shrink rapidly, structures, processes, values, and goals must also change. Downsizing may be incompatible with other organizational change initiatives, such as TQM. And researchers on organizational learning suggest that large-scale staff cuts weaken the social networks that are vital for the organization's capacity to learn and share knowledge (Fisher and White 2000).

The human costs are also extensive and often not anticipated. Medical researchers have tracked individuals who have been downsized, documenting damage to their health. For example, a study in four Finnish municipalities followed workers over 7.5 years, comparing those who experienced no downsizing, minor downsizing, and major downsizing (Vahtera et al. 2004). The group of employees experiencing major downsizing not only had significantly higher rates of sickness absence from work compared with the other two groups, but also had higher rates of cardiovascular death. The term *survivor syndrome* has been coined to describe the negative psychological effects of downsizing on employees who remain behind, although this is only one possible response (Mishra and Spreitzer 1998). Workers experience elevated levels of stress, and generally become demoralized and dissatisfied. Motivation, loyalty, and productivity can suffer as employees avoid taking risks in their jobs and become absorbed in protecting their own interests. This is exacerbated by increased workloads as the organization tries to "do more with less." People get reassigned to new duties without adequate training. Layers of management may be eliminated without authority being delegated downward; departing employees may take with them "organizational memory" and vital skills (Littler and Innes 2003). So it is not surprising that a national survey of Canadian employees conducted in 2000 found that those who experienced downsizing and restructuring in the year prior to the survey had much lower levels of trust in and commitment to their employer, and reported poor communication and less influence on decision making, compared with employees who had not undergone such changes (Lowe and Schellenberg 2001). Building trust relationships and employee commitment is one of the most difficult challenges an employer faces in the wake of downsizing.

Some organizations have reduced staff numbers more humanely. Nova Corporation of Calgary, for example, established the Employment, Transition and Continuity Program to achieve a 10 percent staff reduction. The objective, according to the vice-president of human resources, is "to ensure that those who leave feel good about how they were treated and that those who stay feel energized and ready to move Nova forward."[17] The program underscores the importance of the process by which downsizing is carried out so that morale is not destroyed (Mishra and Spreitzer 1998). It provides employees whose jobs are affected with more options and support, including alternative work arrangements (job sharing, reduced workweek, seasonal employment), financial support for education or working in nonprofit community organizations, small business start-up advice and grants, skills upgrading, leaves of absence, and help with relocation costs. However, staff reductions achieved by employees volunteering to leave under conditions such as those just outlined can create new problems for the organization. This is especially true if the most experienced and productive employees are among those who leave, because it depletes the firm's knowledge assets and reduces its learning capacity. For example, extensive downsizing through voluntary departures at NASA deprived the U.S. space program of the scientific and engineering knowledge it needed to send astronauts on another mission to the moon (DeLong 2004).

Learning Organizations or Organizational Learning?

The concept of a *learning organization* has become popular recently. Employers increasingly are turning their attention to gaining a competitive advantage through effective use and development of employees' knowledge and skills. Not only is this the stated objective of most human resource managers, it is also echoed widely by governments across the industrialized world, which are giving high priority to lifelong learning as the means to continuous skill development (OECD 1996). It is widely assumed that better-trained workers will make an organization more adaptable, innovative, and flexible. So too, entire organizations must learn by continuously incorporating new knowledge, using it to improve their performance, and enabling individual workers to actively engage in learning.

A leading advocate of the learning organization is Peter Senge.[18] He proposes that employees and organizations follow five "learning disciplines": expanding your personal abilities and enabling others to do so; breaking out of old ways of thinking; developing a shared vision of the future; team learning; and holistic or systems thinking. These abstract principles describe a process, not a set of concrete changes. Senge claims that there is no "top 10" list of learning organizations, only individuals, work groups, and teams in various organizations engaged in "mastery and self determination" through learning. Such workers are better able to "embrace change." Senge and other champions of organizational learning present a humanistic image of self-development and self-learning in organizations, quite the opposite to the mechanistic model of organizations created by Taylor or described by Weber.

However, discussions of organizational learning lack clear definitions that could guide implementation strategies (Gherardi 1999; Popper and Lipshitz 2000). A stock definition of a learning organization is one that is "skilled at creating, acquiring, interpreting, transferring, and retaining knowledge, and at purposefully modifying its behaviour to reflect new knowledge and insights" (Garvin 2000: 11). What remains unclear is the relationship between individuals' learning activities in workplaces and organizations as learning systems. Too often, organizational learning initiatives become technical exercises in "knowledge management" using computer-based information systems (Easterby-Smith, Burgoyne, and Araujo 2001). Senge would argue that transforming an organization's culture is the best way of removing the barriers to learning that are rooted in traditional work practices and organizational structures. Learning must become a valued and rewarded process that employees have the autonomy to pursue. According to this view, a learning organization has a *learning culture* that empowers employees to take initiatives to solve problems and learn.[19]

Some firms appear to have moved toward developing a learning culture. At Harley-Davidson Motorcycles, "intellectual curiosity" is now a core value, "Harley University" offers extensive training programs, and the president promotes learning at every opportunity (Gephart et al. 1996: 39). But experts on workplace skill development would argue that Harley-Davidson simply has made greater investments in training, communicating this as a new value commitment. Training-intensive organizations, where skills are increased by regular on-site programs or through support for off-site courses, only go part way to ensuring that individuals engage in learning as they do their job and are able to apply and share that learning throughout the organization (Betcherman, McMullen, and Davidman 1998). Increasingly, organizations are attempting to develop methods for "managing" the knowledge that employees possess. This goes beyond information technology that enables the sharing of information, trying to tap into the "tacit" knowledge or "how-to" learned from doing the organization's tasks. Tacit knowledge is difficult to codify, vital to the organization's success, and often embedded in work-group cultures (Dixon 2000). In organizations with aging workforces facing the retirement of many baby-boom employees this decade, retaining this tacit knowledge is becoming a priority (DeLong 2004).

Thus, creating a learning organization remains an ideal. Achieving this goal will require a more rigorous model of a learning organization that is informed by research, a clearer understanding of the barriers to learning faced by individual employees, and the conditions that can enable knowledge generation and transfer throughout the organization (Argote 1999).

Revolution or Evolution?

To conclude, this brief review of the new management literature raises four pressing issues that researchers and practitioners must consider. The first concerns the barriers to organizational change. The changes heralded above as "revolutionary" suggest a shift from traditional approaches to organizing and managing work. Many managers may be talking about these changes, but, in practice, they are limited. Indeed, new management ideas and their slow diffusion have become a focus of academic research that points to a significant "knowing–doing gap" (Pfeffer and Sutton 2000). Managers may know that "employees are their most valuable asset" but face many barriers to acting on this understanding. Furthermore, fads launched by "management-knowledge entrepreneurs" have a short life cycle and are difficult for senior managers to successfully implement (Carson et al. 2000). A case study of the dynamics of change in a British electric utility, which examined how "managerial modernizers" attempted to change a traditional engineering-dominated culture, made this observation of the modernizers' methods: "The rhetoric intensive practice of talking 'spin' and engaging in elaborate rituals may well, whether in political or organizational life, lead to critical, and sometimes cynical responses" (Carter and Mueller 2002: 1348–49).

Second, the basic assumptions and prescriptions being advocated fall short of a new paradigm. Many of the ideas are recycled from various streams of earlier thinking, some of which are reviewed below. So it would be helpful to have more careful assessments of the intellectual origins and innovative contributions of "new" management approaches. Third, it should be clear by now that most new models for managing and organizing work focus on large organizations, mainly in the private sector. This rather exclusive approach leaves out small and medium-sized businesses and the self-employed, which, when combined, account for far more employment than do large corporations (see Chapter 2). It also tends to overlook the unique operating environments of public-sector and nonprofit organizations.

A fourth issue concerns the likely impact of these changes on employees at all levels of the organization. The research on management fads, just noted, raises the possibility that the effect of new management techniques on employees may be neutral or negative. Probing this further, can the rhetoric of participative management, team work, and employee empowerment now in vogue be taken at face value? Or is it a more sophisticated managerial control strategy aimed at getting workers to make even greater commitments and sacrifices to the organization and to "do more with less"? These questions are taken up in Chapter 6, where we discuss critical perspectives on work. For now, we will show that the belief that people matter in work organizations is firmly rooted in earlier attempts to reform and humanize work.

WORK HUMANIZATION, JOB REDESIGN, AND THE HIGH-PERFORMANCE WORKPLACE

So far, we have documented the problems of bureaucratic hierarchies, assembly-line technology and a fragmented division of labour, and organizational downsizing. The costs to workers, management, and society take the form of job dissatisfaction, stress and anxiety, poor morale, low productivity, and failure to meet the challenges of today's changing economic environment. Central to the new management approaches just reviewed are team work, greater employee participation in decision making, and skill development. But these are hardly new ideas. To help locate them historically, we will take a brief journey through the various theories and some of the actual applications of more democratic forms of work organizations and innovative job designs.

Human relations theory strongly influenced work reform schemes such as job enlargement, job enrichment, participative management, sociotechnical planning, quality circles, and various quality-of-working-life programs. A common ideological thread unites these approaches. Employees should be treated humanely and provided with good working conditions. Even so, beyond a few concessions in the realm of decision making, most of the authority remains with management. However, some reforms in Canada and elsewhere, particularly in Sweden and Norway, have come closer to achieving the dual goals of significantly redistributing power and enhancing the quality of working life. In some instances, the reforms described below may also improve productivity, resulting in mutual gains for employees and employers (Kochan and Osterman 1994).

Swedish Work Reforms

The potential of work humanization has been most fully realized in Sweden, which pioneered work reforms. Decades of social democratic government, a strong labour movement, and legislation giving individuals the right to meaningful jobs provide fertile grounds for the humanistic work reforms that have taken place in Sweden. The widely publicized Volvo Kalmar plant, which opened in the early 1970s, is based on a *sociotechnical work design.* The assembly line was replaced by battery-powered robot carriers, which automatically move car bodies to different work teams, each of which is responsible for a phase of assembly. Productivity and quality improved, but a major limitation of the Kalmar plant's sociotechnical design was that a computer (not teams) controlled the movement of the carriers and short task cycles. Most jobs provided little scope for personal development or for the use of skills and initiative. Workers still complained that their jobs were boring. Volvo's Uddevalla factory, which opened in 1989 and shut in 1993 due to a market downturn, resolved some of these problems by pushing sociotechnical design much further. Small work teams were able to build entire cars at ergonomically designed stationary work locations. By comparison with other Volvo facilities, this plant achieved high levels of quality, productivity, and worker satisfaction.

Another solution to the alienating monotony of assembly-line work can be found in Saab's main auto plant at Trollhattan. The body assembly shop faced problems typical in mass production: high turnover and absenteeism, low quality, widespread dissatisfaction, and a numbingly fast work pace. Reforms initiated in the early 1970s sought to improve the work environment, make jobs intrinsically more satisfying, and boost productivity. As at Volvo, the local union played an active role.

Some remarkable changes resulted. The assembly line was eliminated. Autonomous teams of workers, or *matrix groups,* devote about 45 minutes at a time to completing an integrated cycle of tasks. Robots took over arduous, repetitive welding jobs. Groups of 12 workers control the entire production process in the welding area. This involves programming the computers and maintaining the robots, ensuring quality control, performing related administrative work, and cleaning up their workspace. Buffer zones allow teams to build up an inventory of completed bodies, giving them greater flexibility over how they use their time. Skill development, new learning opportunities, and a broader approach to job design provide the teams with what Saab calls "control and ownership" of their contribution to the production process. Far from being victims of work degradation and deskilling through robotics, these Saab employees have reaped the benefits of upgraded job content and decision-making autonomy.[20]

North American Quality-of-Working-Life Programs

Quality of working life (*QWL*) became a buzzword among managers, public policy makers, academics, and consultants during the 1970s. QWL is an umbrella term covering many different strategies for humanizing work, improving employee–employer cooperation, redesigning jobs, and giving employees greater participation in management. The underlying goal has been to improve employee satisfaction, motivation, and commitment. The expected payoffs are higher productivity, better quality products, and bigger profits. Proponents have argued that employers and employees alike will benefit (Levine 1995). These basic ideas of QWL have been appropriated by the new management literature reviewed earlier, especially notions of employee participation. Often missing, however, is the original QWL goal of improving the quality of the work environment as a worthy goal in its own right.

QWL has diverse intellectual roots. Its core ideas came from Swedish studies on work reform, the Tavistock Institute's sociotechnical systems approach to job design, Norwegian Einar Thorsrud's experiments on self-managing groups, Frederick Herzberg's theory that work is satisfying only if it meets employees' psychological growth needs, and McGregor's "Theory Y," outlined above.[21] In addition, QWL themes are congruent with an emphasis on worker participation in decision making as a means of creating cooperative work relations. *Human resource management,* an influential revision of human relations theory, assumes that employee satisfaction and morale are nurtured in a climate of participative management (Miles 1965). Theoretically, QWL claims to combine both humanistic and economic objectives: challenging, involving, and rewarding work experiences for employees and, for the employer, more productive utilization of the firm's human resources. Yet the lofty rhetoric of QWL advocates often fails to translate into successful applications (see Chapter 6).

A quick overview of some of the major QWL techniques would be useful. *Job enlargement* is meant to expand a job horizontally, adding related tasks to put more variety into the work done by an individual worker. *Job enrichment* goes further by combining operations before and after a task to create a more complex and unified job. For example, in the case of a machine operator in a clothing factory, job enrichment might mean that the operator would now be responsible for obtaining necessary materials, doing the administrative paperwork associated with different production runs, and maintaining the machines. This might not be an enormous change, but it would lead to a somewhat more varied, demanding, and responsible job. *Job rotation* has workers moving through a series of work stations, usually at levels of skill and responsibility similar to their original task. This tactic is frequently used to inject variety into highly repetitive, monotonous jobs. When job rotation is combined with more fundamental redesign strategies (especially the use of work teams), an employee can develop a considerable range of new skills.

An *autonomous work team* consists of roughly a dozen employees who are delegated collective authority to decide on work methods, scheduling, inventory, and quality control. They might also perform what previously would have been supervisory tasks, such as administration, discipline, and even hiring. *Quality circles (QCs),* with a narrower mandate of having workers monitor and correct defects in products or services, are perhaps the best-known application of the team concept. But quality circles often fall short of redistributing authority to rank-and-file workers. By the late 1980s, the "team concept" had become quite common in some industries and Canadian auto production plants were embracing quality or "employee involvement" schemes (Robertson and Wareham 1987). And throughout the 1990s, both researchers and managers in many industries became increasingly interested in various forms of employee participation or "involvement" (Bélanger 2000; Leckie et al. 2001).

Attempts at work reform have generated heated debate between advocates—usually managers and QWL consultants—and critics, who are often trade unionists. As revealed later in this chapter and in Chapter 6 when we discuss "lean production" in the auto industry, team-based production and employee participation do not necessarily humanize working conditions (Wells 1986; Milkman 1997). In fact, the smorgasbord of QWL programs, or what now is referred to as *workplace innovation,* introduced by Canadian employers has resulted in both successes and failures (Lowe 2001). On the negative side, QWL initiatives have resulted in declining work performance, heightened union–management tensions, employee dissatisfaction, and a breakdown in communications. Positive effects can include higher employee satisfaction and commitment, better earnings, improved labour relations, and even productivity gains. The context into which changes are introduced, the level of management commitment to fundamental reform, and the process by which this happens all strongly influence outcomes.

Shell's Sarnia Chemical Plant

The flagship of the North American QWL movement is Shell's chemical plant in Sarnia, Ontario (Rankin 1990; Davis and Sullivan 1980). Here we can see the importance of context and process. What makes the plant unique is that union and management actively collaborated in the sociotechnical planning and design of the new plant. Furthermore, compared with existing facilities, brand-new plants (or "greenfield sites") offer greater scope for innovative work arrangements. Shell's goal was a "post-bureaucratic organization," suitable for the continuous-process technology used in the petrochemical industry, that would facilitate employee control, learning, and participation.

Six teams of 20 workers run the plant around the clock, 365 days a year. Along with two coordinators, a single team operates the entire plant during a shift, and is even responsible for hiring new team members when vacancies occur. Teams are supported by technical, engineering, and managerial personnel, along with a group of maintenance workers, who also teach team members craft skills. The organizational structure is flat, having only three authority levels from top to bottom. Team members have no job titles, and they rotate tasks. Pay is based on knowledge and skills obtained through job training. It takes about six years of training for an operator to reach the maximum pay level.

This design empowers workers and allows them to apply and develop their skills. Continuous-process technology may lend itself more readily to this approach because of the huge capital investment per worker and the enormous losses to the firm should the system malfunction. Thus, management has a strong economic incentive to obtain a high level of employee commitment. By tapping the talents of its employees, Shell created a safer and more productive operation. According to the director of the union, the Shell plant eliminated authoritarian bureaucracy and provided members with opportunities to improve the quality of their work life (Reimer 1979: 5). This sociotechnical design seems to have achieved its objectives: a high level of production efficiency, smoothly functioning teams, and mutually beneficial collaboration between union and management (Halpern 1984: 58–59).

On the whole, autonomous work teams appear to have the greatest potential for significantly re-allocating decision-making power, as well as for creating more interesting, challenging, socially integrated, and skilled work. Why, then, has the Canadian labour movement often been a vocal critic of QWL? The drawback for unions is that management frequently has used QWL to circumvent collective agreements, rationalize work processes, and co-opt workers into solving problems of quality and productivity. QWL has often been used to undermine union bargaining power and frequently spearheads labour relations schemes intended to keep firms union-free. As discussed in Chapter 7, from labour's perspective, gains in the quality of working life are therefore best achieved through collective bargaining. For example, the Canadian Auto Workers (CAW) adopted a policy on work reorganization that supports the involvement and empowerment of workers, provided there is a true partnership and union objectives are not undermined. Rarely does this happen, though.

The Shell experience appears to be atypical. In the Shell chemical plant, the union's involvement came only after guarantees that it would be a full partner in the QWL process, and that its ability to represent the interests of employees would not be undermined. It is noteworthy that members of the same union at an adjacent, older refinery wanted nothing at all to do with QWL. But the story of innovation at Shell's Sarnia plant is now history, because a change in management ushered in more traditional work organization and management systems. It is not unusual for work redesign initiatives to be abandoned when top management changes, especially if the new managers are mainly concerned about costs.

The High-Performance Workplace

Earlier we posed this question: Is a new management paradigm emerging? A tentative answer is yes, and perhaps it is most visible in a human resource management approach called the *high-performance workplace (HPW)*. Despite the downsizing and rationalizations, a small but growing number of organizations have combined various elements discussed above. A high-performance workplace is defined by employee involvement in decision making, team organization and flexible work design, extensive training and learning opportunities, open information sharing and communication, financial incentives for improved performance, support for family responsibilities, and a work environment that improves health and reduces stress. Given these defining characteristics, the working conditions and rewards in such organizations typically are very good, and employers reap the benefits of a productive and loyal workforce.

A major study of human resource practices in Canadian private-sector workplaces in the early 1990s concluded that while an HPW may cost an employer more initially, improved economic performance and worker productivity occurs over time when compared with traditionally run firms (Betcherman et al. 1994). The study also showed that there is no single model of an HPW, just as we noted in discussions of various approaches to work teams. One typical approach to HPW emphasizes QWL-style employee participation and enhanced job quality; another focuses more on providing employee incentives through improved performance-based compensation. According to findings from Statistics Canada's Workplace and Employee Survey (WES), by the end of the 1990s close to three in four Canadian businesses viewed human resource management practices as important to their overall business strategy. However, relatively few had actually implemented employee involvement work designs associated with high-performance workplaces: 9 percent of firms were using self-directed teams, while up to 29 percent reported using some kind of flexible job design (Leckie et al. 2001: 11, 45). The WES also documents that employee involvement is associated with higher job satisfaction. Furthermore, firms using an integrated approach to human resource management—combining training, performance-based pay incentives, and employee involvement in decision making—also tend to be more innovative in their products or services (Therrien and Léonard 2003).

In some respects, the HPW is an updated version of QWL in an internal labour market system (see Chapter 3), but with much greater emphasis on employee participation and human resource development. As such, the HPW approach may contribute to further polarization of the labour market by creating a core of advantaged employees. Yet for this core of employees who have access to good working conditions, high performance appears to mean just that—lots of dedicated hard work and continual upgrading of skills. The results for quality of working life can be mixed (White et al. 2003; Leckie et al. 2001). It is also unclear to what extent, and how, HPW systems actually increase organizational performance (Bowen and Ostroff 2004; Godard 2004). On balance, the overall weight of the evidence would point to some productivity gains when "bundles" of HPW practices are introduced, especially if these changes increase employee involvement and skill levels (Ashton and Sung 2002; Black and Lynch 2000; Appelbaum et al. 2000). This is an important question for future research, along with the conditions that encourage the diffusion of HPW practices (Osterman 2000).

JAPANESE INFLUENCES ON MANAGEMENT AND ORGANIZATIONS

Canadians can learn from Swedish experiences of industrial democracy and work humanization, but the Japanese approach to work has captured most of the attention recently. The global success of Japanese corporations has led many people to conclude that their management systems, industrial organization, and technology are superior. Sociologically, the important issue is whether the Japanese industrial system can be explained by theories of social and economic organization, facilitating its adoption elsewhere, or whether

Japanese industry's distinctiveness owes more to its culture and history, in which case application outside Japan would be difficult (Lincoln and McBride 1987; Jacoby 2005; Morita 2001).

The Japanese Employment System

The four basic elements of the Japanese approach, which typically are found only in the major corporations, include (1) highly evolved internal labour markets (discussed in Chapter 3), with features such as lifetime employment *(nenko)*, seniority-based wages and promotions, and extensive training; (2) a division of labour built around work groups rather than specific positions, a feature that is the basis for the well-known quality circles; (3) a consensual, participative style of decision making involving all organizational levels *(ringi)*; and (4) high levels of employee commitment and loyalty. A closer analysis of these aspects of Japanese firms leads to some interesting conclusions. The differences between North American and Japanese firms and management styles are probably overstated. Moreover, the Japanese system of employment owes more to principles of industrial organization than to any uniqueness in Japan's history or culture (Lincoln 1990; Hamilton and Biggart 1988). Even so, Japan seems to be at the cutting edge of the new approaches to management and work organization discussed throughout this chapter.

The core of the Japanese work organization is the *internal labour market.* Japanese employers strive to maintain a well-functioning internal labour market and to obtain a high level of employee commitment to the firm over the long term (Lincoln and McBride 1987). Teams share responsibility and accountability. When this is coupled with the consensus-building networking process used to make decisions, it is easy to see how individual workers are well integrated in their workplace. Japanese corporate decision making combines centralized authority in the hands of executives who bear final responsibility for decisions with consultations that ensure everyone has some input. On the shop floor, participation commonly takes the form of *quality circles (QCs)*. Made famous by Toyota, QCs are now common in Japanese industry. Some, such as those at Toyota, are essential to maintaining and improving high levels of quality. Yet, others "may be little more than collective suggestion-making exercises, imposed by management, for which workers receive little training" (Lincoln and McBride 1987: 300).

While the contribution of QCs to Japan's industrial success remains a moot point, sociologist Stephen Wood offers an insightful analysis of how they represent a potentially useful approach to worker participation (Wood 1989a). Wood (1989a) notes that the issue of workers' *tacit skills,* or intuitive expertise about how to do their job, figures prominently in debates about the introduction of new technology, deskilling, and management control (also see Chapter 6). Workers' informal expertise is crucial to overcoming the bugs in many new information technology systems. What is innovative about Japanese organization, then, is how it harnesses the tacit skills and latent talents of workers. Wood dismisses arguments that this may be just another power grab by management, or Japanese-style Taylorism. Instead, he argues, Japanese industry has shown that production systems can always be improved and that industrial engineers and other technical experts do not have the final word on these matters. QCs and other forms of participation enlarge workers' overall knowledge of the business and sharpen their analytical and diagnostic skills.

The Japanese Approach in North America

How far have these Japanese management and organizational innovations infiltrated North America? Quality, teams, and participative management are the new management buzzwords. Some organizational design concepts, particularly Toyota's *just-in-time (JIT)* system of parts delivery, have been more readily adapted for use outside Japan than have other aspects of Japanese management or work organization. JIT reduces inventory overhead costs, forges a stronger alliance between a firm and its suppliers, and makes it easier to change production specifications, but does not require major changes in job design. However, JIT assumes a smoothly operating transportation system. In the wake of the September 2001 terrorist attacks in the United States, a number of manufacturing firms, including auto plants in Canada, had to shut down some days because they quickly ran out of parts when the border closed and trucks were held up. So there may be a return to the old practice of stockpiling inventory, a practice that JIT advocates viewed as inefficient. QCs have also become popular among North American firms facing global competition. However, they do not go as far as delegating authority to autonomous work teams. QCs seem to place responsibility for monitoring quality and troubleshooting problems on production workers without a parallel expansion of their authority or increased rewards.[22]

The obvious place to look for the successful adaptation of Japanese employment techniques is in *Japanese transplants* (local plants owned and operated by Japanese firms) or in their joint ventures with North American firms. Even here we find very mixed results. At one end of the continuum is the NUMMI plant in California, a unionized joint venture between Toyota and General Motors. Using methods imported by Toyota, employees work in teams and contribute ideas, but the daily management style is distinctly American (Kanter 1989: 274). However, the norm in transplants or joint ventures may be much closer to the other end of the continuum. As Ruth Milkman's investigation of employment conditions in Japanese transplants in California found, these firms closely resembled American firms employing nonunion labour. Few had introduced QCs, and many local managers were unfamiliar with the principles of Japanese work organization or management approaches (Milkman 1991). These firms' employment strategies mainly emphasized cost reduction, leading them to take full advantage of low-wage immigrant labour.

Canada has numerous Japanese-owned plants, most notably the Honda, Toyota, and CAMI (a Suzuki–GM joint venture) factories in Ontario. The Japanese managers at these three plants have had to develop a hybrid system, modifying aspects of the Japanese approach in a way that is acceptable to Canadian workers (Walmsley 1992). For example, one key to Japanese management's success is "fastidious housekeeping," which emphasizes cleanliness, reduced clutter, and keeping things orderly, as well as consistency in maintaining these goals and mutual respect. Canadian workers at Honda replaced this with their own version of good housekeeping, but dropped the principles of mutual respect and consistency. The major success story among Japanese transplants is Toyota's Cambridge, Ontario, plant, whose Corollas have won the coveted J. D. Power and Associates award for the highest-quality car in North America. From the Toyota workers' point of view, benefits include job flexibility, free uniforms, consensus decision making, teamwork, good pay, and job security. However, there are drawbacks, including regular overtime, open offices for managers, no replacement workers for absent team members, health problems, close scrutiny of absenteeism and lateness, and only selective implementation of employee suggestions. Despite attempts by the CAW to organize this Toyota facility, it remains nonunion.

Probably the peak performance of the Japanese model described above occurred in the 1970s. Even then it faced criticism, such as the stifling of individual creativity through *nenko* and groupism; discrimination against non-permanent employees, particularly women; labour market rigidities that made the horizontal movement of workers among firms difficult; and long hours of work often at an intense pace (Kamata 1983). Japanese industry faced a barrage of new challenges in the 1980s and 1990s, including a prolonged recession, which altered its employment system (Whittaker 1990). Additional stimuli for change included an aging workforce, the different attitudes of young workers, and the growing number of women workers. The seniority principle has given way to other means of rewarding and motivating workers. And the mobility of workers during their careers and across firms is increasing. It will be interesting to see if this restructuring signals a growing convergence of Japanese and North American approaches to organizing and managing work.

Flexibility in Work Organizations

Standardized mass-production techniques were the hallmark of industrial capitalism for over half a century. However, assembly-line mass production of standardized goods is becoming much less economically viable, while *flexible specialization* in production is the way of the future. Attention again turns to Japan, where many firms have moved into more flexible, computerized production systems that can be quickly adapted to produce new product lines. Flexible production systems use computers to link all aspects of production into a coordinated system. Much smaller product runs are possible than with assembly-line systems, and because sales trends and consumer tastes are closely monitored, changes in design or product lines can readily be made. Computers also reduce stock-control costs ("zero inventory") and improve product quality ("zero defects"). Describing the success of northern Italian consumer goods firms, Piore and Sabel (1984: 17) write:

> Flexible specialization is a strategy of permanent innovation: accommodation to ceaseless change, rather than an effort to control it. This strategy is based on flexible (multi-use) equipment; skilled workers; and the creation, through politics, of an industrial community that restricts the forms of competition to those favouring innovation.

This model of flexible specialization in manufacturing firms is based on the use of information technology, highly skilled and versatile workers, and a political environment that favours innovative firms but does not protect ailing industries. The evidence of an increase in flexible, computerized manufacturing systems is indisputable, but it remains unclear whether the full range of manufacturing industries, in all industrialized countries, can and will move in this direction (Wood 1989b). There is, to date, only scattered evidence of improvements in working conditions in specialized flexible production settings. As for predictions of a shifting balance of power between workers and management, and between small firms and huge multinationals, these appear to be more in the realm of hope than of reality. As Stephen Wood notes in a detailed assessment of the flexible specialization model, this part of the theory is really "an intellectual manifesto," a description of the type of workplace that Piore and Sabel believe would be preferable (Wood 1989b: 13). In a way, Piore's and Sabel's optimistic predictions are reminiscent of the postindustrial society theorists, such as Daniel Bell, who were convinced that general skill levels would increase while inequality declined.

Piore and Sabel also overlook the potential for job loss that the adoption of new technologies entails. Nor do they recognize the possibility of its leading to a two-tiered employment system in which a small group of highly skilled workers, primarily male, would be the main beneficiaries of the new production system, while the majority continue to perform routine, unskilled, or semiskilled work. Furthermore, as Jenson (1989) argues, Piore and Sabel fail to recognize the role of gender in the labour market—a serious oversight.

The Flexible Firm

While evidence of a significant shift toward flexible specialization production is limited, there are indications that many employers, in both the manufacturing and service sectors, have begun to reorganize their workforces to gain greater flexibility and to reduce payroll costs (Kalleberg 2001). Frequently, these employer initiatives have involved greater reliance on part-time or part-year workers, and temporary or contract employees. Thus, the economic uncertainties experienced by employers in the past decade have also meant an increase in *nonstandard jobs*. Flexibility is one of the goals of corporate outsourcing and offshoring of jobs in the global labour market. These trends are associated with greater employment insecurity and, potentially, more work-place inequality.

Some observers of these trends suggest that a new type of employment system—the *flexible firm*—is needed, with a core of full-time workers and a periphery of low-cost nonstandard workers whose numbers and functions vary with business conditions (Atkinson 1985; Kalleberg 2001). The employment practices of such firms are thought to give them a competitive advantage in an environment of quickly changing markets and technologies. This adaptability could be achieved three ways: *functional flexibility* (training workers to per-form a variety of different tasks, thus making them more interchangeable); *numerical flexibility* (being able to quickly alter the size of the workforce, or the number of hours worked, through hiring of part-time, temporary, subcon-tracted, and other types of nonstandard workers); and *pay flexibility* (the ability to reduce pay and benefit costs by using alternative wage rates for nonstandard workers and by avoiding traditional collective agreements).

A number of researchers have examined the employment practices of con-temporary private- and public-sector work organizations to see how well they fit the flexible firm model. Despite growing reliance on nonstandard workers, many organizations have not initiated other forms of employment flexibility. Among those that have, however, it is not all that clear that they have done so for long-term strategic reasons (Booth 1997). Many public-sector organiza-tions have relied on nonstandard workers as one way of reducing short-term costs. Furthermore, cross-national differences in the extent to which firms are "flexible" suggest additional political and cultural factors that must be taken into account.[23]

As with the flexible specialization model of production, we will have to wait and see whether flexible firms become the dominant type of work organ-ization in industrial capitalist societies, or whether traditional bureaucratic organizations remain the norm. Both of these theoretical perspectives describe work organizations that, it is argued, will be more successful in a highly com-petitive, rapidly changing economic world. However, each has the potential to increase workplace and labour market inequalities and contribute to the inten-sification and destabilization of work (Smith 1997). As Samuel Rosenberg (1989: 8) cautions, "Behind flexibility—a word with extremely positive con-notations—lies the more serious question of the relative balance of power between different groups in society." Competitiveness and productivity are clearly important goals, but so too are employment security and the quality of working life.

Lean Production

Earlier, we documented that the high-performance workplace model had the potential to be a new paradigm. In general terms, the HPW can be interpreted as the latest reincarnation of the human relations tradition, given its emphasis on good human resource management practices as a key to firm performance. A related management paradigm has emerged in Canada—*lean production* (LP).[24] Three features of lean production distinguish it from HPW. First, it is modelled directly on the Japanese approach to production described above and is well established in Japanese transplants. Second, it is most fully developed in manufacturing, particularly in the North American auto industry. Third, it revolves around an integration of advanced production technology with team organization, based on a detailed analysis of the work process reminiscent of Taylorism. In this respect, LP exemplifies what is called *re-engineering,* or the radical redesign of a firm's entire business process to achieve maximum output and quality with the least labour by tightly integrating technology and tasks (Hammer and Champy 1993: 65–66). As noted earlier, the Toyota Production System exemplifies lean production.

A number of elements define LP: continuous improvement *(kaizen),* continuous innovation, flexible production, work teams, zero downtime, zero defects, just-in-time inventories and production, and employment security (Drache 1994). Lean production is presented by its advocates as a major improvement over mass production and bureaucratic organization. In other words, it has been held up as the production system for a *post-Fordist* work organization. GM's Saturn plant in Tennessee and the GM–Toyota NUMMI factory in California have been examined as models of lean production.

At Saturn, workers and the United Auto Workers union played an active role in all phases of designing the plant (Appelbaum and Batt 1994: Chapter 8). A sociotechnical approach was adopted, integrating technology and the social organization of production. Work teams of 6 to 15 members are self-managed, with responsibility for deciding important issues like workflow and quality. Teams also manage human resources, including hiring, absenteeism policies, and replacement of absent workers. Like the high-performance workplaces described earlier, but unlike traditional North American auto plants, workers at Saturn are paid 80 percent of the industry wage in an annual salary (as opposed to hourly), but they can receive up to an additional 40 percent if production and customer satisfaction goals are met or exceeded. A unique feature is the partnership approach to strategic planning, operating, and problem-solving decisions, whereby managers are "partnered" with elected union representatives. This approach has generated considerable mutual respect, and productivity and quality also seem to have benefited.

NUMMI differs from Saturn in that it was an old GM factory plagued with labour and production problems. When it reopened as a GM–Toyota joint venture, a new approach to work organization and human resource management was applied (Pfeffer 1994: Chapter 3). The NUMMI system included extensive employee training, job security, teams of multi-skilled workers, reduced status distinctions, an elaborate suggestion system, and extensive sharing of what used to be exclusively management's information. Compared with the old GM factory, this new work system resulted in impressive reductions in absenteeism and grievances, and significantly improved quality and productivity. In fact, NUMMI produced high-quality products in less time than other GM plants and was almost as productive as a comparable Toyota plant in Japan. An employee survey showed 90 percent satisfied or very satisfied with work at NUMMI.

Thus, the NUMMI experience shows that the work environment and industrial relations climate can affect the economic performance of a firm. Both the Saturn and the NUMMI cases demonstrate the potential for lean production to positively alter the worker–management relationship. But, as we will see in Chapter 6, the lean production approach to work organization redesign has also been harshly criticized. As with previous new management models that arrived with great promise, implementation often falls short of the mark.

CONCLUSION

We have covered a vast terrain of research and theory in this chapter. We have seen that there are many effective ways to structure and to manage an organization. Yet there is continuity, a set of unifying themes underlying the various theories and models that attempt to explain what happens inside work organizations. Let us reiterate these dominant themes.

Viewed as structures, most work organizations in capitalist societies (1) have been put together according to the principles of bureaucracy; (2) have a specialized and coordinated division of labour; and (3) have been run by professional managers. Technologies, the nature of markets, the type of product or service being produced or provided, management's strategic preferences, and constraints in the social and political environment are among the factors influencing the specific features of an organization's structure. But we must also look beyond these structural aspects of organizations. The actions and reactions of ordinary employees must also be taken into account if we are to understand the dynamics of work organizations.

Human actors create organizations. Even huge bureaucracies consist essentially of social relationships among individuals and groups. Especially fascinating is how these employees, from the night janitor to the president, compete to varying degrees for control of scarce organizational resources such as rewards, opportunities, power, influence, and status. To draw a fixed battle-line between management on one side and subordinate employees on the other would be to caricature the often subtle and complex give-and-take processes at work. This important point will be further explored in Chapter 6. Certainly, worker–management conflict is central to organizational studies. Sometimes conflict erupts, but, typically, truces, negotiations, and tradeoffs create a workable level of stability and cooperation. In sum, the dynamics of conflict and consensus, rooted in unequal power relations, are integral parts of work organizations. But these relationships can assume a myriad of different forms.

Looking ahead, there are signs of a growing emphasis on the human assets of organizations. We can see a shift in the human resource management literature, from the view that an organization's human resource functions merely support its larger goals, to *strategic human resource management,* which emphasizes how human resource activities drive the overall strategy of the organization (Lawler and Mohrman 2003; Becker, Huselid, and Ulrich 2001). Why?

First, the North American economy entered the new century relatively healthy and with unemployment at levels low enough that some industries faced serious labour shortages. Second, the impending retirement of large numbers of baby boomers led some analysts to predict intense competition for talent and labour shortages that could undermine Canada's national competitive advantage (Conference Board of Canada 2000). For a growing number of employers, a top priority is finding effective ways to recruit and retain employees (Lowe 2001). Third, governments, employers, and individuals recognize that skills and knowledge are the key resources in the global economy of the future. This emerging economy is increasingly knowledge-based, so employers are seeking ways to develop what are now referred to as *intellectual assets*—the skills, competence, and commitment of workers (Ulrich 1998). As more jobs become knowledge-intensive, a wide range of experts argue that knowledge workers must be treated as assets, not costs, and must be inspired by managers to learn and be creative (Drucker 1999; Mintzberg 1998; O'Reilly and Pfeffer 2000).

To conclude, the challenges of workplace reform that we have addressed in this chapter are very much alive. How they are addressed by employees, managers, and unions will shape the workplace of the future. To what extent can work be reorganized and tasks redesigned in order to achieve a better balance between the goals of making work satisfying, socially useful, and rewarding on the one hand, and of achieving economic efficiency and profitability on the other? Is this impossible? Are bureaucratic, managerial hierarchies essential for the smooth functioning of capitalism? Can the excesses of Taylorism and Fordism be eradicated? As we have seen, there are some alternative forms of work organization that might be capable of democratizing and humanizing work. The approaches called for in the new management literature—a top-down revolution, TQM, QWL, and Japanese-style employment systems—all contain potential for meaningful change. How far they go in creating non-bureaucratic, democratic, and fulfilling work environments will likely form the script for the drama of workplace change and conflict in coming decades.

DISCUSSION QUESTIONS

1. A few years ago, a senior U.S. executive told a Stanford University MBA class, "All organizations are prisons. It's just that the food is better in some than in others." What arguments and evidence in this chapter would support this (rather cynical) view?
2. The term "flexibility" is used to describe a variety of changes occurring in workplaces. Discuss how workplaces have become "flexible" and explain why managers and workers may view these changes from different perspectives.

3. Critically discuss the following statement, with reference to the changes in management theory/ideology we have observed in North America over the past century: "Worker participation in decision making is neither desirable from the point of management nor is it desired by workers themselves."

4. Assume that within 5 years of graduating, you will be promoted into a management position with responsibility for 50 employees. Discuss how the insights and knowledge gained from this chapter could help you be an effective "people manager."

5. Assessing the major approaches and theories of management discussed in this chapter, which do you think has the greatest potential to improve both productivity *and* the quality of working life?

6. Compare and contrast high-performance work systems, the Japanese approach to management, and Scandinavian approaches to quality of working life.

7. "Many firms still have Taylorist job designs, don't invest in people, are authoritarian, and don't respond to employees' personal needs." Do you agree? Provide evidence and arguments from Chapter 5 to support your position.

8. To what extent can managers create a single, dominant culture in an organization?

9. Given the large number of theories and approaches to management written about in the last century, why has it been difficult to put many of these ideas into practice?

10. The coordination of individuals' activities is one of the most basic problems that any organization must solve to be successful. In your view, what is the most effective way to coordinate work activity?

NOTES

1. Weber (1964: 338); also see Weber's essay on bureaucracy (1946: Chapter 6).

2. Morgan (1997: 15–17). This book gives an excellent overview of the multiple images, or metaphors, of organizations found in the literature.

3. Blau (1963); also see Crozier (1964), whose case studies of French bureaucracies show workers regularly circumventing formal rules.

4. Daft (1995: 25); also see Morgan (1997), Pfeffer (1997), and Clegg, Hardy, and Nord (1996) for overviews of the field of organizational studies.

5. Donaldson (1995, 1996) offers a clear overview and defence of the conventional functionalist brand of organization theory.

6. Trist and Bamforth (1951). The work of Trist and the Tavistock Institute is summarized in Pugh, Hickson, and Hinings (1985: 84–91). Also see Rankin (1990: Chapter 1).

7. The key source for strategic choice theory is Child (1972). Also see Chandler (1962) for a historical discussion of the importance of managerial strategies in shaping modern business enterprises in the United States.

8. Calás and Smircich (1996); Metcalfe and Linstead (2003). The journal *Gender, Work and Organization* is a good resource on this topic.

9. See Kelly (1982: Chapter 2) for an analysis of Taylorism using this typology. Jacoby (1997) examines the historical roots of modern human resource management practices as found in the early 20th century management ideology of welfare capitalism.

10. The application of scientific management in Canada is discussed in Lowe (1984, 1987: Chapter 2), Palmer (1979: 216–22), and Craven (1980: 90–110).

11. Classic studies of auto assembly-line workers include Walker and Guest (1952), Beynon (1984), Linhart (1981), and Meyer (1981). Hamper (1986) and De Santis (1999) offer more personal accounts of this type of work.

12. The research was originally described in Whitehead (1936) and Roethlisberger and Dickson (1939) and has been vigorously debated. See Carey (1967), Franke and Kaul (1978), Sonnenfeld (1985), Jones (1990), and Gillespie (1991).

13. For an assessment of how this book influenced management thinking and practice, see the retrospective on William Ouchi's *Theory Z* in *The Academy of Management Executive* 18(2) 2004: 106–21.

14. For an assessment of how Geert Hofstede's research has influenced management thinking and practice, see the retrospective on Geert Hofstede's *Culture's Consequences: International Differences in Work-Related Values* in *The Academy of Management Executive* 18(1) 2004: 73–93.

15. Kuhn (1970) described scientific revolutions in terms of the shift from old to new paradigms. A paradigm is a set of assumptions, principles, and

guides to action. For useful overviews of shifting paradigms of management knowledge, see Clarke and Clegg (1998). Pfeffer (1997) describes the debates generated by the different theoretical paradigms of organizations.

16. Wood (1989c) reviews nine books representative of what he calls "new wave management." Micklethwait and Wooldridge (1996) offer a scathing critique of management gurus and their ideas. Carson et al. (2000) review 16 management fads over the past 5 decades.

17. *The Globe and Mail* (21 March 1995: B12); *Nova Now* [monthly employee publication] (July–August 1995).

18. Senge (1990); Senge et al. (1994). For a critical assessment of Senge, see Micklethwait and Wooldridge (1996: Chapter 6).

19. See Shieh (1992: Chapter 18) on "learning cultures," and Yanow (2000), Lipshitz (2000), and Garvin (2000) on organizational learning.

20. On Volvo Kalmar, see Jonsson (1980) and Aguren et al. (1985). On Saab's work reorganization, see Logue (1981) and Helling (1985). These accounts of the Volvo and Saab factories are also based on Graham Lowe's personal observations and discussions with union officials, management, and shop-floor workers during visits to both plants in late 1985. The two case studies are meant to highlight different applications of socio-technical principles; the situation in each factory could be quite different now. For a critical view, see Van Houten (1990).

21. Emery and Thorsrud (1969); Thorsrud (1975); Herzberg (1966, 1968); Cherns (1976); Gardell (1977, 1982); Gardell and Gustavsen (1980). On the development of QWL in Canada, see Rankin (1990), Long (1989), and Jain (1990).

22. See Knights, Willmott, and Collison (1985) and Rinehart's (1984) Canadian case study. Hampson (1999) suggests that key features of Toyota's production system are not easily transplanted.

23. See Pollert (1988) and Macdonald (1991) for discussions of the utility of the flexible firm model.

24. On lean production in manufacturing, particularly in the auto sector, see Womack, Jones, and Roos (1990), Rinehart, Huxley, and Robertson (1997), Yates, Lewchuk, and Stewart (2001), Kochan, Lansbury, and MacDuffle (1997), and Green and Yanarella (1996). Nilsson (1996) examines lean production in a white-collar setting.

6

CONFLICT AND CONTROL IN THE WORKPLACE

INTRODUCTION

Throughout the past century, employers sought new ways of managing workers in order to increase efficiency and productivity and to reduce workers' resistance to authority. Scientific management, with its emphasis on complete managerial control and an extreme division of labour, gave way to the softer human relations approach, which tried to motivate workers by making them feel that they were an integral part of the larger work organization.

By the 1970s, management consultants were advocating changes to organizational structures and job design. Japanese-style management strategies, quality-of-working-life programs, and total quality management models, for example, tried to reduce the negative effects of bureaucracy, increase job skills, and enhance worker decision making. Downsizing, organizational restructuring, and new technologies have been the favoured approaches to increasing productivity in the past two decades. Even so, the lean production and high-performance workplace models claim to empower workers, enhance their skills, and involve them as partners in the production of goods and services.

If we were to draw a conclusion about consensus, cooperation, skill development, and fairness in today's workplace from the management literature in Chapter 5, it might be that new approaches to management have solved most of the problems of conflict and worker exploitation observed in the past. But this would ignore a great deal of contradictory evidence.

First, even when the North American economy is performing well, we continue to hear about massive layoffs as employers rely on downsizing and restructuring to compete with aggressive global competitors, deal with declines in market demand for their goods and services, or simply show a higher profit margin for investors (Baumol, Blinder, and Wolff 2003). For example, in June 2005, General Motors announced that it would be cutting 25,000 jobs in North America over the next 3 years. At the same time, Ford was shrinking its workforce by 2,700 positions, Alcoa was laying off 6,500 workers, and Onex was downsizing by cutting 1,800 jobs in its Boeing aircraft plants (Robinson 2005). Meanwhile, over in Japan, the new CEO of SANYO announced her approach to improving the company's performance—cutting 14,000 jobs (Pitts 2005). Along with downsizing, replacement of full-time permanent workers with part-time and temporary employees (Vosko, Zukewich, and Cranford 2003) and the contracting out of work previously done by better paid permanent employees (Shalla 2002) continue to be preferred competitive strategies for many Canadian private- and public-sector work organizations.

Second, despite the rhetoric about the need for better human resource utilization and worker empowerment in order for Canadian firms to be globally competitive, management practices in this regard have been slow to change. Recent studies of large Canadian workplaces continue to show that many still rely on Taylorist management approaches. Most make, at best, limited investments in employee training. Responses to employees' desires for more autonomy and responsibility and their needs for more family-friendly workplace benefits and policies continue to be the exception rather than the rule (Lowe 2000; Jackson 2005).

Third, in response to such employer practices, we regularly see workers resisting and opposing management policies, in all sectors of the economy (Collinson and Ackroyd 2005). In early July 2005, for example, 6,000 of the City of Toronto's outside workers, including garbage collectors, swimming pool attendants, and paramedics, were threatening to strike. Wage increases were among the issues, but the workers' union was also very concerned about the contracting out of formerly unionized jobs, changing job descriptions and overtime requirements, seniority rights, and sick-leave benefits. At the same time, 1,200 container truck drivers in Vancouver were on strike over what they felt were far from adequate payment rates, given the high costs of fuel. While not on strike, B.C. forestry workers were lobbying the provincial government to tighten up safety regulations, in response to yet another fatal accident in the industry, the 21st in 2005. A week earlier, a small group of women in Toronto had protested an employer's decision in a much more provocative manner, by chaining themselves to the door of an executive boardroom. They were drawing attention to the decision to eliminate the jobs of 26 women who had, along with others, been on strike for over half a year over pay equity issues.[1]

These examples demonstrate the need to counterbalance Chapter 5's discussion of management approaches by returning to the themes of power and control, conflict and resistance. These themes are central to an adequate understanding of workplace dynamics. It is also helpful to characterize the management theories and strategies introduced in the previous chapter as sharing a *consensus* perspective. In contrast, this chapter's critical analyses of worker–management relationships have been influenced by the *conflict* perspective in sociological analysis.

It is appropriate, then, to begin by reviewing Karl Marx's critique of work in capitalist society, since his arguments shaped the conflict perspective. We then outline the *labour process perspective,* an attempt to update Marx's ideas to explain contemporary worker–management relationships. This alternative approach begins with the assumption that conflict is to be expected in work settings, where the interests of owners and managers are in opposition to those of workers. It focuses directly on attempts by management to control workers, and on how workers resist such efforts, trying to gain more control over their own labour. While the labour process approach has itself been criticized on various grounds, it does provide a much more critical assessment of quality of working life, lean production, and other contemporary management strategies, as well as the introduction of new information technologies in the workplace.

This chapter is really a debate with the assumptions and conclusions of the mainstream management and organizational literature reviewed in Chapter 5. But we also go further by asking whether workplace health and safety might be one area in which workers and management have come closer to reaching a consensus. We conclude with a discussion of industrial democracy and worker ownership. Both of these approaches to organizing workplaces help workers to gain more control over the conditions of their work. In Chapter 7, we continue the discussion by examining the origins, functions, and future of trade unions, another vehicle through which workers have attempted to look after their interests.

MARX ON EMPLOYMENT RELATIONSHIPS WITHIN CAPITALISM

As we have observed in Chapters 1 and 3, class conflict was central to Marx's perspective on social change. Looking back in history, he argued that feudalism had been transformed into a new mode of production—capitalism, characterized by wage-labour relations of production—because of conflict between different class groupings. Looking forward, he predicted that capitalism would eventually give way to socialism because of the inherently conflictual and exploitative relationships between owners and workers.

Central to Marx's view of social change was a specific economic theory of the capitalist labour market. Beginning with the premise that the value of a product was a direct function of the labour needed to produce it, Marx proposed that, in a capitalist economy, wage labourers produced more than the amount needed to pay their wages. *Surplus value* was being created. Consequently, the relationship between capitalist and worker was exploitative in that the workers who produced the profit were not receiving it.

Unlike feudalism, which greatly restricted the mobility of serfs, workers under capitalism were free. They could always quit and search for another job. However, by purchasing labour, capitalists gained control over the labour process itself. Factory organization and mechanization further increased employer control. Thus, the craftsmen forced into the industrial mills of Marx's time were, according to one commentator, "subjected to inflexible regulations, and driven like gear-wheels by the pitiless movement of a mechanism without a soul. Entering a mill was like entering a barracks or prison."[2]

These relationships of production led, according to Marx, to feelings of *alienation* among workers in a capitalist economy. Lacking control over the process and products of their labour, and deprived of the surplus value (profit) they were creating, workers felt separated (or alienated) from their work. We will discuss alienation in greater depth in Chapter 8, along with other subjective reactions to work. For now, it is sufficient to note that Marx believed that the working class would eventually rise up in revolt against this alienation and exploitation.

But things have changed since Marx's time. The harsh, dangerous, and exploitative working conditions of early industrialization have been largely eliminated in most industrialized and postindustrial economies. The combination of labour legislation, unions and professional associations, and more sophisticated employers has led to higher incomes and standards of living, safer working conditions, and more responsibility and autonomy for workers, at least in relative terms.

However, earlier chapters have documented the growing number of working poor in Canada, a polarization of work rewards, and the rise in involuntary nonstandard employment. Unions continue to resist management attempts to reduce wages while most nonunion workers have little choice but to accept what they are offered. Hundreds of workers are still killed on the job each year, and thousands suffer workplace injuries. Despite the calls for new management approaches to empower workers, some employers continue to act as if control over work belongs to them.

Marx's descriptions of working conditions in early capitalist society may no longer be applicable, but the core themes in his writings can still help us to gain a better understanding of work in today's economy. These themes—power relationships in the workplace, attempts by owners and managers to control the labour process, resistance to these attempts by workers, and conflict between class groups—are central to the labour process perspective on worker–management relationships.

THE LABOUR PROCESS PERSPECTIVE

Harry Braverman on the "Degradation of Work"

In *Labor and Monopoly Capital,* a book written over 30 years ago, Harry Braverman (1974) argued that 20th-century capitalism was considerably different from the mode of production examined by Karl Marx. A relatively small number of huge, powerful corporations now controlled the national and international economies. The role of the state in the production process had expanded. New technologies had evolved, workplace bureaucracies had become larger, and the labour process itself had become increasingly standardized.

Analyzing the transformation of office work in large bureaucracies, Braverman observed that, at the beginning of the 20th century, (male) clerks and bookkeepers had been able to exercise considerable control over their work and had been responsible for a wide variety of tasks. But this was no longer the case. What the feminization of the occupation masked was how the extensive division of labour in offices had narrowed the scope of the work done by one clerk. By the 1970s, office work had been standardized and mechanized. Clerks essentially processed an endless stream of paper on a white-collar assembly line, their routine tasks devoid of much mental activity. Clerical work had been degraded and deskilled.

Braverman attributed these changes to management strategies designed to improve efficiency and gain more control over the office labour process, the same strategies employed in factories many decades earlier. Essentially, Braverman saw Taylorist tendencies at the heart of all modern management approaches, even those used to organize the work of technicians, professionals, and middle-level managers. Thus, both lower- and higher-status white-collar workers were becoming part of the same working class. The deskilling and degradation of work in general was setting the scene for future class conflict.

Braverman's provocative analysis of changing employment relationships in 20th-century capitalism was highly influential, but it was also criticized by researchers sympathetic to its general critical approach.[3] First, Braverman over-generalized from scattered evidence in North America to assert that deskilling was a universal pattern present in all occupations within all industrial capitalist societies. Important cultural differences in the labour process were consequently ignored, as were situations in which automation, work reorganization, and employee relations policies actually provided more autonomy and responsibility to workers. Furthermore, while focusing on declining skills in some occupations, Braverman overlooked the new skills that were required in industrial and postindustrial economies. For example, the management of large numbers of employees involved in different tasks required the development of a wider range of leadership skills, office work in service industries required sophisticated "people skills," and efficient use of new manufacturing and information technologies required computer skills.[4]

Second, Braverman ignored the gendered nature of workplace skills, as have many other researchers (see Chapter 4). Technical skills, frequently more central to "male" jobs, have been valued more highly than people skills, which figure more prominently in jobs typically held by women.[5] Hence Braverman probably underrated the skill requirements of clerical work. In addition, he overlooked how race and ethnicity have played a major part in determining access to better jobs in North American labour markets (see Chapter 3).

Third, Braverman implied that workers passively accepted management assaults on their job skills and autonomy. Seldom in *Labor and Monopoly Capital* does one find mention of workers resisting management, even though power struggles have always been part of the informal side of bureaucracy (see Chapter 5). Furthermore, as we will argue in Chapter 7, workers have organized in unions for well over a century to collectively oppose their employers. The main flaw in Braverman's theory of the degradation of work is its determinism, which suggests that the inner logic of 20th-century capitalism pushed capitalists to devise Taylorism, Fordism, human relations management approaches, and their many variants.

Why, then, was Braverman's book so influential? The answer is that he challenged work researchers to ask new questions about changes in the labour process in contemporary capitalist societies. The ensuing debate has brought into sharper relief how, to what degree, and under what conditions jobs may or may not have been degraded, deskilled, and subjected to various types of management control. Braverman brought the sociology of work and organizations back to issues of class and inequality, power and control, resistance and conflict (Smith 1994). Even in his failings—over-generalizing about deskilling, not taking account of gender, and overlooking worker resistance—he motivated other researchers to ask critical questions about shifting methods of managerial control and about trends in the deskilling or enskilling of work.

Models of Managerial Control

Labour process researchers have documented changing preferences for control strategies among North American employers over the past century. Richard Edwards (1979) and his radical economist colleagues (Gordon, Edwards, and Reich 1982) began this research tradition by extending their theory of labour market segmentation (see Chapter 3) to also describe the evolution of workplace control systems.

They distinguished three basic types of managerial control. With *simple control,* most common in secondary labour markets, employers regulate the labour process with coercive or paternalistic methods, or, in larger organizations, through a hierarchy of authority. *Technical control* is achieved by machine pacing of work and can, in part, replace the direct supervision of simple control methods. For example, Henry Ford's assembly line gave managers a powerful means of controlling the pace of work. As one observer of a Ford plant commented:

> Every employee seemed to be restricted to a well-defined jerk, twist, spasm or quiver resulting in a flivver. I looked constantly for the wire or belt concealed about their bodies which kept them in motion with such clock-like precision.[6]

Bureaucratic control has evolved in large corporations in the core sector of the economy. Good salaries, generous benefits, and pleasant work settings are the inducements provided, usually to middle- and upper-level white-collar employees and some groups of skilled manual workers. The internal labour market, and the prospects of an interesting and rewarding career, are part of an employment package designed to win employees' commitment.

Edwards and his colleagues argued that the evolution of segmented labour markets, with their different control systems, resulted in a socially fragmented and politically weak working class. As they saw it, employers in the United States gained the upper hand over workers in the "contested terrain" of the workplace by developing new and more effective means of control. While impressive in its historical scope, this theory is still quite deterministic, suggesting that changes in the means of production and periodic economic crises forced capitalists to devise new methods for directing, evaluating, and disciplining workers (Nolan and Edwards 1984). Like Braverman, these economists portrayed workers as largely passive.

Other observers of the capitalist labour process have emphasized the agency of workers—how they resist, re-shape, or even actively participate in management control strategies. Andrew Friedman (1977: 82–85), for instance, describes a shifting *frontier of control*, which is influenced alternately by conflict and accommodation between employers and employees. At issue is who sets the hours, pace, and sequence of work tasks, and what constitutes fair treatment and just rewards. Sometimes workers gain a say in these matters through union bargaining. Or management may initiate work reforms, giving workers what Friedman calls *responsible autonomy*, a means of obtaining cooperation by granting workers some scope for making task-related decisions. Some degree of responsible autonomy is found, for example, in the quality-of-working-life (QWL) programs and other participative management schemes described in Chapter 5. The opposite strategy in Friedman's model is *direct control* (like the simple control described by Edwards) which involves strict supervision with very little job autonomy for workers.

Examples of both types of control have been observed in the Canadian mining industry. Wallace Clement's (1981: 204) early study of the International Nickel Company (Inco) describes the introduction of sophisticated "people technology" at the modern Copper Cliff Nickel refinery. Inco's emphasis on a "one big happy family feeling," a flatter job hierarchy, and a new on-the-job training system looked progressive. However, these changes led to a breakdown of traditional job autonomy and the erosion of the bargaining power of the unionized refinery workers. This increased the level of direct control by management. In contrast, Inco's teams of underground miners possessed considerable responsible autonomy. These small, closely knit groups decided among themselves how and when to do each phase of the mining operation.

A more recent study of potash mining in Saskatchewan (Russell 1999: 193–94) revealed that managers in post-Fordist work settings were less likely to employ a "rough" style of management or, in Friedman's terms, traditional direct control. However, it was also apparent that management could not afford to act in this manner, since this might jeopardize the cooperation of workers who were expected to maintain high levels of productivity in a lean-production environment. Ultimately, the researcher concluded that, despite the rhetoric, workers had not really been empowered.

Turning to the service sector, a nightclub in a Western Canadian city provides an interesting example of the shifting frontier of control. Mike Sosteric (1996) describes the strong informal workplace culture that had emerged in the club and the considerable degree of job autonomy enjoyed by workers. As a result, they were loyal and committed, and customers received high-quality, personalized service. But new management saw some of the norms of the informal workplace culture as problematic, and tried to take charge by implementing what might be seen as a system of responsible autonomy. The workers were put off by the training seminars, disliked the new system of job enlargement, and found that elimination of supervisors made their work more difficult. Consequently, they actively resisted the changes. In turn, management adopted a direct and coercive control strategy that led, ultimately, to workers quitting and the quality of service in the nightclub deteriorating.

While Friedman describes responsible autonomy being offered to selected groups of workers in return for their cooperation, Michael Burawoy (1979, 1984) goes even further, suggesting that in many work settings employees actively choose to cooperate. In his study of machine shop workers, he describes how an unspoken agreement with company goals emerged. Some employees adapted to management's control system by simply treating wage bonuses as a game they tried to win, thus exhibiting an individualistic response or adaptation to an otherwise boring job. As long as each worker had a fair chance of "winning" bonuses, management's rules went unchallenged. For these workers, coercion (or the *despotic organization of work,* in Burawoy's terms) was unnecessary since, by accepting the rules of the workplace, they basically began to motivate themselves.

Burawoy also describes the *hegemonic organization of work* in a manner similar to Richard Edwards's bureaucratic control. In large corporations and government departments, employees often see their own futures linked with the success of the organization. Hence, management's goals and values are dominant, or hegemonic. The presence of internal labour markets and responsible autonomy helps to maintain management control. Individual workers are subtly encouraged to adopt the company's or department's value system, and good job conditions foster long-term commitment to the organization. A recent study of how the Klein government in Alberta restructured the department responsible for museums and cultural heritage sites in the 1990s is a good example (Oakes, Townley, and Cooper 1998). By requiring all government departments to develop standardized business plans, an apparently neutral strategy from which everyone would benefit, senior management shifted the department's emphasis from public education and cultural preservation to the logic of business while, at the same time, gaining greater control over professionals employed in the department as they adopted the language and values of business planning.

New forms and places of work, along with the rapid growth of computer-based information technologies, have given rise to yet another type of management control system—*electronic control* (Sewell 1998). Such surveillance may be relatively passive, as when workplaces are monitored by security cameras, or highly active, as when supervisors keep track of work performance electronically and send warning messages to workers who are not producing as much as others. Electronic control is typically also very intrusive, particularly in the case of supervisors monitoring telephone calls, electronic mail, and Internet searches. Consequently, it is not surprising that most workers resent the implementation of such control systems. Nevertheless, most also have to put up with it. A recent study estimated that close to 80 percent of U.S. workers were being monitored by some type of electronic surveillance system (Hansen 2004: 151).

Electronic control has been a common strategy employed by managers in charge of teleworkers (see Chapter 2). In their U.K. study of "managerial control of employees working at home," Alan Felstead and his colleagues (2003) documented the extensive use of electronic mechanisms for recording the amount and type of work being done by off-site employees. However, it also was apparent that this form of control was not always effective. Sometimes managers could not distinguish between productive and unproductive work. They were also unable to get a sense of the employee's frame of mind and their general level of commitment to the work organization. As a result, some managers relied on long, chatty phone calls with their employees. Or they resorted to regular home visits, essentially reverting to traditional methods of direct control. But home visits created their own complications since "the boss" now had to fit into the role of being a guest in the worker's home where others (children, spouses) might not always be very welcoming.

The rapid growth of *call centres* within which telephone and computer systems are linked to generate large numbers of outgoing calls (e.g., for telephone sales or market research) or to handle many incoming calls (typically regarding customer service) has led to extensive electronic control of workers. Computers are used to assign calls to individual workers who may handle hundreds of calls a day. They are also often used to monitor worker performance, both quantitatively (How long did each take?) and qualitatively (Did the customer service agent address the client's problem appropriately? Did the telephone salesperson try hard enough to make a sale?). Supervisors may or may not listen in on individual calls, but the possibility that someone is listening has the same effect. If work teams are part of the management strategy, the electronic monitoring of team performance can be used to push teams to compete with each other against high performance targets.

Some researchers (Fernie and Metcalf 1998) have described these new workplaces from the perspective of the French social theorist Michel Foucault (1977) who wrote about the all-encompassing power of surveillance in modern society. They see call centres as "electronic sweatshops" in which workers have virtually no decision-making opportunities. In fact, because they know they can be monitored at any time, they essentially become their own taskmasters. However, a wide range of subsequent studies have documented considerable resistance by workers to management monitoring and surveillance and have described the varying amounts of control that call-centre workers have, depending on the types of services they provide (Taylor and Bain 2001).[7] For example, customer service agents providing complex financial information to clients have considerably more decision-making authority and are much less likely to be working in a computer-paced environment (Batt 2000). Some labour process researchers have proposed that the combination of technical and bureaucratic control that underlies call-centre management systems should be given a new label—*structural control* (Callaghan and Thompson 2001). While the debate continues on the type and extent of worker control within call centres, it is nevertheless very clear that electronic control is part of the contemporary labour process.

Having reviewed some of the research on managerial control strategies, we can now see the variety of ways in which managers might try to control the labour process. Direct, technical, bureaucratic, hegemonic, and electronic control, along with responsible autonomy, are all possibilities, alone or in some combination. The balance of power and control in any given workplace can range from strictly coercive, with management holding all the cards, to situations where workers have considerable responsibility for regulating their own work, essentially controlling themselves. As John Jermier (1998: 241) wrote when introducing a recent collection of studies on the subject, regimes of control range from those "anchored in the iron fist of power [to] those that rely on the velvet glove." And, as the research discussed above indicates, workers continue to find ways to resist many management control efforts. In Friedman's words, the labour process remains a shifting frontier of control and conflict.

The Deskilling Debate

Questions about changes in the skill level of workers' jobs are central both to our evaluation of Braverman's thesis (as well as the contrasting theories of postindustrial society) and to current public policy discussions about the link between investments in human resources and Canada's economic competitiveness. Are the forces of globalization, technological change, industrial and labour market restructuring, and new management techniques requiring more or less skill from today's workers?

For Braverman, the application of Taylorist-style management along with the adoption of new technologies was systematically deskilling both blue-collar and white-collar work in capitalist society. In contrast, Daniel Bell and other postindustrial society theorists argued that a new economy relying heavily on highly skilled "knowledge workers" was taking shape (see Chapter 1). More recently, other writers have put forward a similar *enskilling* argument about the growing number of multi-skilled workers required with "flexible specialization," "lean production," and other postindustrial modes of production (see Chapter 5) as well as by the proliferation of computers, robotics, and new information technologies in the workplace. And public policy debates about competitiveness have been characterized by warnings that more investments in education and training are needed to increase the skill levels of Canadian workers.

Before we can address the skill debate, we need some agreement on definitions. What exactly is skill and how can we measure it accurately?[8] A first basic question is whether skills reside in workers or whether they are characteristics of jobs. In other words, is deskilling something that happens to individual workers or to the jobs they occupy? The answer to both questions is "yes," since jobs can be redesigned to increase and decrease skill content, and workers can lose skills if they do not have the opportunity to use them regularly.

Second, are skills such as the ability to use advanced computer software, drive a semitrailer, or teach children to read completely objective job requirements that can be reliably measured and ranked in terms of their importance and complexity? Or are skills to some extent socially constructed, reflecting the social status, power, and traditions of a particular occupation? As we noted earlier in our discussion of the Braverman deskilling thesis, the social construction of skill has placed much more value on traditionally male skills and devalued those possessed by women. A useful concept in this regard is *tacit skill,* a term describing the informal knowledge workers have gained through experience or from coworkers about how to do their job. Tacit skills obviously are not part of formal job descriptions and, thus, often go unrewarded.

These observations highlight the importance of thinking about skill in a multidimensional way. The research literature contains a variety of detailed skill classifications, and personnel experts who conduct formal job evaluations have their own lists. But there is considerable consensus that, in general terms, work-related skills have two components: *substantive complexity* (the level and scope of intellectually, interpersonally, and manually challenging tasks performed in a job), and *decision-making autonomy* (the opportunity for individual workers to decide how, when, and at what speed to complete a task).

There are also important research design and measurement issues to be addressed in the skill debate. While case studies of specific workplaces or occupations can illuminate deskilling or enskilling processes, they seldom have the breadth to test Braverman's sweeping conclusion about deskilling within capitalism. Hence, larger-scale studies (both in terms of occupational coverage and the time period examined) are needed. Can we assume that jobs typically held by well-educated people are, in fact, highly skilled? Alternatively, do formal job requirements really indicate the skill levels of the work tasks performed? The answer to both questions is "not necessarily." Recall, for example, that rising levels of educational attainment among Canadians, in the absence of enough jobs requiring the skills acquired through education and training, can lead to significant problems of underemployment (Chapter 2).

Wallace Clement and John Myles (1994: 72) summarize the skill debate by asking if we are faced by "a postindustrial Nirvana of knowledge where everyone will be a brain surgeon, artist, or philosopher (Bell) or, alternatively, a postindustrial Hades where we shall be doomed to labour mindlessly in the service of capital (Braverman)." Some case studies of skill change in specific occupations and work settings have supported the deskilling hypothesis, while others have not. But with larger-scale studies using multiple measures to examine skill shifts over several decades in the complete occupational structure, a more definitive conclusion has emerged. On average, the long-term trend in North America and western Europe has been in the direction of increased skill requirements in the workplace. Although Daniel Bell's overly optimistic predictions of an emerging knowledge society have not been fulfilled, "the net result of the shift to services has been to increase the requirements for people to think on the job" (Clement and Myles 1994: 72).

This conclusion, however, comes with several important caveats. First, Myles and Clement comment on the "net result," acknowledging that deskilling has occurred in some work settings. An example would be the secondary labour market jobs in the contract building-cleaning industry described by Luís Aguiar (2001). In his Toronto-based study, he observed how, in the traditional "zone cleaning" approach, a single individual was responsible for a range of cleaning tasks. But in order to increase efficiency (and profits), supervisors subdivided the tasks into "restroom specialists," "dusting specialists," and "mopping specialists," essentially creating a "mobile assembly line" within which workers had to work harder and faster on a more limited number of tasks.

Second, we also need to recognize that an overall increase in skill requirements need not be accompanied by other improvements in the quality of working life. For example, TQM, lean production, and other recent management innovations may require workers to learn additional skills, but in downsized organizations workers also have to work harder and faster, under more stressful conditions (Russell 1999). Some of the critical assessments of these management strategies (summarized below) distinguish between *multi-tasking* (simply adding more tasks to a worker's job description) and *multi-skilling* (adding to the skill repertoire of workers), arguing that the former merely makes employees work harder, not smarter (Rinehart, Huxley, and Robertson 1997).

Third, the studies showing increased skill requirements over time included data only up until the late 1980s. Until then, growth in higher-skill jobs, most in the upper-tier services and the goods-producing industries, outstripped the expansion of less-skilled positions, many in the lower-tier services. But we have continued to see substantial industrial restructuring and organizational down-sizing, and widespread introduction of automated technologies, the sum of which might have slowed or stopped the enskilling trend. This might account for Michael Smith's (2001: 17) conclusion that "recent technological change has not distinctly raised average skill levels" in Canada.

Finally, we have also witnessed increased polarization between skilled workers and those with less skill over the past several decades. In the United States, writers like Robert Reich (2000) and Richard Florida (2005) worry about this growing inequality and its consequences for social cohesion and the future of Amercan society. In Canada, Clement and Myles (1994: 76) discuss the skill polarization issue in terms of social class, noting that almost all executives and a majority of the new middle class are in skilled jobs, compared to less than a quarter of the working class. Thus, the polarization of skills may be accenting the already pronounced class differences in income, status, and power in the Canadian labour market.

TECHNOLOGY AND THE LABOUR PROCESS

Technology is a central concept in the sociology of work. It figures prominently in theories and debates about industrialization and postindustrial society. Most of the management approaches developed over the past century relied on new technologies, at least to some extent, to improve efficiency and increase productivity. In turn, as we have already seen, labour process researchers have focused on how employers have at times used new technologies to deskill work and increase control over employees.

Lenski on Technology and Social Inequality

Gerhard Lenski (1966), a sociologist analyzing North American society in the 1960s, placed technology at the centre of his theory explaining differences in the level of social inequality. He proposed that a society's technological base largely determines the degree of inequality or the structure of the stratification system within it. In simple hunting and gathering societies, he argued, the few resources of the society were distributed primarily on the basis of need. But, as societies became more complex, *privilege,* or the control of the society's surplus resources, came to be based on power. Ruling elites received a much larger share. In agrarian societies, a governing system had evolved that gave the privileged class control of the political system and access to even more of the society's wealth. However, the arrival of the industrial era reversed this "age old evolutionary trend toward ever-increasing inequality" (Lenski 1966: 308).

Lenski's explanation hinged on the complex nature of industrial technology. Owners of the means of production could no longer have direct control over production. In the interests of efficiency, they had to delegate some authority to subordinates. This led to the growth of a middle level of educated managerial and technical workers who expected greater compensation for their training. Education broadened the horizons of this class of workers, introducing them to ideas of democracy, as well as making them more capable of using the political system to press their demands for a greater share of the profits they produced.

In short, Lenski's theory proposed a causal link between complex industrial technology, rising levels of education, and workers' insistence on sharing in the growing wealth of an industrial society. But why would employers give in to this demand? Lenski believed that the industrial elite needed educated workers; the productive system could not operate without them. Equally important, the much greater productivity of industrial societies meant that the "elite can make economic concessions in relative terms without necessarily suffering any loss in absolute terms" (Lenski 1966: 314). Because so much more wealth was being produced, everyone could have a larger share.

Lenski accurately described the social and economic changes accompanying industrialization. Twentieth-century capitalism had provided a higher standard of living for the working class. Compared to the era observed by Marx, workers were not as exploited and powerless. Social inequality was slowly declining in the decades of economic expansion and full employment immediately following World War II. However, it is virtually impossible to prove that industrial technology was the primary cause. A better explanation might focus on changing social relations of production. Within capitalism's shifting frontier of control and conflict, workers had gained relatively more collective power, through unions, for example. In addition, the state had legislated rules governing the labour market and provided a social safety net that reduced the level of poverty.

But, as we have argued, we are again seeing evidence of growing inequality in North American society. Industrial restructuring, organizational downsizing, the growth of nonstandard jobs, and a reduction in the social safety net are all part of the explanation. Going back to Lenski's starting premise that a society's technological base determines the degree of inequality, is it possible that computers, robotics, and new information technologies have also contributed to the slow increase in social inequality?

The Information Technology Revolution

The new technologies of the industrial age increased productivity largely by reducing the amount of physical labour required in the goods-producing industries. In contrast, today's computer-based information and automated production technologies are re-shaping both the physical and mental requirements of work, in both the goods-producing and service sectors (Burris 1998). Factory robots, globally linked office computer networks, e-commerce in the retail sector, automated materials-handling systems in transportation, computer-assisted design in engineering firms, computer-assisted diagnostics in health care, and the Internet, cell phones, and fax machines in workers' homes are all part of a rapidly changing technological context for work at the beginning of the 21st century.

It would not be an exaggeration to describe the changes occurring as an information technology (IT) revolution. In fact, some observers argue that the arrival of the "information society" has had a larger impact on society than the shift to a service-dominated economy (Aoyama and Castells 2002). The change has been extensive and rapid. Twenty years ago, only a very small minority of Canadian workers used computers, in any form, in their jobs. By 1989, more than one-third (35%) were using computers, and by 2000 well over half (57%) were doing so (Marshall 2001b). The 2001 census showed almost 400,000 Canadians, close to 3 percent of the total workforce, directly employed in information technology (IT) as computer engineers, programmers, and technicians, software and systems analysts, and web designers, for example (Habtu 2003).

Information technology innovations have clearly contributed to increased productivity (Harchaoui et al. 2002). At the same time, the potential for electronic control of workers has been enhanced by new forms of IT, as we noted earlier in this chapter. But have these new technologies also led to improvements in other aspects of job quality? Automation's potential to eliminate dirty, dangerous, and boring factory jobs has been partly realized through robotics.[9] Robots are ideal for work in cramped spaces, in extreme temperatures, or in otherwise hazardous situations, and they have been used to eliminate dangerous jobs in many factories (welding and painting automobiles is an example). It is possible to organize production around robots in a way that offers workers considerable opportunity for skill improvement and job autonomy. The Saab factory described in Chapter 5 is an example.

But what is possible is not always what happens. The retooling of a U.S. automobile plant provides an example (Milkman and Pullman 1991). Until the mid-1980s, the GM factory in Linden, New Jersey, produced luxury-sized cars using traditional assembly-line technology. Around 1985, the factory was retooled in order to start producing smaller cars. Over 200 robots were added to the production system, along with more than 100 automated guided vehicles (AGVs) which moved the cars in production through the system. In addition, management introduced organizational innovations like just-in-time delivery and quality control circles. However, the basic hierarchy separating production workers from skilled trades workers (electricians and machinists) remained unchanged, and the automated system required 25 percent fewer production workers. This decline in production jobs could be traced to the replacement of welders by robots as well as the reorganization of several production processes. Previously, a group of workers with a broad range of skills relieved others, one at a time, from their various positions on the line. With the new technology, everyone took breaks at the same time, so the multi-skilled relief workers were no longer needed. Under the old system, production workers frequently repaired defective component parts. With the just-in-time delivery system, defective parts were returned to the suppliers.[10]

Computer numerical control (CNC) machines, used to make metal parts and tools, have also led to an erosion of workers' knowledge requirements and responsibilities. These machines, run by semiskilled operators, have replaced tool and die makers, who traditionally were among the most skilled manufacturing workers. As one skilled Canadian machinist described the impact of CNCs on his job:

> "You don't even need a man to monitor [the CNC], the machines will monitor themselves. They've got all these electronic scans that tell when the tool edge is wearing, what horsepower the machine is using. They've got all these tool change systems, so they even can change tools whenever they want, so you don't even need a man there."[11]

As for white-collar employees, the impact of IT has been positive more often than negative.[12] Many of the dreary filing and typing chores of office work have disappeared, and opportunities to acquire word-processing, graphics, database management, and Internet skills have provided more challenging and interesting jobs for some clerical staff. But the full potential of IT for enhancing the quality of work has seldom been realized. On the basis of her early case studies of computer use in the workplace Shoshana Zuboff (1988: 57) concluded that IT "creates pressure for a profound *reskilling.*" Yet, as her studies showed, old-style management control and bureaucracy often prevented workers from learning how to effectively interact with computers. Fearing loss of power, many middle managers in the firms she studied were reluctant to give workers the training and responsibility they needed to make the leap to *informated* (as opposed to simply *automated*) work. A more recent overview of research on the impact of IT on job quality makes a similar point. Rubery and Grimshaw (2001) conclude that the effects are not predetermined. It is how IT is implemented in the workplace, and how new work patterns are regulated within society, that really matters.

From the broader societal perspective, the widespread adoption of IT may have led to greater polarization of skills (Burris 1998). Workers who already are highly skilled (professionals, managers, and technicians) are more likely to have the opportunity to work with IT and to receive the necessary training (Lowe 1997; Marshall 2001b). As Thomas Idle and Arthur Cordell concluded a decade ago, "We are beginning to see the creation of a workforce with a bimodal set of skills. Highly trained people design and implement the technology, and unskilled workers carry out the remaining jobs" (1994: 69).

And what about job creation and job loss? Have new technologies created more jobs than they eliminated? Or will the further spread of IT result in the "end of work" as Jeremy Rifkin has predicted:

> Within less than a century, "mass" work in the market sector is likely to be phased out in virtually all of the industrialized nations of the world. A new generation of sophisticated information and communication technologies is being hurried into a wide variety of work situations. Intelligent machines are replacing human beings in countless tasks, forcing millions of blue and white collar workers into unemployment lines, or worse still, breadlines.[13]

As we have noted in our discussion above, new automated technologies have frequently led to workforce reductions within specific organizations. However, there is little evidence, at least to this point, that the introduction of IT has led to an overall increase in unemployment in Canada (Sargent 2000). Instead, to the extent that job losses have occurred in different industries and regions, corporate downsizing, the shrinking of the government sector, and global business systems that have moved jobs overseas have been much more directly responsible. Thus, there is insufficient evidence to support Rifkin's apocalyptic view of the future and the targeting of IT as the primary cause of job loss. As Charley Richardson (1996) wrote some years ago, "computers don't kill jobs, people do."

Technological, Economic, or Social Determinism?

Technology is a tool that humans have used to their advantage historically, to create jobs and to improve the quality of life. Technology has also been used to eliminate jobs, to destroy the natural environment, and to kill people. There is evidence that today's new IT can create jobs, increase skill levels, and provide workers with more control over their working environment. New technologies also have the potential to do the opposite, as our quick review of the research literature demonstrates.[14] Hence, as a society, our challenge is to find ways to use technology to provide economic opportunities for as many citizens as possible.

Writing in the 1960s, Gerhard Lenski viewed industrial technology positively, arguing that it had led to reduced inequality. At the same time, Marshall McLuhan presented a far more pessimistic analysis of the impact of technology on social life. He concluded that human beings were at risk of becoming the servants of technology. His actual words were stronger; he suggested that if trends continued, we might simply become "the sex organs of the machine world . . . enabling it to fecundate and to evolve ever new forms."[15]

We do not agree with either of these positions since they attribute too much independent power to technology. *Technological determinism,* the belief that the developmental pattern and effects of a given technology are universal and unalterable, removes the possibility of human agency, the potential for people to shape technology for the greater social good. Furthermore, this position ignores the strong evidence that new technologies have been used in very different ways, with different outcomes, in different work settings and societies.

We also reject *economic determinism,* the belief that the "market knows best" how to choose and implement new technologies. Such a perspective would argue that, despite current problems with underemployment, deskilling, and loss of worker autonomy, new technologies will in time lead to more positive than negative outcomes. Behind the "free hand of the market" are real people, making decisions about how to implement the new technologies. These decisions typically have been made by only a very small minority of citizens—owners and managers guided by the profit motive.

Instead, we prefer a *social determinism* perspective, advocating education, wide-ranging discussion, and open decision making about how technologies will be used, by whom, and for whose benefit. We believe that technology should be used to serve individual, community, and societal needs, not merely the needs of a particular company or work organization. Consequently, workers need to be able to participate in decisions about the choice and implementation of new workplace technologies that affect them directly, and citizens need a voice in shaping broader industrial technology strategies.

Because the quest for increased competitiveness through technology has frequently meant job loss or job downgrading for workers, labour movements in North America and Europe generally have opposed the introduction of new technology by management without any input from unions. But this does not mean that unions oppose technological change in principle. Instead, they have insisted that employees be consulted so that the negative effects can be minimized and opportunities for upgrading jobs and improving working conditions maximized. Unions also have argued that any productivity gains should be shared equitably. But while European unions have been successful in obtaining some of these goals, similar gains for organized labour have yet to be achieved in Canada (see Chapter 7).

Involving unions and employees in the technological change process in Canada would constitute a major step toward balancing economic and social goals. Research indicates that in countries with strong labour movements, consultation on technological change is more often the norm. Negative effects such as deskilling and income polarization have been less pronounced, and workers have not been as opposed to technological change.[16] Because Sweden and Norway have powerful unions that take an active role in shaping social policy, legislation requires employers to consult with employees prior to automating. Workers have direct input regarding the reorganization of work around new technologies. Widespread layoffs due to technological change have generally been avoided, and, in cases of redundancy, employers have retrained or found other jobs for those affected. In the past few years, however, some of these policies have been weakened and others are being questioned (Smith et al. 1995: 713). It remains to be seen whether Sweden can maintain the system that has provided a model for North American trade unionists.

NEW MANAGEMENT APPROACHES: A CRITICAL PERSPECTIVE

Chapter 5 introduced some of the recent management approaches that have promised to flatten workplace hierarchies, provide workers with additional job autonomy, and enhance workers' skills. While we have already questioned some of these promises earlier in this chapter, it would be useful to further highlight the specific labour process criticisms of several prominent contemporary management approaches.

Quality of Working Life (QWL) and Total Quality Management (TQM)

The quality-of-working-life (QWL) model of management grew out of several different traditions, including the British sociotechnical systems approach, early Swedish work reforms, and Herzberg's writings on job satisfaction. The emphasis in QWL programs was primarily on workers' immediate job tasks, with job enrichment, job enlargement, job rotation, and autonomous work teams being among the most common innovations. In some settings, management hierarchies were also reduced as work teams took on some of the basic management tasks. The QWL package was typically presented to workers as an effort on the part of management to improve the quality of working life, but obviously managers were hoping to improve productivity as well.

Some QWL initiatives resulted in model "high-performance" workplaces, such as the Shell Sarnia refinery and GM's Saturn factory. Unions participated in the design of the program, workers received extensive training, and considerable autonomy and decision-making responsibility was given to self-directed work teams. But more often, critics charge, changes in workers' job tasks were minimal, they did not gain the opportunity to influence decisions on larger workplace changes (the introduction of new technology, for example), and management imposed the programs rather than involving the workers in the decision.[17] In some cases, QWL programs were introduced in a cynical attempt to counter the organizing efforts of unions—if management could provide these benefits, who needs a union?

Like QWL approaches, total quality management (TQM) programs have been heralded as a "win–win" approach to labour–management relations: workers benefit through improved working conditions and productivity increases. TQM emphasizes customer satisfaction and continuous improvement in addition to job redesign. It also places more emphasis on the need to develop a strong organizational culture. By the 1980s, when TQM was becoming popular, concerns about global competitiveness were becoming more pronounced. Hence, the need for workers to buy into the goals of the program and the values of the corporate culture took on a larger significance. The message for workers was that winning—being part of an ever-improving and more productive company or government workplace—was important not only for the workers and management, but also for the economy. By the same logic, losing meant more than just being part of an unsuccessful work organization. In the new competitive economy, it might also mean losing one's job.

But TQM efforts have seldom invited workers to participate in decisions about whether to introduce new technologies, or to restructure or downsize organizations. Instead, critics point out, workers have been invited to find ways to work harder, often in an atmosphere of anxiety about possible job loss. Consequently, promises of improved working conditions have frequently translated into only superficial changes. In fact, there is strong evidence linking the large increase in occupational injuries and illnesses (carpal tunnel syndrome, for example) in the United States over the past two decades to the increased prevalence of lean production, quality circles, and just-in-time delivery systems (Askenazy 2001; Brenner, Farris, and Ruser 2004).

As in the case of QWL, some TQM initiatives were presented to workers as alternatives to unions. Systems put in place to monitor quality, and to encourage teamwork and product/service improvement, have also been effective ways to control workers (Sewell 1998). Ironically, many middle managers also find it stressful to constantly reorganize work processes. Thus, while TQM may have increased productivity in some settings (see Chapter 5), it has done so by requiring employees to work harder. All too often, working conditions have not improved and workers have not been empowered (Zbaracki 1998).[18]

Flexible Specialization and Lean Production

Flexible specialization joined the list of new management approaches with considerable fanfare. It promised revolutionary change in Fordist manufacturing systems (computer-coordinated, small-run, flexible production) and the return of craft forms of production (highly skilled, autonomous workers involved in all aspects of the production process). Consequently, bureaucracy should be reduced and workplace conflict should subside. In addition, the power of multinational corporations employing traditional production methods would decline. Essentially, flexible specialization was seen as the production system of a postindustrial economy characterized by less conflict and inequality.

Unlike QWL and TQM, flexible specialization has limited applicability as a management model outside of the manufacturing sector. In fact, it remains unclear just how applicable it is to the full range of manufacturing industries. More important for our discussion here, there is little evidence of working conditions improving and workplace conflict subsiding in firms of this type. Multinational corporations have certainly not become less powerful over the past decade.

Thus, one critique of the flexible specialization model is that it is more a blueprint for a desirable workplace than a description of an emerging trend. However, if it were to become a trend, this model proposes no solution to the social problem of workers made redundant by technological change. Instead, it offers a vision of a severely segmented labour market in which a small, elite group of highly skilled workers are the main beneficiaries, while the majority continue to perform routine, unskilled, or semiskilled work.[19]

The *lean production* system of manufacturing also relies heavily on computer-based and automated technology, but without the commitment to the small-scale, highly flexible production processes of flexible specialization. Nevertheless, it goes beyond the rigid and standardized mass production of Fordism by encouraging a flexible approach to product redesign and maintaining an emphasis on quality, continuous improvement in production methods *(kaizen)*, and just-in-time delivery. Like QWL and TQM, lean production claims to rely on highly skilled and thinking workers, and promises a reduction in workplace conflicts through its teamwork model of decision making. With all of these features, it is little wonder that lean production has been portrayed by its advocates as the final solution to Taylorism and Fordism (Womack, Jones, and Roos 1990).

Nevertheless, many critics have been harsh in their assessments of lean production's impact on the labour process.[20] They point out that multitasking (one person doing more jobs) is not the same as multi-skilling. Instead, it is one of the ways in which management has squeezed more work out of a smaller number of workers. The elimination of replacement workers has also had this effect, as has *kaizen,* the consensus-based approach to continuous improvement. By emphasizing how workers in a particular factory need to outperform their competition elsewhere, management essentially harnesses peer pressure to speed up production.

By way of example, James Rinehart describes a California automobile factory that, before lean production, managed to keep its workers busy for 45 out of every 60 seconds they were on the job. Lean production brought this up to 57 out of 60 seconds. But, as he puts it, "[t]o call kaizen a democratization of Taylorism is to demean the concept of democracy" (2006: 162). Other critics agree, noting that management continues to tightly control the topics discussed in kaizen sessions (Graham 1995). Essentially, lean production has not led to significant skill enhancement or to worker empowerment (Yates, Lewchuk, and Stewart 2001). It has meant a faster pace of (still repetitive) work (Schouteten and Benders 2004), higher stress for employees (Anderson-Connolly et al. 2002), and, in some cases, higher injury rates (Brenner, Farris, and Ruser 2004). Hence, lean production has also led to resistance by workers—the 1992 strike at the CAMI plant in Ingersoll, Ontario, is an example. Thus, as Rinehart (2006: 200) concludes, "lean production constitutes an evolution of rather than a transcendence of Fordism," and therefore can be called neo-Fordism.

High-Performance Workplaces

The *high-performance workplace* (HPW) is one of the latest additions to the list of new management strategies (Godard 2001, 2004). This model shares a number of features with other recent management approaches, including an emphasis on reducing bureaucratic hierarchies and rigid job descriptions, increasing workers' decision-making opportunities, and offering them more training to enhance their skills. But the HPW approach goes further by advocating human resource management programs that are family friendly and emphasizing the importance of a healthy, non-stressful work environment. It also recommends that employers share profits from productivity increases with workers through variable-pay plans that link some compensation to performance.

Like flexible specialization, the high-performance workplace is more a model of a desirable future than a description of current practice. Only a minority of work organizations have moved in this direction. Among those that have, a majority have adopted only some of its core features (Osterman 2000). Furthermore, as studies evaluating the effectiveness of the HPW accumulate, it is becoming apparent that in many situations there are only negligible payoffs, for both employers and workers (Godard 2004). Perhaps we should not be surprised, given the inherent conflict of interests between employers and workers in a capitalist economy (Godard 2001). Nevertheless, for workers in an HPW, the potential benefits (skill upgrading, decision-making opportunities, profit sharing, improved working conditions) are considerable. But there are also drawbacks, since the HPW model assumes that employment security can no longer be guaranteed, and that profit sharing also means risk sharing (if the company loses money, so do the workers). In fact, downsizing and layoffs have frequently accompanied the implementation of HPW systems (Osterman 2000; Danford et al. 2004).

Critics influenced by a labour process perspective are skeptical of the commitment asked of workers in return for these benefits (Danford et al. 2004), pointing to previous examples of companies using QWL, TQM, and other similar approaches to speed up work and sideline unions. Nevertheless, there are some examples of unions and management working together to build high-performance organizations. Clarke and Haiven (1999) describe how, in Saskatchewan in the 1990s, the Communications, Energy and Paperworkers Union (the same union involved in the Sarnia Shell plant described in

Chapter 5) participated in redesigning the Saskatoon Chemicals work organization, including profit sharing and payment for skills. As in other successful union–management projects, what made the difference was a very strong union that was centrally involved in all aspects of the change process. In addition, both parties agreed to a process of *continuous bargaining*, in contrast to the traditional approach of seeking a collective agreement that would remain unchanged for several years.

There have been few attempts to introduce HPW practices in the lower-tier service industries. However, a study of a handful of large U.S. retail companies like Wal-Mart and Home Depot that have attempted such organizational changes concludes that workers have not really benefited (Bailey and Bernhard 1997). Wages did not rise appreciably, skill sets were not significantly enhanced, and only a few workers found new career opportunities. High-performance work organizations, while not widespread, are more likely to be found in the upper-tier services and in some of the goods-producing industries. Thus, the advent of this form of work organization may signal greater labour market segmentation. The result might be a society in which a small, privileged elite of high-skill, relatively autonomous workers are well rewarded while the majority of workers face greater risk of low-skill, less-rewarding work with greater employment insecurity. In short, the high-performance workplace model does not offer a solution to the larger social problem of growing labour market inequality.

New Management Approaches: A New Labour Process?

The management literature reviewed in Chapter 5, while recognizing some of the flaws and inconsistencies in these approaches, uses labels like *participative management* and *employee empowerment* to describe recent innovations. More critical observers have described them as *neo-Fordist* or *post-Fordist*, implying that the positive changes have not been nearly as significant as their proponents claim. By offering some small concessions to workers, these approaches have maintained the basic production framework and power structure of industrial capitalist society. Some critics go even further, using terms like *hyper-Taylorism* (Russell 1997: 28) or *neo-Taylorism* (Pruijt 2003). They argue that work intensification signals a new era of "management by stress" in which workers push themselves and each other to make more profits for their employers (Kunda and Ailon-Souday 2005: 209). A good example is provided by Vivian Shalla (2004: 363) who describes how Air Canada has been "squeezing more time and output from employees" as it attempts to become a more flexible and competitive global airline.

But labels are not as important as real outcomes. What do we see when we look back at more than three decades of new management approaches in North America? All of them offer some potential for skill upgrading, increasing worker participation in decision making, and reducing bureaucracy. In some cases (Shell's Sarnia refinery, the Tennessee Saturn plant, and the Saskatoon Chemicals workplace, for example), the implementation of these approaches has gone a considerable way toward improving the quality of working life for employees. None of these management models, however, really provides permanent and effective avenues for workers to have ongoing input into larger organizational decisions, such as the introduction of new technologies or a decision to restructure the organization.

The two approaches that go the furthest to counter Taylorism and Fordism—flexible specialization and the high-performance workplace—have not been implemented often. Other approaches typically do not deliver nearly as much as they promise. Furthermore, when we examine them more critically, we frequently find that they have also been used to increase control over workers and to speed up work (Burchell, Ladipo, and Wilkinson 2002). In addition, the advent of the new management approaches has been accompanied by widespread organizational restructuring and downsizing. The outcome, for society as a whole, has been a polarization of the labour force in terms of skill, income, work intensification, and job security, leading to greater social inequality.

As we concluded about the impacts of new technologies on the labour process, the problem does not lie in the management system so much as in how it is implemented, by whom, and for what purposes. Overall, new management strategies—we might call them *social technologies*—have almost always been introduced with productivity and profit as the main goals. For example, based on his in-depth study of four U.S. manufacturing plants that introduced a wide range of teamwork and other new management approaches in the 1990s, Steven Vallas (2003) concludes that the overriding emphasis on increasing profit margins led managers and corporate executives to maintain or re-introduce standardized (i.e., Taylorist) production methods that clearly

stood in the way of job enrichment and increased worker involvement in decision making. In short, improvements in quality of working life and worker
empowerment have been secondary motivations, while reductions in social
inequality have seldom figured in the decision. And workers have rarely been
involved in the decision to implement new social technologies, either directly
or through their unions. With a few exceptions, when workers have been
invited to participate, their input has been restricted to smaller decisions about
changes in their immediate work tasks. Consequently, we should not be surprised that much of the promise of the new management approaches has not
been realized.

A CRITICAL PERSPECTIVE ON WORKPLACE HEALTH AND SAFETY

Our profile of good jobs and bad jobs in Chapter 3 documented the high costs
to workers, employers, and society of workplace health and safety risks. Every
year, hundreds of Canadian workers are killed and thousands are injured on the
job. Acknowledging that it is impossible to put a price tag on human suffering
and death, the magnitude of the problem is nevertheless indicated by the
$4.65 billion paid as compensation for work-related injuries and fatalities in
Canada in 1998 (Human Resources Development Canada 2000a: 10). In addition, the future health of many workers is regularly placed at risk. Every year, a
sizable minority of all Canadian workers is exposed to workplace health hazards,
including air pollution, excessive noise, physically demanding repetitive work,
and stress-inducing work environments (Bachmann 2000; Williams 2003).

Workplace health and safety is highly relevant to our discussion of conflict, control, and resistance in the workplace. Historically, dangerous
machinery, unsafe work sites, polluted air, and exposure to carcinogenic substances and other risks have taken their toll on the working class in terms of
shorter life expectancies and higher illness and disease rates. Most of the
progress made over the years in reducing workplace health and safety risks has
been the result of workers fighting for improvements, often against the strong
opposition of employers. Furthermore, workplace health and safety continues
to be a contentious issue in union–management contract disputes and in the
larger political arena. At the same time, the area of workplace health and
safety provides one example where, at least to some extent, workers and their
employers have managed to identify and work together toward common goals.

The Politics of Workplace Health and Safety

In the early years of Canada's industrialization, workers had little protection from what were often extremely unsafe working conditions. Work in the resource industries, in construction of canals and railways, and in factories was hazardous, and the risk of injury and death was high. As in other industrializing countries, an *administrative model of regulation* slowly developed, in which the government set standards for health and safety and tried to enforce them (Lewchuk, Robb, and Walters 1996: 225). Employers frequently opposed these efforts, arguing that their profits were threatened and that the state had no right to interfere in worker–employer relationships. In turn, unions fought for change and the public was mobilized behind some causes, including restrictions on the employment of women and children in factories. Thus, the Factory Acts of the 1880s led to improvements such as fencing around dangerous machines, ventilation standards for factories, and lunchrooms and lavatories in large workplaces. The employment of women and children was also curbed on the grounds of protecting their health.

Over the years, additional safety standards were introduced. In unionized workplaces and industries, *collective bargaining* also made a difference as unions negotiated for better working conditions and the elimination of specific health and safety hazards. By the second decade of the 20th century, provincial workers' compensation boards had put in place *no-fault compensation systems* that still exist today. Injured workers are provided with some money, the amount depending on the severity of their injury, and with partial compensation for lost wages, no matter whose fault the accident. In return, they give up their right to sue for compensation if the employer was at fault. The system is funded by contributions from employers, with the amount based, in part, on their safety record. Thus, there is a monetary incentive for employers to reduce workplace health and safety hazards and to encourage safe working practices among employees.

All of these approaches to dealing with health and safety issues—standards setting and enforcement by the state, collective bargaining between unions and employers, and the no-fault compensation system—offer little room for direct involvement on the part of workers who are most directly affected by unsafe and unhealthy working conditions. But the introduction of the *internal responsibility system (IRS)* has changed this. A guiding principle of this approach is that workers' personal experience and knowledge of work practices and hazards are an integral part of any solution to health and safety problems. In addition, workers should have the right to participate in the identification and elimination of workplace hazards. Furthermore, health and safety is also management's responsibility. Overall health and safety is a workplace issue—an internal responsibility—too important to be left to government alone. In the IRS system, employees are directly involved with management in monitoring and inspection, and in education and health promotion in their workplaces.

Direct worker involvement in health protection and promotion in the workplace increased dramatically in Europe in the 1970s. This trend grew out of demands for greater industrial democracy (discussed later in this chapter), and government reviews of traditional and unsatisfactory approaches to occupational health and safety (Tucker 1992). In Canada, the first major initiative of this sort was the 1972 Saskatchewan Occupational Health Act. It broke new ground by broadly defining occupational health as "the promotion and maintenance of the highest degree of physical, mental, and social well-being of workers" (Clark 1982: 200), and by making joint health and safety committees (JHSCs) mandatory. Other provinces and the federal government (about 10% of Canadian workers are covered by federal labour legislation) followed. Today the internal responsibility system, built around JHSCs, is part of the legislated workplace health and safety system across the country, although there are significant differences in the extent to which it has been implemented and promoted by provincial governments (Tucker 2003).

JHSCs range in size from 2 to 12 members, half of whom must be non-managerial employees (either elected or appointed by a union if one exists), and are required in any workplace employing 20 or more. In most provinces, workplaces with at least 5 but fewer than 20 employees must also have at least 1 non-managerial safety representative. The committees keep records of injuries, participate in safety inspections, make recommendations to management about health and safety concerns, inform employees about their rights with respect to such concerns, and develop safety enhancement and educational programs (Gordon 1994: 542–43).

Along with giving JHSCs the *right to be involved* in health and safety issues, legislation also allows workers the *right to refuse unsafe work,* the definition being based on the individual employee's own assessment that a particular task presents a genuine risk. If subsequent inspections determine otherwise, the legislation protects workers from reprisals from employers.

Workers also have the *right to be informed* about potentially hazardous materials with which they might be working.

The IRS system appears to be effective. Research has shown that the recommendations of JHSCs to management are usually heeded (Gordon 1994: 548). An Ontario study examining data on claims for lost time due to accidents between 1976 and 1989 concluded that "where management and labour had some sympathy for the co-management of health and safety through joint committees, the new system significantly reduced lost-time accidents" (Lewchuk, Robb, and Walters 1996: 225). But if an organization reluctantly introduced JHSCs, the system had little effect. The success of JHSCs in Canada has led to recommendations that the Canadian system be copied in the United States. The more traditional U.S. system, which still relies primarily on legislation and, where unions are present, on collective bargaining, is less effective (Gordon 1994).

However, there has also been some movement backward in Canada over the past decades (Tucker 2003). During the 1990s, the Klein government in Alberta shifted to more of a "voluntary compliance" approach to worker health and safety issues, and also reduced its expenditures on occupational health and safety programs. As a result, many fewer employers were prosecuted for non-compliance with health and safety legislation during this era. The fact that the number of government inspections and prosecutions rose again in the late 1990s suggests that, despite its ideological appeal, the employer self-regulation approach may have been seen to be ineffective by the provincial government. Interestingly, the Conservative government in Ontario in the mid-1990s did not follow the Alberta lead. Instead, provincial government enforcement of health and safety legislation increased, and a number of changes were made to legislation that reduced risks to workers. The larger presence of industrial unions in Ontario, as well as the greater power of opposing political parties, may be part of the explanation for these variations (Tucker 2003).

The Labour Process and Workplace Health and Safety

Does the relative effectiveness of JHSCs and the IRS system indicate that workers and management are no longer in conflict over health and safety issues, that this is a win–win situation? We would argue that workers and management have moved further in this direction on health and safety than they have elsewhere. In part, this is due to the economic incentives built into the system for employers—a better safety record means lower costs of production. More important, legislation has reduced some of the power differences between management and workers and directly involved each in addressing health and safety issues.

Even so, workplace health and safety remains a contested terrain. Critics of the IRS system have pointed out that, in many JHSCs, non-management members continue to have much less influence, particularly in non-unionized workplaces. Some employers largely ignore the joint committees, despite legislation (Lewchuk, Robb, and Walters 1996: 228). Rather than seeking ways to make their workplace safer and healthier (and by so doing, reducing their costs), some employers instead focus on challenging workers' claims of injuries and pushing them to return to work sooner (Thomason and Pozzebon 2002). Research has also shown that workers are often unaware of their rights and lack the knowledge needed to address complex health hazards, and that scientific and medical experts give the appearance of impartiality to a system that continues to be dominated by employers for whom profits come first.[21] In some tragic cases like the Westray mine explosion in 1992 in Nova Scotia, the system failed completely (Comish 1993; McCormick 1998). Local miners had warned that the mine site was unstable, and the initial mining plan was rejected for safety reasons. But the mine still opened, and remained open even after repeated problems with methane gas, coal dust, and roof falls. Eventually, 26 miners died in an explosion. In other Canadian industrial settings, workers die more slowly, but not necessarily less painfully, from long-term exposure to carcinogens like asbestos (Mittelstaedt 2004).

While Canadian unions have fought for safer workplaces for well over a century, they have sometimes found themselves balancing concerns about their members' health and well-being with equally strong concerns about potential job loss. Robert Storey and Wayne Lewchuk (2000) describe how, in the early 1980s, the United Automobile Workers of America tried to force Bendix Automotive in Windsor, Ontario, to deal with the deadly problem of asbestos dust in its brake shoe factory. Having already closed one of its plants because of this problem, the company announced that it now planned to close the remaining factory. Some of the workers wanted their union to insist that the workplace be made safer, whatever the cost. Others, fearful that the company would close the factory and they would lose their jobs, wanted the union to back down. The factory was eventually closed anyway. While highlighting how little power the workers really had, compared to the company, Storey and Lewchuk also describe the limits of provincial health and safety legislation, which is incapable of stopping an employer from closing a workplace rather than dealing with its health hazards.

Along with disputes over management's and worker's rights, the definitions and sources of work-related health problems also continue to be contested. For example, there is mounting evidence that job-related stress takes a toll on employees' health. Stress-induced health problems can range from headaches to chronic depression and heart disease. But stress is frequently rejected as a legitimate concern by workers' compensation boards who administer claims.[22] They tend to use a narrow definition of health and illness, focusing mainly on physical injuries and fatalities. Employers and compensation boards sometimes also attempt to place the blame for workers' health problems on their lifestyles or family situations rather than on their jobs. Such a "blaming the victim" ideology assumes that workers are careless, accident-prone, or susceptible to illness.

Rather than accepting a narrow definition of workplace health and safety issues, Bob Sass (1986) has argued that we need to

"stretch" the present legal concept of risk, which covers dust, chemicals, lighting, and other quantifiable and measurable aspects of the workplace to cover all work environment matters: how the work is organized, the design of the job, pace of work, monotony, scheduling, sexual harassment, job cycle, and similar work environment matters of concern to workers.

Sass wrote this two decades ago. Since then, he has gone even further to argue that the "weak rights" workers now have within the IRS system need to be extended. "Strong rights," as he sees it, go beyond co-management and would involve "*worker control* in the area of the work environment."[23] In such a democratized work environment, workers would not have to accept the constant tradeoff between health risks to themselves and the efficiency and profit demanded by employers. Sass sees unions as the only vehicle for such change since they are worker-controlled, even though most unions continue to work within the current IRS system and accept its basic premises and goals.

However, with some exceptions, it has not been unions that have been pushing for a broader and more inclusive definition of worker health and well-being. Instead, corporate health promotion, or "workplace wellness" programs, emerged in the 1980s as yet another human resource management policy that would help to motivate workers and raise productivity (Conrad 1987). The orientation of these wellness programs has typically been to change workers' lifestyles or health-related behaviours (exercising more and quitting smoking, for example). Some have gone considerably further to promote family-friendly and less-stressful employment policies and practices.[24]

Workplace wellness programs have good intentions, and can benefit employees who participate in them. But they have not addressed issues of job redesign and have not had an impact on the unequal distribution of power within work organizations which contributes to health and safety problems.[25] Furthermore, because such programs are typically only available in larger work organizations, many workers are unable to participate. Thus, while offering assistance to some workers, corporate wellness programs do not improve working conditions for the working class as a whole.

Terrance Sullivan (2000) introduces a collection of essays about "injury and the new world of work" by observing that we find ourselves, once again, in a period of revolutionary change, with new information and communication technologies, new types of work organization, and new management approaches. He argues that the legislation and the internal responsibility system that was developed to address workplace health and safety problems in the last century may not be adequate for the workplaces of the 21st century, an argument echoed by O'Grady (2000). Not only can we expect to see new types of injuries, we also know more today about long-term effects of workplace hazards and stressors, and about the wider array of physical and mental disabilities that might be caused in the workplace. Thus, like Bob Sass (1986), Sullivan (2000) argues for a broader definition of workplace health and safety issues, and calls for a re-examination of the legislation and programs developed to address the problem.

INDUSTRIAL DEMOCRACY: RETHINKING WORKERS' RIGHTS

Our evaluation of new management models concluded that none really offered workers input into decisions beyond their own jobs. New technologies generally have been imposed by management and then, perhaps, workers have been given more freedom to use the technology. Decisions about downsizing and organizational restructuring are rarely negotiated, and unions are rarely consulted about policies that replace permanent, full-time jobs with nonstandard jobs. The IRS system attempts to go further, acknowledging the need to involve workers in the co-management of health and safety risks. But it does not give workers and employers equal control over the work environment.

Within capitalism, ownership carries with it the ultimate right to control how work is organized and performed. We have seen some employers relinquishing some of these rights, allowing workers greater autonomy over their immediate work tasks in the belief that efficiency and productivity could be improved and conflict reduced. *Industrial democracy*, in contrast, attempts to involve workers in a much wider range of decisions within the organization. Industrial democracy applies the principles of *representative democracy* found in the political arena to the workplace. Workers have a voice at the work-group level, as well as indirectly through elected representatives on corporate boards and other key policy-making bodies. A greater degree of open discussion and consensual decision making involving all employees is the goal of democratized work environments.

Even so, as we will argue below, industrial democracy in its various forms has not completely eliminated workplace power differences. The shared decision-making structures of industrial democracy still give owners and managers more influence. Furthermore, industrial democracy does not guarantee less bureaucracy, reduced income inequality, skill upgrading, or even more task-related autonomy for workers. Even if workers are involved extensively in decision making through elected representatives, the debates might not be around these issues. Job security, for example, or health and safety might be the major concern. Thus, in some settings, industrial democracy has led to job redesign and profit sharing. In others, the results may have been fewer layoffs or more consultation about technological change. Overall, however, industrial democracy has been implemented to a greater degree in European countries than in North America (Frege 2005).

Industrial Democracy in North America

Attempts to involve workers in management, at least to some extent, actually have a long history in North America. Following World War I, *works councils* including elected workers and management representatives were set up in large workplaces in a number of industries in both Canada and the United States. The councils met to discuss health and safety, workers' grievances, efficiency, and sometimes even wages. In a number of settings, particularly the coal industry, the impetus for these initiatives was a series of long, bitter, and violent strikes.[26]

Mackenzie King, who later became prime minister of Canada, was a labour consultant to some of these companies at the time. He was a strong advocate of the principles of industrial democracy, believing that works councils and related initiatives would reduce industrial conflict and help usher in a new era of social harmony (King 1918). Critics argued that mechanisms to encourage more cooperation between workers and management continued to favour the latter, and that works councils were just a way to deter workers from joining unions. Clearly, in many companies this was the case. With the onset of the Depression at the end of the 1920s, interest in these earlier forms of industrial democracy waned, perhaps because unemployment reduced labour unrest.

Since then, calls have periodically been made for a revival of such worker–management decision-making systems. Not coincidentally, industrial democracy has attracted more interest during periods of labour unrest—immediately after World War II, for example, and again in the 1970s. During those strike-filled eras, employers and politicians looked to Germany and other European countries where industrial democracy had been implemented more widely. North American unions have generally opposed the idea, believing that works councils and other similar forms of industrial democracy undermine collective bargaining, the more traditional means by which workers have negotiated with management for improved working conditions (Guzda 1993: 67).

The spread of industrial democracy in North America has been slow. Most employers have been more interested in avoiding strikes than in really sharing power, and unions have not been supportive. One estimate from the early 1990s indicates that four out of five Fortune 500 companies have set up "worker involvement programs" (Guzda 1993: 67). However, this general term probably includes a wide variety of QWL, TQM, lean production, and other management programs that may provide only limited opportunities for worker involvement in decision making beyond their immediate work tasks.

Institutionalized Industrial Democracy in Germany

North American experiments in industrial democracy have usually been introduced by management, occasionally with some pressure from unions. In contrast, principles of industrial democracy have been institutionalized via legislation in many northern European countries.[27] Workers are given the right to elect representatives to sit on works councils or corporate boards, the right to consultation regarding technological changes, and more grassroots control over health and safety matters (Beaumont 1995: 101–8).

The German model of *codetermination* has a century-long history. In the 1890s, concerns about widespread labour–management conflict motivated Bismarck to bring in legislation giving works councils the right to advise management on workplace regulations. This legislation was broadened in the 1920s as the Weimar government sought ways to counter the threat of the Russian Revolution spreading into Germany. After World War II, concerned that labour strife would hold back the reconstruction of Germany's devastated economy, the British, American, and French administrators of occupied Germany extended the codetermination legislative framework. A number of additional changes were made in the 1970s and 1980s (Beaumont 1995).

In its current form, German legislation makes works councils mandatory in any workplace with more than five employees. Elected representatives of the workforce (their number proportional to the size of the firm) share decision making with management representatives. The legislation stipulates that in larger work organizations where unions are present, a proportion of the elected members must be from the union. But non-unionized white-collar workers must also be represented. In larger companies, there are several layers of shared decision making. The most powerful management board, a small group of three or four individuals, has only one worker representative, selected from within the larger works council.

The codetermination legislation requires employers to advise and consult with works councils about plans to introduce new technologies, restructure work, re-allocate workers to different tasks or locations, or lay off workers. Works councils can demand compensation for workers negatively affected by new technologies or organizational restructuring. Consequently, the North American strategy of downsizing has been much less common. Elected representatives are jointly involved with management in determining hours of work, pay procedures (bonus rates, for example), training systems, health and safety rules, and working conditions. However, collective bargaining between employers and unions about overall pay rates takes place outside the works councils on an industry-wide level. Consequently, the right to strike remains with the larger unions, not with the local works councils. The unions, however, have much less influence in the joint decision-making process at the local level.

The German *dual representation system* of industrial democracy—industry-wide unions and local works councils—is viewed relatively positively by both employers and workers. There are some concerns that unions may be weakened by the dual system of representation (in the long term, this could jeopardize workers' employment security), but union members still constitute a majority of elected representatives on the works councils. Overall, joint decision making has meant better communication between workers and management, greater protection of workers' rights, improved working conditions, and a reduction in labour–management conflict. The latter was, of course, one of the original goals of the codetermination legislation, and this is also what has frequently impressed North American observers concerned about strikes and reduced productivity.

During the 1990s, the German economy began to lose some of its momentum, in part because of the costs of reuniting West and East Germany. Employers began to call for changes in the dual system of representation. In particular, there were pressures to reform the industry-wide collective bargaining system that had meant similar pay rates and conditions of employment in different-sized firms throughout the country. Employers argued that they needed more flexibility to remain competitive (Beaumont 1995: 107–8). But so far, major changes have not been made to Germany's labour legislation, and works councils continue to protect workers' rights. They also remain active and effective in other northern European countries (Looise and Drucker 2003; Poutsma, Hendrickx, and Huijgen 2003).

Sweden: Industrial Democracy as a National Goal

In Germany, concerns about strikes and social unrest led to codetermination legislation. In Sweden, the motivation to implement industrial democracy was somewhat different. An exceptionally strong union movement and the governing Social Democratic Party worked together to improve employment conditions and give workers more rights. In fact, tripartite cooperation and consultation among large and centralized unions, the employers' federation, and the state has been a hallmark of Swedish democracy (see Chapter 7). Thus, in Sweden in the 1970s and 1980s, industrial democracy was elevated to a national goal within a much broader policy of commitment to reduced inequality and full employment.[28]

The full-employment policy has three key elements. A wage policy, based on the concept of equal pay for equal work, removes pay differentials among regions, industries, and firms by pushing low-wage firms to pay decent wages, and by allowing profitable enterprises to pay less than they would under conventional collective bargaining. The reasoning is that weaker enterprises will either become more efficient or go out of business. The relatively lower wages in expanding industries should stimulate economic growth. Active labour market interventions, including heavy investments in training, assist new labour force entrants, as well as workers displaced from declining industries, in finding work. Selective employment policies also attempt to create jobs in regions or industries affected by structural shifts in the economy.

As in Germany, Swedish law mandates employee representation on corporate boards of directors in all but the smallest firms. Elected members are involved in joint decision making with management in a wide range of areas, but cannot participate in discussions regarding negotiations with unions. But the 1977 Act on Employee Participation in Decision Making extended Swedish employees' rights beyond the rights of German workers. Specifically, in addition to the obligation to inform and consult with workers on major decisions such as factory closures or introducing new technologies, Swedish employers must negotiate with unions prior to making such decisions. Unions must also be given complete access to information on the economic status of the firm, its personnel policies, and so forth. In 1984, Sweden went further by setting up Wage Earner Funds that redirected corporate taxes into share purchases on behalf of employees in manufacturing and related industries (Whyman 2004).

The 1978 Work Environment Act goes beyond the goal of industrial democracy—joint decision making throughout a work organization—to address problems of worker satisfaction and personal fulfillment, a subject we examine in more detail in Chapter 8. The act aims to achieve "working conditions where the individual can regard work as a meaningful and enriching part of existence." It is not enough for work to be free of physical and psychological hazards; it must also provide opportunities for satisfaction and personal growth, and for employees to assume greater responsibility. Thus, unlike QWL and other recent management models that view improved productivity and less conflict as desirable outcomes of worker satisfaction, this legislation makes worker satisfaction and individual growth a high priority.

During the 1970s and 1980s, Sweden's impressive economic performance, low unemployment, and reduction of wage inequalities were seen as strong evidence of the success of its model of society-wide industrial democracy. But the following two decades brought changes. Some of the active labour market policies encouraging worker mobility and maintaining full employment were eliminated, and unemployment was allowed to rise. Employers sought more flexibility in the allocation of labour in the production process, and were able, as in Germany, to reduce the extent of nationwide centralized wage bargaining. Lean production made its appearance in Swedish factories, some of the most progressive factories (in terms of workers' autonomy and skill enhancement) closed, and employers introduced new technologies to cut jobs.[29] But the framework legislation that supports the Swedish system of industrial democracy is still in place, and the Swedish version of codetermination continues to be successful, although it is at the local level, rather than the national level, where unions continue to most effectively represent their members' interests (Levinson 2000).

WORKERS TAKING OVER OWNERSHIP

A basic capitalist principle is that ownership carries with it the right to control how work is performed and how the organization is run. By offering some degree of co-management or codetermination, the various forms of industrial democracy have given workers a few of the rights traditionally attached to ownership. However, in some settings workers have gone much further, taking over partial or complete ownership of the means of production. How economically viable is worker ownership within a global capitalist economy? And does worker ownership really provide more opportunity for control over the labour process?

Employee Share Ownership Plans

In North America, *employee share ownership plans (ESOPs)* have been widely promoted as a way for workers to share in the profits of production and as a means of generating more consensus in the workplace. If workers are part-owners, the reasoning goes, they should be more likely to identify with the company, work harder, and cooperate with management. In the past few decades, ESOPs have been part of the package offered to employees in many private-sector high-performance workplaces.

In some companies with ESOPs, shares in the company are purchased for employees and provided as part of a benefit package. In other workplaces, employees are allowed to purchase shares at reduced rates, or their employer pays for a portion of the cost if workers choose to buy shares (Luffman 2003). Depending on the details of the plan, some categories of employees might receive (or be allowed to buy) more shares than others. Typically, higher-paid employees receive or can buy the most shares. Thus, employees receive some of the company profits, in the same way as would other shareholders. However, when shares available through the ESOP are provided as benefits in lieu of wages, employees will also lose when the company is losing money, since their shares then lose value.

A recent estimate (Brohawn 1997: 1) suggests that there are more than 10,000 ESOPs in existence in the United States, in firms representing over 11 million employees (not all of these employees would necessarily take advantage of the stock option plan). In Canada, about two-thirds of the firms listed on the Toronto Stock Exchange have ESOPs. About half of these plans are available to all employees, with the company paying at least some of the cost. However, when we focus on the full Canadian labour force, we find that only 1 in 10 private-sector workers have the opportunity to buy shares in their company. And only 8 percent are employed by companies that subsidize such share purchases (Luffman 2003: 29). Thus, as with many other fringe benefits (see Chapter 3), the opportunity to participate in an ESOP is much more often available to workers who are already working in better jobs.

Similar plans are in existence in a number of European countries, although they are not nearly as widespread as various forms of institutionalized industrial democracy. A recent survey of large workplaces in 10 European Union countries showed ESOPs present in about 1 in 10 private companies. ESOPs were somewhat more common in Britain than in other continental EU countries with histories of industrial democracy (Poutsma, Hendrickx, and Huijgen 2003). In Sweden, the government-initiated Wage Earner Funds have been used to make share purchases on behalf of employees in some industries since 1984.[30]

In North America, ESOPs have been set up primarily to provide additional benefits to workers or to allow them to purchase shares, but not to promote joint decision making. In workplaces where employees participate in an ESOP as well as in some form of joint decision making, it is usually because some other managerial model (QWL, for example) has also been implemented. A British study of the recently privatized bus industry shows that firms with ESOPs are somewhat more likely to have workers on the board of directors, although there is no evidence that share ownership by employees led to this weak form of industrial democracy. The study also reveals that workers participating in ESOPs were no more likely to have control over their own work tasks, although management was concerned that workers might begin to demand more control. Unions representing the workers were more interested in using their seats on the board of directors to preserve jobs than to gain control over the labour process (Pendleton et al. 1996).

An example of a company offering profit sharing, but going much further in terms of worker participation, is the fascinating Brazilian company SEMCO (Semler 1993). The financially successful company with 300 employees (another 200 workers run "satellite" businesses that subcontract to SEMCO) also allows workers to set their own production quotas, redesign products, develop marketing plans, and determine salary ranges. Every six months, workers evaluate their managers. For big decisions, such as buying another company or relocating a factory, every worker gets a vote. Thus, SEMCO appears to have gone far beyond the basic principles of industrial democracy, and also the ESOP model, to give workers rights and obligations normally reserved for owners.

Employee Buyouts

Economic restructuring and factory closures over the past several decades have made worker ownership of a different kind a fairly frequent news item in Canada (Long 1995). Workers have bought out their employers (or attempted to do so) in an effort to save their jobs when the company was about to close. In some cases, huge firms were bought out, while in others, much smaller enterprises were involved. *Employee buyouts* or attempted buyouts have occurred in both the service industries (the attempted buyout of Canadian Airlines in the late 1990s, for example) and the manufacturing sector.

One of the most prominent examples of the latter was the purchase of Algoma Steel in Sault Ste. Marie in 1992. The company had been operating in the community since 1901 and employed about 5,000 workers at the beginning of the 1990s. If the company had folded, the community would have been decimated.[31] Through their union (the United Steel Workers of America), and with the assistance of loan guarantees from the New Democratic provincial government, workers negotiated the purchase of a majority of shares in the company. Cost savings were obtained by a voluntary pay reduction and through a plan to eliminate several thousand jobs through attrition. A decade after the buyout, Algoma Steel ran into severe financial difficulties but, with the participation of the union, refinanced and restructured itself once more. As a result of the public sale of shares to finance the upgrading of the steel mill, workers now own only about one-quarter of Algoma's shares. However, they still appoint 5 of the company's 13 board members, and the participative management structure set up years earlier continues to operative effectively.

But the Algoma Steel experiment is an exception. Typically, new worker-owners do not implement industrial democracy or innovative participative management schemes. For example, when forestry workers and managers bought the mills in which they had been employed for many years in Kapuskasing, Ontario (1991) and Pine Falls, Manitoba (1994), they initially experimented with some forms of enhanced worker participation in decision making. But within a year or two the traditional system of management had been reinstated (Krogman and Beckley 2002).[32] Thus, after a buyout has occurred and production resumes, the worker–management relationship is generally similar to that in a company with an employee share ownership plan (ESOP). But, there are still two critical differences. With majority ownership, workers can manage their own job security (they will decide if the company continues to function or not). And, as owners, workers retain the option to change the management system in the future, if they wish to do so.

Research has shown that most ailing companies bought by workers become successful once again. For example, a recent meta-analysis of 43 published studies (Doucouliagos 1995) showed that profit sharing (ESOPs), worker ownership, and institutionalized industrial democracy are all associated with higher levels of productivity, with the largest impact occurring in worker-owned firms. The chances of a worker buyout being successful are increased when all the participants, including the union if one is present, support the buyout plan. In addition, workers typically need expert advice to assist them in deciding how to restructure the organization, since major changes (often in wage rates and staffing) are usually needed. Government support is also critical, particularly with respect to financing, since many lenders are reluctant to gamble on worker buyouts, despite their better-than-average track record.

Producer Cooperatives

Producer cooperatives have a long history in North America and Europe (Oakeshott 1978). The first examples appeared in Britain over a century ago as workers looked for alternatives to the exploitative excesses of early capitalism. However, the producer cooperative movement was overshadowed by another form of collective response by workers, the formation of trade unions (see Chapter 7). Nevertheless, the cooperative alternative continues. In Atlantic Canada, for example, records show that 221 worker cooperatives were founded between 1900 and 1987. Three-quarters came into existence after 1975. Recent estimates indicate that about 300 producer cooperatives are functioning in Canada, with a total of about 6,000 members.[33] About 60 percent of these organizations are in Quebec, in large part because the provincial government has put in place programs that provide financing for producer cooperatives. While producer cooperatives are not a widespread phenomenon, and those that exist are quite small, they nonetheless offer another alternative to conventional employer–employee relationships. Examples in Canada include everything from construction companies and daycare centres to taxi firms and fish farms, as well as small manufacturing companies, consulting firms, restaurants, and forestry management firms.

Producer cooperatives must be distinguished from other collective enterprises, including consumer cooperatives (such as co-op food stores) or housing co-ops, in which a number of individuals or families jointly own and maintain a dwelling or housing complex. They are also different from marketing co-ops (the wheat pools set up by Western farmers are the best example) and financial services co-ops, such as credit unions. While sharing with these other organizations a general commitment to collective ownership and shared risk taking, *producer cooperatives* are distinguished by their function as collective producers of goods or services.[34]

Distinguishing producer cooperatives from other types of worker-owned enterprises is not quite as easy. However, there are several characteristics that do stand out. First, to a much greater extent, producer cooperatives have typically been established because of a desire to collectively solve problems faced by their members. For example, discontent with low pay and erratic work schedules have led some truckers to leave the large companies they worked for and set up their own trucking cooperatives. On a much larger scale, some employee buyouts that resulted in the formation of producer cooperatives were instigated by imminent job loss and the possibility of a community's economic base disappearing. Thus, unlike most enterprises in a capitalist economy where profit making is the first priority, producer cooperatives have a social purpose interwoven with the need to be successful in the marketplace.[35] Second, central to the philosophy of producer cooperatives is a commitment to the democratic principle of "one person—one vote." This defining characteristic would eliminate companies with employee share ownership plans (ESOPs) and many examples of employee buyouts. Third, and related to the second point, producer cooperatives typically do not allow non-members to own shares. As a result, members cannot benefit financially from the success of the cooperative by selling shares. Instead, some arrangements are usually made to reimburse members for their own contributions when they leave the organization.

It has often been assumed that producer cooperatives, like employee buyouts, are typically a response to difficult economic times. In other words, workers decide to set up a cooperative, or take over a company and re-shape it into a cooperative, because their jobs are at risk. However, analysis of the cooperative experience in Atlantic Canada over the course of the last century indicates no relationship between economic cycles and the formation of producer cooperatives (Staber 1993). This suggests that the social goals of producer cooperatives' members may be as important as their economic goals.

Small producer cooperatives can survive in a capitalist economy. The study of Atlantic Canada showed an estimated median lifespan of 17 years for rural producer cooperatives and 25 years for those in urban settings (Staber 1993: 140). But producer cooperatives face many challenges. It is difficult to finance such enterprises, since banks are often skeptical of their ability to survive. This problem is also faced by other small businesses and by employees attempting to buy out their company. In situations where producer cooperatives are set up as a solution to high unemployment caused by economic downturns, the new worker-owned enterprises are at a disadvantage from the outset. And, as with any attempts to democratize workplaces, participants in producer cooperatives have to learn how to work collectively toward common goals, something that is not taught or encouraged in a competitive capitalist society.

Given their small size and, in particular, their social goals, it is not surprising that producer cooperatives tend to offer workers greater control over the labour process. Leslie Brown (1997) argues that producer cooperatives actually incorporate many of the democratizing and work-humanizing features of workplace reform that have been promised, but not always delivered, by the many new management approaches of the past decades. She also concludes that producer cooperatives are a viable solution to problems of community economic development that have not been solved by attempts to encourage traditional forms of capitalist enterprise in underdeveloped regions of the country.

Worldwide, the most successful producer cooperative is found in Mondragon, a city in the Basque region of Spain, where over 50,000 worker-owners run more than 200 highly diverse manufacturing and service companies with annual sales (in 2004) of more than 10 billion Euros.[36] Each worker has an economic stake in the enterprise where she or he works and, as a member of the firm's general assembly, establishes policies, approves financial plans, and elects members to a supervisory board, which, in turn, appoints managers. Mondragon co-ops have replaced the private ownership of industry in this region with a system of collective ownership and control. By all accounts, these co-ops have achieved high levels of growth, productivity, and employment creation, strong links with the community, harmonious labour relations, a satisfying and non-alienating work environment, and a close integration between workplace and community (Whyte and Whyte 1988).

Mondragon's success is linked to, among other things, the Basque region's decades-long struggle for greater autonomy from Spain, the destruction of the area's industrial base during the Spanish Civil War in the 1930s, and support from the local Catholic church and unions. This unique historical mix of political, cultural, and economic circumstances has sometimes been used as an argument that producer cooperatives could not be successful in other settings such as North America. Granted, it is never easy to transplant production systems from one cultural and political context to another. But, if we are willing to accept that Japanese management techniques or German codetermination, for example, offer alternatives to the North American approach to organizing workplaces, we should not ignore the Mondragon-style producer cooperative alternative (Macleod 1997).

CONCLUSION

In this chapter, we have critically examined the assumptions and conclusions of the mainstream management literature outlined in Chapter 5. Influenced by the labour process research tradition, we questioned whether new management models really do empower workers and allow them to enhance their skills. We then shifted our focus to an examination of alternative ways of organizing work that might reduce conflict and provide workers with more control over the labour process and their work organizations.

Several general conclusions sum up this discussion. First, concentration of power in the hands of owners and managers continues to characterize workplaces within our society, and the desire to increase profits continues to overshadow any commitment to empowering workers. Managers continue to systematically control workers, although their methods have changed. Second, employees' expectations of reasonable wages and working conditions, and their desire for more control over the labour process, are frequently at odds with the goals of management. Consequently, conflict and resistance are often present in the workplace, manifested in many different ways.

Third, it is apparent that the new management models intended to empower workers and enhance their skills, while reducing conflict and improving productivity, have frequently not lived up to their promises. In fact, in some cases, the rhetoric of new management models has been used to gain additional control over workers. New technologies have the potential to raise workers' skill levels and humanize the workplace, but they have also been employed to control workers and to eliminate jobs. Fourth, new technologies and management models are created and implemented by human beings. The social problems generated by these "hard" and "soft" technologies are a result of human choices. By the same reasoning, it is our responsibility to try to use these technologies to improve working conditions, empower workers, create jobs, and reduce social inequality.

Workplace health and safety continues to be a contested issue in the workplace. Even so, the systems put in place to address health and safety concerns provide some evidence of workers and management finding common ground. Different forms of industrial democracy and worker ownership also offer considerable potential for worker empowerment and conflict reduction. Hence, our last and most general conclusion is that there are alternatives to traditional employment relationships within capitalism that deserve serious consideration.

Despite our emphasis on the need to examine worker resistance when analyzing the labour process, such resistance has been only one of many themes in this chapter. In fact, we have talked much more about management control systems, the deskilling debate, the impact of new technologies, and alternative approaches to organizing the workplace. In the next chapter, we examine the primary collective vehicle for worker resistance in Canada over the past century, namely, the organized labour movement.

DISCUSSION QUESTIONS

1. Some social theorists have argued that conflict is inevitable in the workplace in industrial capitalist societies. Others believe that some kind of consensus might be possible. Critically discuss these two positions.

2. Outline the core aspects of the "labour process" perspective on work in industrial capitalist societies. In your opinion, does this perspective have any relevance in 21st-century Canada?

3. Discuss the various methods of workplace control used by employers and managers. Which types of control would be most effective, in your opinion? Why?

4. What is meant by the terms *technological, economic, and social determinism?* Use the terms to discuss the replacement of workers with robots on assembly lines, the outsourcing of call-centre jobs to India, and the viability of worker buyouts.

5. Assess the major impacts of technology on work, and use this assessment to speculate about the future impact of technology on work.

6. Critically discuss the following statement: "Producer cooperatives have been successful in places like Mondragon, Spain, and in some isolated small Canadian communities, but they are irrelevant for modern, urban Canada."

7. Workplaces are organized differently, workers are managed differently, and industrial relations have different rules in some European countries. Can

we learn anything from these societies that would lead to higher productivity and less conflict in North American workplaces?

8. The authors of this text propose that "workplace health and safety provides one example where, at least to some extent, workers and their employers have managed to identify and work together toward common goals." On what basis do they reach this conclusion? Do you agree with their assessment?

NOTES

1. These management–labour disputes were described in *The Globe and Mail* (5 July 2005: A8), (5 July 2005: B5), (4 July 2005: S1), and (29 June 2005: A12), respectively.

2. Paul Mantoux, quoted by Beaud (1983: 66).

3. See Smith (1994), Warhurst (1998), and Thompson and Smith (2000/1) for discussions of Braverman's contributions to the sociology of work.

4. See Spenner (1983), Attewell (1987), Gallie (1991), Burris (1998), Gallie et al. (1998), de Witte and Steijn (2000), Thompson and Smith (2000/1), and Handel (2003) for contributions to the deskilling debate. Canadian studies addressing this issue include Myles (1988), Clement and Myles (1994: Chapter 4), Bob Russell (1999), and Hughes and Lowe (2000).

5. See Horrell, Rubery, and Burchell (1990), Steinberg (1990), Wajcman (1991), Hughes (1996), Clarke (2000), and Belt, Richardson, and Webster (2002) on the gendered nature of occupational skills.

6. Ewan (1976: 11), quoted by Swift (1995: 30).

7. Also see Frenkel et al. (1995), Buchanan and Koch-Schulte (2000), Taylor et al. (2002), Grimshaw et al. (2002), Belt, Richardson, and Webster (2002), Taylor and Bain (2003), Russell (2004), and Beirne, Riach, and Wilson (2004) on work in call centres.

8. See Spenner (1983, 1990), Form (1987), Vallas (1990), Gallie (1991), Smith (2001), and Handel (2003) on the definition and measurement of skill.

9. On robotics, see Robertson and Wareham (1987, 1989), Block (1990: 100–3), and Suplee (1997).

10. See Milkman (1997) for an excellent study of technological change, new management strategies, and downsizing and layoffs in the same factory.

11. Robertson and Wareham (1987: 28); also see Noble (1995) and Block (1990: 100–3).

12. See, for example, Hughes (1989, 1996), Zuboff (1988), Lowe (1997), Maclarkey (1997), and Belt, Richardson, and Webster (2002).

13. Rifkin (1995: 3); for equally concerned but somewhat less pessimistic views, see Idle and Cordell (1994), Noble (1995), Swift (1995), and Richardson (1996).

14. For additional discussions of how IT does not have deterministic effects on working conditions and work rewards, see Rubery and Grimshaw (2001), Hunter and Lafkas (2003), and Mishel and Bernstein (2003).

15. McLuhan (1964: 56), quoted by Menzies (1996: 44).

16. Mahon (1987); several recent Canadian studies have also demonstrated how a strong union (Clarke and Haiven 1999) and a positive labour relations climate (Smith 1999) can influence workers' willingness to accept change.

17. See Wells (1986), Robertson and Wareham (1987), and Rinehart (2006: Chapter 6) on QWL.

18. For additional critiques of TQM see Fox and Sugiman (1999), McCabe (1999), and Knights and McCabe (2000). For a study that concludes that TQM benefits workers, see Rosenthal, Hill, and Peccei (1997).

19. See Neis (1991), Macdonald (1991), and Fox and Sugiman (1999) for critiques of flexible specialization.

20. Graham (1995), Rinehart, Huxley, and Robertson (1997), Kochan, Lansbury, and MacDuffle (1997), Bob Russell (1999), Schouteten and Benders (2004), Rinehart (2006: 159–63); see Nilsson (1996) for a somewhat more positive assessment of lean production.

21. Walters and Haines (1988), Walters (1985); also see Sass (1995).

22. See Karasek and Theorell (1990), Jex (1998), and Williams (2003) on work stress. Kompier et al. (1994) note that Sweden, the U.K., and the Netherlands recognize work-related stress in their health and safety legislation, while Germany and France do not.

23. Sass (1995: 123); elsewhere, Sass (1996) criticizes the organized labour movement for accepting the status quo in terms of legislation and policies regarding health and safety.

24. For a useful overview of workplace wellness programs, see the report for Health Canada by Graham S. Lowe (2003) on "healthy workplaces and productivity" at http://www.grahamlowe.ca/documents/48/.

25. For a more critical assessment, see Hansen (2004), who labels employee assistance programs as the "caring cousin" of workplace surveillance strategies,

since employers can use such programs to identify workers who may potentially become less productive.

26. See Russell (1990: Chapter 3) and Rinehart (2006: 48–49) on early work councils in Canada. Also see Guzda (1993) on the U.S. situation.

27. See Poutsma, Hendrickx, and Huijgen (2003) for an overview of different approaches to employee participation in Europe, including works councils.

28. This discussion of Sweden draws on Milner (1989), Smith et al. (1995, 1997), Smucker et al. (1998), Levinson (2000), and Whyman (2004).

29. Sandberg (1994); Smith et al. (1995, 1997); Smucker et al. (1998); Nilsson (1996).

30. See Gunderson et al. (1995: 418), Marshall (2003), and Luffman (2003) for Canadian data on ESOPs, and Bruno (1998) and Logue and Yates (1999) for U.S. data.

31. This discussion of worker buyouts is based on Gunderson et al. (1995), Long (1995), and Nishman (1995).

32. Bacon, Wright, and Demina (2004) studied employee buyouts in Britain in the 1990s and observed somewhat more emphasis on human resource management and on getting workers involved in decision making.

33. Staber (1993) provides the Atlantic Canada figures while Quarter (1992) presents the national count. Also see Co-operatives Secretariat (2000) for a similar estimate of the prevalence of producer cooperatives in Canada and a dozen case studies of successful organizations of this type.

34. This discussion of producer cooperatives is based on Quarter (1992: 27–32), Staber (1993), Lindenfeld and Wynn (1997), and Brown (1997). For additional information on producer cooperatives in Canada, see http://www.canadianworker.coop. Information about producer cooperatives internationally can be obtained from http://www.ica.coop.

35. Employee buyouts, on their own, are not necessarily accompanied by other progressive social changes. For example, the workers and managers who took over the forestry mills in Pine Falls, Manitoba, and Kapuskasing, Ontario, were no more likely than the previous corporate owners to engage in ecologically friendly forestry practices (Krogman and Beckley 2002).

36. For more information on the Mondragon cooperatives, see http://www .mondragon.mcc.es/.

UNIONS, INDUSTRIAL RELATIONS, AND STRIKES

INTRODUCTION

We begin this chapter with three prominent stories from Canada's industrial relations scene. While vastly different, these examples highlight key issues that likely will dominate labour–management relations in the coming years.[1]

- In the spring of 2004 on opposite sides of the country, provincial governments did what only government employers can do to end a labour dispute: they introduced legislation to force their striking employees back to work. British Columbia was facing a province-wide public-sector strike, as teachers, transit workers, and other public employees rallied in support of 43,000 striking health care workers who had been legislated back to work with a 15 percent wage cut. In this pressurized environment, unions and the government were able to reach a last-minute agreement, from the union's perspective, providing at least some protection against the contracting out of members' jobs. Meanwhile on Canada's east coast, 20,000 striking public-sector workers returned to their jobs after the Newfoundland government imposed back-to-work legislation. Strikers who defied the legislation faced dismissal. This legislation imposed a new four-year collective agreement, which included wage freezes for two years followed by modest increases and, more contentiously from the workers' perspective, cut sick-leave allotments in half.

- Wal-Mart is the world's largest corporation, with $288 billion in sales and 1.6 million employees. The giant retailer's approach to employee relations is firmly nonunion. So the certification of a union at a Wal-Mart store in Windsor, Ontario, in 1997 was a first. No other Wal-Mart store in the world was unionized. However, the union representing the Windsor store employees gave up its bargaining rights, unable to negotiate a first collective agreement. This did not stop employees at other Wal-Mart stores in Canada from seeking union representation. Wal-Mart in 2005 closed a store in Jonquière, Quebec, just months after workers voted to unionize. The closure came in the wake of a Quebec Labour Commission ruling to consider an application by UFCW Canada (United Food and Commercial Workers Canada) to keep the store operating until decisions on a series of unfair labour practice changes were reviewed by the commission. Within weeks of the Jonquière store closing, employees at Wal-Mart's Brossard, Quebec, store voted against joining UFCW Canada. The union accused the company of intimidation; the company accused the union of coercing employees into signing union cards and using pressure tactics. This battle continues.

- In July 2005, the longest labour dispute in the history of professional sports ended when the National Hockey League Players' Association reached a new collective agreement with team owners. While the public may not feel much empathy for professional hockey players who earn 7-figure salaries, the 301-day NHL lockout illustrates labour–management dynamics in a relatively new arena of collective bargaining: professional sports. Powerful players' unions are the norm in professional team sports. However, it was the team owners who "locked out" the players by cancelling the NHL season, and the deal reached included a salary cap and a 24 percent salary rollback. This distinguishes the NHL lockout from other union–management disputes, because it was the union that wanted to preserve a free market for salaries. In other industries, it would be management accusing the union of putting artificial constraints on the labour market. In the aftermath of the collective agreement, there is much speculation as to how "labour peace" will be restored and what the long-term damage to league revenues and its fan base will be.

These examples go to the heart of contemporary *employee–employer relations,* a topic we raised in Chapter 6 and explore here. The public-sector disputes raise questions about employees' rights to negotiate a decent wage, fair treatment, and economic security. The ongoing battle between Wal-Mart and UFCW Canada highlights the rights of workers to join a union, widely considered a basic human right. The NHL lockout shows that collective bargaining has extended far beyond its origins among manual labourers in mines and factories into high-profile occupations, including athletes, musicians, actors, judges, and doctors. The above examples raise important questions we will address in this chapter: When is it justifiable for market pressures, or "the public good," to take precedence over the rights and interests of employees? What responsibilities do employers have to their employees? Is there an inherent conflict between workers' desires for good jobs and employers' goals of profits, flexibility, and competitiveness? Conversely, to what degree are interests between workers and employers shared? Are there circumstances under which labour and management can cooperate to their mutual benefit? What is the role of government in regulating the relations between labour and management? Why do workers join a union in the first place? Finally, what improvements in wages and working conditions have unions achieved through collective bargaining?

These are the sorts of industrial relations issues that will be examined in this chapter. From a sociological perspective, we will focus on power, conflicts, and compromises over the distribution of resources in the workplace, as well as on employee collective action. These themes are highlighted in the study of unions, the major organizations representing the interests and aspirations of employees in capitalist societies. As we concluded in Chapter 6, much of the resistance to employers by North American workers has been channelled through their unions. We will, therefore, explore the nature and development of the labour movement in Canada, as well as how union–management relations are regulated through a complex legal framework. The causes and implications of industrial conflict are the subject of later sections in this chapter.

THEORETICAL PERSPECTIVES OF THE LABOUR MOVEMENT

We can begin to understand why unions developed in the first place by reviewing some of the early theories of the labour movement. No single issue stands out as the driving force of unionization. Rather, workers historically have rallied collectively to oppose the imposition of arbitrary management power, to protect their jobs and retain some semblance of control over the labour process, and to obtain improvements in rights, wages, and working conditions.

Why Do Workers Unionize?

In a classic work on British trade unionism, Sidney and Beatrice Webb (1894/1911) suggested that workers' pursuit of higher wages expressed a more basic desire to reduce employer domination. Collective action could improve working conditions and reduce the competition for jobs that drives wages down. Furthermore, the Webbs argued that it could curb an employer's authority by instituting common rules governing the employment relationship. Selig Perlman (1928) later stressed the role of unions in controlling jobs. After studying the International Typographical Union, the oldest union in America, Perlman concluded that workers develop an awareness that jobs are scarce and must, therefore, be protected through unionization. Michael Poole sheds additional light on the emergence of the labour movement in his discussion of the various goals it pursued.[2] Unions can thus be viewed five different ways: as moral institutions, fighting against the injustices and inequities of capitalist industrialization; as revolutionary organizations intent on overthrowing capitalism; as psychological or defensive reactions against the threat early capitalism posed to workers' jobs; as responses to economic realities, aimed at achieving better wages and working conditions; and as political organizations extending workers' rights further into the industrial arena.

Generally speaking, the harsh working conditions endured by industrial workers in the late 19th century sparked the first major surge of unionization. Yet the basic idea that workers are motivated to unionize by the desire to gain greater control over their jobs, as well as fair treatment and just rewards, remains a dominant theme in industrial relations. Mainstream industrial relations theory views the *system of job regulation* as the core of worker–management relations.[3] In other words, the rules and regulations that form the basis of collective agreements are assumed to inject stability into employment relations by tilting the balance of power slightly away from management and toward workers.

However, this perspective overplays the importance of predictable and harmonious industrial relations, and downplays the frequency of conflict. That is, by focusing on the system of rules, regulations, and institutions governing industrial relations, the mainstream perspective does not question the existing distribution of power between workers and management. It also presents an incomplete picture of industrial relations, due to a narrow focus on the formal aspects of the system, such as legislation and collective bargaining. An equally important informal side involves the daily interaction between workers and employers, as "the rules of the game" are constantly being negotiated. It is, therefore, helpful to acknowledge that work is a power relationship in which conflict is always a possibility. As Richard Hyman (1975: 26) explains, "in every workplace there exists an invisible frontier of control, reducing some of the formal powers of the employer: a frontier which is defined and redefined in a continuous process of pressure and counter-pressure, conflict and accommodation, overt and tacit struggle." Echoing themes from Chapter 6, Hyman proposes a more critical perspective on industrial relations, defining it as the process of control over work relations—a process that especially involves unions.

Conflict and Cooperation in Union–Management Relations

It is important to understand the role of conflict in union–management relations. Collective bargaining is the process by which a union, on behalf of its members, and an employer reach a negotiated agreement (called a collective agreement or a contract) that defines for a specific time period wages, work hours and schedules, benefits, and other working conditions, and procedures for resolving grievances. Collective agreements are negotiated and administered under provincial and federal labour laws and are designed to reduce conflict. Indeed, the main thrust of modern industrial relations practice is the avoidance of conflict. Thus, for the system to operate with some degree of fairness and equity for workers, who on the whole are in the weaker bargaining position, there must be the threat of conflict that could disrupt the employer's business. A stock criticism of the Canadian industrial relations scene concerns its adversarial nature. A federal government task force on labour relations, responding to these concerns, explained the underlying conflict in these words:

> Paradoxical as it may appear, collective bargaining is designed to resolve conflict through conflict, or at least through the threat of conflict. It is an adversary system in which two basic issues must be resolved: how available revenue is to be divided, and how the clash between management's drive for productive efficiency and the workers' quest for job, income and psychic security are to be reconciled. (Canada 1969: 1, 19)

We must also view industrial relations as a continuous process. New problems regularly confront the parties in collective bargaining. Solutions for one side may create difficulties for the other, as in cases of wage rollbacks. Or an agreement may be based on compromises that both sides have trouble living with (for example, when a third-party arbitrator imposes a solution). Part of the difficulty is that workers and management have different definitions of social justice and economic reality.[4]

As pointed out in Chapter 5, work organizations typically have a division of labour that requires collective interdependence. Management, therefore, must balance the need to control employees with the necessity of achieving a workable level of cooperation and commitment from them. Stephen Hill (1981) has explored this basic tension and concludes that workers and managers must cooperate to provide goods or services. Yet in doing so, Hill notes, each side also strives to maximize its own interests. Workers aim for higher wages, better working conditions, and more autonomy in their jobs. Employers pursue higher profits, lower costs, and increased productivity. The chronic tension between these opposing interests forces tradeoffs on both sides, and may also generate open conflict. Of course, we must recognize that not all union–management negotiations are a *zero-sum game:* that is, a situation in which one side can gain something only if the other gives up something. Indeed, on some issues, such as improved health and safety conditions (see Chapter 6), negotiations can produce a win–win situation in which both workers and management benefit. This approach to negotiations is called mutual gains bargaining (Weiss 2003).

Unions as "Managers of Discontent"

There is a general consensus among scholars that the industrial relations systems in modern capitalist societies "keep the lid on conflict." Having observed the oppressive conditions under which mid-19th-century factory workers toiled, Marx concluded that eventually their misery and poverty would ignite a revolution. But, as Stephen Hill (1981: Chapter 7) points out, even Marxists, committed to the belief that capitalism pits the workers and bosses against each other in constant struggle, acknowledge that collective bargaining integrates workers into the existing economic system. For many Marxists, unionism itself embodies a basic contradiction by striving to solve workers' problems within the confines of capitalism. This is why Lenin, father of the Russian Revolution, dismissed trade unions as capable only of reform, not revolution. But to lay all the blame on capitalism is to ignore the basic dilemma of the distribution of scarce resources, which all societies must face (Crouch 1982: 37–38).

In some respects, then, unions function as *managers of discontent*.[5] Unions in Canada and other advanced capitalist societies channel the frustrations and complaints of workers into a carefully regulated dispute-resolution system. Unions help their members to articulate specific work problems, needs, or dissatisfactions. Solutions are then sought through collective bargaining, or through grievance procedures. In Canada, for example, labour legislation prohibiting strikes during the term of a collective agreement puts pressure on union leaders to contain any actions by rank-and-file members that could disrupt the truce with management. This may seem ironic, since unions developed in response to the deprivations suffered by workers in the early stages of industrial capitalism. But unions today operate in ways that contribute to the maintenance of capitalism, seeking reforms that smooth its rough edges.

In order to channel the discontent of their members, unions must legitimately represent the views of the majority of members. This raises the issue of internal *union democracy*. Generally speaking, unions are democratic organizations whose constitutions allow members to elect leaders regularly. Theoretically, this should make leaders responsive and accountable to rank-and-file members, translating their wishes into tangible collective bargaining goals. But, despite this, there has long been a sociological debate about the pitfalls on the road to union democracy.

Robert Michels was the first to investigate the problems of union bureaucracy and democracy. His study of German trade unions and the Social Democratic Party prior to World War I concluded that leaders in working-class organizations always dominate members. Michels's famous *iron law of oligarchy* draws on technical, organizational, and psychological explanations. According to this theory, leaders develop expert knowledge, which gives them power; once in office, leaders can control the organization to maintain their power; and, finally, the masses tend to identify with leaders, and expect them to exercise power on their behalf. But in Canada and elsewhere, we have recently witnessed the emergence of strong rank-and-file movements challenging entrenched union leadership cliques and, in effect, opposing oligarchic rule. Most unions today espouse democratic principles, but some have been more successful in putting these into practice than others. In fact, union officials seem to be constantly trying to increase the participation of rank-and-file members in union activities. Images of corrupt and autocratic leaders, while headlined by the media and reinforced by the history of a few unions, are far from typical. Thus, what Michels discovered was not a universal trait of unions, but a potential problem faced by all large bureaucratic organizations.[6]

WHAT DO UNIONS DO?

How do unions set goals and devise strategies to attain them? Of great interest to sociologists is why unions, rather than espousing revolutionary or even reformist aims, have engaged in largely conservative and defensive actions. Typically, the daily activities of unions focus on two types of goals: control over work and the rewards of work (Crouch 1982: Chapter 4). It is not inconceivable that demands for control over the work process could, if they escalated, spark radical challenges to the capitalist system of business ownership. Yet, historically, such demands often involved skilled craftworkers fighting to defend their relatively privileged position in the labour market against the onslaught of modern production methods. Their goal, in short, was to preserve the status quo.

There are frequently compromises between control over the labour process and economic rewards. Workers may be forced by their immediate economic needs to pursue the latter to the exclusion of the former. In addition, employers are sometimes willing to give up more of their profits rather than concede to workers' greater decision-making authority. The emphasis on material gain rather than job control, known as *business unionism,* has become a hallmark of the North American labour movement. In the last decade, we can detect a shift away from business unionism toward what can be called *social unionism.* Public-sector unions, along with some leading private-sector unions such as the Canadian Auto Workers, have taken on a much broader agenda of reform, entering public debates about issues such as globalization, international human rights, and health care reform.[7]

Public Opinion about Unions

How does the public view the activities of unions? Canadian public opinion has become less favourable toward the role of unions in our society. Gallup polls in the 1950s asked respondents if they thought unions were good or bad for the country: 12 to 20 percent said unions were bad, while 60 to 69 percent said they were good. However, polls during the 1980s found an increase in "bad" responses to between 30 and 42 percent. Generally, Canadians have become less trusting of major institutions, including unions. For example, a 1996 national poll found considerable cynicism toward corporations, the federal government, and unions, which can be interpreted as part of the shift in social values in a postindustrial society.[8]

Public attitudes toward unions are, however, usually more ambiguous than suggested by public-opinion polls. For example, in a study of Winnipeg and Edmonton residents, we found that individuals held both positive and negative images of unions, depending on the specific issues at stake (Krahn and Lowe 1984; Lowe and Krahn 1989). Many respondents believed that unions contributed to inflation, while also recognizing that they had achieved material gains for members. At a general level, this study identified two different images of unionism: a *big labour* image that sees unions as too powerful, believes that they harm society, and thinks, therefore, that they require greater government regulation; and a *business unionism* image that focuses on the positive gains in wages and working conditions unions have made through collective bargaining. Both images are reinforced by the media.

Researchers also have linked nonunion employees' attitudes toward unions to their level of support for joining a union (Barling, Fullagar, and Kelloway 1992). This perspective has direct relevance for unions' future success. According to a national survey conducted by Canadian Policy Research Networks in 2000, 25 percent of nonunion employees in Canada said it was likely or very likely they would join a union if one existed in their workplace or profession (Kumar, Lowe, and Schellenberg 2001). Younger workers and members of visible minorities (including Aboriginal people) expressed higher than average interest in joining a union. Lack of trust in one's employer, not having a supportive and healthy work environment, and dissatisfaction with pay and job security also are associated with support for joining a union. The relatively strong support for union joining among young workers is corroborated in other studies (Lowe and Rastin 2000). This bodes well for the labour movement, although it appears that workplace socialization reduces pro-union attitudes as these young workers gain more experience.

The Economic Impact of Unions

Unions benefit their members financially. Research by Canadian economists suggests that, on average, unionized workers earn about 10 percent more than nonunion workers do, with the impact being greater for women than men. However, this *union wage premium* has declined since the 1970s. Union members also are more likely than comparable nonunion workers to have pension plans provided by their employer, more paid vacations and holidays, dental and medical plans, and better job security. A study of Canadian child-care workers, a low-wage service industry, found that unions raised wages by 15 percent, had a positive impact on benefits, and contributed to professionalization through financial incentives for workers to improve their qualifications and skills (Cleveland, Gunderson, and Hyatt 2003). Unions also contribute to reducing overall wage inequality in a nation's labour market, although less so than in the past. This is largely because the union impact on wages is greatest for workers in the lower and middle ranges of the income, education, and skill distributions. In other words, without unions, income inequality in Canada would most likely rise. Indeed, in Canada, the United States, and Britain, growing income inequality has been linked, in part, to declining union membership.[9]

Access to these economic advantages of union membership depends mainly on where you work. In this regard, workplace size and type of employment matter. Unionization is significantly higher in large workplaces. Recall from earlier chapters that large firms make up much of the "core" sector of the economy, providing better wages and benefits, and are more likely to introduce new technologies and innovative forms of work organization. They are in a better competitive position than are smaller firms to provide decent wages and working conditions. In terms of employment type, recall from Chapters 2 and 3 that nonstandard jobs are often referred to as "bad jobs" because of their inadequate wages, security, and benefits—outcomes related to a lack of union representation (Kalleberg, Reskin, and Hudson 2000). Barriers to unionization include restrictive legislation and, until very recently, lack of interest among unions. The lack of union representation for temporary, part-time, and self-employed workers—groups that have been increasingly marginalized in the labour market—has become a recent research focus (Carré et al. 2000: Section IV; Vosko 2000).

Does the conventional wisdom that unions further their members' interests at public expense have a basis in fact? The most thorough analysis of what trade unions actually do is a study by two Harvard economists, Richard Freeman and James Medoff (1984: 5–11), who make a useful distinction between the two "faces" of unionism. The *monopoly* face represents unions' power to raise members' wages at the expense of employers and other workers. The *collective voice* face shifts attention to how unions democratize authoritarian workplaces, giving workers a collective voice in dealing with management. Freeman and Medoff admit that unions do impose some social and economic costs, but think that these are far outweighed by their positive contributions. Unions significantly advance workers' economic and political rights and freedoms. And, to the chagrin of their opponents, unions also boost productivity. As these researchers explain, unions improve productivity through lower employee turnover, better management performance, reduced hiring and training costs, and greater labour–management communication and cooperation. But because of higher wage costs, productivity gains do not necessarily make unionized firms more profitable.

Even though they are examining American unions, Freeman's and Medoff's conclusions deserve careful attention. They show that, in addition to raising members' wages relative to those of non-members, unions alter nearly every other measurable aspect of the operation of workplaces, from turnover to productivity to profitability to the composition of pay packages. The behaviour of workers and firms and the outcomes of their interactions differ substantially between the organized and unorganized sectors. On balance, unionization appears to improve rather than to harm the social and economic system.[10]

THE DEVELOPMENT OF THE CANADIAN LABOUR MOVEMENT

The Canadian labour movement has long been a major force in the economic life of the nation. Its history is tied to the development of capitalism. But, as workers in the late 19th and early 20th centuries reacted to the inequities and deprivations of their employment by organizing unions, they also had to fight to gain the right to do so. The story of unions is a human drama in which groups of workers often struggled against strong-willed employers and a reluctant state to gain higher wages, better working conditions, and some control over their daily working lives.

Craft Unionism

Skilled craftworkers were the first to unionize. The earliest union in Canada was organized by printers in York (now Toronto). Carpenters, bricklayers, masons, cabinetmakers, blacksmiths, shoemakers, tailors—these were the pioneers of the union movement before the days of modern industry. A strike by Toronto printers in 1872 resulted in the Trade Unions Act, which for the first time legalized union activity. Prior to that, unions had been considered a conspiracy against the normal operations of business. Other significant events that laid the foundations for trade unionism in Canada were the Nine-Hour Movement in the 1870s, involving working-class agitation for shorter working hours, and the creation of the Trades and Labour Congress (TLC) in 1883 as the first central labour body.[11]

Craft pride based on the special skills acquired through a long apprenticeship, solidarity with fellow artisans, and a close integration of work and communities were the hallmarks of these early *craft unions*. Craftsmen (there were no women among them) were the aristocrats of the working class. Printers, for example, reinforced their status by referring to their work as a "profession." Like other crafts, shoemakers drew on a long heritage to bolster craft pride. "From medieval craft lore, the shoemakers brought forth St. Crispin as a symbol of their historic rights and their importance to the community," reports historian Greg Kealey (1980: 292).

Craft unions were organized according to specialized craft skills, and served their members in several important ways. They were benevolent societies, providing members with a form of social insurance years before the rise of the welfare state. They also protected their members' position in the labour market by regulating access to the craft, thus monopolizing its unique skills. Today, this would be referred to as a *labour market shelter* (see Chapter 3). And, as small local enterprises of the 19th century gave way to the factories and large corporations of the 20th, unions provided craftsmen with a defence against the erosion of their way of life. Artisans opposed scientific management, the mechanization and reorganization of craft production in factories, and other attempts by employers to undermine the skills and responsibilities on which their craft traditions were based. Moreover, the tightly knit social relations of working-class artisan communities—reinforced by educational institutes, parades, picnics, and other neighbourhood events—bolstered these reactions to the march of modern industry.[12]

Craft unions dominated the young Canadian labour movement well into the 20th century. These *international unions* were American-based and affiliated with the conservative American Federation of Labour (AFL). The iron moulders' union was the first international union to set down permanent Canadian roots, in 1859. This was a time when the labour market for many skilled trades spanned both sides of the Canada–U.S. border. The internationals quickly came to control the Canadian labour scene. At the 1902 convention of the TLC, the AFL unions purged their Canadian-based rivals by successfully moving that no organization duplicating one already in the AFL could be a member of the TLC. They thus stole power from more radical Canadian labour leaders (Babcock 1974).

Industrial Unionism

The craft principle underlying the early AFL unions contrasts with *industrial unionism,* where all workers in an industry are represented by the same union, regardless of their occupational skills. The Knights of Labour, the earliest industrial union in Canada, organized its first local assembly in Hamilton, Ontario, in 1875. For a brief period in the 1880s, they challenged the dominance of the AFL craft unions. Driven by an idealistic radicalism, the Knights' immediate goal was to organize all workers into a single union, regardless of sex, skill level, craft, or industry of employment. They saw their efforts ultimately leading to the abolition of the capitalist wage system and the creation of a new society. Their membership peaked in 1887, with more than 200 local assemblies representing workers in 75 occupations. A combination of factors led to the Knights' demise: rapid membership growth that made it difficult to maintain an idealistic philosophy; regular defeats at the hands of employers; internal political rivalries; the union's growing hostility with the AFL unions; and a recognition among craftworkers that uniting with unskilled workers in industrial unions would jeopardize their privileged position. The Knights survived longer in Canada than south of the border, although by the early years of the 20th century they had all but disappeared (Kealey 1981).

Only a handful of industrial unions surfaced in the early 20th century. Several were part of the rising tide of labour radicalism that reached a crest with the 1919 Winnipeg General Strike. There was a distinctive regional flavour to these working-class protests, as most were rooted in Western Canada. The Western Federation of Miners, an international union but the archrival of the AFL, gained a foothold in British Columbia mines. The radical ideology of the Chicago-based Industrial Workers of the World (the "Wobblies") attracted unskilled immigrants employed in lumbering, mining, agriculture, and railways in the West prior to World War I. The Wobblies advocated a form of *syndicalism,* which held that international industrial unions and general strikes are the main vehicles for working-class emancipation.

Our discussion would be incomplete without mention of the One Big Union (OBU). This revolutionary industrial union received widespread support in the Western provinces, particularly among miners, loggers, and transportation workers, around the time of the Winnipeg General Strike. The OBU called for secession from the conservative AFL and its Canadian arm, the TLC. Its support for the Russian Revolution brought counterattacks from employers, governments, and craft unions, and caused the union's eventual defeat in the 1920s.

Not until the 1940s did industrial unionism become firmly established in Canada. The breakthrough was the United Auto Workers' (UAW) milestone victory in 1937 against General Motors in Oshawa, Ontario. The UAW sprang up under the banner of the left-leaning Congress of Industrial

Organizations (CIO) in the 1930s to organize unskilled and semiskilled workers in mass-production industries. Many of the initial forays into the auto, electrical, rubber, and chemical factories of corporate North America were led by communist organizers. Craft union leaders opposed the CIO largely on political grounds, despite the fact that Canadian workers embraced the CIO form of industrial unionism (Abella 1974; Moulton 1974).

In 1956 Canadian craft and industrial unions buried their differences, uniting skilled and unskilled workers within a single central labour organization. This marked the merger of the craft-based TLC and the industrial unions of the Canadian Congress of Labour, resulting in the creation of the Canadian Labour Congress (CLC). A similar merger of the AFL and the CIO had occurred one year earlier in the United States. The CLC remains Canada's "house of labour." Its affiliated unions represent 69 percent of union members in the country (Human Resources Development Canada, Workplace Information Directorate 2000: 68).

The CLC promotes the economic, political, and organizational interests of affiliated unions by providing research, education, and organizational and collective bargaining services, as well as by eliminating jurisdictional conflicts and organizational duplication. It sometimes becomes embroiled in inter-union disputes over jurisdictions and complaints about unions *raiding* each other's members. Prominent examples of such rivalries include the Canadian Auto Workers' (CAW) recruitment of Newfoundland fishery workers who were represented by the United Food and Commercial Workers International Union, and the raid on the Canadian Union of Postal Workers by the International Brotherhood of Electrical Workers. The CLC also has strong ties to the New Democratic Party, having taken a leading role in founding the party in 1961. The CLC has become active on the national political stage, for example, by forming coalitions with other community-based groups to oppose free trade. This more broadly based political action increased with the election of Bob White, the high-profile former head of the CAW, as CLC president in 1992.

Quebec Labour

The history and present character of the labour movement in Quebec contribute to the distinctiveness of social and economic life in that province (Boivin and Déom 1995; Lipsig-Mummé 1987). Quebec labour's history is a fascinating topic in its own right, and deserves more than the brief treatment we can afford to give it here. Many of the issues central to Canada's ongoing constitutional debates are amplified in the arena of labour relations. For example, industrial relations have been shaped by a different legal framework in Quebec (its laws are derived from the Civil Code of France, rather than British-based common law). A more interventionist state role has created a higher degree of centralized bargaining than in other provinces, as well as resulting in some innovative legislation such as the 1977 anti-strikebreaking law. And nationalist politics have left an indelible mark on union policies and priorities. Indeed, the question of special representation for Quebec on the executive of the Canadian Labour Congress at the 1992 CLC convention raised the spectre of a split between the CLC and the Quebec Federation of Labour, which represents CLC affiliates in that province.

The development of unions in Quebec followed a different path from that of the rest of Canada. For example, in the early 20th century the Catholic Church organized unions that, unlike their counterparts elsewhere, emphasized the common interests of employers and employees. During the years that Premier Maurice Duplessis and the ultraconservative Union Nationale party held power (from 1936 to 1960), repressive state actions attempted to stifle union development. Worker militancy flared up in response, the most violent manifestations being the 1949 miners' strike at Asbestos and the 1957 Murdochville copper miners' strike. Some analysts view these strikes as major catalysts in Quebec's Quiet Revolution, which ushered in major social, economic, and political reforms in the 1960s.

During this period, the old Catholic unions cut their ties with the church, becoming one of the main central labour organizations in the province (the CNTU, or Confederation of National Trade Unions) and adopting an increasingly radical stance. Hence the Quebec Federation of Labour, made up of CLC-affiliated unions, differs from its counterparts in other provinces; it represents a minority of union members in the province and operates in a more independent manner. Another influential union organization is the Quebec Teachers' Federation. Thus, there is no unified Quebec labour movement, although these central organizations have banded together on occasion, the most notable example being the 1972 Common Front strike by some 200,000 public-sector workers against government policies.

THE ROLE OF THE CANADIAN STATE IN INDUSTRIAL RELATIONS

Canada's contribution to the quest for industrial peace has been the development of a legislative and administrative framework that casts the state into the role of *impartial umpire,* mediating between labour and capital. The architect of this system was William Lyon Mackenzie King, the first federal minister of labour and later a Liberal prime minister. King's 1907 Industrial Disputes Investigation Act (IDIA) became the cornerstone of Canada's modern industrial relations policy. The act provided for compulsory *conciliation* (fact finding) in disputes during a "cooling off" period, a tripartite board of *arbitration,* and special treatment of public interest disputes involving public services. At first the act was applied to disputes in coal mines and railways. Its scope was extended during World War I, and in the 1950s its principles were incorporated into provincial legislation. In some instances, the state used the powers of the act to legislate an end to strikes. There is little question that this sort of intervention has shaped the pattern of industrial conflict in Canada—some would argue, in the interests of employers (Craven 1980; Huxley 1979).

State intervention in industrial conflict is not a 20th-century invention, however. Claire Pentland, for example, documents the role of the British army in suppressing unrest among colonial labourers building the Rideau Canal in 1827. He concludes: "Intervention by the Canadian state on behalf of employers in more recent years should not be regarded as novelties, but as fruits of what employers and officials learned about labour problems in the middle of the nineteenth century" (Pentland 1981: 196). Yet, it was the 1907 IDIA that marked the first major step toward the institutionalization of industrial conflict through the control of law.[13]

We can identify four major phases in the development of the legal framework for industrial relations in Canada (Riddell 1986). In the first, or pre-Confederation phase, common law prohibited collective bargaining. The second phase was entered with the 1872 Trade Unions Act, passed by the Conservative government of Sir John A. Macdonald. This neutralized the legal restrictions on unionism, but did not grant workers positive legal rights or protections that facilitated collective bargaining. The National War Labour Order, regulation P.C. 1003, launched the third, or modern, phase in 1944. Modelled on the 1935 U.S. National Labour Relations Act (Wagner Act), P.C. 1003 granted employees in the private sector collective bargaining rights, set down union certification procedures, spelled out a code of unfair labour practices, and established a labour relations board to administer the law. These measures paved the way for a postwar labour–management pact designed to maintain industrial peace. This truce was enshrined in federal legislation in 1948 and in subsequent provincial legislation.

Another milestone in the legal entrenchment of *collective bargaining rights* came out of a 1945 strike by the United Auto Workers at the Ford Motor Company in Windsor, Ontario. The *Rand Formula,* named after Justice Ivan Rand of the Supreme Court, whose ruling was instrumental in settling the strike, provided for union security through a *union shop* and *union dues checkoff.* According to the Rand Formula, while no one should be required to join a union, because a union must act for the benefit of all employees in a workplace it is justifiable to automatically deduct union dues from the paycheques of all employees in a workplace regardless of whether or not they actually belong to the union.

The fourth phase involved the rise of public-sector unions. The movement toward full-fledged public-sector unionism began in Saskatchewan in 1944. The real push, however, started when Quebec public employees were granted collective bargaining rights in 1964. Another major breakthrough was the 1967 Public Service Staff Relations Act, which opened the door to unions in the federal civil service.

The intent of all these legislative changes was to establish industrial peace. It is obvious that one of their unintended effects was to spur the growth of unions. But the goal of conflict reduction should not be underestimated. The centrepiece of the 1907 IDIA was the regulation of work stoppages through compulsory postponement of strikes and lockouts, provisions for mediation or conciliation, the banning of strikes or lockouts during the term of a collective agreement, and alternative means of dispute resolution. The act also made strikes or lockouts illegal in key industries (transportation, communications, mines, and public utilities) until a board of inquiry had studied the problem and its conciliation report had been made public. These methods of *institutionalizing conflict* have been fine-tuned over the years. But work stoppages are still illegal during the term of a collective agreement (although they do occur) and the state has taken away the right to strike altogether from certain public employees who perform essential services.

Some analysts would argue that industrial relations entered a coercive phase during the 1980s. Leo Panitch and Donald Swartz view government imposition of wage controls, more restrictive trade union legislation, and the use of courts to end strikes as signals of the end of "free collective bargaining" as established by the industrial relations system set out in the 1948 federal legislation. Panitch and Swartz characterize this new era as *permanent exceptionalism*, reflecting how the suspension of labour's rights and more heavy-handed state intervention that were once the exception have become the rule.[14] As we shall see below, there is now considerable debate over whether Canadian industrial relations have entered troubled times.

UNION MEMBERSHIP TRENDS

Canadian Union Membership Growth and Stabilization

Figure 7.1 traces union membership growth in Canada since 1911. In 2004, there were approximately 4.3 million union members, representing 30 percent of the non-agricultural paid workforce.[15] Considering that only 4.9 percent of

FIGURE 7.1 *Union Membership in Canada, 1911–2004*

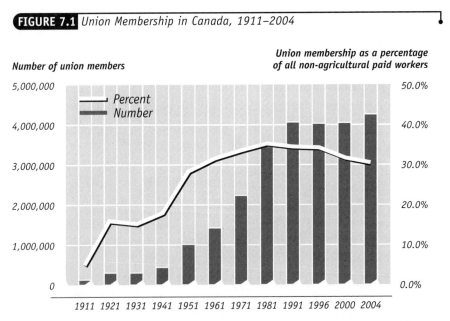

Source: "Union Membership in Canada, 1911–2004," *Workplace Gazette* 7, 3 (2004): 44; Henry, Manon. URL: http://www110.hrdc-drhc.gc.ca/millieudetravail_workplace/gazette/articles//gaze_articles.cfm Human Resources and Skills Development Canada, 2003. Reproduced with the permission of the Minister of Public Works and Government Services Canada, 2005.

all non-agricultural paid employees belonged to unions in 1911, this is an impressive record of expansion. This percentage measures the unionization rate—or *union density*—and reflects the proportion of actual union members to potential members. The exclusion of agriculture, where most workers are self-employed and are, therefore, ineligible for union membership, allows more accurate comparisons with earlier periods when agriculture was a much larger sector, and with other countries at different levels of industrialization. However, some employees in agriculture belong to unions. And unlike some countries, labour legislation in Canada usually prohibits managers and supervisors from joining.[16]

It is also important to note that collective bargaining coverage extends beyond union members. In 2004, collective bargaining coverage was 32.4 percent of paid employees, or about 2 percentage points higher than the unionization rate (*Perspectives on Labour and Income* 2004: 60). This reflects the fact that some non-members in unionized workplaces are entitled to the wages and benefits negotiated by the union in a collective agreement. Among them are supervisory employees who are excluded because of their management role, new hires on probation, and individuals who because of their religious or other personal beliefs opt not to join. Furthermore, Canadian labour law permits nonunion forms of collective representation, such as staff associations, which represent 5 percent of all employees. Finally, another 9 percent of employees belong to professional associations, and some of these set wages and working conditions for members.[17]

Scanning the chart, we can identify three major spurts in membership growth. The first two coincided with the two World Wars. This is not surprising because national mobilization for these wars resulted in economic growth, labour shortages, and the need for a high level of cooperation between employers and employees, all of which are key ingredients for successful union recruitment.

In the years during and immediately following World War II, Canada's contemporary industrial relations system took shape. Massive changes in the size, composition, legal rights, and goals of unions have occurred since the 1940s. The third growth spurt took place in the 1970s. The rise of *public-sector unions* in that decade, largely facilitated by supportive legislation, brought many civil servants, teachers, nurses, and other public employees into organized labour's fold.

In the three decades since 1971, union density hit a peak of 37.2 percent in 1984. In part, the decline since that year occurred because two recessions and industrial restructuring cut deeply into the traditional membership strength of unions in manufacturing and other blue-collar occupations. Similarly, budget cuts and downsizing in the public sector during the 1990s eroded membership ranks. Furthermore, employer pressures for concessions and the whittling away of collective bargaining rights by governments and the courts have contributed to a more hostile climate for labour relations. Despite these adversities, Canadian unions have held their own by adapting to these harsher realities. While membership numbers levelled off in the 1990s and increased in the early 21st century, recent membership increases have lagged behind labour force growth, resulting in a decline in union density.

The labour movement has undergone organizational changes as well. A notable trend is consolidation, resulting from mergers and membership growth since the 1960s (Chaison 2004). In 1968, there were 14 large unions (30,000 or more members), accounting for just under half of total union membership. By 2004, 32 unions had memberships of 30,000 or more, comprising over 75 percent of total union membership in the country.[18] One of the defining features of the Canadian labour movement is the large number of *locals*. In 2004, the 267 national and international unions comprised 16,267 locals, the basic self-governing unit of the labour movement and the legal entity for collective bargaining. Membership in locals ranges from just a few members to more than 45,000. In short, despite consolidations, Canadian labour remains fragmented and, consequently, collective bargaining is very decentralized. Unlike some European nations, where industry-wide national bargaining is the norm, the Canadian pattern of single-establishment, single-union bargaining results in thousands of collective agreements in effect at any one time.

A Comparative Perspective on Canadian Unions

A comparative perspective on Canada's unions provides insights about the nature of union–employer relations, and how these are institutionalized into national systems of industrial relations. Figure 7.2 provides two key measures for 11 countries: recent rates of union membership and the overall trend in union density since the early 1990s. Keep in mind that these comparisons are approximate because countries collect these data in somewhat different ways. As we noted above, density measures the extent of union membership among wage and salary earners. Through their unions, these workers engage in

FIGURE 7.2 *Unionization Rates in 2003 and Trends Since 1990**

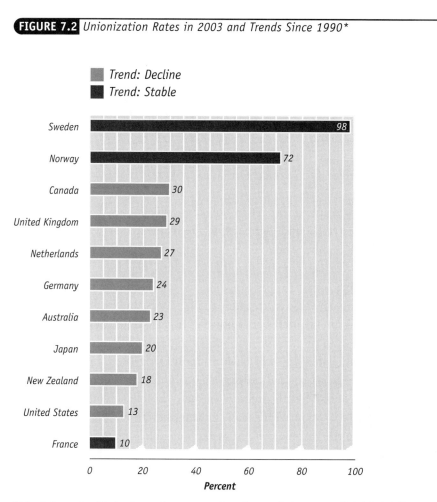

*Refers to the number of trade union members as a percentage of wage and salary earners. Methodological differences in how countries collected these data make comparison difficult. Data are for 2003, except for the United States (2004), New Zealand (1998), Germany (2000), and France (2000). Union density going back to 1990 or earlier was examined to determine trends.

Sources: Compiled from various sources, none of which uses identical measures of union density. Unpublished data on trade union membership from the International Labour Office, 2005; Carley, Mark 2005 *Industrial relations in the EU, Japan, and United States, 2003–04.* European Industrial Relations Observatory online (http://www.eiro.eurofound.ie); Carley, Mark 2004 *Trade union membership 1993–2003.* European Industrial Relations Observatory online (http://www.eiro.eurofound.ie); Bamber, Greg J., Russell D. Landsbury, and Nick Wailes, editors, *2004 International and Comparative Employment Relations: Globalisation and the Developed Market Economies,* 4th ed. (London: Sage, 2004), p. 379.

collective bargaining, the decision-making process in which union and management negotiate wages, benefits, hours, and other employment conditions for a group of workers, resulting in a *collective agreement* (or contract) (OECD 1994: Chapter 5; Lipsig-Mummé 2005).

In terms of union density, Canada is higher than many European countries, Australia and New Zealand, the United States, and Japan, but lower than the Scandinavian countries of Norway and Sweden. However, union density reveals only part of the strength of unions; collective bargaining coverage also must be considered. In Europe, it is more common than in Canada, the United States, or Japan for workers who are not union members to have terms and conditions of employment set by collective agreements. For example, in France, the Netherlands, and Germany, where relatively few wage and salary earners are union members, between 80 and 90 percent are covered by the provisions of collective agreements that these unions negotiate (International Labour Office 1997).

What accounts for these cross-national differences? In North America, key factors are management's traditional opposition to unions, labour laws that make the certification process for new bargaining units difficult, and a decentralized industrial relations system based in local workplaces. In Japan, where density and coverage are similar, collective bargaining also is decentralized and firm-based. But, unlike North America, in many European countries there is much greater coordination within each industrial sector. With the exception of Sweden, where the vast majority of workers eligible for union membership actually belong, countries with even moderate levels of union density can have extensive collective bargaining coverage because centrally negotiated contracts apply to most workers in a sector.

The western European pattern is a result of highly coordinated and centralized collective bargaining systems, supported by law, in which national unions negotiate with large employers' federations. Pay and other basic working conditions are set nationally or by industrial sector, greatly reducing competition among individual workplaces on these issues (unlike Canada, where a firm may resist union pay demands because it would increase its wage costs relative to local competitors). So, in France, for example, despite a union density of only 10 percent, employer associations are legally required to regularly negotiate broad agreements with unions, the benefits of which are extended to virtually all employers and employees. In addition, industrial democracy systems at the firm level in some European countries (discussed in Chapter 6) give workers and unions a voice in a wide range of local issues, from technological change to work arrangements. European systems of collective bargaining have recently become more decentralized, in some cases bringing greater industrial democracy to workplaces and providing more scope for worker participation in management decisions.

We noted earlier that Canadian union density has been in slow decline over the last several decades. Of 92 industrialized and developing countries, only 20 saw increases in the number of union members, with the largest growth seen in Asia (International Labour Office 1997). This is reflected in Figure 7.2, which indicates ongoing declines in union membership density in 8 of the 11 countries listed. Declines in both density and coverage were more common in nations with employer-centred, decentralized industrial relations systems—notably the United States and Japan. Drops in coverage were less likely in countries with strong sectoral or national bargaining regulated by the state to balance employers' and employees' interests. Britain and New Zealand, two countries with previously centralized systems, stand out with declining density and coverage. In each case, market-oriented governments deregulated and decentralized industrial relations systems in attempts to weaken the power of unions—basically, a shift to a more Americanized model of industrial relations (Freeman 1995). There are also various signs in other European countries, Australia, and New Zealand of a shift in collective bargaining from national or sectoral levels down to firms, largely due to the introduction of more flexible forms of work organization, although the extent and implications of this decentralization process vary considerably (International Labour Office 1997).

The field of *comparative industrial relations* examines whether labour and employment relations are undergoing similar transformations across advanced industrial nations (Bamber and Lansbury 1998; Godard 1997). Both these issues generate much debate among researchers. In the context of economic globalization, there is also growing interest in unionization trends in Asia, the former Soviet Union, and central and south America. At issue is how economic development affects human rights, with workers' freedom to join unions and engage in collective bargaining viewed as an extension of these basic rights. For example, in China economic reforms have reduced direct state regulation of labor relations, but because unions are integrated with management, they are unable to consult or negotiate on behalf of workers (Clarke, Chang-Hee, and

Qi 2004). In Russia, former communist trade unions have pursued social partnerships to reduce industrial conflict and protect members' interests during the transition to a market economy, but this has been ineffective because of the history of union collaboration with government and employers during the Soviet era (Ashwin 2004). In Mexico, union density has declined in the face of increasing employer resistance to unions and the growing use of what are called "protection contracts," whereby employers pay unions not to represent their members' interests in a workplace (Fairris and Levine 2004).

The Decline of U.S. Unions

Canada's union density and coverage have surpassed those of Japan and the United States, the two other countries with employer-centred industrial relations. It is significant that Canadian union membership remained stable while other countries, especially the United States, saw a decline. Canadian unions appear to have weathered the economic storms of the last quarter of the 20th century, although the future looks far less certain (Lowe 1998). U.S. unions, in contrast, are engaged in a struggle for survival. Membership has plunged from one-third of the workforce in the mid-1950s to 12.5 percent in 2004, and below 8 percent in the private sector (Carley 2005). Observers attribute this decline to a number of factors, but in particular to fierce anti-union campaigns launched by private-sector employers, labour laws that permit these coercive tactics, and ineffective responses from unions to these changes. Also significant is the shift of employment away from the union strongholds of the northeastern industrial regions to southern states, where *right-to-work laws* undermine union security; under these laws, neither union membership nor payment of union dues may be required as a condition of employment.[19]

Some industrial relations scholars have concluded that the American industrial relations system, which grew out of the New Deal in the 1930s, was replaced in the 1970s and 1980s by a nonunion approach to industrial relations. The breakdown of the old union-based industrial relations system can be traced to three trends: (1) *concession bargaining* (that is, unions agreeing to rollbacks in wages, benefits, and collective rights); (2) management innovations such as teamwork and greater employee participation; and (3) explicitly nonunion human resource management strategies in the private sector, in part a response to American unions' greater positive impact on wages compared with other countries, giving employers a strong incentive to oppose them. Taking a more historical view, Goldfield links the decline of American unions to changing class relations. He argues that the failure to form a solid working-class base for the labour movement during its period of greatest membership strength, from the 1930s to the 1950s, seriously weakened it as a political force. Hence, U.S. labour lacked the strategies, leadership, and membership solidarity necessary to survive in a more hostile environment.[20]

Another perspective is offered by Seymour Martin Lipset, an expert on Canadian–American differences. Lipset's (1990) thesis is that a core difference between the two societies is the historically stronger commitment among Americans to individualistic values, and Canadians' greater sense of collective rights, which are more amenable to unionization. A comparison of attitudes and values of workers on both sides of the border (Lipset and Meltz 1997) found an interesting paradox. On the one hand, American workers were slightly more approving of unions than Canadian workers, and considerably more nonunion workers would vote for a union if they had the opportunity. Yet on the other hand, the actual level of union membership in the United States, as we noted, is one-third that of Canada. This paradox reflects the difference between deep-rooted *values*, which underlie the institutions of labour relations, and more immediate *attitudes* toward unions, which reflect current labour market conditions and union strength. So in the case of America, unions are so weak today that workers believe they should be stronger, while in Canada, the greater power of unions ends up being a source of criticism among many. This debate is far from finished, however. Two American academics (Freeman and Rogers 1999: 36) have argued that if one looks beyond unions to any collective form of representation, in both countries there is a gap between workers' desire for more say in decisions that affect their work lives and actual opportunities for participation.

The U.S. experience shows the decisive role of legislation and management opposition in encouraging or inhibiting free collective bargaining. Some analysts argue, pessimistically, that unions lack the resources and supportive public policies required for the massive new organizing successes needed to grow their membership. However, other scholars detect signs of union revival in the United States, which may reverse the trends described above.[21] The main evidence of revival in the late 1990s includes increased public approval of unions (seen, for example, in the widespread support for striking UPS workers in

1997); new leadership in the AFL-CIO (the equivalent of the CLC) committed to vigorous organizing; increased success in union certification elections; and innovative strategies among local unions that have appealed to low-wage workers, women, and racial and ethnic minorities.

The Justice for Janitors campaign in Los Angeles exemplifies the latter. As in many North American cities, janitorial work in Los Angeles is done under short-term contracts by companies who employ immigrant workers. Trying to counter a decade of restructuring and de-unionization, the Service Employees International Union launched in the 1990s a successful public campaign to re-unionize the industry and improve the lot of the Cental American immigrants. Many of these workers were women, and many had entered the United States without documentation. The campaign focused on getting building owners to help provide health and welfare benefits to janitors and their families (Cranford 2004). In 2005, two major unions—the Teamsters and the Service Employees International Union—voted to leave the AFL-CIO, claiming that the national labour federation was not devoting sufficient resources to new organizing (*The Globe and Mail* 26 July 2005: B10). Whether this move results in more campaigns like Justice for Janitors, or more direct recruitment of new members, will be closely watched by researchers—and employers.

THE CURRENT STATE OF UNIONS IN CANADA
Membership Patterns

The likelihood of a Canadian worker being a member of a union, or joining one in the future, varies according to province of residence, industry, and occupation. Figure 7.3 shows that in 2004, 30.6 percent of employees (excluding the unemployed and self-employed) belonged to a union, a decline of 7 percentage points since 1981. Provincial differences in labour legislation, industrial mix, and economic performance influence union membership. Newfoundland has the highest rate of unionization, at 39 percent, because of the disproportionately large size of the unionized public sector in a historically weak economy and the strength of unions in the fishing industry. Quebec is next, for reasons that largely reflect the unique political context noted above. At the opposite end of the scale, Alberta's union density is only 22 percent. A critical perspective would suggest that Alberta's industrial relations have moved closer to the American model, given this province's more restrictive labour legislation, the

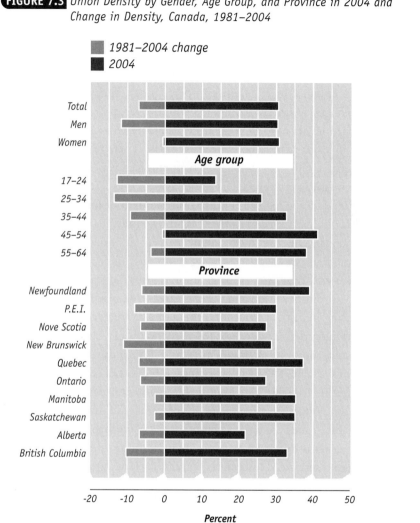

FIGURE 7.3 *Union Density by Gender, Age Group, and Province in 2004 and Change in Density, Canada, 1981–2004*

Source: Adapted from the Statistics Canada publication *Perspectives on Labour and Income,* Diverging trends in unionization, Catalogue 75-001, Summer 2005, vol. 17, no. 02.

decimation of construction industry unions in the early 1980s, and American-style nonunion human resource policies found in key industries such as oil and gas (Panitch and Swartz 1993; Noël and Gardner 1990; Reshef 1990). To view this from a broader perspective, labour laws are one aspect of labour standards that provide workers with a range of protections. According to a comparison of legislated labour standards in all U.S. states and Canadian provinces and territories, Alberta ranks close to the bottom (Block and Roberts 2000).

We also can see from Figure 7.3 that the once large gender gap in levels of union representation has disappeared, reflecting other labour force changes discussed in Chapters 2 and 4. However, age is a major source of variation in union membership rates. In 2004, only 14 percent of young workers (aged 17 to 24 years) belonged to unions, compared with 26 percent of 25- to 34-year-olds, and 41 percent of those aged 45 to 54. This pattern of aging membership is the main reason why unions need to bolster their recruitment efforts (Lowe and Rastin 2000). Yet this is not happening, given the decline since 1981 in union density among workers under age 35.

Figure 7.4 provides a detailed industrial breakdown of unionization rates, highlighting how industry of employment is an important factor in explaining membership trends. The five most highly unionized industries, each with 65 percent or more of employees belonging to unions, are in the public sector. The only private-sector industry to have union density substantially above the national average is transportation and warehousing, at 42 percent. Construction and manufacturing industries are around the national average in their levels of union density. Union density does not tell the whole story, so we also need to consider the actual number of union members. Five industries—two in the public sector (health care and social assistance; educational services) and three in the private sector (durable and nondurable manufacturing; transportation and warehousing)—together account for 60 percent of all union members in Canada.[22] In contrast, rates of union membership are less than half the national average in service industries such as retail, finance and insurance, and accommodation and food services. This reflects both the lack of opportunity to join a union and the difficulties in organizing one. Unions' lack of success in these industries does not bode well for the future of unions, given our discussion in Chapter 3 about the contributions of these sectors to job growth.

Another way of examining variations in rates of union membership is by occupation. Figure 7.5 shows that individuals in wholesale, food and beverage, managerial, and retail occupations are least likely to be union members. In contrast, the most highly unionized jobs are in the public sector: teachers, nurses, professors, and health care technical workers. Between 58 percent and 89 percent of workers in these occupations belong to unions. The unionization of these well-educated knowledge workers has given a whole new character to the

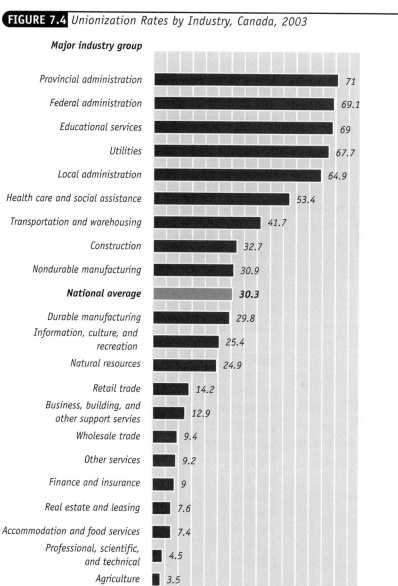

FIGURE 7.4 *Unionization Rates by Industry, Canada, 2003*

Major industry group

Provincial administration	71
Federal administration	69.1
Educational services	69
Utilities	67.7
Local administration	64.9
Health care and social assistance	53.4
Transportation and warehousing	41.7
Construction	32.7
Nondurable manufacturing	30.9
National average	**30.3**
Durable manufacturing	29.8
Information, culture, and recreation	25.4
Natural resources	24.9
Retail trade	14.2
Business, building, and other support servies	12.9
Wholesale trade	9.4
Other services	9.2
Finance and insurance	9
Real estate and leasing	7.6
Accommodation and food services	7.4
Professional, scientific, and technical	4.5
Agriculture	3.5

0   20   40   60   80

Percentage of employees belonging to unions

Source: Adapted from the Statistics Canada publication *Perspectives on Labour and Income,* Catalogue 75-001, August 2004, vol. 5, no. 8.

FIGURE 7.5 *Unionization Rates by Occupation, Canada, 2003*

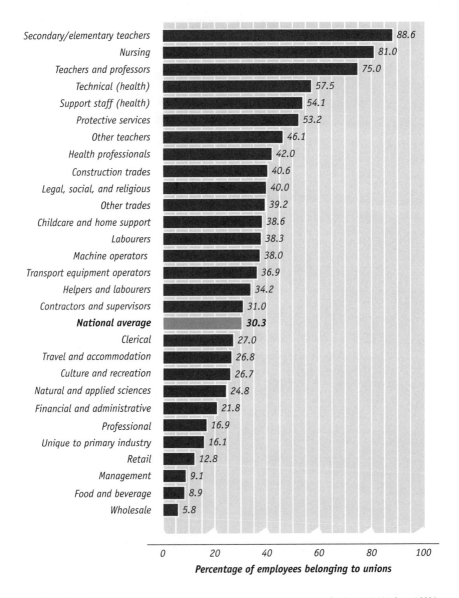

Occupation	Percentage
Secondary/elementary teachers	88.6
Nursing	81.0
Teachers and professors	75.0
Technical (health)	57.5
Support staff (health)	54.1
Protective services	53.2
Other teachers	46.1
Health professionals	42.0
Construction trades	40.6
Legal, social, and religious	40.0
Other trades	39.2
Childcare and home support	38.6
Labourers	38.3
Machine operators	38.0
Transport equipment operators	36.9
Helpers and labourers	34.2
Contractors and supervisors	31.0
National average	**30.3**
Clerical	27.0
Travel and accommodation	26.8
Culture and recreation	26.7
Natural and applied sciences	24.8
Financial and administrative	21.8
Professional	16.9
Unique to primary industry	16.1
Retail	12.8
Management	9.1
Food and beverage	8.9
Wholesale	5.8

Percentage of employees belonging to unions

Source: Adapted from the Statistics Canada publication *Perspectives on Labour and Income,* Catalogue 75-001, August 2004, vol. 5, no. 8.

labour movement. The traditional, blue-collar union strongholds in manual labour, construction trades, machine operation and assembly, and transport equipment operation now have union densities ranging from 34 to 41 percent. In terms of actual numbers of union members, postsecondary teachers and professors have the most union members (450,000), followed by clerical occupations, machine operators and assemblers, secondary- and elementary-school teachers, and travel and accommodation workers. These five occupations account for almost half of all union members in Canada.

Other employment characteristics also help to predict union membership. Workplace size is one of the strongest predictors of union membership: 54 percent of employees in workplaces with over 500 employees are unionized, compared with 13 percent of those in establishments with under 20 employees. Furthermore, being a full-time, permanent worker improves your chances of being a union member, or, in other words, of having the opportunity to join. Unions represent 24 percent of part-time and 25 percent of temporary workers, compared with approximately 31 percent of full-time workers or permanent workers.

In short, the composition of the labour movement has been transformed since the 1960s. The typical unionist today is a white-collar worker employed in one of the service industries, quite likely in the public sector. Women have joined the ranks of organized labour faster than men, so chances are that new union recruits will be women. There also is the potential for growing union representation among immigrants and racial minorities. As noted in Chapter 3, wages for immigrants and visible minorities (often the same people) tend to be lower than average. With the exception of black immigrant women, immigrants and racial minorities are less likely to be union members than the white majority (Reitz and Verma 2004). This gap in unionization is somewhat reduced the longer that immigrants are in Canada. Union campaigns targeting the garment industry, meatpacking, building services, and other sectors with high concentrations of immigrants could speed up this process.

The Rise of Public-Sector Unions

In marked contrast, membership grew in both relative and absolute terms during this time throughout the service sector. An example of this expansion is that of the Canadian Union of Public Employees (CUPE), the largest union in Canada since the early 1980s. The key to CUPE's success is membership diversification: the union has moved into new areas such as universities, airlines, and nursing homes in this period. This strategy also has paid off for some industrial unions. The Canadian Auto Workers has recruited new members in food services and casinos, and the United Steel Workers has organized security guards and taxi drivers.[23]

The present character of the labour movement is reflected in its largest organizations. Table 7.1 lists the 10 largest unions in 2000 and in 2004,

TABLE 7.1 *Ten Largest Unions, Canada, 2000 and 2004*

	Membership (000s) in 2000	*Membership (000s) in 2004*	*Change in Membership (000s) 2000–2004*
Canadian Union of Public Employees	485	535	50.0
National Union of Public and General Employees	325	337	12.0
National Automobile, Aerospace, Transportation and General Workers Union of Canada (Canadian Auto Workers or CAW)	220	260	40.0
United Food and Commercial Workers International Union	210	188	−22.0
United Steel Workers of America	190	180	−10.0
Public Service Alliance of Canada	147	153	6.4
Communications, Energy and Paperworkers Union of Canada	144	150	5.7
International Brotherhood of Teamsters	100	110	9.8
La Fédération de la santé et des services sociaux (formerly la Fédération des affaires sociales)	97	101	4.2
Labourers' International Union	60	85	25.0
Membership of the 10 largest unions	1,978	2,099	121.1
Total union membership	4,058	4,261	203.0
Percentage of total membership in the 10 largest unions	48.7%	49.3%	

Source: "Union Membership in Canada—2004," *Workplace Gazette* 7, 3 (2004): 44; Henry, Manon. URL: http://www110 .hrdc-drhc.gc.ca/millieudetravail_workplace/gazette/articles//gaze_articles.cfm Human Resources and Skills Development Canada, 2003. Reproduced with the permission of the Minister of Public Works and Government Services Canada, 2005.

which, as a group, represent about half of all union members in Canada. The top two, CUPE and NUPGE, represent government and other public-sector employees. CUPE members work in a wide spectrum of jobs in municipalities, electrical utilities, social services, child-care centres, schools, libraries, colleges and universities, hospitals, nursing homes, and many other public institutions. Typical of many government employee unions, CUPE evolved from a traditional and rather docile staff association. Established in the 1960s, it grew to 97,000 members by 1967. CUPE's fourfold growth in three decades made it the first union in the country to break the 400,000 membership mark. In 2004, it represented close to 13 percent of all union members in Canada. NUPGE is the umbrella organization for the various provincial government employee unions. Two other public-sector unions are among the top 10—one is based in Quebec, and the other, PSAC, represents federal government workers. The largest private-sector union is the CAW, which is Canadian-based. The largest U.S.-based union is the United Food and Commercial Workers. Facing declining membership, the UFCW has been actively trying to organize new members in service industries but has not gained much ground against formidable opponents, such as Wal-Mart.[24]

A watershed in the Canadian labour movement was reached in 1967 when the federal government passed the Public Service Staff Relations Act. This opened the door to collective bargaining for federal civil servants. Provincial government employees were already moving in this direction, with the granting of collective bargaining rights to Quebec public employees in 1965. Unionism soon spread into municipal governments, hospitals, schools, prisons, social services, and other expanding public institutions. Consequently, the international unions representing mainly male craft and industrial workers have lost their once-dominant position in the Canadian labour movement. Thus, the large unions in Table 7.1 represent the wave of public-sector unionism that began in the late 1960s. They also signal a new brand of worker militancy. Nurses, teachers, librarians, social workers, clerks, university professors, and other white-collar employees have taken to the picket lines to pressure their employers (ultimately, the government) into improving wages and working conditions and maintaining the quality of public services.

"Canadianizing" Unions

The rise of public-sector unions has also helped to *Canadianize* the labour movement. At the beginning of the 20th century, U.S.-based international unions represented about 95 percent of all unionized workers in Canada. By 1969, this proportion had dropped to 65 percent and, with the groundswell of nationalism since the 1970s, has continued to decline to 31.9 percent in 1990.[25] Not surprisingly, the recent upsurge in union growth is largely because of the efforts of national public-sector unions. Equally important in explaining the Canadianization trend is the push for greater autonomy, or outright independence, within Canadian sections of international unions.

The vulnerability of Canada's branch-plant economy taught growing numbers of workers the need for greater local control of union activities. A strong argument in support of international unions is that they are labour's best defence against the global strategies of multinational corporations. Yet the internationals have not always been effective in dealing with the sorts of problems multinational corporations created for Canadian employees. For example, officials at the U.S. headquarters of these unions sometimes equated layoffs or plant closures in Canada—associated with multinational firms shifting production to their U.S. facilities—with more jobs for their much larger American memberships. The North American Free Trade Agreement and the rising tide of protectionism south of the border have compounded these problems.

Different bargaining agendas also tended to arise in the two countries, reflecting their distinctive industrial relations environments. Canadian auto workers, for instance, roundly rejected the concessions made to employers by the U.S. wing of their union. The issue of national autonomy came to a head in the 1984 strike against General Motors by the Canadian division of the United Auto Workers. These Canadian autoworkers found themselves pitted against not only GM, but also the UAW leadership in Detroit, who wanted Canadian workers to accept the concessions agreed to by their American counterparts. While autonomy was not the goal of the Canadian workers going into the strike, it became an inevitable result.[26]

Beyond having different bargaining priorities and strategies, some Canadian branches of international unions felt that their dues were flowing into the American headquarters with few services flowing back. These and other factors have prompted a growing number of separations. In addition to the Canadian Auto Workers Union, the Communications, Energy, and Paperworkers Union, the Communications and Electrical Workers, the Energy and Chemical Workers, and the Canadian Paperworkers Union were all created in this way.[27] Usually, however, good working relations are maintained with their former U.S. parents. What will be interesting to watch in future is how unions create new forms of international cooperation—beyond Canada and the United States—to address the impact of economic globalization on workers.

WOMEN AND UNIONS

Why historically has the level of unionization been lower among female workers than among male workers? At one time, it was assumed that women would be less interested in unions because of their family responsibilities and presumed lower commitment to paid employment. Because of their domestic roles, the argument went, women's ties to the labour force were weak. Hence, work problems that might prompt unionization were of secondary concern. Old myths die hard, but close scrutiny of recent membership trends totally shatters this one. Women made up only 16.4 percent of all union members in 1962, yet by 2003, this figure had jumped to 48 percent. In fact, women have been joining unions at a much faster rate than men. Charlotte Yates's analysis of union organizing campaigns in Ontario during the 1980s and 1990s found that female-dominated workplaces were far more likely to vote in favour of union representation than were male-dominated workplaces. Clearly, women are not apathetic, passive, or indifferent to unions, if in fact they ever were. This is an important point that Yates believes unions have not adequately considered in their organizing campaigns.[28]

Indeed, women are often more pro-union than men because of their inferior positions in firms.[29] Kate Purcell rejects the "passive woman worker" argument on the grounds that industrial patterns of unionism are most decisive in who joins. Men and women alike join unions and engage in militant action according to the established patterns of their industries or occupations (Purcell 1979: 122–23). Miners and forestry workers are highly unionized, as we noted above, and have a history of militancy. But because these are non-traditional areas of women's employment, we would not expect to find many female unionists in these industries.

Thus, gender segregation in the labour market (discussed in Chapter 4) is the key factor in women's unionization patterns. We have seen how, historically, women's employment has been restricted to a few predominantly female jobs. The oppressive and unrewarding character of this type of work, not the fact of being a woman, underlies the lower rate of female unionization. Such conditions in themselves can undermine collective action as a way of solving problems: workers, be they male or female, may be more likely to quit than to stay and fight for change.

What has been the role of male-dominated unions in keeping women out of certain jobs and, therefore, out of unions? There are numerous examples of craft and industrial unionists adopting policies that restricted female access to their jobs, mainly due to fears about having their wages undercut. In the early 20th century, craft unions lobbied with middle-class reformers to keep women out of the industrial labour force (allegedly to protect them). Such efforts discouraged union initiatives by women, thereby defining the union movement as a male institution. In Canada, for example, the failure of the 1907 strike by female Bell Telephone operators in Toronto was partly due to the lack of support their unionization campaign received from the International Brotherhood of Electrical Workers, an exclusively male craft union (Sangster 1978).

Consequently, for many women the option of joining a union has not existed. Only in the last 20 years, for example, were major organizing efforts launched in the largely female retail and financial industries. Massive counterattacks were mounted by management in banks and in retail stores. Despite a huge investment of organizing resources since the late 1970s, the CLC has achieved only limited success in unionizing banks. Breakthroughs during the 1980s in achieving union certification in big department stores, such as Eaton's, initially augured well for union expansion in the retail sector. Yet the 1987 *decertification* of unions by employees in five Ontario Eaton's stores was a major setback.[30] While some areas of the retail industry, such as grocery stores, have moderate levels of unionization, vast areas of the service sector still pose large hurdles, as witnessed in recent failed unionization attempts at McDonald's and Wal-Mart.

Feminist scholars have cast new light on the collective struggles of women to achieve fairness and equality through unions. In predominantly female occupations, women have a long tradition of collective action. As professionals, female teachers have been organized for decades. While not based on trade union principles, the 1918 founding of the Ontario Federation of Women Teachers' Associations showed the growing commitment of women to organize for improved working conditions (Strong-Boag 1988: 69). Also, a history of union activism exists among women in the garment and textile industries, particularly in unions such as the International Ladies' Garment Workers (Gannagé 1986, 1995; Lipsig-Mummé 1987; Frager 1992). Women also have exerted pressure on the male-dominated unions to which they belong to have their concerns addressed. As Pamela Sugiman shows in her study of women's struggle for gender equality within the Canadian wing of the United Auto Workers during the 1960s, feminists active in the union used "gendered strategies of coping and resistance," drawing on women's combined experiences as autoworkers, unionists, wives, and mothers. Yet while formally espousing this goal, male UAW members had difficulty getting beyond their patriarchal ideology (Sugiman 1992: 24).

This research underscores a prominent theme in women's growing influence in unions, namely the importance of coalitions with the organized women's movement outside the workplace. *Pay equity* has become a major new workplace objective, largely due to a broad alliance of feminists and some major unions. Indeed, the central thrust of pay equity is to challenge the gendered division of labour as enshrined in many union-negotiated job classifications. It also places equality squarely on the collective bargaining agenda, especially given that legislation in Ontario and elsewhere requires unions to negotiate pay equity with employers. Perhaps the most prominent example of how pay equity for women workers has become a rallying point for unions is the Public Service Alliance (PSAC) national strike of federal civil servants in 1991. However, it took until the end of the decade to reach a pay equity settlement with the government for these clerical and administrative workers. As Gillian Creese's (1999) case study of the white-collar workers' union at BC Hydro documents, public-sector unions' quest for pay equity in the 1990s was made more difficult by the fact that it coincided with employers downsizing, restructuring, and contracting out work. These organizational changes affected a wide range of workers, raising complex issues about standards of fairness in how workers are treated. Furthermore, even though the job hierarchy at BC Hydro did become more open to women and members of racial minorities, union traditions and culture continued to reflect the values and interests of white males.

In sum, women's demands for equality of opportunities and rewards in the workplace have already had a major impact on Canadian unions. The CLC elected its first female president, Shirley Carr, in 1986, and women are making up a growing share of elected union officers and paid staff, especially in public-sector unions. The largest union in Canada, CUPE, elected Judy Darcy as its president in 1991. And a few unions, such as the Canadian Union of Postal Workers, have given top priority to concerns most often expressed by women (for example, paid maternity leave, sexual harassment, childcare, and fair treatment of part-time workers) (Julie White 1990). Gradually, "women's issues"—which really are issues for all workers—have come to have higher stature in labour's collective bargaining and social action. Important bargaining items, at least for large unions, now include gender-neutral contract language and clauses dealing with discrimination, sexual harassment, family-related leave, childcare, technological change, and rights for part-time workers (Kumar and Acri 1991; Briskin and McDermott 1993). Labour's success in achieving this bargaining agenda is obviously crucial to its future.

MANAGEMENT OPPOSITION TO UNIONS

Despite laws giving each employee the right to join a union and to participate in free collective bargaining, employer opposition to unionism has frequently been a major obstacle to putting these rights into practice. Workers who encounter strong management opposition to their union organizing activities may legitimately fear for their jobs, and this will have a chilling effect on the union campaign. Moreover, if an employer chooses not to recognize a union after it has been legally certified in the workplace, it is very difficult for members of that union to exercise their legal rights. Prior to the introduction of a legal framework for *union certification* and *collective bargaining* procedures during World War II, many industrial disputes occurred over the employer's refusal to recognize the workers' union. Surveying the historical record, Pentland observes: "It is sad but true that Canadian employers as a group and Canadian governments have never taken a forward step in industrial relations by intelligent choice, but have had to be battered into it."[31]

The New Human Resource Management

White-collar workers in the private sector remain the largest unorganized group in the labour force. Strenuous efforts by employers to remain union-free, with the help of the new human resource management practices already mentioned, and government inaction with regard to promoting private-sector union recognition, have created an inhospitable climate for unionism. There are several additional reasons for this climate (Bain 1978: 23). White-collar employees often have direct contact with management and, therefore, may identify more closely with the company ideology than do blue-collar workers. When management openly disapproves of unions, white-collar workers may be reluctant to join one. Workers join unions for the benefits, such as better working conditions, higher wages, job security, or access to a grievance proce-dure. Concrete proof that unions can "deliver the goods" is obviously lacking in industries that have never had high levels of unionization. The difficulty of overcoming these obstacles discourages unions from launching recruitment drives. Consequently, the option of joining a union has not been available to many private-sector white-collar workers.

Management usually prefers to deal with each employee individually. This creates a "David and Goliath" situation, pitting individual employees against the power of a corporation. However, the once virulent anti-unionism of Canadian business turned into grudging acceptance in the decades immedi-ately following World War II (Smucker 1980: Chapter 7). Instead of fighting unions head on, managers now often criticize them as being unrepresentative organizations under misguided leadership. An underlying concern is what constitutes *management's rights.* There are still some employers today who zeal-ously believe that the owners of a business should unilaterally define the employment relationship. In such cases, the right to manage takes absolute priority.

Some leading corporations have adopted a more subtle strategy. This is the *new human resource management,* which aims to maintain high-quality human resource policies and good working conditions, and encourages employee par-ticipation and commitment. Both lean production and high-performance workplace models, discussed in Chapter 5, have been adapted by some employers to maintain a union-free workplace. In some cases, a *company union* or employee association creates the appearance of democratic employee repre-sentation. The basic assumption, derived from the human relations manage-ment tradition, is that if management treats employees well and listens to their concerns, a union will be unnecessary. IBM is a good example of this approach in action. In Britain, the corporation's 13,000 employees were balloted at the request of the government's industrial relations board to find out if they desired union recognition. Only about 4 percent of the staff favoured unionization. This vote was not the result of corporate coercion or intimidation, as is becoming increasingly common in the United States (although not at IBM). Rather, it reflected the attention and resources that the company had devoted to employee relations. These human resource management policies produce a corporate culture based on strong employee loyalty and identification with management's goals (Keenoy 1985: 98–102). Some Canadian firms, most notably the Hamilton steel producer Dofasco, have developed their own unique brand of nonunion labour relations to obtain the cooperation of employees (Storey 1983).

Workplace reforms, particularly lean production and high-performance work systems, pose major dilemmas for unions. However, some of the most successful innovations of this kind suggest that union involvement is a prerequisite (P. Kumar 1995: 148). The examples of Saturn, Shell, and NUMMI (see Chapter 5) make this point. But in most cases, unions have tended to respond with either passive acceptance of these initiatives or active rejection.[32] The CAW, after extensive experience with lean production, adopted a policy of very carefully assessing the benefits and costs of participation (CAW 1993: 12). Only if workers' interests can be advanced and the union as an organization strengthened will the CAW enter into such change programs with management. Crucial ingredients are worker education and openness on management's part to share information (Martin 1995: 118–21). Pradeep Kumar (1995: 148–49) explains why "workplace innovation" poses difficulties for unions as well as management:

> . . . unions and management have different views of the goals of workplace innovations. Management goals are primarily efficiency-oriented. The general mission of HRM [human resource management] is to improve organizational effectiveness and achieve competitive advantage through a more efficient utilization and development of human resources. . . . Unions, on the other hand, are voluntary organizations of workers that serve as their collective voice in order to improve their physical, economic, social, and political well-being. The underlying values of the unions are inherently pluralistic and collective. Union activities extend beyond the workplace to legislative and political action to effect economic and social change. Consequently, if the unions are to have an independent role in workplace change, the objective of HR innovations must be more than simply enhancing organizational

effectiveness: it must include improving the lot of the individual worker and promoting societal well-being.

One of the most insightful studies of a union's experience with new systems for organizing and managing examines CAMI, a joint venture between General Motors and Suzuki (Rinehart, Huxley, and Robertson 1997). The firm built a new factory in Ingersoll, Ontario, and carefully recruited a relatively young workforce. The CAW represented the workers, making CAMI one of the few North American "transplants" to be unionized. The plant used *lean production,* a hybrid of Japanese and North American industrial organization and management systems. CAMI claimed its competitive advantage derived from its emphasis on four core values: *kaizen* (continuous improvement); team spirit; open communications; and worker empowerment. However, rather than representing a paradigm shift from old-style mass production to post-Fordism, CAMI basically just fine-tuned Fordism. Work under lean production was "arduous and intense" (204). The biggest problem was chronic understaffing, mainly due to CAMI management's intent to operate very lean from the start. "CAMI values" declined as worker discontent rose—the reality of conflict replaced the promise of "win–win." CAMI jobs were not challenging; most were low-skilled, repetitive, and routinized, with a high risk of injury. The logic of the system was a relentless push to reduce the labour component of each vehicle—hardly conditions that foster more humane work. Workers reacted to these conditions by going on strike for five weeks in 1992. As a result of the strike, the union gained more voice in setting production standards, in addition to higher wages. Not surprisingly, the researchers quote many workers who label CAMI "just another car factory." It is experiences like this that provide unions good reason to be skeptical and cautious when dealing with management strategies to reorganize work.

Unions in an Era of Continuous Restructuring

Crystal-ball gazing is not a very scholarly activity, but we should at least contemplate likely future scenarios in Canadian industrial relations. The collective bargaining agenda is being transformed. Declining productivity growth since the mid-1960s, recessions in the early 1980s and early 1990s, privatization and cuts in government spending, high taxes and shrinking real incomes, and the increasing use of legislation and courts to restrict labour's rights have forced unions to rethink their negotiating priorities. For organized labour, the challenge has been to protect members' jobs and incomes. Concession bargaining began seriously in 1983, and since then unions have faced employer demands for wage and benefit concessions and greater flexibility in hiring practices— and the NHL lockout is another example of this.

In an era of industrial restructuring, downsizing, and public-sector cutbacks, employers are increasingly using work reorganization, flexible employment systems, and contracting out ("privatization" in the public sector) to their advantage. There is little doubt that these will be major collective bargaining challenges in the future. When 26,000 members of the Canadian Auto Workers union went on strike in October 1996 against General Motors Canada, the key issue was "outsourcing" of work (using other companies to make parts previously made by GM workers). The union gained some ground in its fight to restrict GM's ability to sell plants or to contract out "non-core" jobs. In the public sector, similar issues arose in November 1995 when 120 laundry workers at two Calgary hospitals staged an illegal strike, protesting the contracting out of their work to a private Edmonton company. They had previously taken a 28-percent wage cut that they thought would preserve their jobs. Other unionized hospital workers joined the picket line and there was considerable support from citizens fearful about the general deterioration of health services. Within a few days, the regional health authority and the provincial government had backed down on this particular privatization initiative, but the larger trend was unchecked.[33]

This realignment in collective bargaining in favour of management reflects underlying economic trends. Many factors are motivating employers to slash labour costs, boost employee productivity, intensify the work process, or even relocate jobs outside of Canada. Included among these are globalization; pressures from NAFTA to bring Canadian employment standards into line with lower standards in the United States, Mexico, or elsewhere; the acceleration of technological innovations; government deregulation of industry; the desire to obtain greater flexibility in staffing and production systems; and pressures to cut government deficits. Some analysts argue that, in response, organized labour must develop truly international strategies. Not surprisingly, the impact of globalization trends on national and workplace industrial relations is a growing area of research. A case in point is maritime shipping, a truly global industry (Lillie 2005). Unions representing seafarers and port workers have formed a global alliance, the International Transport Workers' Federation's Flag of Convenience campaign. Based in London, England, the association coordinates actions to maintain minimum pay standards, negotiate collective agreements with ship owners, and prevent union-busting at ports.

THE NATURE OF COLLECTIVE ACTION

So far, we have assumed that union members' common interests and objectives make it likely that they will bargain collectively with their employer. But how do workers come to act collectively? We recognize the importance of work-group dynamics, informal norms, and social relations, topics already discussed in Chapter 5. But this is only a partial answer. A more complete understanding of collective action requires us to examine the underlying basis for unionization, which is the predominant form of collective worker action in industrial society. What makes this particularly interesting is that the very idea of collective action seems contrary to the individualistic norms of capitalist employment relations.

The Dilemmas of Collective Action

Drawing on the insights of informal group dynamics, sociologists portray unions as *social movements* that unite members so they can actively seek shared goals. But, as with bureaucracies, unions also require rules and sanctions to regulate the behaviour of members. As Allan Flanders (1970: 44) writes, "trade unions need organization for their power and movement for their vitality." Both qualities are essential for unions to achieve their social and economic objectives. At the individual level, this perspective focuses on why employees would organize themselves into unions to collectively pursue common goals. In addition, it incorporates an analysis of the economic and social conditions that encourage or inhibit unionism and, moreover, raises questions about the internal dynamics of unions.

It is helpful to consider two dilemmas that a worker must resolve if she or he is to seriously consider joining a union or, if already a member, becoming an activist. The first is the *free-rider problem*. Mancur Olson asserts that individuals will not naturally organize to further their collective interest. "The rational worker," explains Olson (1965: 88), "will not voluntarily contribute to a union providing a collective benefit since [s]he alone would not perceptibly strengthen the union, and since [s]he would get the benefits of any union achievements whether or not [s]he supported the union." Just like social movements concerned with environmental, peace, or feminist issues, unions provide *collective goods*. In other words, all potential members have access to the organization's achievements—a sustainable environment, a world with fewer weapons, employment equity programs, or a grievance procedure and negotiated regular wage increases—whether they have assisted in achieving these objectives or not (Crouch 1982: 51–67). This is one reason that unionization levels remain low in European countries such as France, the Netherlands, and Germany. Given that the centralized, state-regulated industrial relations systems in these countries guarantee many non-members the full benefits of union-negotiated wage agreements, the incentive to join is considerably reduced. In contrast, in the United States where such legislation is absent, "right to work" laws weaken unions by taking advantage of the "free rider" dilemma.

The second dilemma involves Albert Hirschman's (1970) *exit* and *voice* methods of expressing discontent. According to Hirschman, a dissatisfied employee can either leave her or his employer or stay and push for changes. The presence of a union increases the chances that the employee will pursue the latter strategy. But in nonunion workplaces, the poor employment conditions that could spark an organizing drive also increase the chances of an individual opting to quit. This is a major obstacle to unionization in low-wage job ghettos in the service industries today.

However, as we have already noted, many people do not have employment options. Unemployment may be high in their community, or their particular skills may not be in demand. Consequently, as we document in Chapter 8, such individuals adapt to their limited opportunities to find a better job, perhaps becoming increasingly apathetic and alienated from their work. But, while it is true that apathy and alienation stifle collective action, it would be wrong to assume that these conditions cannot be overcome. Instead, we should ask why some groups of employees mobilize while others do not.

The Mobilization Process

The process of mobilizing initial support for a union requires obtaining signed union cards from the majority of employees in the workplace. It also requires the financial assistance and organizational expertise of an established union.

Once organized, a union faces the problem of rallying members in support of collective bargaining goals. Ultimately, this could involve strike action. According to Charles Tilly's (1979) research on the causes of social protest, what is essential here is a sense of shared identity with the group. Collective action in the workplace will be easier when the work group is an important part of each employee's life. Strong social ties inside and outside the workplace integrate employees into the group. Individual interests become synonymous with group interests, acting as a springboard to collective action. Good examples of groups possessing this *solidarity* are printers, miners, fishers, and members of other occupations in which the occupational culture encompasses all of a worker's life.[34] Levels of unionization and industrial militancy tend to be high under these conditions.

Assuming the existence of a cohesive group, what pushes its members into common action? The role of the group leader is obviously central. Especially important are *organic leaders* who, by virtue of being part of the group, are best able to tap its potential for collective action. This is because they understand the experiences of group members and can gain their trust better than an outsider could. In short, an effective leader will build up group solidarity, create an awareness of common interests, map out a realistic program of action, and seize opportunities to launch the plan.

But we are not suggesting that all union activity can be reduced to the characteristics of a particular group, its members, and its leaders. Also important is the environmental context, which might involve a supportive community, a hostile employer, or fair labour legislation—all factors that can either nurture or dampen union activity. Even if such conditions are favourable, employees ripe for unionization may still not act. Similarly, unionized workers facing a deadlock in negotiations with management over a new collective agreement may not strike. Often missing is one or more precipitating factors: an arbitrary change in work practices; the denial of a long-awaited salary increase; the dismissal of coworkers; or a pent-up sense of being treated unfairly by management. Such perceived injustices could serve as the catalysts for mobilization.

Research on the inner workings of two Canadian unions gives us glimpses of how some of these factors actually contribute to *militancy*. Julie White's (1990: 147) study of the Canadian Union of Postal Workers (CUPW) quotes a woman worker in a Saint John post office: "We're a militant union, and I really believe that Canada Post has made us that way." This view is typical of CUPW members, White argues. The union has developed a "culture of struggle," based on the belief that management does not have the interests of workers at heart and, further, that any past improvements have been extracted from management by militant action (Langford 1996). Against a background of unsuccessful attempts to resolve disputes through negotiation, mediation, and conciliation, this culture of struggle has been cultivated by an openly democratic structure based on rank-and-file involvement.

The importance of internal union dynamics was also a crucial factor in the militancy of Canadian autoworkers in the 1980s. Charlotte Yates attributes CAW's success at holding back on concessions demanded by North American auto firms to its greater membership solidarity, compared with the United Auto Workers union in the United States, which agreed to concessions. Some of the key features of union organization that contributed to CAW's militancy were members' influence over union decision making, including the rank-and-file's ability to mobilize opposition to the leadership if unpopular positions were being imposed, and effective communication channels linking all levels in the union (Yates 1990: 77).

In summary, a solid organizational base, strong leadership, union democracy that encourages rank-and-file participation, a shared understanding of who is responsible for employees' grievances, and support from other groups in the community are key ingredients for collective action. Equipped with this understanding, we can now turn our attention to the most common manifestation of collective action—strikes.

STRIKES

According to conventional wisdom, unions in this country are strike-happy, and our entire industrial relations system is far too adversarial. In this section, we will evaluate the popular complaint that Canadian workers are strike-prone. We will also probe beneath the surface of these conflicts, locating strikes within the context of the larger political arena. Are strikes a sign of worker militancy? Does industrial conflict have any connection with working-class politics? After all, many of the grievances fuelling strikes could be blamed on the economic system, but do workers make such connections? And what role do employers and governments, through legislation, play in industrial disputes?

What Are Strikes?

Strikes are high drama on the stage of industrial relations. The public views strikes as an inconvenience or even as a major social problem. Many politicians and business leaders argue that strikes harm the economy. The participants seldom want strikes, least of all the union members, who will never recoup the wages lost should the dispute drag on. But if conflict has been largely *institutionalized,* as described earlier, why do strikes occur at all? What motivates workers to strike, and what are the larger social and economic implications of their actions?

Let us begin with a widely accepted definition of a *strike* or *lockout:* "A temporary stoppage of work willfully effected by a group of workers, or by one or more employers, with a view to enforcing a demand" (Lacroix 1986: 172). Workers initiate strikes typically after a legally required strike vote, while lockouts (which account for relatively few work stoppages, recent examples being the NHL and the CBC disputes) are initiated by employers. In the 1990s, close to 90 percent of collective agreements in Canada's private sector were negotiated without a strike, although some did rely on third-party conciliation or mediation to reach this goal. The public sector is a different story. Some 70 percent of agreements are settled between the parties. However, strikes are actually less frequent than in the private sector, because governments have increasingly relied on legislation to prevent or terminate strikes—as they did in 22 percent of the collective agreements concluded with public-sector workers in the 1990s (Gunderson et al. 2005: 357).

The B.C. and Newfoundland strikes described at the start of the chapter are examples of this kind of government action. Overall, strikes are infrequent events. Contrary to public opinion, settlement of union–management disputes without recourse to work stoppages is clearly the norm in Canada. There is evidently some truth to the argument that conflict has been institutionalized through the elaborate legal framework that regulates union–management relations.

Strikes are, nonetheless, central to the wage-labour relationship. If workers are selling their labour power to an employer in return for wages, their ultimate bargaining lever is to withdraw that labour. Even before the emergence of industrial capitalism in Canada, strikes erupted as workers protested against harsh treatment. For example, carpenters imported to build ships in colonial New France staged a slowdown to win living expenses above basic wages

(Pentland 1981: 27). We have already mentioned that the struggling 19th-century labour movement mounted several landmark strikes, most notably the 1872 Toronto printers' strike and the Nine-Hour Movement agitation that happened around the same time.

But strikes are only one possible form of workplace conflict. Richard Hyman distinguishes between *unorganized* and *organized conflict* (1978: Chapter 3). The former is not a calculated group action, and typically involves workers responding to oppressive situations by individual absenteeism, quitting, or sabotage. The latter is a planned collective strategy, the aim of which is to change the source of the discontent. Recall that strikes in Canada are only legal after the collective agreement has expired and specific conditions, such as a strike vote, have been met. Unauthorized strikes during the term of an agreement, often spontaneous responses by rank-and-file workers to an immediate problem in the workplace, are known as *wildcat strikes*.[35] The Calgary laundry workers that we discussed earlier provide a recent example.

Not all strikes are over economic issues, nor do they exclusively involve working-class unionists. They can be political, as in the case of the Canadian Labour Congress's 1976 "national day of protest," when more than 1 million employees marched on the streets to register their opposition to the federal government's imposition of wage and price controls. The British Columbia Federation of Labour's concerted opposition in 1983 to new legislation that would seriously weaken unions is another example. So too are the 1996 "days of action" initiated by the Ontario Federation of Labour to protest the newly elected Conservative government's massive public-sector funding cuts. Industrial disputes also can involve work stoppages by high-status professionals. Examples include teachers, doctors, university professors, airline pilots, and professional athletes. Finally, strikes do not necessarily entail all members of the union leaving the work site as a group. Under certain situations, *rotating strikes* across a number of work sites, *working-to-rule* (doing the minimum required or refusing overtime work, which has been used recently by health care workers), or staging work stoppages by "sitting down" on the job are variants of strike activity that can also communicate workers' demands to management.

Our emphasis on strike activity may leave the impression that workers are the disruptive factor in what otherwise would be harmonious labour relations. The key word here is *relations,* for there are two parties in every strike—three if the government intervenes, and possibly four if the public is directly affected. Seldom is the employment relationship a balanced one, however, since employers inevitably wield greater power. After all, management holds the final bargaining chip—the keys to the factory gate or office door, and, therefore, to the workers' jobs. Nonetheless, the public usually sees unions as the cause of strikes and labour unrest.

Canadian Strike Trends

On an annual basis, work stoppages rarely account for more than 0.5 percent of all working time in Canada (Table 7.2). This indicates that the seriousness of the strike "problem" often gets blown out of proportion. But, historically, we can identify eras when strikes were more common and widespread. There have been four particularly stormy periods of industrial conflict in Canada during the 20th century, gauged by the percentage of total working time lost due to strikes and lockouts.

A number of issues were at stake in early-20th-century strikes. Skilled artisans in the 19th century had been able to retain much of their craft status, pride, and economic security through their control of the production process. This privileged position was eroded by industrialization after 1900, as advancing technology and scientific management techniques undermined craftworkers' autonomy. Thus, craftworkers angrily resisted rationalization of their work, sparking many of the 421 strikes and lockouts that occurred between 1901 and 1914 in southwestern Ontario manufacturing cities (Heron and Palmer 1977).

Prior to union recognition and compulsory collective bargaining becoming encoded in law during World War II, many strikes were precipitated by an employer's refusal to recognize the existence of a union, much less bargain with it. The historic peak in labour militancy occurred at the end of World War I. Workers across the country were agitating against oppressive working conditions, low wages, and declining living standards due to soaring wartime inflation. Most of all, they wanted recognition of their unions. Western Canadian unions were far more militant and inclined toward radical politics than those in the rest of the country. Thus, in 1919 when Winnipeg building and metal trades employers refused to recognize and negotiate with unions over wage increases, the Winnipeg Trades and Labour Council called a *general strike*.

A massive display of working-class solidarity erupted, bringing the local economy to a halt. Sympathy strikes spread to other cities across Canada, and

TABLE 7.2 *Work Stoppages* Involving One or More Workers, Canada, 1976–2004*

Period	Number in existence during year	Total number of workers involved	Average number of workers per stoppage	Total person-days not worked	Average days lost per worker involved	Days lost as a percentage of estimated total working time
1976	1,040	1,586,221	1,525	11,544,170	7.3	0.53
1977	806	217,647	270	3,320,050	15.3	0.15
1978	1,057	400,622	379	7,357,180	18.4	0.32
1979	1,049	462,386	441	7,819,350	16.9	0.33
1980	1,028	439,003	427	9,129,960	20.8	0.37
1981	1,050	341,852	326	8,850,560	25.9	0.35
1982	680	464,181	683	5,712,500	12.3	0.23
1983	645	329,472	511	4,440,900	13.5	0.18
1984	716	186,916	261	3,883,400	20.8	0.15
1985	829	162,333	196	3,125,560	19.3	0.12
1986	748	484,255	647	7,151,470	14.8	0.27
1987	668	581,882	871	3,810,170	6.5	0.14
1988	548	206,796	377	4,901,260	23.7	0.17
1989	627	444,747	709	3,701,360	8.3	0.13
1990	579	270,471	467	5,079,190	18.8	0.17
1991	463	253,334	547	2,516,090	9.9	0.09
1992	404	149,940	371	2,110,180	14.1	0.07
1993	381	101,784	267	1,516,640	14.9	0.05
1994	374	80,856	216	1,606,580	19.9	0.06
1995	328	149,159	455	1,582,321	10.6	0.05

1996	328	283,744	865	3,339,560	11.8	0.11
1997	284	257,664	907	3,610,206	14.0	0.12
1998	381	244,402	641	2,443,876	10.0	0.08
1999	413	158,612	384	2,445,770	15.4	0.08
2000	379	143,795	379	1,656,790	11.5	0.05
2001	381	221,145	580	2,198,850	10.0	0.07
2002	294	168,002	571	3,033,430	18.0	0.15
2003	266	79,526	299	1,736,312	22.0	0.03
2004	300	259,150	864	3,257,719	13.0	0.14
1976–79 average	988	666,719	654	7,510,188	14.5	0.33
1980–89 average	754	364,144	501	5,470,714	16.6	0.21
1990–99 average	394	194,997	512	2,625,041	13.9	0.09
2000–04 average	324	174,324	539	2,376,620	14.8	0.09

*Refers to strikes and lockouts.

Source: "Chronological Perspective on Work Stoppages 1976–2004." URL: http://www110.hrdc-drhc.gc.ca/millieudetravail_workplace/chrono/index.cfm/doc/english. Reproduced with the permission of the Minister of Public Works and Government Services Canada, 2005.

even into the United States. The battle lines of open class warfare (one of the few instances of this in Canadian history) were drawn when, fearing a revolution, Winnipeg's upper class fought back with the help of the state. For several days, strikers squared off against police and employer-sponsored armed vigilantes. The confrontation ended in violence after the Royal Northwest Mounted Police, sent in by the federal government, charged a crowd of demonstrators. Strike leaders were arrested and jailed, while the workers' demands were still unmet (Bercuson 1974; Jamieson 1971).

Strike activity declined with rising unemployment during the Depression of the 1930s. As a rule, unions are less likely to strike in tough economic times. Conversely, when industry is booming and there is a relative shortage of labour, reflected in low unemployment rates, a strike becomes a more potent bargaining lever. The World War II era marked the rise of industrial unionism in manufacturing industries. Organizing drives accelerated as military production demands helped to restore the ailing economy. Again, union recognition was a dominant issue, driving workers in automobile factories, steel plants, and mines onto the picket line. As mentioned earlier in the chapter, the 1945 Ford strike led to the introduction of the Rand Formula to protect union security (Colling 1995). Subsequently, this became a standard feature of collective agreements, setting the tenor of labour relations by legitimizing unions while simultaneously ensuring orderly collective bargaining (Russell 1990).

Canada experienced a series of strikes in the mid-1960s. The fact that about one-third of these work stoppages involved wildcat strikes (mainly over wages) led the government to perceive a serious crisis in industrial relations. A task force, chaired by Professor H. D. Woods of McGill University, was set up to investigate the causes of industrial unrest and to recommend ways of achieving labour peace. Yet rampant inflation during the 1970s, and an increasingly militant mood among public-sector workers, escalated labour–management confrontations.

The most recent strike wave reached its apex in 1976. The Trudeau government's imposition of wage and price controls in 1975 as part of its anti-inflation program made strikes over higher wages a futile exercise (Reid 1982). But in the 20 years following 1976, lost time due to work stoppages dropped to post–World War II lows. The recession-plagued 1980s and 1990s dampened strike activity. Ironically, the fact that Canada's lost working time due to strikes and lockouts had dropped by the late 1980s was used by the federal government to attract foreign investors looking for a stable industrial relations environment (*The Globe and Mail* 4 April 1987: A3).

Table 7.2 profiles the labour relations patterns of the past three decades by reporting four characteristics of strike activity: *frequency* (number of strikes), *size* (number of workers involved), *duration* (days lost), and overall *volume* (days lost as percentage of total working time). For example, in 2004 there were 300 work stoppages involving an average of 864 workers, with each striking worker on the picket line an average of 13 days. In comparison, 1980 witnessed more than three times the number of strikes, each involving fewer workers (an average of 427) but lasting considerably longer (20.8 days on average). A more systematic comparison over time can be found in the bottom three rows of Table 7.2. Here we provide averages for the last four years of the 1970s, the decades of the 1980s and 1990s, and the 2000–2004 period. Note two key trends over these four time periods: a drop in both the number of work stoppages and the days lost because of these stoppages. Changing economic and industrial relations conditions in Canada have resulted in fewer strikes and lockouts with diminishing impact on economic productivity, at least as measured by lost time.

However, the greater number of public-sector workers who have faced legislated restrictions on strike action contributes to this trend. Most governments' deficit-cutting strategy during the 1990s involved imposing wage rollbacks through legislation, effectively suspending their employees' collective bargaining rights (Swimmer 2001; Reshef and Rastin 2003). So the declining numbers in Table 7.2 do not necessarily signal the dawn of a new era of labour peace. Clearly the system of public-sector industrial relations needs reform. Taking stock of labour–management relations in the federal government, the Advisory Committee on Labour Management Relations (2001), led by John Fryer, recommended a new framework, based on collaborative problem solving, that would be better able to address pressing human resource management issues. Extending the Committee's diagnosis across the public sector, what's needed is a fundamental shift to a less adversarial approach based on cooperation and mutual trust.

A Comparative Perspective on Strikes

How does Canada's strike record compare with that of other industrial nations? Recognizing the difficulties involved in making international comparisons of strike activity, due to differences in how strikes are defined and measured (Ross, Bamber, and Whitehouse 1998), we can, nonetheless, get a rough idea of where Canada stands in this regard. Based on the annual averages of working days lost in nine industrialized countries between 1970 and 1992, Canada's rate was the second highest after Italy (Adams 1995: 512). But the earlier years in this period were historically high for Canada, inflating the overall average. Looking at the mid-1990s, Canada has significantly lower strike volume than Italy, France, or Australia, but considerably higher than other countries, in particular the United States and the Netherlands (International Labour Office 1997: 249–50). More importantly, the strike volume in highly industrialized countries has declined, in large part because of difficult economic times and high unemployment through much of the 1990s. Some experts suggest that while strikes are less frequent, when they do happen they are now more intense confrontations (Aligisakis 1997). However, other analysts argue that cross-national patterns in industrial disputes reveal considerable diversity, reflecting the unique institutional contexts in which employment relationships are shaped (Adams 1995; Ross, Bamber, and Whitehouse 1998).

There are major institutional differences that account for international variations in strikes. For instance, in Italy, disputes, while short, are frequent and involve many workers. By contrast, the Swedish system of centralized bargaining, along with the country's powerful unions and their huge strike funds, mean that a work stoppage could quickly cripple the economy. This imposes considerable pressure to peacefully resolve potential disputes. In Japan, unions are closely integrated into corporations in what really amounts to a type of company unionism. Strikes are infrequent; workers voice their grievances by wearing black armbands or by making other symbolic gestures calculated to embarrass management. German union–management collective bargaining is centrally coordinated, but, at the individual workplace, employee-elected works councils frequently negotiate employment conditions (see Chapter 6). German laws require these participatory councils; they also make it very difficult to strike. Even though the strike rate in Canada is relatively low now, the fact that employees in this country have to confront employers when seeking collective bargaining has created a far more adversarial system (R. J. Adams 2000).

Thus, legislation governing the structure of collective bargaining is vitally important. The North American system of bargaining is highly decentralized and fragmented, involving thousands of separate negotiations between a local union and a single employer, each of which could result in a strike. Low-strike nations such as Austria, Germany, the Netherlands, and Sweden avoid these problems by having national or industry-wide agreements, and by legislating many of the quality-of-working-life issues that Canadian unions must negotiate with employers on a piecemeal basis. And as discussed in Chapter 6, laws in these countries give employees greater participation rights in enterprise-level decision making. This provides unions with more information about the firm's operations and opens up more communication channels with management, thereby encouraging more cooperation.

There are also major industrial and regional variations in strikes, which may be the result of how industries are organized. The least strike-prone Canadian industries are finance, trade, and services, which makes sense given their low unionization levels. Of the highly unionized industries, mining has historically been the most strife-ridden. Similarly, Newfoundland and British Columbia have significantly higher than average strike rates because of their high concentration of primary and other strike-prone industries.[36]

Explaining Strikes

One prominent but widely challenged interpretation of inter-industry differences in strike behaviour is Kerr and Siegel's *isolation hypothesis*. These writers propose that isolated industries have higher strike rates than those where workers are integrated into the larger society. Isolated workers ". . . live in their own separate communities: the coal patch, the ship, the waterfront district, the logging camp, the textile town. These communities have their own codes, myths, heroes, and social standards."[37] Canada has always had large numbers of remote, resource-based, single-employer towns (see Chapter 2), where a combination of social isolation and a limited occupational hierarchy moulds workers into a cohesive group in opposition to management.

Research by Eric Batstone and his colleagues in a British auto plant adds to our understanding of strikes by highlighting the day-to-day processes that may culminate in a strike (Batstone, Boraston, and Frenkel 1978). Strikes, argue these researchers, do not just happen. Rather, as a form of collective action, strikes require a high degree of *mobilization*. How production is organized, the type of technology used, and the formal institutions regulating union–management relations shape the context for this mobilization. But a crucial (yet largely ignored) ingredient of mobilization is how the social relations within the union allow certain individuals and groups to shape the course of events leading up to a strike. This mainly occurs through the identification of potential strike issues and through the use of rhetoric, by which the escalation of collective action is justified.

At the time of Batstone's study, most British strikes were unofficial stoppages initiated on the shop floor, not by top union leaders or by a democratic vote by all members. Laws more strictly regulate Canadian strikes, although recent British legislation has also moved in this direction. But the Batstone study offers some important insights about strikes in general. Typically workers are reluctant to strike, so there will not be much support for a strike until a vocabulary has developed that justifies such action. This involves the translation of specific grievances into the language of broad principles and rights. Furthermore, disputes over wages also typically involve *effort bargaining* (Gannagé 1995). That is, even if management refuses to grant wage increases, workers can informally shift their bargaining tactics to the other side of the "wage-effort equation" through slowdowns, refusal of overtime, and unauthorized work breaks. Finally, the study refutes the common view that the most powerful unions throw their weight around by striking. A basic contradiction of trade unionism, especially when examined in the workplace, is that ". . . the groups of workers most able to strike (in terms of bargaining awareness and collective strength) may rarely have to resort to strike action" (Gannagé 1995).

Industrial conflicts in recent decades were more likely than earlier disputes to occur in the public sector and to involve women, immigrants, and other workers new to collective bargaining. Jerry White's (1990) analysis of the 1981 illegal strike in Ontario hospitals highlights how the pressures of cutbacks and the resulting work reorganization in health care have the potential to push workers onto the picket line. White dismisses the "agitator" explanation of strikes, because the union leadership generally did not support the strike. Rather, the strike was a grassroots protest by nonprofessional service workers such as orderlies, housekeepers, food handlers, registered nursing assistants, lab technicians, and maintenance workers. In the end, the strike won the workers slight wage improvements, but at a price: thousands of strikers were suspended, 34 were fired, and 3 union leaders were jailed.

Interestingly, the motivations to strike were different for men and women in this dispute. At issue was the transformation of the labour process associated with patient care due to massive cost-cutting. Male workers wanted compensation in the form of higher pay, reduced workload, and guaranteed benefits. Women, in contrast, were far more concerned with how cutbacks had led to deteriorating relationships with patients. Jerry White (1990: 124–25) explains:

> There was a complex of rewards [for women], including formal *recognition* of the usefulness of the work, an *interaction* with other workers, and an intrinsic sense of a *job well done*. The hospital work provided this complex of rewards. Care-giving was a key aspect of this. Labour process changes broke this care-giving bond for many women workers. The desire to reestablish this bond was one consideration in the decision to strike.

WORKER MILITANCY AND CLASS POLITICS

Social scientists often use strikes as a measure of worker *militancy*. Are strikes actually deeply rooted in class antagonisms, reflecting basic working-class discontent? Or do unionized workers act simply out of economic self-interest? General strikes have temporarily disrupted the economies of several advanced capitalist societies in recent years, and there is debate over how to interpret these events. Are they signs of heightened *class consciousness,* or are they merely pressure tactics to force changes in specific government or employer policies? Economists have calculated that strikes have only a modest impact on the economy (Gunderson et al. 2005). Sociologists are equally interested in the broader implications of strikes for class consciousness and for working-class political action.

Comparative Perspectives on Working-Class Radicalism

Michael Mann's (1970) comparative study of industrial conflict addresses this question. Workers in France and Italy, he found, were more politically radical than those in Britain or the United States, who identified more closely with management. But the political potential of the French and Italian working classes was frequently limited by their failure to articulate an alternative social system and to chart a course toward it. Mann argues that strikes can be *explosions of consciousness*. But the working-class solidarity they generate rarely gathers momentum beyond the immediate event. For example, a test of the "explosion of consciousness" hypothesis among Hamilton, Ontario, postal workers who had participated in a strike found increased positive identification with fellow workers and the union, but little impact on class consciousness (Langford 1996). Ultimately, then, strikes by even politically radical workers are transformed into tactical manoeuvres to obtain concessions from employers.

Most workers, according to Mann, possess a *dual consciousness.* Instead of having a unified understanding of how their work dissatisfactions are organically linked to the operations of capitalism—real class consciousness in the Marxist sense—workers tend to compartmentalize their work experiences from the rest of their lives. They develop a "pragmatic acceptance" of the alienation and subordinate status they endure at work. Satisfactions in life are found in the basic pleasures of family, friends, and community, not in the quest for a new society.

Several other comparative studies have carried this discussion further, trying to identify which factors may transform varying degrees of working-class consciousness into collective action. This question gets at what perhaps is the weakest link in Marxist theories of class (Form 1983: 175). Duncan Gallie (1983) investigated the nature and determinants of worker attitudes to class inequality in France and Britain. At the time, these two countries had similar levels of socioeconomic development, but the French working class was radical, while its British counterpart was moderate. Indeed, Gallie (1983) points out that the May 1968 general strike in France was "arguably the most powerful strike movement unleashed in the history of capitalist society, shaking its very foundations." Why is it, asks Gallie, that British workers accept major social inequities as inevitable, while for the French they are a source of resentment and a catalyst for industrial militancy and political radicalism?

Gallie's evidence shows that workers in both France and Britain recognize class-based inequities in opportunity, wealth, and privilege. They differ, however, in their attitudes toward these inequities. For French workers, class position is an important part of personal identity. They identify with the larger working class, resent the system that puts them at the bottom, and believe that political action can improve their situation. British workers, in contrast, are more concerned about changing things in the workplace rather than in society as a whole.

Gallie's explanation of these differences hinges on the role of left-wing political parties in translating workplace experiences into a radical critique of society. French employers exercise greater unilateral power than do their British counterparts, and workplace industrial relations in France are not governed by mutually-agreed-upon rules. Thus, social inequality is more pronounced than in Britain. Furthermore, France has a revolutionary political tradition going back to the late 18th century. Hence, French workers' grievances with their employers are more intense and more readily carried outside the workplace, where they are moulded into a broad counter-ideology by exposure to trade unions and radical politics (especially the Communist Party).

Scott Lash (1990) provides another perspective on the determinants of worker militancy, with conclusions echoing those of Duncan Gallie. He compares the radical French with the conservative American working classes. Data collected at industrial sites in both countries show that differences in objective employment conditions cannot account for variations in militancy. Lash's alternative explanation focuses on how ideological and cultural factors shape militancy. Trade unions and political parties in a society or region are vehicles for socialization, transmitting to workers ideologies of natural rights in the workplace and in the political arena. Exposure to left-wing political parties determines *societal radicalism*. Socialization by trade unions, "industry's alternative rule-makers," is the most important cause of *industrial radicalism* (Lash 1984). Both kinds of radicalism aim to achieve the kinds of broader legal rights that would give power to individual workers. Militant workers do not, therefore, necessarily try to overthrow existing institutions; rather, they use them to their collective advantage.

Class and Politics in Canada

What about the Canadian situation? Gallie and Lash conclude that unions and political parties are the main vehicles for channelling general working-class discontent into collective action. Have Canadian unions, which are basically reformist in orientation, had the same role (Smucker 1980: Chapter 11)? The Canadian Labour Congress helped to found the New Democratic Party in 1961, and remains closely affiliated with it. However, if all union members voted for the NDP, it would have formed the federal government by now. And, despite the tough economic times that many workers have endured since the early 1980s, there are few signs that widespread discontent is on the rise. When New Democratic Party governments were elected in Ontario, British Columbia, and Saskatchewan in the 1990s, the victories actually involved little change from previous elections in the NDP's share of the popular vote. And the 1990–95 Ontario NDP government's attempt to impose a "social contract" on unionized public-sector workers to reduce the government deficit backfired, creating what then premier Bob Rae later called a "social contract tunnel of doom" that contributed to the NDP's electoral defeat (Rose 2001: 73–74). Nor have two major recessions made the unorganized members of the working class more inclined to seek union representation. If anything, they have tended to blame their economic woes partly on unions (Baer, Grabb, and Johnston 1991).

Despite theories suggesting that involvement in strikes may have a radicalizing effect on workers' political consciousness, our earlier discussion showed that this is seldom the case. Because strikes involve the organized and better-off members of the working class, resulting improvements in wages and working conditions may further divide the working class. Michael Smith (1978) claims that if strikes bring gains for the unionized workers involved, the more disadvantaged, unorganized workers may see this as proof of the unfairness of the system. Cynicism, political alienation, and anti-union feelings may result. So, even though strikes may briefly kindle the flames of class consciousness among strikers, their overall political impact within the Canadian working class appears small.

One of the few Canadian studies to probe these theoretical issues examined the ideologies of male manual workers in an aircraft repair plant where skill levels were high, and in a home insulation factory employing low-skilled labourers. No significant differences in political attitudes or class ideologies were found between the two groups (Tanner 1984; Keddie 1980). Furthermore, among those who were committed to left- or right-wing political beliefs, these beliefs were not related to other attitudes about work and society (Tanner and Cockerill 1986). The researchers concluded that a coherent and integrated set of political views was absent among the male blue-collar workers they studied. Worker consciousness regarding class and politics tended to be fragmented and often contradictory. For example, some of the same individuals who believed that corporate profits should be more equally distributed (a left-wing position) also agreed that trade unions are too powerful (a right-wing stance). But according to the researchers, this does not demonstrate an acceptance of the capitalist status quo or, alternatively, a rejection of socialism; rather, it suggests that workers' limited encounters with alternative ideologies lead them to question the system only on certain issues.

Other attitude surveys reinforce this assessment of class consciousness among Canadian workers. Even the economic crisis of the late 1970s and early 1980s, marked by high inflation and rising unemployment, did not further polarize class ideologies (Baer, Grabb, and Johnston 1991). However, a more recent Alberta study did reveal that individuals personally affected by the recession were somewhat more likely to be critical of the distribution of wealth in society (Krahn and Harrison 1992). Whether or not there is a class basis to it, there are recent expressions in Canada and the United States of collective consciousness. This is best seen in the Living Wage movement, which in a number of cities has brought together coalitions of citizens and community groups, including unions, to have local bylaws passed that ensure all workers receive a wage high enough to live on (Luce 2004; Living Wage 2005).

The Future of Worker Militancy

Despite the absence of a strong working-class radical ideology and, moreover, a decline in strike activity in Canada, there have been several examples of worker militancy that deserve careful scrutiny. Can we expect to see an upswing in industrial conflict in the future, coupled perhaps with a leftward shift in working-class politics? Some observers believe that future working-class mobilization could be provoked by employer and government demands for greater productivity, economic concessions, and more management rights.

A graphic illustration is the British Columbia Solidarity Movement, which sprang up in opposition to the Social Credit government's 1983 public-sector restraint program. Proposed legislation would have limited the scope of public-sector bargaining, allowed the firing of employees at will, cut back on social services, ended rent controls, and abolished the B.C. Human Rights Commission. The launching of Operation Solidarity by the B.C. Federation of Labour was a delayed reaction, which followed the more militant responses by the grassroots Lower Mainland Budget Coalition.

Bryan Palmer's (1986) analysis of the B.C. Solidarity Movement raises some provocative questions. Class relations in Canada have begun to realign themselves, argues Palmer, and the 1983 events in British Columbia signal that class conflict may be on the rise. Palmer goes as far as calling the protest marches, public rallies, picket lines, and community-based resistance during the summer of 1983 "class warfare." These actions reflected mass opposition to the proposed legislation, rather than being merely the response of union leaders. In fact, the moderate leadership of Operation Solidarity rejected calls for a general strike that would have deepened the conflict. Negotiations between the government and mainstream union leaders eventually resulted in minor revisions to the legislation.

Not everyone would agree with Palmer's Marxist interpretation of the events in British Columbia. In hindsight, it may have been premature to draw conclusions about realignments in class relations on the basis of this case alone. It is interesting to speculate about other possible outcomes of this confrontation if organized labour had not become involved. What seems clear, though, is that government offensives against employee rights and social programs can provoke the labour movement into more broadly based political action. Even though these reactions are often defensive, with unions scrambling to preserve past gains or members' jobs, public opinion can sometimes be rallied against unpopular government decisions, as in the case of the Calgary hospital laundry workers. Another example of widespread protests occurred in the mid-1990s when Ontario unions forged a broad coalition with community-based organizations to oppose policies proposed by the Harris Conservative government, which they saw as threatening social programs and undermining workers' rights. The political goal of this short-lived initiative was practical and immediate (and, we might add, unsuccessful): to pressure the Harris government into rethinking its agenda. The actions had no clear impact on class relations.

In general terms, however, the environment of labour relations became more restrictive after the recession of the early 1980s. Canadian employers increasingly resorted to lockouts. Several provincial governments introduced public-sector restraint legislation during this period, and the use of court injunctions and legislation to end strikes has escalated. Major private-sector strikes have been accompanied by the increased use of strikebreakers. On the other hand, there have been positive breakthroughs for labour, such as Quebec's anti-strikebreaking legislation. But these are exceptions, and some analysts believe that the new developments may undercut the established labour–management accord by placing severe restrictions on free collective bargaining, especially by public employees.[38] These backward steps violate international labour conventions, resulting in the International Labour Organization (ILO) investigating several provincial laws in the 1980s.

CONCLUSION

To return to the themes of conflict and cooperation introduced at the start of this chapter, do the new economic realities of the opening decade of the 21st century require that unions and management create new relationships? To answer this question, researchers have expanded the study of industrial relations far beyond just unions. In response to the relentless pressures of economic globalization and restructuring, industrial relations research has expanded to encompass employment relations, management, and work organization issues (Giles 2000). This interdisciplinary perspective has raised key questions about the future of unions that hinge on cooperative relationships and new forms of participatory work organization.

Today there are still many observers who believe that labour–management relations in Canada are too adversarial. Unions interpret this as employers wanting cooperation to be on their terms. Management sees it as a way of improving competitiveness and efficiency and lowering costs. But there is a growing recognition that the legislative framework regulating labour–management relations is in need of major renovation. As one study of Canada's future industrial relations challenges concluded, "In the ideal world, the legislative framework governing employer-employee relations would have a higher level of consistency across jurisdictions, be balanced and respectful of both labour and management perspectives, and be exempt from modifications when the political regime changes"(Lendvay-Zwickl 2004: ii). Unions do not dismiss the importance of productivity, competitiveness, or organizational flexibility. They simply want cooperation and consultation on these issues to be fully open and democratic. As documented elsewhere in the book (Chapters 5 and 6), there are models of a more equal partnership between labour and management. But this approach requires a fundamental shift away from adversarialism and toward full recognition of the rights of workers. This will require greater trust between workers and management. The end result could be significant benefits for labour, management, and society as a whole.

DISCUSSION QUESTIONS

1. In your opinion, what are the greatest challenges Canadian unions face to their future survival, and what strategies will most likely help them to meet these challenges in the coming years?

2. Public opinion in Canada is divided over the role of unions in the economy and society. Do you think that unions have outlived their usefulness? Explain the reasons for your position.

3. Some sociologists argue that conflict is inevitable in the workplace. Others believe that labour–management cooperation is not only possible, but essential. Critically discuss and evaluate these two positions.

4. Discuss how unions contribute to reducing overall wage inequality in society. In what ways might they contribute to other sources of inequality?

5. Is there ever a clear "winner" after a strike or lockout?

6. In the early 21st century, why would workers be motivated to join a union? How have the reasons for joining unions changed historically?

7. Based on the material presented in Chapters 5, 6, and 7, what do you consider to be the most effective ways to give workers a "voice" in their workplace today?

8. If you were a union organizer, on which group or groups of workers would you focus you efforts? Draw on material from Chapter 7 and earlier chapters to develop a membership recruitment strategy.

9. "Historically, Canadian governments at the federal and provincial levels have played the role of 'impartial umpire' in industrial relations." Critically discuss this statement.

NOTES

1. The examples are based on the following reports:
   - B.C. and Newfoundland disputes: Centre for Industrial Relations, University of Toronto, *Weekly Work Report* for the week of 4 May 2004 (http://www.chass.utoronto.ca/cir/library/wwreport/ wwr2004_05_04.html)
   - Wal-Mart dispute: Centre for Industrial Relations, University of Toronto, *Weekly Work Report* for the weeks of 4 April 2005 and 11 April 2005; the UFCW Canada Wal-Mart campaign website

(http://www.walmartworkerscanada.com); Wal-Mart Canada's website (http://www.walmartcanada.ca/CA-CustServ-PeopleFAQs.html); and Serwer (2005)

- NHL lockout: Jim Kelley, "Winners and losers in the NHL agreement," 14 July 2005 (http://sports.espn.go.com/nhl/news/story?num=0&id=2105489); "Sides will have to ratify new CBA," ESPN.com, 13 July 2005 (http://sports.espn.go.com/nhl/news/story?id=2106776); and "NHL deal reached," *The Globe and Mail* (13 July 2005: A1)

2. Poole (1981: Chapter 1). Godard (1994: Chapter 7) provides a useful discussion of the role and function of unions as institutions.

3. Bain and Clegg (1974). The systems model of industrial relations was pioneered by John Dunlop (1971). For critical assessments, see Bamber and Lansbury (1998), Poole (1981: Chapter 2), Crouch (1982: Chapter 1), and Hyman (1975: Chapter 1). A systems approach is contrasted with a political economy approach to Canadian industrial relations by Taras, Ponak, and Gunderson (2005).

4. See Poole (1981: Chapter 8) for a discussion of the importance of workers' values and perceptions (which are the basis of their actions) in the study of industrial relations.

5. The phrase "managers of discontent" was coined by C. W. Mills (1948). An especially interesting discussion of the mobilization and management of discontent is provided by Batstone, Boraston, and Frenkel (1978: Chapter 16).

6. Michels (1959); also see Lipset, Trow, and Coleman (1956) and Freeman (1982).

7. On the social unionism of the CAW, see Gindin (1995). Also see the issues and campaigns on the CAW website (http://www.caw.ca) and the Canadian Union of Public Employees' (CUPE) magazine, *Organize* (http://www.cupe.ca) for that union's perspective on broad social and political issues.

8. The data in this paragraph are from Riddell (1986: 5), Coates, Arrowsmith, and Courchene (1989: 119), and The Angus Reid Group (1996). On value shifts, see Nevitte (2000).

9. Card, Lemieux, and Riddell (2004) Also see Jackson et al. (2000: Chapter 4); for an international comparison, see Gustafsson and Johansson (1999).

10. Freeman and Medoff (1984: 19); also see Gunderson and Hyatt (2005) on how unions affect compensation, productivity, and management practices in Canada.

11. Basic sources for historical accounts of the rise of the Canadian labour movement include Heron (1989), Morton (1989), Smucker (1980: Chapters 7 and 8), and Godard (1994: Chapter 4). Historical and contemporary research on Canadian labour is published in *Labour/Le Travail, Relations Industrielles/Industrial Relations, Studies in Political Economy,* and *Just Labour: A Canadian Journal of Work and Society* (http://www .justlabour.yorku.ca).

12. On labour radicalism during the early 20th century, see Robin (1968), McCormack (1978), Bercuson (1974), and Frager (1992).

13. The institutionalization of conflict thesis was put forward by sociologists to explain the relative peacefulness of union–management relations in the immediate postwar period. Some U.S. academics went so far as to suggest that the decline in industrial disputes signalled a lessening of class distinctions and class conflict. See Crouch (1982: 106–9) and Hill (1981: 124–27).

14. Panitch and Swartz (1993). Also see Swimmer (2001); Russell (1990); and Kettler, Struthers, and Huxley (1990). Similar concerns about collective bargaining in the United States and the United Kingdom are discussed in Freeman (1995).

15. Note that the data in Figure 7.1 for 1981 and later are based on revised labour force estimates, to reflect changes to the Labour Force Survey in 1995 (see Chapter 2). This results in higher estimates of non-agricultural paid employment, and, therefore, lower levels of union density than obtained using earlier labour force data series. Also note that the data in Figures 7.3 to 7.5 use a slightly different method of reporting union membership, calculating it as a percentage of all employees, which is why the membership level for 2000 varies from that in Table 7.1.

16. See Murray (2001), OECD (1994: Chapter 5), and Chang and Sorrentino (1991) on measurement issues regarding union membership.

17. On nonunion forms of representation, see Taras (2002) and Lowe and Schellenberg (2001: 25–29).

18. The 2004 data in this paragraph are from Henry (2004).

19. For an assessment of right-to-work legislation from a Canadian perspective, see Ponak and Taras (1995); for a critical U.S. analysis, see American Federation of Labor (1995).

20. See Goldfield (1987). On the breakdown of the old industrial relations system see Kochan, Katz, and McKersie (1986) and Bamber and Lansbury (1998: 18).

21. Those detecting signs of union renewal include Voss and Sherman (2000), Clawson and Clawson (1999), and Cornfield (2001). The pessimistic view is presented by Rose and Chaison (2001).

22. Additional analysis of union density in this section is based on Akyeampong (2004) and Statistics Canada's annual update on unionization published in *Perspectives on Labour and Income* (2004).

23. These unions have informative websites: Canadian Union of Public Employees (http://www.cupe.ca), Canadian Auto Workers (http://www .caw.ca), and the United Steel Workers of America (http://www.uswa.ca).

24. See the websites for these unions: NUPGE (http://www.nupge.ca), PSAC (http://www.psac.ca), and the UFCW Canada (http://www.ufcw.ca).

25. The website of Human Resources and Skills Development Canada, Labour Program, Workplace Information Directorate (http://206.191 .16.130/millieudetravail_workplace/index.cfm/doc/english) has data on membership according to type of union. See Roberts (1990) and Chaison and Rose (1990) on the Canadianization trend.

26. See Gindin (1995) and Wells (1997) for an analysis of the CAW's shift away from principled militancy to more pragmatic cooperation with management. Yates (1998) examines the challenges posed by the growing diversity of the CAW's membership.

27. A history of the Energy and Chemical Workers Union and its drive for a Canadian approach is found in Roberts (1990). The Communications Workers, the Energy and Chemical Workers, and the Canadian Paperworkers Union later merged to form the Communications, Energy, and Paperworkers Union of Canada.

28. Yates (2000: 662–64). On women in unions in Canada, see Canadian Labour Congress (1997), Briskin and McDermott (1993), Creese (1999), Gannagé (1995), and Julie White (1990).

29. Based on a major U.S. study of employee voting in union representation elections; see Farber and Saks (1980) and Antos, Chandler, and Mellow

(1980). Once members, women are no less committed to their union than are men; see Wetzel, Gallagher, and Soloshy (1991).

30. Decertification refers to unionized workers voting to terminate their relationship with the union representing them. On the bank organizing drive, see Lowe (1981) and Ponak and Moore (1981). For an account of the unsuccessful 1948–52 Eaton's organizing drive, see Sufrin (1982).

31. Pentland (1979: 19); also see Jamieson (1971: 51–52). See Bendix (1974) on U.S. employers' anti-union tactics.

32. Downie and Coates (1995: 177). For differing perspectives on this issue, see Rinehart, Huxley, and Robertson (1997); Clarke and Haiven (1999); Lowe (2000); Wells (1986); Appelbaum and Batt (1994); Nissen (1997); and Payette (2000).

33. Sources for this paragraph are *The Globe and Mail* (23 October 1996: A1, A6) on the GM strike and *The Globe and Mail* (23 November 1995: A13; 25 November 1995: A3) on Alberta.

34. Tilly's (1979: 64) example is Lipset, Trow, and Coleman's (1956) study of union democracy, which found that printers' union locals "have both distinct, compelling identities and extensive, absorbing interpersonal networks," and incorporate much of the members' lives. Clement's studies of miners (1981) and fishery workers (1986) provide Canadian examples. See Conley (1988) for a useful theoretical treatment of working-class action. Crouch's (1982) rational-choice model of trade unionism focuses even more directly on the attitudes and beliefs of individual employees, as well as on the larger contextual constraints on their actions.

35. Gouldner (1955) presents the classic study of a wildcat strike. Jamieson (1971: Chapter 7) discusses the wildcat strikes that characterized the turbulent industrial relations climate in Canada during the 1960s. Also see Fisher (1982) and Zetka (1992) on wildcat strikes.

36. See Gunderson, Hyatt, and Ponak (1995) for a discussion of these trends. For historical perspectives on militancy and industrial conflict in mining, see Craven (1980: Chapter 8), Frank (1986), and Seager (1985).

37. Kerr and Siegel (1954: 191). For critiques, see Shorter and Tilly (1974: 287–305) and Stern (1976). Fisher (1982) finds empirical support in Canada for the Kerr and Siegel thesis.

38. Compare Panitch and Swartz (1993), Russell (1990), and Godard (1997).

8

THE MEANING AND EXPERIENCE OF WORK

INTRODUCTION

In previous chapters, we presented a structural analysis of work in Canada, discussing, among other topics, labour markets, the occupational structure, work organizations, labour unions, and gender, ethnic, and other forms of stratification. Some of this material addressed individual reactions to work—workers' problems in balancing work and family responsibilities, and resistance and conflict in the workplace, for example. But in this final chapter we get to the very core of the individual–job relationship by examining the meaning of work in our society and workers' subjective response to their work.

We begin with a broad overview of work values. What does "work" really mean in our society today? Have work values changed over time, and are there cultural differences in work values? We then focus our analysis on the work orientations (or preferences) held by individual Canadians. How are these preferences shaped? Are alternative work orientations emerging in response to changing employment relationships and demographic shifts in Canadian society? The last section of the chapter discusses job satisfaction, alienation, and job-related stress. Overall, how satisfied are Canadian workers with their employment situation? Is this changing? What factors influence job satisfaction and dissatisfaction, and what are the underlying causes of job-related stress?

DEFINING WORK VALUES, WORK ORIENTATIONS, AND JOB SATISFACTION

Values, as sociologists use the term, are the benchmarks or standards by which members of a society assess their own and others' behaviour. We could talk about how personal attributes such as honesty and industriousness are valued in our society, about how we value freedom of speech, or about the value placed on getting a good education. We might also ask how work is valued, or, in other words, what the meaning of work is in a particular society.

Having identified work values as societal standards, we can define *work orientations* more narrowly as the meaning attached to work by particular individuals within a society. Blackburn and Mann (1979: 141) define an orientation as "a central organizing principle which underlies people's attempts to make sense of their lives." Thus, studying work orientations involves determining what people consider important in their own lives. Does someone continue to work primarily for material reasons (a desire to become wealthy or because of economic necessity), because income from the job allows him or her to enjoy life away from work, or to get enjoyment and personal fulfillment from work? Even more specifically, what types of work and work arrangements do individuals prefer?

The distinction between work values and work orientations is not always clear, since it hinges on the extent to which the latter are broadly shared within a society. As we will argue below, several different sets of work values can coexist within a society, influencing the work orientations of individuals within that society. In fact, as our earlier discussion of managerial ideologies and practices showed, particular work values have often been promoted by employers to gain compliance from workers.

An individual's work orientations are also shaped by specific experiences on the job. Indeed, a worker whose job or career is dissatisfying may begin to challenge dominant work values. Thus, over the long term, shifting work orientations on the part of many workers might also influence societal work values. *Job satisfaction* (or dissatisfaction), then, is the most individualized and subjective response of an individual to the extrinsic and intrinsic rewards offered by her or his job.[1]

WORK VALUES ACROSS TIME AND SPACE

Historical Changes in the Meaning of Work

The meaning attached to work has changed dramatically over the centuries.[2] The ancient Greeks and Romans viewed most forms of work negatively, considering it brutalizing and uncivilized. In fact, the Greek word for "work," *ponos,* comes from the root word for "sorrow." According to Greek mythology, the gods had cursed the human race by giving them the need to work. Given these dominant work values, the ruling classes turned their attention to politics, warfare, the arts, and philosophy, leaving the physical work to slaves.

Early Hebrew religious values placed a different but no more positive emphasis on work. Hard work was seen as divine punishment for the "original sin" of the first humans who ate from the "tree of the knowledge of good and evil." According to the Old Testament, God banished Eve and Adam from the Garden of Eden, where all of creation was at their easy disposal, to a life of hard labour, telling them that "In the sweat of thy face shalt thou eat bread, till thou return unto the ground."[3] This perspective on work remained part of the early Christian worldview for many centuries.

A more positive view of work was promoted by Saint Thomas Aquinas in the 13th century. In ranking occupations according to their value to society, Aquinas rejected the notion that all work is a curse or a necessary evil. Instead, he argued that some forms of work were better than others. Priests were assigned the highest ranks, followed by those working in agriculture, and then craftworkers. Because they produced food or products useful to society, these groups were ranked higher than merchants and shopkeepers. A comparison of this scheme to contemporary occupational status scales (see Chapter 3) reveals some interesting reversals. The status of those involved in commercial activity—bankers, corporate owners, and managers, for example—has increased, while farmers and craftworkers have experienced substantial declines in occupational status.

During the 16th-century Protestant Reformation in Europe, Martin Luther's ideas marked a significant change in dominant work values. He argued that work was a central component of human life. Although he still had a negative opinion of work for profit, Luther went beyond the belief that hard work was atonement for original sin. When he wrote, "There is just one best way to serve God—to do most perfectly the work of one's profession" (Burstein et al. 1975: 10), he was articulating the idea of a "calling," that is, industriousness and hard work within one's station in life, however lowly that might be, as the fulfillment of God's will. Whether or not peasants or the urban working class actually shared these values is difficult to determine. However, to the extent that the ruling classes could convince their subordinates that hard work was a moral obligation, power and privilege could be more easily maintained.

The Protestant Work Ethic

The Industrial Revolution transformed the social and economic landscape of Europe, and also generated a new set of work values. In his famous book, *The Protestant Ethic and the Spirit of Capitalism,* Max Weber (1958) emphasized how Calvinists, a Protestant group that had broken away from mainstream churches, embraced hard work, rejected worldly pleasures, and extolled the virtues of frugality. Weber argued that such religious beliefs encouraged people to make and reinvest profits and, in turn, gave rise to work values conducive to the growth of capitalism. Other scholars have questioned whether these early Protestant entrepreneurs really acted solely on religious beliefs, and whether other groups not sharing these beliefs might have been equally successful (Dickson and McLachlan 1989). Nonetheless, Weber did draw our attention to the role of work values in capitalist societies. He also recognized that such an *ideology of work,* while justifying the profit-seeking behaviour of capitalists, might also help control their employees who, for religious reasons, would be motivated to work hard (Anthony 1977: 43).

Freedom and equality are additional secular values that fit into the belief system of capitalist democracies. A central assumption is that workers are participants in a labour market where they can freely choose their job. If this is so, the fact that some people are wealthier and more powerful than others must be because of their hard work and smart choices, most specifically, investments in higher education. Hence, within this ideology of work, equality does not refer to the distribution of wealth and power, but to access to the same opportunities for upward mobility in a competitive labour market. But, as demonstrated in Chapter 3, some workers are advantaged by birth, and some occupational groups are more protected in the labour market. Consequently, for many workers, the daily realities of the labour market often contradict the dominant set of work values.

Work as Self-Fulfillment: The Humanist Tradition

The importance of hard work and wealth accumulation are only one perspective on work in a capitalist society. The belief that work is virtuous in itself gave rise to another set of values in the 17th and 18th centuries. The *humanist tradition* grew out of Renaissance philosophies that distinguished humans from other species on the basis of our ability to consciously direct our labour. A view of human beings as creators led to the belief that work should be a fulfilling and liberating activity, and that it constituted the very essence of humanity.

Karl Marx fashioned these ideas into a radical critique of capitalism and a formula for social revolution. He agreed that the essence of humanity was expressed through work, but argued that this potential was stifled by capitalist relations of production. Because they had little control over their labour and its products, workers were engaged in alienating work. For Marx, capitalist economic relations limited human independence and creativity. Only when capitalism was replaced by socialism, he argued, would work be truly liberating.

Marx's theory of alienation has had a major impact on the sociology of work, as we explain later. Interestingly, a similar set of beliefs about the centrality of work to an individual's sense of personal well-being underlies a number of contemporary management approaches, which are decidedly non-Marxist in their assumptions (see Chapter 5). It is often argued, for example, that workers would be more satisfied if they were allowed to use more of their skills and initiative in their job. But these management perspectives do not see capitalism as the problem. Rather, they advocate improved organizational and job design within capitalism.

Humanistic beliefs about the essential importance of work have been espoused by many contemporary workplace researchers.[4] In fact, as should be apparent by now, such beliefs clearly underlie our own perspective on the role of work in people's lives. Humanistic work values have also been promoted by various 20th-century theologians, including the Catholic priest who founded the system of Mondragon producer cooperatives in the 1950s (Chapter 6). More recently, in response to rising levels of unemployment in Canada in the early 1990s, a group of Catholic bishops prepared a discussion paper on "the crisis of work." In their call for greater government involvement in the economy and more corporate responsibility for job creation, they argued that:

> It is through the human activity of work, that people are able to develop their sense of self-identity and self-worth, acquire an adequate income for their personal and family needs, and to participate in the building-up of one's community and society. (Canadian Conference of Catholic Bishops 1991: 2)

Experiencing Unemployment: Identifying the Meaning of Work in Its Absence

Many studies tell us that unemployment is extremely traumatic for jobless individuals and their families.[5] But the Catholic bishops suggest that the unemployed are disadvantaged in ways that go beyond lost income, a lower standard of living, personal and family stress, and health problems. They take the humanist position that work gives meaning in our lives. Thus, by examining the experience of unemployment, the gaps left by the loss of a paid job, we can gain further insights about the personal and social functions of work.

Based on her research among the unemployed during the Depression of the 1930s, Marie Jahoda (1982) identified some of the *latent functions* of work which people miss if they lose their jobs. While the *manifest function* of work is, primarily, maintaining or improving one's standard of living, the latent (less obvious) functions contribute to an individual's personal well-being. Work can provide experiences of creativity and mastery, and can foster a sense of purpose. It can be self-fulfilling, although, clearly, some jobs offer much less fulfillment than others. When hit by unemployment, an individual loses these personal rewards, as illustrated by the following comments by a nutritionist about her 14 months of unemployment:

> I feel like I'm wasting time. I feel like I'm not accomplishing anything. I don't really feel like I'm contributing to the marriage, to society, or anything like that.[6]

Work also provides regularly shared experiences and often enjoyable interactions with coworkers (Hodson 2004). When the job is gone, so are such personally satisfying routines. An unemployed person quickly comes to miss these social rewards, and may also find that relationships away from work are no longer the same. Such feelings are expressed in the following way by an unemployed teacher:

> The whole feeling that I had . . . [was] that nobody really understood where I was, what I was going through, what I cared about, what was important to me. Because I was no longer talking about my job as a teacher, in fact I wasn't talking about my job as anything. I had a strong sense of not fitting in.

In addition, work structures time. Individuals who have lost their job frequently find that their days seem not only empty, but disorienting. A divorced mother of a four-year-old, living with her own mother to make ends meet, explains:

> If you went to work, at least you're coming home in the evening. When I'm home all day . . . I get confused. My whole metabolism's gone crazy, because it's, like, I'm coming back from . . . where I have been. Or I've gone and I'm still waiting to come home.

Being unemployed often requires that one seek financial assistance, sometimes from family members, but more often through government agencies. Dealing with the bureaucracy can be frustrating. Even more problematic, receiving social assistance carries a great deal of *stigma*. As a former factory worker, the mother of two small children, explains:

> When I went down there, I felt that I just stuck right out. I thought, "Oh, my God, people think I'm on welfare . . . You used to think "It's those people who are on welfare," and now you discover you're one of those people.

The same value system that rewards individuals for their personal success also leads to what Jean Swanson (2001) calls "poor-bashing," the perpetuation in the media, in our legislation and social programs, and in our everyday discourse, of myths and stereotypes about the poor and the unemployed being lazy, unmotivated, and undeserving of assistance. This, in turn, can lead unemployed people to blame themselves for their own problems, even if they recognize the structural barriers they are facing. Swanson (2001: 10) quotes an unemployed former bank employee who exemplifies such contradictory feelings:

> I never chose to be poor. I'm ashamed of what I am now, but it's beyond my ability to change things. I'm still alive. I haven't committed suicide. I'm living with hope. Although I know with all my heart that it's not my fault, the system makes you feel guilty. Society makes you feel guilty.

These glimpses into the lives of unemployed Canadians reflect the value placed on work in our society as a potential source of self-fulfillment and social integration. At the same time, a set of more materialistic work values are highlighted through the stigma and self-blame that can result from joblessness, seen by some unemployed people as a mark of their own labour market failure. Thus, while career success and wealth accumulation are central to the value systems in capitalist society, the humanist perspective on work as a source of self-identity and personal satisfaction reflects an equally important alternative set of work values.

Cultural Variations in Work Values?

Max Weber argued that the Protestant work ethic provided a set of work values that were conducive to the emergence of a new capitalist mode of production. Today, when we compare the economic performance of different countries in the global economy, we might be tempted to make a similar "cultural differences" argument. Specifically, do Japan, the Southeast Asian tiger economies, China, and India owe some of their success to a different set of work values?

Japan's economy outperformed the rest of the world in the 1980s. A common explanation was that Japanese workers had a much stronger work ethic and a higher level of commitment to their employers. Underlying such explanations was an assumption of powerful cultural differences. The Japanese, it was argued, had always exhibited strong patterns of conformity and social integration. In modern times, the corporation had come to assume the once-central roles of the family and the community. From the vantage point of North America, the presumably stronger work ethic of Japanese employees appeared to be a key ingredient in the Japanese "economic miracle."

As the economies of Singapore, South Korea, Taiwan, and Hong Kong expanded rapidly in the 1990s, similar cultural explanations about different underlying work values were proposed. Like descriptions of the presumably more-motivated Japanese workers, this "Confucian work ethic" argument attributed the economic success of these economies to traditional habits of hard work, greater willingness to work toward a common social goal, and employees' ready compliance with authority. While we have not yet seen similar cultural explanations of rapid economic growth in China and India over the past decade, we would not be at all surprised if they surfaced.

Undoubtedly, there are cross-national differences in the way people respond to work in general, to new technologies, and to employers' demands for compliance. North American workers, for example, may be somewhat less accepting of Japanese-style management approaches that expect workers to show unwavering loyalty to company goals (Graham 1995). Workers in some Asian countries may think of their work organization in less individualistic and more family-like terms (Jiang et al. 1995). That said, it is overstating such differences to argue that a Confucian work ethic explains the rapid growth of the East Asian economies. If it does, why did the economic growth in these countries not begin decades earlier? India is a largely Hindu country with a great diversity of ethnic and linguistic groups, so why is its economy so hot today? As for China, after more than half a century of communist dictatorship, it is unlikely that Confucianism is still the psychological fuel for its economy.[7]

Conceding that there are some cultural differences in work values (Super and Šverko 1995), we argue that they are not central to explanations of national differences in economic growth in today's global economy. Far more crucial are the production decisions of firms regarding technology, employment practices, and research and development, and the extent to which governments actively participate in economic development strategies. In fact, research suggests that the cause-and-effect relationship may frequently operate in the opposite direction—work values and, in turn, employee behaviour can be influenced by employment practices and labour market institutions (Schooler 1996).

Japan is a good example. Its labour market is highly segmented, even more so than in Canada. Only a minority of Japanese workers are employed in the country's huge, profitable, high-technology corporations. Most people work in smaller businesses that subcontract to make parts or provide services for the giant firms. The major advantages of this arrangement for the large corporations are the flexibility of being able to expand and contract their labour force without hiring permanent employees, and the reduction in inventory costs through just-in-time delivery of component parts from the subcontracting businesses.

Until the 1990s, major Japanese firms generated worker loyalty by offering lifetime employment guarantees and opportunities for upward advancement within the corporation, along with higher wages and good benefit packages (Hill 1988). Critics argued that company loyalty was actually built on fears of job loss, and that the economic benefits provided to workers in major firms simply made it easier for their employers to demand compliance and hard work (Kamata 1983). Whatever the source of employee commitment, job security and access to internal labour markets were severely eroded in the 1990s as major Japanese corporations struggled to adapt to a changing global economy. By contrast, in small firms and family businesses, low pay, little job security, and long working hours had long been the norm. Thus, for both core- and periphery-sector Japanese employees, the "willingness" to work long and hard may always have had much more to do with organizational and economic factors than with unique cultural values (Lincoln and Kalleberg 1990).

Time for a New Set of Work Values?

During the 1980s and 1990s, unemployment rates rose steeply in most Western economies. They declined again over the past decade, although not to the levels of the post–World War II era. In contrast, the growth in nonstandard jobs that began several decades ago has continued, placing more workers in precarious employment and financial situations. At the same time, a growing minority of workers have chosen or been required to work longer hours per week (Chapter 2). Given this polarization of employment experiences—not enough work for some and too much for others—perhaps we need to promote a new set of work values? Specifically, if more citizens would "get a life" beyond the workplace, perhaps our society could get back on track toward a better future.

This is not a new idea. More than two decades ago in his provocative book, *Farewell to the Working Class*, Andre Gorz (1982) argued that higher levels of unemployment also meant more free time for individuals to participate in non-paid work and leisure activities. Recognizing that people still need to work to make a living, he suggested that new technologies have reduced the time it takes to produce what is needed for a decent standard of living. Consequently, Gorz called for a new set of values that would replace hard work, labour market competition, and consumption of material goods with more emphasis on the personal fulfillment that comes from non-paid work activities and leisure pursuits.

Similar ideas have been promoted in a number of more recent books about employment patterns and the future of work in Western societies. For example, Jeremy Rifkin (1995) argued that contemporary industrial transformations would be more traumatic for workers than were previous industrial revolutions. While manufacturing jobs replaced agricultural jobs and, in turn, service jobs replaced manufacturing jobs, new technologies and global production patterns essentially mean the "end of work" as we know it for many people. Rifkin advocated the promotion of new work values that would encourage more people to participate in the voluntary sector of the economy where they could find personal fulfillment in caring for others, improving the environment, and making other contributions to society.

Jamie Swift's (1995) concerns were less about the "end of work" than about growing labour market inequality in Canada. He observed that, along with a growing number of working poor, many well-paid labour force participants were also overworked and stressed from the long hours they put into their jobs. Swift reflected on "the lunacy of lives driven by the compulsions of work, speed, and consumption," and called for more attention to "the good life" rather than "the goods life" (Swift 1995: 221, 224). Like Gorz and Rifkin, Swift also recommended a new set of work values that emphasize working less, consuming less, and seeking self-fulfillment not only in paid work but also in healthy leisure pursuits and in (non-paid) caring for others.

Clearly, labour market polarization and growing social inequality are social problems that need to be addressed. Emphasizing the value of non-paid work and reducing inequities in access to paid work are part of the solution. So, too, are stressing the importance of caring for others and the environment and emphasizing that there is more to a good life than merely earning a high income and acquiring and consuming material goods. But writers like Gorz, Swift, and Rifkin failed to tell us how such a transformation of social values can be achieved. Nor did they explain how high levels of inequality between those with paid jobs and those without could be avoided.[8] In a society where women have traditionally been expected to do most of the "caring work" (see Chapter 4), would non-paid voluntary work continue to be seen as "women's work"? There is also an intergenerational equity concern. Since older workers are often more economically secure, is it fair to encourage young Canadians to change their values and reduce their career aspirations, arguing that they are the ones who will face the "end of work" as we know it today?

While it is clearly important to rethink dominant work values, an overemphasis on the value of unpaid work as *the* solution to labour market polarization diverts our attention from the sources of growing inequality within the labour market. Today's more precarious economy is not simply the product of natural market forces. Real people in positions of power make decisions to change technologies, to move factories, and to push workers to accept less income, fewer hours of work, and less job security. Consequently, there is a need for individuals, unions, and the government either to resist these trends or to re-shape them (Noble 1995). In fact, the work values of employers need to be re-shaped as well, with commitments to profits at all costs being balanced with greater responsibility to employees and the larger community (Lowe 2000; Lowe and Schellenberg 2001).

In her description of "how the overwork culture is ruling our lives," Madeline Bunting (2004) continues the debate about the need for new work values.[9] Like other observers of contemporary labour market trends, Bunting recognizes how industrial restructuring and organization re-engineering have deprived some people of jobs and forced others to work longer and harder to make a living. But she also documents how managers, professionals, and other advantaged workers are willingly participating in the new culture of overwork. While leisure used to be a sign of status for the middle and upper classes, today overwork has become a status symbol as professionals take pride in the many hours they work, how they "multitask" while on the job, and how they continue to work (using laptop computers and sophisticated communication technologies) while on vacation.

Bunting argues that a consumer-focused culture and economy (work hard to spend more) and new managerial ideologies (work hard to show how committed you are to the company) make many of us into "willing slaves" in an era where "narcissism and capitalism are mutually reinforcing" (Bunting 2004: xxiv).

What is required, she concludes, is a return to a traditional work ethic that respects human dignity and autonomy, the promotion of a "care ethic" that places high value on looking after the needs of others and that adequately compensates the women and men in "caring" occupations, and the emergence of a new "wisdom ethic" that recognizes that building a better society is more important than simply heating up the economy by working harder and spending more (Bunting 2004: 312–24). Thus, while reinforcing many of the points made by the other writers discussed above, Bunting places more emphasis on the need to retool our work values, in contrast to replacing them with non-work values.

WORK ORIENTATIONS

Having discussed how societal work values have changed over time, might vary across cultures, and need to be re-examined in light of today's economic realities, we now shift our analysis to individuals' job expectations—the type of work orientations they bring to their jobs. Are most Canadians motivated primarily by a desire to be successful and become wealthy? Do they seek to work as hard as possible, whatever the costs, as Madeline Bunting (2004) suggests British workers do? Or is their work a source of personal satisfaction and fulfillment? Do women and men have similar orientations to work? Are specific groups of labour force participants beginning to exhibit different types of work orientations, and if so, why?

Instrumental Work Orientations

Four decades ago, David Lockwood discussed the differences he observed in male British working-class "images of society." Lockwood was interested in the issue of *working-class consciousness*—how British workers perceived social inequality. He identified three distinct worldviews originating in different community and workplace experiences. *Proletarian workers,* argued Lockwood, saw the world in much the way Marx had predicted: they perceived themselves to be in an "us against them" conflict with their employers. Lockwood (1966) concluded that this image of society was more pronounced in industries such as shipbuilding and mining where large differences between management and workers in terms of income, power, and opportunities for upward mobility had produced heightened class consciousness.

Deferential workers also recognized class differences, but accepted the status quo, believing that wealth and power inequities were justified. Provided they were treated decently, deferential workers were unlikely to engage in militant actions against employers. Lockwood located this worldview in the traditional service industries and in family firms where paternalistic employment relationships foster acceptance of the existing stratification system.

Both of these traditional orientations to work, Lockwood argued, were found primarily in declining industries. The third type, *instrumental work orientations,* was more typical of the attitudes of the contemporary (1960s) working class, whose members Lockwood labelled *privatized workers.* They viewed society as being separated into many levels on the basis of income and possessions. Rather than expressing opposition or attachment to their employers, the dominant feeling of these workers was one of indifference. According to Lockwood, the "cash nexus" (the pay received), and not feelings of antagonism or deference, constituted the employment relationship for such workers. Work was simply a way to obtain a better standard of living, an "instrument" used to achieve other non-work goals.

Lockwood and his colleagues went on to conduct a major study of the work orientations of a group of male autoworkers in Luton, a manufacturing town north of London, England. These "affluent workers" had chosen to move to this new industrial town because the jobs available would provide them with the money and security needed to enjoy a middle-class standard of living. The fact that work in the Vauxhall automobile factory was not intrinsically rewarding did not seem to matter. In their book describing this study, the researchers essentially concluded that these workers exemplified the modern worker in Western industrialized economies (Goldthorpe et al. 1969).

Critics argued that British workers in the 1960s appeared to be instrumental only in comparison with an idealized proletarian worker of the past, and that the conclusions drawn from this study were over-generalizations, at the very least applicable only to men. Based on their study of the male labour market in another British city, Robert M. Blackburn and Michael Mann found that most workers reported a variety of work orientations. In addition, there was often little congruence between an individual's expressed work preferences and the characteristics of his job. Blackburn and Mann (1979: 155) argued that the major flaw with an *orientations model* of labour market processes is that many people have few job choices, taking whatever work they can get, even if it is not what they would prefer.

Individuals' work preferences can also be modified in the workplace. If a job offers few intrinsic rewards or little opportunity to develop a career, workers may adjust their priorities accordingly. This might explain why studies have shown instrumental work orientations to be more common among workers in monotonous, assembly-line jobs (Rinehart 1978). Richard Sennett and Jonathan Cobb (1972) examined this theme in their book, *The Hidden Injuries of Class.* Their discussions with American blue-collar workers revealed a tendency to downplay intrinsic work rewards, few of which were available to them, while emphasizing pay and job security. Workers interviewed by Sennett and Cobb also re-defined the meaning of personal success, frequently talking about how hard they were working in order to provide their children with the chance to go to college.

Thus, we should not be surprised if some workers do have instrumental work orientations, given the strong emphasis on material success in our society, and current fears about job security. By way of example, in 1996, 35 percent of a national sample of employed Canadians agreed that "[m]y current job is a way to make money—it's not a career." In another study conducted several years earlier, almost 2,500 private sector and crown corporation employees across the country were asked to rank the relative personal importance of a number of job rewards. This 1991 survey showed that about one in four workers (23 percent) ranked pay higher than any other work reward.[10]

But most employed Canadians are not motivated primarily by instrumental work orientations. Instead, along with a desire for good pay, extensive benefits, and job security, workers also want intrinsically satisfying work. A national survey of 2,500 Canadian workers conducted in 2000 revealed, for example, that 85 percent or more felt that good pay, good benefits, job security, work that was interesting, work that provided a sense of accomplishment, and work that allowed the development of skills and abilities were "important" or "very important" to them personally. However, it is also noteworthy that the intrinsic work rewards (interesting work, accomplishment, and skill development) were rated as important by 95 percent or more of the sample members, compared to 85 to 90 percent for the extrinsic rewards (pay, benefits, security).[11]

Gender and Work Orientations

The early British studies of work orientations focused only on men. When the work orientations of women were mentioned, it was often assumed that their interests were directed primarily toward the home, and less toward paid employment. The argument that *gender role socialization* might encourage women to place less value on intrinsic and extrinsic job rewards and more on social relationships in the family, the community, and on the job seemed plausible in the 1960s and early 1970s when many of these studies were completed.

However, there is another more compelling structural explanation. Women and men work in very different labour market locations, and women typically receive fewer job rewards. Hence, gender differences in work orientations might also be adaptations to their employment situation. Using the same logic as in our explanation of why unskilled blue-collar workers might have more instrumental work orientations—their low-level jobs discourage other types of work preferences—we could also argue that female work orientations, to the extent that they differ from those of men, are a function of workplace differences (Rowe and Snizek 1995: 226).

In addition, the domestic and child-care responsibilities that many women continue to carry force some to readjust their work and career goals. As a part-time nurse explained:

> My career is important to me. I don't want to give that up. But my main concern is the children . . . So I give about 100 percent to the children and maintain a career at the same time. (Duffy and Pupo 1992: 116)

Similarly, in her study of young university-educated women, Ranson (1998) observed that, particularly for those who had chosen non-traditional careers in science and engineering, the pressures of balancing family and job demands required a re-assessment of either their parenthood or career aspirations, or both. As one young woman explained when asked about her parenthood plans:

> Actually I've thought, would I work or would I stay home? . . . I've spent a lot of time going to school. I've worked my way up. Like, do I want to give all that up? [Because] if you come back to the work

force in five years you're not going to work where you left off. Then the other part of me thinks, well, you have to make that choice, it's one or the other. (Ranson 1998: 529)

Once again, these examples point to structural determinants—the gendered household division of labour and the frequent absence of family-friendly employment policies—rather than to female personality characteristics as the source of gender differences in work orientations.

The Work Orientations of Youth

Socrates is reputed to have once said that "[c]hildren today are tyrants. They contradict their parents, gobble their food and terrorize their teachers." If he had lived today, he might also have added that "they don't want to work hard" since such concerns about deficient work orientations have frequently been part of the public stereotype of youth. In the 1960s and 1970s, for example, the counterculture activities of some North American youth led to fears that the work ethic of this new generation was declining. Most members of that cohort are now settled into their careers, thus demonstrating that these fears were largely unfounded. It seems that each generation, when observing the generation following, sees lifestyle experimentation, forgets its own similar experiences, and concludes that values are slipping.

What are the concerns about the work orientations of today's youth? At first glance, debates about the possibly inadequate *employability skills* of young Canadians seem to have little to do with the subject of work orientations. The Conference Board of Canada, an organization representing large public- and private-sector employers, published an *Employability Skills Profile* (ESP) in 1993. In this pamphlet, the Conference Board (1993) listed a set of "critical skills required for the Canadian workplace" as identified by employers. Included in the list are *academic skills* (communication, thinking, ability to learn), *personal management skills* (positive attitudes and behaviours, responsibility, adaptability), and *teamwork skills* (the ability to work with others). The ESP has been influential, focusing debate on the types of skills that schools should be developing in students in order to prepare them for employment in the rapidly changing labour market. At times, it also has fuelled public criticisms of the educational system.

Clearly, all of the skills identified in the ESP are important, not only for young workers, but for all labour force participants, even though not all of these skills would be equally necessary in all jobs. However, more careful scrutiny of the public education system suggests that, with respect to teaching academic skills, Canadian schools are performing quite well.[12] Furthermore, research indicates that when hiring young people for entry-level positions, many employers focus mainly on motivation, perceived work ethic, and the personality of the applicant—in other words, on work orientations rather than on skills. Young job seekers are told that "proper attitude earns a job."[13]

Are today's youth less willing to "work hard" and to make a commitment to an employer? Unfortunately, there is little reliable data documenting shifts in the "soft" employability skills of youth. But a majority of high-school students work part-time at some point while completing school, suggesting that there is a strong desire to work, even if only for the money. Continued high rates of participation in postsecondary education (see Chapter 2) indicate that youth are willing to work hard to obtain the credentials needed for better jobs. And studies of youth continue to show high levels of general work commitment. For example, in 1985 we conducted a survey of approximately 1,000 Edmonton Grade 12 students about their work and education plans and experiences. In 1996, we repeated the study. Only 9 percent of the first cohort agreed that "I'd rather collect welfare than work at a job I don't like," compared to 7 percent of the 1996 cohort. Similarly, only one in eight respondents in both surveys agreed that "I would not mind being unemployed for a while" (Lowe and Krahn 2000). Such facts leave us skeptical about the supposed "work attitude" problem of today's youth.

The employment difficulties faced by Canadian youth are real, as we have demonstrated in Chapters 2 and 3. Corporate and public-sector downsizing, along with an increase in nonstandard jobs, have made it much more difficult for young people to cash in their educational investments for good jobs. It may also have led to a prolonged entry into adult roles, such as living on one's own, marriage, and parenthood.[14] Recognizing these trends, in 1993, Douglas Coupland wrote a fictional account of the experiences of today's youth moving between jobs and relationships. *Generation X: Tales for an Accelerated Culture* was a bestseller. Its portrayal of the post-baby-boom generation as cynical and alienated, with few career goals and largely instrumental work orientations, became part of popular culture.

The Generation X view of changing work orientations differs from public concerns about employability skills. It focuses on somewhat older, more educated youth rather than on recent high-school graduates, and diagnoses "the problem" more specifically, pointing to a devaluing of careers and intrinsic work rewards, rather than simply a reluctance to work hard. It points to structural changes in the labour market—a lack of satisfactory employment opportunities—as the source of the problem, rather than to schools and their presumed failure to socialize appropriate work attitudes in youth.

Is there evidence of the emergence of Generation X work orientations? The answer appears to be "no." A majority of young Canadians continue to invest in higher education, despite rising tuition fees, hoping to get good jobs. Many continue to pursue additional work-related training after obtaining formal qualifications.[15] Surveys of youth fail to reveal a decline in commitment to work or a reduced interest in intrinsic work rewards. For example, in 1985 we surveyed graduates of the five largest faculties at the University of Alberta, asking them about their work goals and career ambitions. In 1996, we repeated the study, asking identical questions of graduates from the same faculties. In response to a question asking "How important would the following be to you when looking for a full-time job after leaving school?" 92 percent of the class of 1985 agreed that "work that is interesting" would be important to them, compared to 96 percent of the class of 1996. Similarly, over 90 percent of both cohorts agreed that "work that gives a feeling of accomplishment" would be important to them. For both questions, the proportion who answered "very important" increased over time (Rollings-Magnusson, Krahn, and Lowe 1999). Other findings from these studies lead to the same conclusion: young Canadians continue to aspire to good jobs and satisfying careers.

The 1996 survey also showed 74 percent of the university graduates agreeing that "it will be harder for people in my generation to live as comfortably as previous generations," a sentiment echoed by 65 percent of Alberta Grade 12 students (Rollings-Magnusson, Krahn, and Lowe 1999). Thus, while the basic work orientations of youth may not have changed, and while their work ethic has not declined, there is considerably more anxiety about future employment prospects. A generation ago, an expanding economy allowed a larger proportion of youth to launch satisfying careers. Today, labour market restructuring has made career entry more difficult for Canadian youth and heightened their employment anxiety.

At the same time, reduced loyalty and commitment by many employers to their workers, as reflected in greater reliance on part-time and temporary workers, may also translate into less commitment and loyalty to the work organization by employees, both young and old (Lowe and Schellenberg 2001; Gallie et al. 1998: 306). For example, a national survey conducted by Canadian Policy Research Networks in 2000 showed 50 percent of the 2,100 paid employees in the sample (the self-employed were excluded in this analysis) agreeing that "I feel very little loyalty to this organization." Another 23 percent agreed strongly with this statement.[16] However, rather than indicating a decline in the work ethic, this points to a possible change in the relationship between workers and their employers. It also indicates a problem that employers, not schools, need to address. How do you motivate workers without offering them some degree of employment security and satisfaction in return? As nonstandard work becomes more common, employers will be forced to face this issue more directly.[17]

Welfare Dependency and an Emerging Underclass?

The poor also have been the frequent target of public concern about a "declining work ethic," since it is often easier to blame them for their plight than to try to understand the structural conditions that create poverty (Wright 1993; Swanson 2001). During the 1970s, growing awareness of extensive poverty, particularly among Black Americans, prompted fears that fewer people were willing to work. By way of example, in 1972, U.S. president Richard Nixon stated that American society was threatened by the "new welfare ethic that could cause the American character to weaken."[18]

However, in both Canada and the United States, national surveys conducted in the 1970s failed to reveal a declining work ethic. Very few Canadian respondents agreed that they would rather collect unemployment insurance than hold a job. Most stated that, given the choice, they would prefer working to not having a job, and that work was a central aspect of their lives. Paradoxically, these same respondents doubted the work commitment of other Canadians. Four out of five agreed that "[t]here is an atmosphere of welfare for anybody who wants it in this country."[19] People seemed to be saying that they personally wanted to work, but that social institutions made it easy for others to avoid working.

Has anything changed over the past several decades? A national survey of Canadian workers conducted in 2000 led to the conclusion that the work ethic remains strong (Lowe and Schellenberg 2001: 40).[20] But negative public opinion about social assistance programs and the work orientations of recipients also remains relatively unchanged. National surveys conducted in 1994 showed 68 percent of Canadians agreeing that "the existence of our social programs makes it too easy for people to give up looking for work." A smaller majority (55%) agreed that "if people just took more responsibility for themselves and their families we wouldn't need all of these social programs." Sixty-three percent supported lower unemployment insurance benefits for frequent users of the UI system, presumably because they felt they were less deserving (Peters 1995: 115–16).

What has changed, however, are the social assistance programs themselves. During the 1990s, the Unemployment Insurance program (now called Employment Insurance) was altered, making it more difficult for seasonal workers to obtain benefits. In several provinces, social assistance benefits have been cut back for "able-bodied" recipients, and *workfare* programs that require recipients to work in public projects have been introduced.[21] Thus, in the past decade, long-standing beliefs that many of the poor are reluctant to work, perhaps because the *social safety net* is too comfortable, have had a significant influence on public policy.

In fact, with the number of long-term unemployed increasing in many Western industrialized countries, and with income inequality rising, some observers have begun to label the most marginalized members of society as an *underclass,* even though the term has been used very imprecisely (see Chapter 3). Some discussions of this new underclass are sympathetic, arguing that this level of social inequality is unacceptable. Other commentators treat the trend toward greater inequality as largely inevitable, worrying instead about welfare dependency. The argument, simply put, is that prolonged receipt of social assistance and unemployment insurance benefits leads to changed attitudes: both the stigma of receiving assistance and the desire to seek paid employment decline. Even more troubling, such accounts suggest, these welfare dependency orientations can be passed on to the children of the poor (Buckingham 1999).

Is there any basis to such fears about a declining work ethic within the poorest members of society, or are we simply seeing ideological justifications for reducing social assistance benefits to the poor and unemployed? The best test would be to determine whether the presumed weak work ethic of the poor translates into little effort to seek paid work or other sources of income. Research that has taken this approach invariably leads to a different conclusion—a weak work ethic and a lack of effort are seldom the problem. Much more often, the problem is one of not enough good jobs and inadequate social and labour market policies (Schwartz 1999; Jaynes 2000).

In fact, the majority of Canada's poor are not welfare-dependent but "working poor," seeking to maintain their standard of living on jobs with low pay and short hours. There is actually considerable movement in and out of poverty as individuals lose jobs, move from social assistance to a low-paying job, become divorced, or enter retirement without an adequate pension.[22] In regions or cities where unemployment is high, a great deal of productive work continues to take place in the informal economy, including subsistence work performed by street people and the homeless.[23] Such widespread initiative is hardly evidence of a failing work ethic.

Furthermore, when offered jobs that pay only a few dollars more than the minimum wage, welfare recipients have demonstrated a great willingness to work.[24] As for the argument that welfare dependency attitudes are passed from one generation to the next, research shows little of this happening (Duncan, Hill, and Hoffman 1988).[25] Thus, while we are concerned about rising social inequality, we are not convinced by accounts of welfare dependency and a declining work ethic among the poor and unemployed.

JOB SATISFACTION AND DISSATISFACTION

We began this chapter with an overview of societal work values and then examined the work orientations or preferences of individuals. Now we will focus on job satisfaction and dissatisfaction, the subjective reactions of individual workers to the particular set of rewards, intrinsic or extrinsic, provided by their job (Blackburn and Mann 1979: 167).

What do Canadians find most and least satisfying about their jobs? Are most of them satisfied? Do people's work orientations influence what they find satisfying in a job? Can job satisfaction or dissatisfaction affect workers off the job? Some researchers assume that productivity is a function of job satisfaction. Their goal has been to discover how to organize work and manage employees in a way that leads to increased satisfaction, productivity, and profits. Others, adopting the humanistic view that work is an essential part of being human, view satisfying and self-fulfilling work for as many people as possible as a desirable societal goal in itself.

The Prevalence of Job Satisfaction

The standard measure of job satisfaction in North American survey research is some variation of "All in all, how satisfied are you with your job?" In response, a very large majority of workers typically report some degree of satisfaction. Just how large that majority is depends on the response categories offered to survey participants. If they cannot choose a neutral midpoint, a larger proportion indicate that they are satisfied. For example, according to a 2002 EKOS Research national survey, more than 80 percent of Canadian workers said they were satisfied with working conditions in their main job and 40% reported being very satisfied when the other two response options were "somewhat dissatisfied" and "very dissatisfied."[26] But when provided with a neutral midpoint on a five-point job satisfaction scale, a significant minority choose this response, leading to a somewhat lower estimate of job satisfaction. Using such a scale, an EKOS Research Associates–Graham Lowe Group national survey in 2004 found that 70 percent of Canadian workers were satisfied with their job (31% were very satisfied), 21 percent were neither satisfied nor dissatisfied, while only 9 percent expressed dissatisfaction (see Figure 8.1).[27]

These results, and many similar findings over the past few decades (Firebaugh and Harley 1995), imply that job dissatisfaction is not a serious problem. However, there are several reasons why we could question such a conclusion. First, as James Rinehart (1978) wisely observed, workers' behaviours— strikes, absenteeism, and quitting—all indicate considerably more dissatisfaction with working conditions than do attitude surveys. Rinehart (1978: 7) proposed that answers to general questions about job satisfaction are "pragmatic judgements of one's position vis-à-vis the narrow range of available jobs." Most workers look at their limited alternatives and conclude, from this frame of reference, that they are relatively satisfied with their work.

Second, responses to general job satisfaction questions may be similar to replies to the question, "How are you today?" Most of us would say "fine," whether or not this is the case. Hence, more probing questions may be needed to uncover specific feelings of job dissatisfaction. We should also recognize that in a society with highly individualistic work values, people may be unwilling

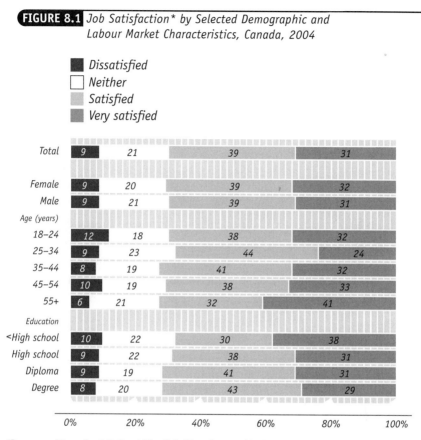

FIGURE 8.1 *Job Satisfaction* by Selected Demographic and Labour Market Characteristics, Canada, 2004*

■ *Dissatisfied*
□ *Neither*
▨ *Satisfied*
▨ *Very satisfied*

	Dissatisfied	Neither	Satisfied	Very satisfied
Total	9	21	39	31
Female	9	20	39	32
Male	9	21	39	31
Age (years)				
18–24	12	18	38	32
25–34	9	23	44	24
35–44	8	19	41	32
45–54	10	19	38	33
55+	6	21	32	41
Education				
<High school	10	22	30	38
High school	9	22	38	31
Diploma	9	19	41	31
Degree	8	20	43	29

0%   20%   40%   60%   80%   100%

*Responses of "very dissatisfied" and "dissatisfied" have been combined.

Source: *Rethinking Work,* EKOS Research Associates and Graham Lowe Group national survey (n = 2002). Reprinted by permission of the Graham Lowe Group.

to express dissatisfaction with their jobs because it could reflect negatively on their own ability and efforts.

This is why more indirect measures of job satisfaction are useful. We have already noted that, according to the 2000 CPRN survey, three out of four (73%) paid employees in Canada agree that they feel little loyalty to their work organization. A 1987 Environics survey asked Canadian workers whether they would choose the same occupation if they had another chance to choose. Almost half (46%) said that they would stay in their occupation, 42 percent said they would change occupations, and the rest were uncertain. A 1996 Angus Reid poll showed 32 percent of employed Canadians agreeing that they "would take a comparable job at another company if it were offered."[28] Thus, *behavioural intention* measures typically reveal somewhat more widespread job dissatisfaction. As Sandy Stewart and Bob Blackburn (1975: 503) put it, "satisfaction is expressed within a framework of what is possible, liking is expressed within a framework of what is desirable."

Age and Job Satisfaction

A consistent finding in job satisfaction research is that older workers generally report somewhat more satisfaction (Firebaugh and Harley 1995; Brisbois 2003: 58). Figure 8.1 documents this pattern, with the proportion answering "very satisfied" increasing from 32 percent for the youngest respondents in this 2004 study to 41 percent for the oldest survey participants. Typical of the research on differences in job satisfaction, explanations for this age-related pattern take two basic forms. Some are individualistic, focusing on motivations and work orientations, while others are more structural, pointing to characteristics of the job and the workplace (Ospina 1996: 181).

Perhaps older workers have reduced their expectations, becoming more accepting of relatively unrewarding work. This is an *aging* effect. Alternatively, a *cohort* explanation emphasizes lower expectations (and consequently higher satisfaction) on the part of a generation of older workers, assuming that older workers grew up in an era when simply having a secure job was all one desired and when little self-fulfillment at work was expected. Third, a *life-cycle* effect argues that older workers are more likely to have family and community interests that might compensate for dissatisfying work. A fourth explanation emphasizes the nature of the work performed (a *job* effect), suggesting that older people have been promoted into more rewarding and satisfying jobs, or have changed jobs until they found one they liked. A final explanation points to *self-selection,* arguing that less satisfied workers will drop out of the labour force as they get older, leaving behind the more satisfied workers.

Which is the best explanation? No clear answer emerges from the many studies on this topic, indicating that all of these processes may be involved. Regardless, we know that older workers are more likely to report satisfaction with their work, and that this is due mainly to the better jobs they have obtained over time and to their cumulative social experiences and adaptations in the workplace and society.[29]

The data we have displayed in Figure 8.1 also support several studies that have suggested that the relationship between age and job satisfaction is U-shaped (Clark, Oswald, and Warr 1996: 75; Birdi, Warr, and Oswald 1995). Specifically, satisfaction is moderately high among the youngest workers, lower in the next cohort, and then higher again in each successive cohort. The explanation is that the youngest workers are usually reasonably satisfied with their entry-level jobs, perhaps because many are still attending school. But if they remain in the student labour market after completing their formal education, the mismatch between low-level jobs and higher aspirations begins to have an effect. Hence, job satisfaction is lower among workers in their late twenties and early thirties. With time, however, the combination of better jobs and declining expectations again leads to higher levels of job satisfaction. This argument points to both individualistic and structural determinants of job satisfaction. It also raises an interesting question for future research. Because today's youth seem to be facing many more barriers to entry into good jobs, will we continue to observe the overall strong positive relationship between age and job satisfaction in the future?

Gender and Job Satisfaction

Given that women are more likely to work part-time in the secondary labour market, are they less satisfied with their jobs? The research evidence offers a clear answer. Despite large differences in work rewards, there is typically little difference between men and women in self-reported job satisfaction (de Vaus and McAllister 1991; Brisbois 2003: 58). Figure 8.1 tells the same story. Again, as with age differences in job satisfaction, an explanation of this non-difference brings us back to the subject of work orientations. Some researchers suggest that women have been socialized to expect fewer intrinsic and extrinsic work rewards. Hence, the argument goes, women are more likely to be satisfied with lower-quality jobs, focusing instead perhaps on satisfying social relationships within the workplace (Phelan 1994).

However, we should not explain the job satisfaction of women with a *gender model* (differences due to prior socialization) while employing a *job model* (differences due to the nature of the job) to account for the satisfaction of men (de Vaus and McAllister 1991). In fact, it makes more theoretical sense to explain the satisfaction of both women and men with reference to the type of jobs they hold and their roles outside the workplace. While, on average, men might report high satisfaction with their relatively good jobs, women might be equally satisfied with less rewarding jobs, having modified their expectations because of the time spent in these same jobs. And as Chapter 4 discussed, men's and women's involvement in family roles differ greatly. This also needs to be taken into account when explaining the non-differences in job satisfaction between women and men.

The critical test would compare women and men doing similar work, as in a study of post office employees in Edmonton where men and women performed identical tasks (Northcott and Lowe 1987). The researchers found very few gender differences in job satisfaction or in work orientations after accounting for both job content and family roles. Thus, while gender differences in work orientations may exist, characteristics of the jobs women typically hold and family responsibilities, rather than prior socialization, are probably responsible (Hodson 2004).

Educational Attainment and Job Satisfaction

Educational attainment and job satisfaction might be linked in several different ways. Following human capital theory (see Chapter 3), we might hypothesize that higher education should lead to a better job and, in turn, more job satisfaction. A second, more complex hypothesis begins with the assumption that better-educated workers have higher expectations regarding their careers but recognizes that not all well-educated workers will have good jobs. Hence, well-educated workers in less-rewarding jobs would be expected to report low job satisfaction. As one commentator wrote several decades ago, "the placing of intelligent and highly qualified workers in dull and unchallenging jobs is a prescription for pathology—for the worker, the employer, the society" (O'Toole 1977: 60).

Job satisfaction studies completed in the 1970s and 1980s (Wright and Hamilton 1979; Martin and Shehan 1989) typically found that education had little effect on job satisfaction, perhaps because the two hypothesized effects cancelled each other out. Even when comparing well-educated and less-educated workers in the same blue-collar jobs, few differences in job satisfaction were observed. Perhaps the more educated workers anticipated future upward mobility, and so were willing to tolerate less rewarding work for a time. The 2004 data featured in Figure 8.1 show that job satisfaction (the percent "satisfied" and "very satisfied") is marginally higher among better-educated Canadian workers but the overall pattern is not all that different from what we observed in earlier decades.

In previous chapters we have speculated that as education levels have been rising, the mismatch between workers' skills, credentials, and aspirations, on one hand, and their job content, on the other, may be rising. Although our single-point-in-time analysis (of 2004 survey data) cannot answer this question, we might hypothesize that levels of job dissatisfaction will slowly increase in the future as more well-educated young workers find themselves unable to find jobs that meet their aspirations. Researchers have yet to compare the relationships among age, education, and job satisfaction today with the situation several decades ago. However, one study from the 1990s did show that workers who felt underemployed in terms of their education and skills were much more likely to be dissatisfied with their jobs. In addition, several more recent studies have linked perceptions of job insecurity to reduced job satisfaction.[30] Given the extent to which youth, including well-educated young workers, have been negatively affected by increased labour market polarization and reduced employment security, this is a trend that needs to be monitored.

Work Rewards, Work Orientations, and Job Satisfaction

Our review of the effects of age, gender, and education on job satisfaction indicates that work orientations are frequently part of the explanation. Nevertheless, as we have also argued, differences in job characteristics and rewards are typically more directly linked to job satisfaction and dissatisfaction. In fact, hundreds of studies have attempted to identify the specific job conditions that workers are most likely to find satisfying. Such studies have focused on pay, benefits, promotion opportunities, and other *extrinsic work rewards;* autonomy, challenge, social relationships in the workplace and a range of *intrinsic work rewards; work organization features* (for example, bureaucracy, health and safety concerns, and the presence of a union); and *job task design* characteristics, including the use of new technologies.

A popular theory developed decades ago by Frederick Herzberg drew on all of these traditions, emphasizing both the extrinsic and intrinsic rewards of work. *Hygiene factors* like pay, supervisory style, and physical surroundings in the workplace could reduce job dissatisfaction, Herzberg argued. But only *motivators,* such as opportunities to develop one's skills and to make decisions about one's own work, could increase job satisfaction (Herzberg 1966, 1968). Herzberg also insisted that the presence of such motivators would lead, by way of increased job satisfaction, to greater productivity on the part of workers. We will return to that issue below.

Influenced by Herzberg's *two-factor theory,* most job satisfaction researchers now use multidimensional explanatory frameworks incorporating both intrinsic and extrinsic work rewards, as well as organizational and task characteristics. Although researchers categorize the specific features of work in somewhat different ways, there is considerable consensus at a broader level. Arne Kalleberg (1977), for example, identified six major dimensions of work. The first, an intrinsic reward dimension, emphasizes interesting, challenging, and self-directed work that allows personal growth and development. Career opportunities form a second dimension, and financial rewards (pay, job security, and fringe benefits) a third. Relationships with coworkers, convenience (the comfort and ease of work), and resource adequacy (availability of information, tools, and materials necessary to do a job) complete the list.[31]

It is also clear from the research we have reviewed that work orientations or preferences must remain part of the job satisfaction equation. Indeed, Kalleberg argues that it is the specific match between work rewards (characteristics of the job) and work orientations that determines one's degree of job satisfaction. This would apply, he suggests, to both extrinsic and intrinsic dimensions of work. In his classic study, Kalleberg (1977) found that job rewards had positive effects on expressions of job satisfaction in the U.S. labour force, as predicted. But he also observed that, other things being equal, work preferences (orientations) had negative effects on satisfaction. In other words, the more one values some particular feature of work (the chance to make decisions, for example), the less likely it is that one's desires can be satisfied. However, job rewards had substantially greater effects on satisfaction than did work preferences. In addition, Kalleberg's analysis led him to conclude that intrinsic job rewards were more important determinants of job satisfaction than were extrinsic rewards.

Canadian Research on Intrinsic Job Rewards

What does Canadian research tell us about intrinsic and extrinsic work rewards and their fit or mismatch with the work orientations of workers?

Going back more than 30 years, respondents in the 1973 national Job Satisfaction Survey were given a list of more than 30 specific job characteristics and asked to assess, first, how important each was to them and, second, how their own job rated on these characteristics. The analysts then compared the average "importance" and "evaluation" scores, and examined the gap between them. The largest discrepancies appeared in the areas of "opportunities for promotion" and "potential for challenge and growth" (Burstein et al. 1975: 31–34). These are not unusual findings since most workers seek intrinsic rewards from their work and wish to "get ahead" in their careers.

But has labour market restructuring, particularly downsizing and greater reliance on nonstandard workers, changed this pattern? Figure 8.2 presents selected findings from the CPRN 2000 national survey of employment relationships. For each of five intrinsic and five extrinsic dimensions of work, sample members were asked to indicate how important it was to them personally (their own work orientation) and how much it was provided by their job (an evaluation of the specific job reward). Thus, these "job quality deficit" comparisons replicate the contrasting of "importance" and "evaluation" scores in the 1973 study. They are, in a sense, the ingredients for the contemporary patterns of job satisfaction and dissatisfaction that we have described above.

It is clear from this chart that there still are sizable gaps between what Canadian workers desire and what they get from their jobs, in both the intrinsic and extrinsic dimensions, with the largest gap (34 percentage points) for "good chances for career advancement." Thus, with respect to this specific finding, very little has changed since 1973 when "opportunities for promotion" was one of the two areas in which workers' desires far exceeded what their jobs provided. But unlike the situation in 1973, the work orientation–work reward gap for all of the extrinsic dimensions considered in this analysis (pay, benefits, job security, balancing work and family) is larger than the gap for the five intrinsic dimensions (interesting work, feeling of accomplishment, freedom to decide how work will be done, chance to develop skills and abilities, receive recognition for work well done). Perhaps what we are seeing here reflects the growth in nonstandard employment and, more generally, the labour market restructuring and polarization that has occurred over the past decade or two.

That said, we still see that there are sizable gaps between workers' orientations and their job rewards on the five intrinsic dimensions of work featured in Figure 8.2. For example, 94 percent of Canadian workers report that having the

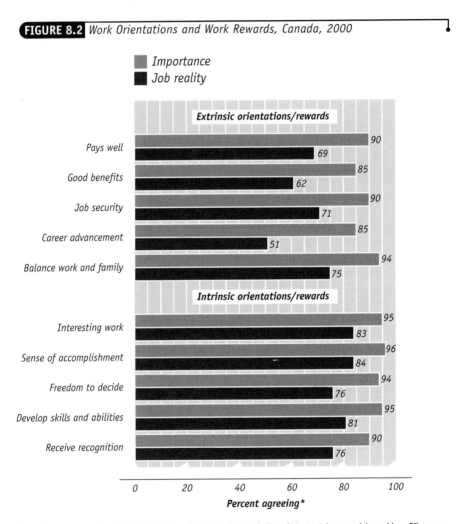

FIGURE 8.2 *Work Orientations and Work Rewards, Canada, 2000*

■ *Importance*
■ *Job reality*

Extrinsic orientations/rewards

Pays well — 90 / 69
Good benefits — 85 / 62
Job security — 90 / 71
Career advancement — 85 / 51
Balance work and family — 94 / 75

Intrinsic orientations/rewards

Interesting work — 95 / 83
Sense of accomplishment — 96 / 84
Freedom to decide — 94 / 76
Develop skills and abilities — 95 / 81
Receive recognition — 90 / 76

0   20   40   60   80   100
Percent agreeing*

*For both work orientations ("how important is …?") and work rewards ("to what extent does your job provide …?"), survey respondents were asked to agree or disagree on a five-point scale. The percentage agreeing or agreeing strongly with each statement is presented in this chart.

Source: CPRN=EKOS 2000. *Changing Employment Relationships Survey.* Reprinted by permission from Canadian Policy Research Networks.

freedom to decide how their job will be done is important to them, but only 76 percent feel that their job actually provides such freedom. Almost all (95%) value opportunities to develop their skills and abilities, but only 81 percent are in jobs that provide such opportunities. Similarly, 95 percent want an interesting job, but only 83 percent consider their current job to be interesting.

Researchers agree that, for most workers, jobs that provide considerable autonomy, complexity, and variety are more personally fulfilling. The opportunity to make decisions about how a job should be done, to develop and use a wide range of skills, and to do interesting work can also increase a worker's satisfaction with her or his job. This would explain the very large differences in job satisfaction reported by self-employed women (86% satisfied) and women who were paid employees (68%) in a 2000 national survey of Canadian workers (Hughes 2005: 86). In this study, the self-employed women were much more likely than paid employees to indicate that their jobs were interesting and provided decision-making and skill development opportunities.

In contrast, repetitive jobs with little variety often have the opposite effect. In the following quotations, compare the variety and challenge that are so satisfying for the male oil-field service company manager and the sense of accomplishment reported by the professional engineer (also male) with descriptions of repetitive, low-skill work by a female food-processing factory worker and a male assembly-line worker.[32]

> Every morning even now, I'm happy to come to work; I look forward to it. I think it's because this business is, well, you never know when you come in here in the morning what's going to be asked of you. It could be different every day; generally it is.
>
> ***
>
> I love it. I think I have one of the best jobs I know of . . . I enjoy working on projects from start to finish as opposed to just having a small piece of them . . . I have found over the years that I am very easily bored and in this field and this environment you don't get bored very often.
>
> ***
>
> Basically, I stand there all day and slash the necks of the chickens . . . The chickens go in front of you on the line and you do every other chicken or whatever. And you stand there for eight hours on one spot and do it.
>
> ***
>
> Your brain gets slow. It doesn't function the way it should. You do the same thing day in and day out and your brain goes. I'm like a robot. I walk straight to my job and do what I have to do.

Fortunately, the jobs held by the majority of Canadians appear to be more challenging than those described in these last two quotations. The 2000 CPRN survey reveals, for example, that about three-quarters or more of Canadian workers report that their jobs are interesting, provide them with a sense of accomplishment, and offer some decision-making opportunities as well as chances to enhance skills and abilities (Figure 8.2). In addition, 77 percent of the respondents in this survey reported that their job required high skill levels, while a similar proportion (75%) agreed that "you get the training needed to do your job well" (results not shown in Figure 8.2).

Nevertheless, when asked to agree or disagree that "your job requires that you do the same tasks over and over," 59 percent of the 2000 survey respondents agreed or agreed strongly. These findings suggest that repetitious work (a lack of variety) is more common than non-challenging, low-skill work, or work with little autonomy. The 2000 survey also asked workers to respond with a "yes" or "no" to the question: "Considering your experience, education, and training, do you feel that you are overqualified for your job?" One in four workers (27%) said "yes." An earlier (1994) national survey of literacy skills also indicated that about one in five employed Canadian workers are underemployed with respect to their reading and writing skills (Krahn 1997: 18). As we also have noted, there is strong evidence that underemployment may lead to job dissatisfaction (Johnson and Johnson 1995).

Consequences of Job Satisfaction and Dissatisfaction

"If the job's so bad, why don't you quit?" Many of us may have thought this about an unsatisfying job, but fewer have actually done so. There may be some features of the job—pay, hours, location, friendly coworkers—that make it palatable, despite the absence of other work rewards. In addition, unless other jobs are available, most employees simply cannot afford to quit. Even during a period of relatively low unemployment in 2000, the CPRN national survey found 41 percent of Canadian workers agreeing that it would be difficult to find another job as good as their current job. Two-thirds (67%) agreed that it would be difficult to cope financially if they lost their job. Thus, job dissatisfaction will not necessarily translate into quitting behaviour (Hammer and Avgar 2005).

But feelings of job dissatisfaction may allow workers to rationalize coming in late, calling in sick, or generally not working as hard as they could

(Hulin 1991). Studies have also shown a relationship between job dissatisfaction and overt acts of employee deviance, such as theft of company property, or the use of drugs and alcohol on the job and away from work (Mars 1982; Martin and Roman 1996). Dissatisfaction with work is also correlated with the number of complaints and grievances filed in unionized work settings. As for nonunion workers, surveys have shown that those dissatisfied with their work are more likely to view unions positively and are more likely to join a union, if given the chance.[33]

From an employer's perspective, however, the critical issue is whether increases in job satisfaction will boost productivity. If dissatisfaction can lead to tardiness, absenteeism, deviance and, in some situations, quitting, surely improvements in the quality of working life will lead to a more satisfied and, hence, more productive workforce? This seems like an obvious relationship (Fisher 2003). But although one can find examples of studies showing such a pattern, there are also many studies that reject this hypothesis. Overviews of research on the topic typically show that the relationship between satisfaction and productivity is weak, or is only present in some work settings (Iaffaldano and Muchinsky 1985).

There are several possible explanations for this. First, productivity is more often a function of technology and workers' skills than of their attitudes. Thus, even if high levels of satisfaction are evident, low skill levels, inadequate on-the-job training, or obsolete technology will limit opportunities for productivity increases. Second, work-group norms and expectations must be taken into consideration. Managers and consultants who have introduced job enrichment programs or high-performance work practices, and have perhaps even found higher levels of satisfaction as a consequence (Berg 1999), have frequently been disappointed when productivity increases did not follow. They failed to realize that workers might view an improved quality of working life as their just reward, or that informal work norms and long-standing patterns of behaviour are difficult to alter (Macarov 1982).

Finally, it may be that productivity can be influenced by job satisfaction, but only under certain conditions. For example, our earlier discussion suggested that workers in low-level jobs might report job satisfaction because they were assessing their work with a limited set of alternatives in mind. Workers in higher-status jobs might, on the other hand, report satisfaction because of tangible work rewards. If so, perhaps productivity increases due to job satisfaction might only be expected in the latter group. In fact, the most extensive overview of research on this topic—a meta-analysis of findings from 254 previous studies—concludes that the relationship is much stronger in high-complexity jobs (Judge et al. 2001: 388).[34]

WORK AND ALIENATION

William Faulkner once wrote that "You can't eat for eight hours a day nor drink for eight hours a day nor make love for eight hours a day—all you can do for eight hours is work. Which is the reason why man makes himself and everybody else so miserable and unhappy."[35] Faulkner didn't use the term *alienation,* but he might have since his cynical observations about the misery of work capture some of the meaning of the term as used by social philosophers. Specifically, referring back to our earlier discussion of the meaning of work, we can define alienation as the human condition resulting from an absence of fulfilling work.

Karl Marx and Alienating Work within Capitalism

The verb "alienate" refers to an act of separation, or to the transfer of something to a new owner. Marx used the term in the latter sense when he discussed the "alienating" effects of capitalist production relations on the working class. The noun "alienation" refers to the overall experience of work under these conditions.

Marx identified a number of sources of alienation under capitalism. Products did not belong to those who produced them. Instead, ownership of the product remained with those who owned the enterprise and who purchased the labour of workers. Decisions about what to produce and about the sale of the finished products were not made by the workers, and profits generated in the exchange remained with the owners of the enterprise. In fact, given an extensive division of labour, many of the workers involved might never see the finished product. Thus, workers were alienated from the product of their own work.

Marx also emphasized alienation from the activity of work. Transfer of control over the labour process from individual workers to capitalists or managers meant that individual workers lost the chance to make decisions about how the work should be done. In addition, extensive fragmentation of the work process had taken away most intrinsic work rewards. Alienation also involved the separation of individual workers from others around them. Obviously, bureaucratic hierarchies could have this effect. But more importantly, because capitalist employment relationships involve the exchange of labour for a wage, work was transformed from a creative and collective activity to an individualistic, monetary activity. Work itself had become a commodity. As a consequence, Marx argued, workers were alienated from themselves. Capitalist relations of production had reduced work from its role as a means of human self-fulfillment to being merely a market transaction.[36]

This structural perspective on alienation from work rests on several key assumptions. First, alienation occurs because workers have little or no control over the conditions of their work, and few chances to develop to their fullest potential as creative human beings. Second, the source of alienation can be traced to the organization of work under capitalism. Third, given that alienation is characteristic of capitalism, it exists even if workers themselves do not consciously recognize it.

Alienation, then, is a "condition of objective powerlessness" (Rinehart 2006: 11–20). Whether individual workers become aware of the cause of their discontent with work depends on a variety of factors. In the absence of a well-defined and legitimate alternative to the capitalist economic system, we would not expect most North American workers to be able to clearly articulate their alienation, or to act on it. Apathy, or an attempt to forget about work as soon as one leaves it behind, are the common responses of many workers to a situation to which they see no viable alternative. But as we have seen in Chapter 7, discontent does exist, as demonstrated by the frequency of strikes, slowdowns, and other behaviours showing frustration and unhappiness with the conditions of work.

The Social–Psychological Perspective on Alienation

The social–psychological perspective on alienation shares with the structural perspective an emphasis on the powerlessness of workers, and the conclusion that many jobs offer limited opportunities for personal growth and self-fulfillment. While the structural perspective emphasizes separation from the product and the activity of work, from coworkers, and from oneself, the social–psychological approach attempts to measure feelings of powerlessness, meaninglessness, social isolation, self-estrangement, and normlessness (Seeman 1967, 1975). Thus, in a sense, the main focus of the social–psychological perspective is the absence of intrinsic job rewards.

Social–psychological studies of alienation, moreover, are unlikely to lay all or much of the blame on capitalism itself. Instead, technologies that allow workers few opportunities for self-direction, bureaucratic work organizations, and modern mass society in general are identified as the sources of alienation. This perspective also differs from the structural approach in its emphasis on feelings of alienation—the subjective experience of alienating work conditions. Researchers within this approach have relied primarily on the self-reports of workers, in contrast with the Marxist focus on the organization and content of work.

Consequently, social–psychological accounts of alienation resemble job dissatisfaction, although their explanations of the sources of alienation have pointed more explicitly to the negative consequences of work fragmentation and powerlessness. The general job satisfaction literature examines a broader array of causal variables (work orientations, for example). But since some social–psychological studies of alienation actually define the phenomenon as an "extrinsic orientation to work" (Seeman 1967: 273), there is really more congruence between the two research traditions than first impressions would suggest. Hence, it is not surprising to find studies that examine alienation and job dissatisfaction simultaneously, or writers who use the terms interchangeably.

Robert Blauner on "Alienation and Freedom"

Over four decades ago, Robert Blauner (1964) published *Alienation and Freedom,* a classic study of industrial differences in workers' responses to their jobs. Following the social–psychological tradition, he defined alienation in terms of powerlessness, meaninglessness, isolation (or social alienation), and self-estrangement. The centrality of powerlessness in Blauner's theory of alienation is clear from his use of the word *freedom* to refer to the ability to choose how one does one's work.

Blauner compared work in four different industries to test his theory that technology (and the attendant division of labour) is a major determinant of the degree of alienation experienced at work. He argued that the traditional printing industry, which at the time still operated largely in a craftwork mode, produced low levels of alienation because of the considerable autonomy of workers and the high levels of skill required. The textile manufacturing industry represented an intermediate step in the process of technological development, while the culmination occurred with the assembly lines of the automobile industry. Here, alienation was most acute because of low-skilled, repetitive tasks that deprived workers of control over their actions.

Blauner argued that new automated technologies reversed this trend. In work settings such as oil refineries, skill levels were higher, tasks were varied, and individuals could make a number of decisions about how they would do their work. Since this industry was providing us with a glimpse of the future of industrialized societies, he reasoned, the prospects for individual freedom were good. By predicting that alienation would decline as fewer workers were employed in mass-production settings, Blauner was rejecting Marx's argument that work in a capitalist society was always alienating.

Alienation and Freedom has been both influential and controversial. Blauner was undoubtedly correct about the negative consequences of continued exposure to low-skill, routinized work. But critics have argued that a social–psychological definition does not address alienation in the broader way that Marx had defined it.[37] Furthermore, Duncan Gallie's (1978) comparative study of refinery workers in France and England has highlighted cross-cultural differences in workers' responses to technology, an important point Blauner overlooked. And as we have argued earlier, effects of technology are not predetermined. A position of *technological determinism* conveniently overlooks why different technologies are designed or chosen by those who control an enterprise, and how these technologies are built into the organizational structure of the workplace.

Richard Sennett on "The Corrosion of Character"

Writing in the middle of the 20th century, before the era of industrial restructuring, downsizing, and privatization began, and before the advent of widespread nonstandard employment, Robert Blauner was optimistic in his assessment of the future of work in Western industrialized economies. Alienation would decline and freedom (to decide how work should be done) would increase. In contrast, Richard Sennett's recent book on "the corrosion of character" in the 21st-century economy is extremely pessimistic in its predictions. While Sennett (1998) does not reference the social–psychological research literature on alienation, his analysis of the negative impact of contemporary work organizations and relationships on the minds and souls of American workers is firmly rooted in this tradition.

Sennett describes how today's labour market is characterized by the constant restructuring of work organizations, an emphasis on flexibility in production and the delivery of services, a greater reliance on nonstandard workers, and a glorification of risk taking. Together, these new values and practices have combined to get rid of the order and routine that traditionally defined the lives of workers. Why is this a problem? Because, Sennett argues, order, routine, and stability lead to the formation of personal and occupational identities which give meaning to our lives. In a working world where constant change is glorified and the positive aspects of structure and routine are lost, individual identities are also lost or fail to take shape. Even the teamwork that is central to so many new management approaches is criticized by Sennett, since he sees it as merely a way of gaining compliance from workers, as "the group practice of demeaning superficiality" (1998: 99). He concludes that:

> The contradictions of time in the new capitalism have created a conflict between character and experience, the experience of disjointed time threatening the ability of people to form their characters into sustained narratives. (1998: 31)

Sennett recognizes that these fleeting personal identities and the resulting concerns about "who in society needs me?" (1998: 146) are shaped in a capitalist economy. But he does not blame the capitalist relations of production that separate workers from owners and that leave control of the labour process with the latter. Instead, the postmodern alienation that he describes is attributed to the forms of work organization and the much less stable divisions of labour that characterize the 21st-century workplace. In this sense, Sennett's critique of the meaninglessness, social isolation, and normlessness of work today fits centrally within the social–psychological tradition of alienation research.

But where it fits is ultimately less important than whether he is correct in his analysis. The management theorists discussed in Chapter 5 see the changes taking place in today's workplace as largely positive, empowering workers and increasing their job satisfaction. Sennett is almost uniformly negative in his assessment of the alienating impact of contemporary work. These divergent perspectives on work experiences raise fundamental questions about whether, compared to earlier generations of workers, today's workers, individually and through their own organizations, are in a better or worse position to negotiate greater control over the labour process. However, as we have concluded a number of times in earlier chapters, there is not a single answer. Instead, increased labour market polarization may mean greater control (and less alienation) among well-paid and empowered "knowledge workers" and professionals, and less control (and more alienation) among poorly paid workers in low-complexity jobs and precarious employment situations.

WORK AND STRESS

If we take a broader *work and well-being* approach to the subject of how workers experience and react to their jobs, both physical and psychological reactions to work become part of our subject matter. This approach also encourages us to consider how paid work might affect an individual's life in the family and the community. Consequently, workplace health and safety issues, discussed in Chapter 3, are also central to discussions of work and well-being, as are concerns about balancing work and family responsibilities (see Chapter 4). Here, we look more specifically at work and stress.

Defining Work-Related Stress

Defining *work-related stress* independent of job dissatisfaction is not easy. In fact, some researchers studying stress and its consequences rely on measures that might, in a different context, be considered indicators of job dissatisfaction. But there are important distinctions to be made between job dissatisfaction, anxiety and tension about work, job stress, and job burnout (Humphrey 1998: 4–9). It is possible to feel dissatisfied with work, perhaps even a little anxious about some aspects of a job, without experiencing a great deal of stress. But work-related stress, with its physical and mental symptoms, can also accumulate to the point of *burnout,* where an individual is simply unable to cope in the job (Maslach and Leiter 1997). It is useful, then, to conceive of work-related stress as a many-sided problem—with job dissatisfaction as one of its components—that can lead to serious mental and physical health problems.

It is also helpful to distinguish *stressors* (or *strains*) from an individual worker's reactions to them. Stressors are objective situations (for example, noisy work environments or competing job demands) or events (a dispute with a supervisor or news that some workers are about to be laid off) that have the potential to produce a negative subjective or physical response. Thus, work-related stress is an individually experienced negative reaction to a job or work environment. Obviously, the absence of stress does not imply the presence of job satisfaction. What distinguishes stress reactions are the wide range of ill health (physical and psychological) symptoms.

Causes and Consequences of Work-Related Stress

Research in many different settings has shown that physical reactions to stress can include fatigue, insomnia, muscular aches and pains, ulcers, high blood pressure, and even heart disease. Depression, anxiety, irritation, low self-esteem, and other mental health problems are among the documented psychological reactions to stressful work. The research literature has also shown that the effects of work-based stressors can be conditioned by individuals' psychological coping mechanisms and by the amount of social support they receive from family, friends, and coworkers (Jex 1998).

In their study of workers who had lost well-paying manufacturing and resource sector jobs in five rural Ontario communities, and who were now struggling with unemployment or low-paying, part-time service-sector jobs, Anthony Winson and Belinda Leach (2002: 127–30) provide graphic examples of one form of work-related stress. These workers and their families worried constantly about paying bills, making do with less, losing the family home, having to rely on food banks, and dealing with medical expenses that used to be covered by fringe benefit packages. But such stress is not restricted to workers in small communities faced with factory closings (Reynolds 1997). In 1994, a nationwide survey of Canadian workers revealed that almost 1 in 4 (22%) had experienced stress over possible job loss or layoff during the previous 12 months. By 2000, when unemployment rates had declined considerably, the proportion of Canadian workers experiencing stress due to possible job loss had declined to 13 percent (Williams 2003: 24).

As Chapter 4 documented, equally stressful for some workers, especially women, are the pressures of looking after family responsibilities while trying to devote oneself to a job or career. As one of the young, university-educated women interviewed by Gillian Ranson (1998: 527) lamented:

> In this job, in this particular job, I don't think I could do justice to the child and I've made the decision that . . . either after maternity leave or in the near future, I'm quitting, because it's not fair. I come home, I can't even talk to my husband because I'm so wound up, stressed out.

But most research in this area has concentrated on stressors in the work environment (Barling, Kelloway, and Frone 2005).[38] Continual exposure to health and safety hazards, working in a physically uncomfortable setting, shift work, or long hours can all be stressors. Similarly, fast-paced work (especially when the pace is set by a machine) and inadequate resources to complete a task can generate considerable stress for workers. So too can the experience of constant organizational restructuring and/or the introduction of the latest management fad. In addition, working at tasks that under-utilize one's skills and abilities, that do not meet one's expectations for the job, or that allow little latitude for decision making are stressful for many workers. Finally, there is the stress that an unreasonable and overly demanding supervisor can create, as well as the stress resulting from sexual harassment or from discrimination in the workplace based on gender, race, sexual orientation, religion, or disability.

Statistics Canada's 1994/95 national Population Health Survey asked employed Canadians (aged 15 to 64) about a number of sources of stress in their job. Forty-four percent of men and 37 percent of women indicated that their job had high physical demands. Seventeen percent of men and 31 percent of women reported high job strain, that is, "they were in hectic jobs, and had little freedom in controlling the pace of work or in deciding how to carry out their duties" (Statistics Canada 2000a: 22–24). The national General Social Survey conducted in 2000 revealed that one-third (34%) of Canadian workers had experienced stress in the previous year due to "too many demands or too many hours." Fifteen percent of the employed survey participants reported stress from "poor interpersonal relations" (with supervisors or coworkers), and 13 percent had felt stressed because of the risk of accident or injury (Williams 2003: 24). More recently, a 2004 national survey estimated that 24 percent of Canadian workers had experienced job stress "often" in the previous 12 months while 14 percent experienced job stress "always."[39]

The 2004 study examined the consequences of job stress for the 38 percent of working Canadians who experienced job stress "often" or "always." Compared with their peers who reported less stress, this high-stress group was more likely to report mental and physical health problems, seek medical help for such health problems, be absent from work, quit their jobs, and report less satisfying family and personal lives due to job stress. In short, a very large number of working Canadians experience a significant amount of job stress which, in turn, has negative consequences for themselves, their families, and their employers.

The "Demand–Control" Model of Work-Related Stress

A useful perspective for understanding workplace stress is the *demand–control model,* which re-defines stressors as *job demands* but also introduces the concept of worker control.[40] It distinguishes between active jobs where individual decision-making potential is high, and passive jobs where it is largely absent. If psychological demands on a worker are high, but she or he can do something about it, stress is less likely to result. If demands are high and control is low, stress and the health problems that can follow are far more often the outcome.

From this perspective, we can see why researchers have repeatedly found highly routinized, machine-paced work to be extremely stressful. Assembly-line jobs are often considered to be among the most stressful (Hamilton and Wright 1986: 266). Demand is constant, the physical work can be extremely taxing, and worker control is virtually absent. Solange De Santis describes her reaction to the assembly line after several weeks of work in General Motors' Scarborough, Ontario, assembly plant:

> The line went down for a few minutes, and I sat on the stool attached to the job station, next to my table full of red and yellow lenses, ropes of black putty, boxes of screws and light bulbs. Maybe it was because I had a few moments to think, maybe it was the sensory overload of the past few weeks, but for the first time an abrupt, raw hatred of the place rose in my craw, a loathing of the noise, the dirt. All the people

are stupid, I thought. The work is meaningless and exhausting. There is no privacy. The line never stops. The last was a bit irrational, since I was thinking this during the instant that the line had stopped. (De Santis 1999: 85)

Earlier we suggested that instrumental work attitudes may be a coping mechanism—the paycheque becomes the only relevant work reward. Some assembly-line workers rely on alcohol or drugs to get through a shift (De Santis 1999: 111–14). Others adapt to assembly-line work by "tuning out" the boredom and waiting for the chance to get away from the job. Ben Hamper (1986: 88), better known as "Rivethead," describes how he was introduced to his new job on a truck-assembly line in Flint, Michigan:

"Until you get it down, your hands will ache, your feet will throb and your back will feel like it's been steamrolled . . . "

"Are there any advantages to working down here?" I asked pitifully.

The guy scratched at his beard. "Well, the exit to the time clocks and the parking lot is just down these stairs. Come lunchtime or quittin' time, you can usually get a good jump on the rest of the pack."

But psychological problems can also arise. Several decades ago, Robert Linhart (1981: 57) described an extreme reaction to the assembly line in a French automobile factory:

He was fixing parts of a dashboard into place with a screwdriver. Five screws to fix on each car. That Friday afternoon he must have been on his five hundredth screw of the day. All at once he began to yell and rushed at the fenders of the cars brandishing his screwdriver like a dagger. He lacerated a good ten or so car bodies before a troop of white and blue coats rushed up and overcame him, dragging him, panting and gesticulating, to the sick bay.

Highly routine, monotonous, mechanized, and closely supervised jobs are also found in the service industries. A 1980s study of Canada Post mail sorters and letter carriers in Edmonton revealed that the most stressful job in the organization involved "keying" postal codes using automated machinery. The relentless pace of the machinery, conflicting demands imposed by management, constant repetition, and lack of challenges and job autonomy were associated with diminished mental and physical health among coders. The use of pain relievers and tranquilizers was significantly higher among the coders using automated technology compared with those sorting mail by hand (Lowe and Northcott 1986).

The widespread adoption of electronic communications systems in the contemporary workplace has created some highly rewarding and complex jobs, but also some new types of work that can be extremely routinized and, hence, very stressful. Most prominent here are jobs in call centres that require workers to repeatedly attempt to contact members of the public, either to sell products or services or to solicit their opinions on various topics (Taylor et al. 2002). The same technologies have also increased the scope for electronic surveillance, yet another new source of stress (see Chapter 6). This practice might involve monitoring the conversations of teleworkers and others who deal with the public by telephone, keeping track of Internet searches made by employees, recording output in settings where workers use computers or electronic communications systems, and even delivering electronic warnings to those performing below certain levels. According to one recent study, almost four out of five U.S. workers are being monitored by some type of electronic surveillance system (Hansen 2004: 151).[41]

Frederick Taylor would have applauded such high levels of work routinization and worker control, but employees do not. A data entry clerk whose work group had been told that they had not met management productivity goals the previous week comments:

> I feel so pressured, my stomach is in knots. I take tons of aspirin, my jaws are sore from clenching my teeth, I'm so tired I can't get up in the morning, and my arm hurts from entering, entering, entering. (Nussbaum and duRivage 1986: 18)

A Canadian call-centre employee described the workplace in similar terms:

> They count on high turnover because of the way this business is run, there is a very high burnout level. People burn out quickly because of the stress, because of the pressure, because of the way people are treated, because of the degrading nature of the work. (Buchanan and Koch-Schulte 2000: 24).

It is not difficult to see how such working conditions—high demands and virtually no worker control—can be stressful, and can lead to physical and psychological ill health.

The "Person–Environment Fit" Model

A different theoretical model of work-related stress emphasizes the *person–environment fit*. According to this model, stress results when there is a significant gap between an individual's needs and abilities and what the job offers, allows, or demands (Johnson 1989). To take a specific example, reports of stress and burnout among social workers, teachers, and nurses are common. Individuals in these helping professions have work orientations and expectations (the desire to solve problems and help people) and useful skills (training in their profession) that are frequently thwarted by the need to deal with an excessive number of clients, limited resources, and administrative policies that make it difficult to be effective.[42]

We might also use this perspective to help us understand the stress many workers feel when they attempt to juggle their work and family roles. A paid job requires one to be present and involved; at the same time, family responsibilities demand time and attention. Because women continue to carry a large share of domestic and child-care responsibilities, they are much more likely to experience role conflict and stress, as this individual interviewed by Duffy and Pupo (1992: 134) reported:

> When I worked full-time, I felt very guilty about the children. If they were sick and I went to work, I felt guilty. If I stayed home with them, I felt guilty about work. Part-time work could offer the flexibility that full-time work cannot.

This woman chose part-time work to try to handle her problem, but gave up additional income and career options (which are limited in most part-time jobs). From a person–environment fit perspective, she was attempting to improve the fit between demands of her family and her job (or the financial need to work). But that fit could also have been improved if structural, rather than just individual, changes were implemented.

Thus, while the person–environment fit model of stress broadens our explanatory framework by bringing in work orientations, its individualistic focus advocates primarily personal solutions. From the demand–control perspective, we are more inclined to seek solutions to the lack of worker control. In this case, if more men accepted an equal share of the responsibilities in their homes, if more employers offered flexible work schedules to their workers, and if affordable, quality childcare were more accessible, would the woman quoted above have been forced to make these choices? With respect to professional burnout, how could organizational structures and administrative policies be changed to allow social workers and nurses to be more effective? How can we use new technologies to reduce stress and increase job satisfaction, rather than merely to control workers and increase the speed and intensity of work?

CONCLUSION: THE "LONG ARM OF THE JOB"

Early in this chapter, we described how individuals are exposed to at least two competing work-value systems: the first suggests that monetary rewards are paramount; the second emphasizes how work can be personally self-fulfilling. This raises the possibility that, because of different socialization experiences, some workers would be primarily instrumentally oriented while others would be more motivated by the intrinsic rewards of a job. In short, it could be argued that work orientations or preferences are brought to the job, and will influence feelings of satisfaction or alienation.

But most research tends to support an alternative explanation, namely, that participation in low-skill or routinized work can produce instrumental work attitudes. For many workers, such attitudes may be a means of adapting to employment that has little but a paycheque to offer. Even so, studies of job satisfaction, alienation, and stress highlight the negative consequences of work that is repetitive and routine, offers few chances for individual decision making, and does not develop a person's skills and abilities. While it is important to recognize that work orientations affect how an individual feels about his or her job, the nature of the work itself is likely to have a stronger impact, both on an individual's feelings of satisfaction, alienation, and stress and on her or his work orientations.

We have suggested that work-related stress does not get left behind at the end of the workday. Chronic work pressures, as well as anxiety about potential job or income loss, can undermine an individual's overall quality of life. At the same time, it is apparent that satisfaction with work translates into a broader sense of well-being. But are there any other more long-term, perhaps even permanent, psychological effects a job can have on an individual? The answer appears to be "yes"—continued exposure to some kinds of work can have long-lasting effects on non-work behaviours and on one's personality.

In a study conducted over three decades ago, Martin Meissner argued that the "long arm of the job" has an impact on one's life away from work. He studied male British Columbia sawmill workers to test rival hypotheses about how the nature of work might affect after-work behaviours. The *compensatory leisure* hypothesis proposes that people will look for activities away from work that will compensate for what is absent in their jobs. Thus, workers who have little opportunity to develop their skills and abilities on the job might seek these opportunities away from work. The *spillover* hypothesis suggests that the effects of work will influence one's choice of after-work activities. Meissner concluded that there is a spillover effect. Workers who had little chance to make decisions about how their work should be done were considerably less likely to engage in free-time activities that required or allowed this kind of individual discretion. Similarly, those who had few opportunities for social interactions at work were more inclined toward solitary leisure-time activities (Meissner 1971: 260).

One could argue that the subjects in Meissner's study had chosen their jobs in order to satisfy their personal preferences. Such an emphasis on work orientations would explain, for example, that individuals with a preference for solitary activities would choose both their jobs and their leisure activities with this in mind. While plausible, this explanation is not convincing for several reasons. First, it suggests that individuals have a much greater range of job choices than is typically the case. Second, subsequent studies comparing these two hypotheses have agreed with Meissner (Martin and Roman 1996). Third, there is research demonstrating even more conclusively that work with limited scope has a negative effect on personality.

Melvin Kohn and his associates have examined this relationship for many years using the concept of *occupational self-direction*. Kohn argues that work that is free from close supervision, that involves considerable complexity and independent judgment, and that is non-routine will have a lasting positive effect on one's personality and psychological functioning. Specifically, individuals whose work allows self-direction are more likely to develop a personality that values such opportunities and a more self-confident, less fatalistic, and less conformist approach to life. They also exhibit greater flexibility in dealing with ideas. Alternatively, jobs allowing little self-direction are more likely to lead to psychological distress, a finding we have already documented from stress research.

Kohn presents a convincing case, since he used longitudinal data in his studies. By comparing the jobs and personalities of workers at two points in time (up to 10 years apart), he clearly demonstrated the personality changes in those who had and those who did not have the opportunity to work in jobs allowing self-direction. Hence, he has also been able to demonstrate that "both ideational flexibility and a self-directed orientation lead, in time, to more responsible jobs that allow greater latitude for occupational self-direction."[43] The overall conclusion of Kohn's studies and related research[44] is that intrinsically rewarding work, particularly work that allows self-direction, can have important positive long-term consequences for the personalities and careers of those fortunate enough to participate in it.

But the "long arm of the job" extends even further, beyond feelings of satisfaction and dissatisfaction, experiences of alienation and stress, difficulties balancing work and family, and actual personality change. A growing number of population health studies have documented that poor jobs and problematic working conditions are simply bad for your health. We have already seen how frequently workers are injured or killed on the job, or become ill because of hazardous working conditions (Chapter 3). But strong evidence linking heart and other diseases, as well as mental illness, to work settings is also accumulating. In fact, a number of excellent longitudinal studies have demonstrated that, other things being equal, workers in low-skill and poor-paying jobs don't live as long.[45] That, surely, must be a sufficiently convincing argument for increasing our efforts to improve the quality of working life for as many working Canadians as possible.

DISCUSSION QUESTIONS

1. Compare and contrast the following concepts: work values, work orientations, job satisfaction, work-related stress, and alienation.
2. Based on discussions in this chapter, as well as Chapters 2 and 3, what are the key factors that distinguish "good jobs" from "bad jobs" in the Canadian labour market?

3. Do you think that, compared to their parents and grandparents, young people today have different work values and orientations? How would you design a study to determine whether your opinion is correct?

4. Critically discuss the following statement: "The majority of Canadians go to their (paid) jobs to get the money they need to enjoy life away from work. If they won a lottery, most Canadian workers would quit their jobs."

5. When asked, a large majority of workers say they are satisfied with their job. So why should we worry about changing working conditions, management approaches, or organizational structures?

6. Older workers tend to be more satisfied with their jobs. What might explain this phenomenon? Do you think this long-standing research finding will continue to be observed in the future?

7. What kinds of jobs generate the most stress for workers? Why? What might be done to reduce the stress created by such jobs?

8. What do the authors of this text mean by "the long arm of the job?" Are you convinced by their argument? Why or why not?

NOTES

1. See George and Jones (1997) for a somewhat different classification of work values, attitudes, and moods.

2. Bernstein (1997); also see Anthony (1977; Chapter 1) and Byrne (1990: Chapter 3).

3. Genesis 2:17, 3:19 (King James Version).

4. Recent examples include Hodson (2001), Rayman (2001), and Bunting (2004).

5. See Burman (1988, 1996), Newman (1989), Feather (1990), Gallie and Russell (1998), and Winson and Leach (2002: Chapter 6) on social and psychological consequences of unemployment.

6. The quotations in this section are from Burman (1988: 161, 113, 144, 86).

7. See *The Economist* (1996) for a critique of cultural explanations of economic growth.

8. Rifkin (1995: 255–57) proposes tax deductions for voluntary work and a "social wage" for the unemployed who participate in voluntary work, but such policies would still produce a very unequal income

distribution. In a more recent book, Gorz (1999) goes further in proposing a guaranteed annual income and recommending cooperative economic organizations as solutions to the inequality problem. Others have emphasized the need for greater sharing of jobs (e.g., Human Resources Development Canada 1994; Hayden 1999; Huberman and Lanoie 2000).

9. See Schor (1991, 1998), Hochschild (1997), and Rayman (2001) for similar critiques of overwork and over-consumption in the United States. Green (2001) and Kemeny (2002) document the extent of overwork in Britain and Canada, respectively. Harpaz and Snir (2003) present a useful discussion of the varied meanings of "workaholism."

10. The 1996 survey was conducted for the Royal Bank by Angus Reid (*The Edmonton Journal* 8 October 1996: F1). Wyatt Canada (1991: 11) completed the 1991 survey.

11. Unpublished findings from a survey conducted by Canadian Policy Research Networks Inc. (http://www.cprn.org). See Lowe and Schellenberg (2001) for discussions of the same data analyzed in a slightly different manner.

12. Barlow and Robertson (1994: 25–38); Osberg, Wein, and Grude (1995: 161–71); Taylor (2001: 23).

13. National Center on Education and the Economy (1990: 24); Cappelli (1992: 6); Ainley (1993: 24); Holzer (1996).

14. Meunier, Bernard, and Boisjoly (1998); Boyd and Norris (1999); Côté (2000); Molgat (2002); Mitchell, Wister, and Gee (2004).

15. See Statistics Canada, *The Daily* (30 July 2004) on high postsecondary enrollments and Jackson (2005: 47) on training received by young workers.

16. Unpublished research findings; see Lowe and Schellenberg (2001) for discussion of the same data analyzed in a different fashion.

17. See Feldman, Doerpinghaus, and Turnley (1994), Nollen and Axel (1996), Booth (1997), and Sennett (1998) on reduced loyalty and the dilemmas of managing contingent workers.

18. Nixon's comment was quoted in *Time* (7 September 1987: 42).

19. Burstein et al. (1975: 12, 22, 60); also see Hamilton and Wright (1986: 288) and Furnham (1990: 196–208), who review a range of similar U.S. studies and draw the same conclusion.

20. Gallie et al. (1998: 303) draw the same conclusion about the U.K., namely, that the work ethic has not declined over the past several decades.

21. See Burman (1996: 42–45) and Broad and Antony (1999) on welfare state restructuring, Richards et al. (1995) for debates about "workfare" programs, Swanson (2001) on stereotypes about the poor, and Black and Stanford (2005) on "welfare reform" in Alberta. Battle and Torjman (1999) and Riddell and St-Hilaire (2000) discuss more progressive public policies for dealing with unemployment and poverty.

22. See Laroche (1998), Schecter and Paquet (1999), Winson and Leach (2002), and Morissette and Zhang (2005) on movement in and out of poverty in Canada; also see Newman (1989), Morris and Irwin (1992), Payne and Payne (1994), Rank and Hirschl (1999), and Ehrenreich (2001).

23. See Felt and Sinclair (1992), MacDonald (1994), and Leonard (1998).

24. In Canada, the Social Research and Demonstration Corporation has been conducting research for over a decade showing how incentives added to social assistance and other government transfer payments lead to more, rather than less, work by the poor and underemployed. See http://www.srdc.org/.

25. In contrast, Page (2004) presents a U.S. study showing that children of parents who received welfare are more likely, as adults, to receive social assistance themselves. Even so, this researcher is reluctant to conclude that a culture of dependency is being passed from one generation to the next, and points out that a majority of children of welfare recipients did not become recipients themselves (Page 2004: 242).

26. The 2002 survey is described in Brisbois (2003: 57–58), a Canadian Policy Research Networks document that can be downloaded at http://www.cprn.com/en/doc.cfm.

27. Unpublished data from the 2004 *Rethinking Work* survey of a random sample of 2002 Canadian workers conducted by EKOS Research Associates Inc. and the Graham Lowe Group Inc.

28. Maynard (1987: 115); *The Edmonton Journal* (8 October 1996: F1); also see Burstein et al. (1975: 29) and Levitan and Johnson (1982: 76–79) on indirect measures of job satisfaction.

29. Kalleberg and Loscocco (1983) conclude that all of the explanations have some relevance; Hamilton and Wright (1986: 288) favour the "job effect,"

while Firebaugh and Harley (1995: 97) believe that "aging" and "life-cycle" explanations fit the data best.

30. See Johnson and Johnson (1995) on the link between underemployment and job dissatisfaction, Hellgren and Sverke (2001) on downsizing and job dissatisfaction, and Preuss and Lautsch (2002), De Witte and Näswall (2003), and Gallagher and Sverke (2005) on employment insecurity and job dissatisfaction.

31. Ospina (1996) focuses on workers' perceptions about the justness of the opportunity structure in addition to work rewards and work orientations as determinants of job satisfaction.

32. In order, the quotations are from House (1980: 335), Bailyn and Lynch (1983: 281), Armstrong and Armstrong (1983: 129), and Robertson and Wareham (1987: 29). For additional graphic "tales from the assembly line," see Hamper (1986) and De Santis (1999). Hodson (2004) provides an excellent content analysis of workers' reactions to their jobs as reported in 149 published workplace ethnographies.

33. See Bender and Sloane (1998) on job dissatisfaction and filing grievances. Kochan (1979: 25) links dissatisfaction to willingness to join a union. Gordon and Denisi (1995), Bryson, Cappellari, and Lucifora (2004), and Hammer and Avgar (2005) reverse the question by asking about the effects of union membership on job satisfaction.

34. Petty, McGee, and Cavender (1984) reached a similar conclusion in an earlier, smaller-scale meta-analysis; also see Altman (2001). Lowe (2003) reviews a parallel debate about whether implementation of workplace wellness programs (see Chapter 6) leads to increased productivity.

35. Faulkner is quoted by Studs Terkel (1971: xi).

36. See Archibald (1978: 35–43), Erikson (1990), Hodson (2001: 23–25), Grabb (2002: 19–21), and Rinehart (2006: 11–20) for additional discussion of Marx's writings on alienation.

37. Archibald (1978: 124–30); however, see Hodson (1996), who links Blauner's thesis to labour process debates about management control over workers.

38. Also see Karasek and Theorell (1990), Roberts and Baugher (1995), Cartwright and Cooper (1997), Jex (1998), Statistics Canada (2000a), and Williams (2003).

39. Unpublished data from the 2004 *Rethinking Work* survey conducted by EKOS Research Associates Inc. and the Graham Lowe Group Inc.

40. See Karasek (1979), Karasek and Theorell (1990), and Sauter, Hurrell, and Cooper (1989) on the "demand–control" explanation of workplace stress; for recent studies using this approach, see Soderfeldt et al. (1997), Carayon and Zijlstra (1999), de Jonge et al. (2000), and Schouteten and Benders (2004).

41. Also see Frenkel et al. (1995), Sewell (1998), Fernie and Metcalf (1998), and Felstead, Jewson, and Walters (2003) on workplace electronic surveillance. Mann and Holdsworth (2003) describe the stress generated among teleworkers, as well as the negative physical health symptoms they report as a result of their working conditions.

42. Leiter (1991); Lait and Wallace (2002). Also see Wallace and Brinkerhoff (1991) and Maslach and Leiter (1997) on "burnout."

43. Kohn and Schooler (1983: 152); also see Kohn (1990), Miller et al. (1979), and Miller, Kohn, and Schooler (1985).

44. Krahn and Lowe (1997) propose that under-use of literacy skills at work can lead to the loss of such skills. Schooler (1984) develops a more general theory of "psychological effects of complex environments" that has applicability beyond the workplace.

45. Mustard, Lavis, and Ostry (2005) provide an excellent, up-to-date summary of the linkages between work and health outcomes. Also see Health Canada (2004) on the relationship between social inequality and health outcomes.

CONCLUSION

Our discussion of work, industry, and Canadian society has ranged widely over many different topics. We have reviewed and dissected key theoretical debates in the sociology of work and industry and the sociology of organizations, and we have documented emerging industrial, labour market, and workplace trends. Instead of summarizing the many issues we have covered, we conclude by highlighting some of the major themes in current debates about changing labour markets and workplaces.

Debates about the emerging postindustrial society and global economy have been part of sociological discourse for several decades, and will no doubt continue. But it is very clear that, like all major industrialized nations, Canada has become a service-based society with a globally linked economy. Natural resource, construction, and manufacturing industries will continue to play an important, if diminished, role in the Canadian economy, but it is the service sector, and the wide range of jobs it provides, that now dominates. Three out of four employed Canadians depend on the service sector for their livelihood. Some service industry jobs are highly skilled, well paying, and intrinsically rewarding, yet a substantial number are relatively unskilled and less rewarding, both financially and psychologically. One of the most clearly defined characteristics of this postindustrial economy is the distinction between "good jobs" and "bad jobs." In particular, while a majority of working Canadians are still employed in permanent, full-time jobs that have a Monday-to-Friday daytime schedule, a growing proportion of the millions of jobs in the Canadian labour market no longer fit this post–World War II definition of a "standard" job. Employment relationships and work arrangements have become much more varied and, in some key ways, less secure and more precarious.

Equally important demographic changes have been quietly transforming the Canadian workforce. The population is slowly aging, and the implications of this trend are only beginning to be understood. The oldest members of the baby-boom generation are approaching retirement. Having lived through the mid-20th-century decades of economic expansion and rising personal incomes, but also through the turbulent labour market restructuring of the 1990s, this large and privileged cohort is looking ahead with mixed feelings. Some baby boomers eagerly anticipate retirement, planning to enjoy a leisurely lifestyle paid for by their pensions and savings. Others, equally financially secure, love their job and can't imagine life without it. Still others would prefer to retire, but must keep working due to inadequate pensions and savings. How these trends, fuelled both by demographic shifts and an inequality-producing economy, eventually play out remains to be seen. Will the normal age of retirement rise past 65? Will the government maintain or decrease its pension commitments? Will more employers shift their hiring preferences to contingent workers who are typically not eligible for private pensions?

At the other end of the age distribution, in the 1980s and 1990s successive cohorts of younger workers experienced considerable difficulty finding good jobs, despite their increased investments in higher education. Since then, postsecondary tuition rates have increased dramatically, while the number of well-paying, full-time, secure jobs awaiting university- and college-educated youths has not kept pace with the number of postsecondary graduates. What are the long-term implications for Canadian society, for worker–employer relationships, and for individual feelings of job satisfaction, of this kind of intergenerational inequality?

These questions highlight the issue of growing social inequality in Canadian society, clearly the most disturbing trend we have observed. While unemployment rates today are not nearly as high as they were during the recessions of the 1980s and 1990s, we still have more than 1.2 million people seeking jobs in our society. Nonstandard jobs have become much more common, and the ranks of the working poor have been growing. The causes of the observed increase in social inequality are complex and intertwined, but involve the following: transformation of the global economy, including outsourcing of jobs; the desire for greater hiring flexibility and the continued popularity of downsizing in both the private and public sectors; the use of new technologies as labour replacements; shrinking employment in the goods-producing sector; polarization within the growing service sector; a decline in the strength of labour unions; and governments distancing themselves from their 20th-century role of labour market regulators. Together, these trends are creating a more segmented labour market in terms of income, benefits, job security, and skill requirements, and increased anxiety among Canadians about their overall economic security (Lowe 2000).

Overlaid on the trend toward a more polarized labour market are inequities based on gender, age, race and ethnicity, disability, and region. With respect to the latter, some regions of the country remain underdeveloped, largely bypassed by the economic growth experienced by others. We have already commented on the labour market advantages enjoyed by baby boomers in contrast to the generations following them into the Canadian labour market. The lack of opportunities, outright discrimination, and other employment barriers faced by women, people with disabilities, racial and ethnic minorities, and Aboriginal Canadians are still very evident in Canadian society. When we look back at the long and difficult battles fought by organized labour to obtain the employment rights and standard of living we take for granted today, we are reminded that social inequities such as these are not easily eliminated. As we have argued at various points throughout this book, there is clearly a need for activist governments, both federally and provincially, to tighten and improve employment standards (e.g., minimum wages, working hours, health and safety issues), and to revamp employment insurance, pension, child-care, and other work-related policies and programs so that more Canadian workers benefit. The Canadian economy would benefit as well, since workers who are healthy and secure in decent jobs are likely to be more productive. Equally important, there is a continued need for a strong organized labour movement and for socially and environmentally responsible corporate leaders.

There are a few signs of positive change. A small number of employers have begun to take their community and social responsibilities more seriously. In particular, some more enlightened employers have begun to address work–life balance problems faced by workers and to support workplace wellness initiatives. Some Canadian unions have continued to play a strong role in promoting workers' rights and in seeking ways to improve the quality of jobs and reduce workplace inequalities. Some politicians are beginning to recognize the vital role of caregiving in our society, and the negative consequences, for the economy and society as a whole, of growing social inequality.

Over the past few decades, women have made progress in their quest for labour market equality. Acknowledging the present flaws of employment equity, pay equity, and work–family policies, their very existence has begun a slow process of institutional change which has benefited many, though not all, women. Although members of visible minorities continue to encounter labour market barriers, there is evidence that discrimination against them is becoming less acceptable. Unfortunately, there is also evidence that recent immigrants to Canada, most of whom are non-white, are not faring as well in the labour market as did immigrants several decades earlier. Even more troublesome is the continued extremely disadvantaged situation of Canada's First Nations. The levels of unemployment and underemployment in Aboriginal communities are simply unacceptable. Recently, governments and employers have (once again) begun to concern themselves with "workforce diversity" issues, which could lead to improved opportunities for racial and ethnic minorities. But expressed concerns do not necessarily lead to actions and, sometimes, even allow those in positions of authority to avoid actions. Critics also point out that relatively low unemployment rates, and the resulting greater competition for skilled workers in some sectors of the economy and some regions of the country, may lie behind these heightened concerns. If so, will these concerns about increasing workforce diversity be maintained if the economy slows down?

Our examination of new management approaches within Canadian work organizations also identifies conflicting trends. On the one hand, we have documented how downsizing and an over-reliance on nonstandard workers have frequently been the substitute for innovative approaches to increasing productivity. Similarly, despite their promise for humanizing work and empowering workers, new management schemes have frequently been used to push workers harder with no compensation in terms of autonomy, skill enhancement, or improved incomes. Yet, on the other hand, some of the management literature advocates major workplace reforms that put people first, and there are positive examples of the implementation of such reforms. Furthermore, there are important alternatives, including legislated industrial democracy and incentives to encourage worker ownership, that could lead to more worker autonomy, increased job security, and reduced social inequality. The challenge will be to convince employers of the benefits of real workplace reform, to assist workers in their efforts to obtain it, and to encourage governments to remove legislative and other barriers to such improvements.

Canadian unions have reacted to new management schemes with some uncertainty. Some have strongly resisted change while others have been willing to accept it, so long as job and income security have been maintained. Others have insisted on being active partners in the change process, in return for their cooperation. Where equal-partner status has been obtained, and where real consultation took place, the outcomes have often been positive. Recognizing that unions have been responsible for many of the gains made by working Canadians over the past century, it is extremely important, in our opinion, to maintain and strengthen the labour legislation that has allowed this to happen.

Conclusion

Even in the new "knowledge economy," there are large numbers of low-wage, low-skilled workers who have virtually no bargaining power or employment security. Canada's working poor could clearly benefit from a stronger union presence in their workplaces.

The issues we have been raising here are ultimately about our societal goals. Is economic competitiveness at any social cost our objective? Or would we prefer that the primary goal be a better quality of life, achieved through a reasonable standard of living, employment security, and improved opportunities for personally rewarding work for as many Canadians as possible? If it is the latter, and we strongly believe it should be, then a political response is also needed. As we have argued at various places in this book, a "let the market decide" approach (the position taken by most provincial and federal governments during the past two decades) is unlikely to assist us in reaching these social goals.

So what is the alternative? For a start, no single group—particularly employers—should have a monopoly on solutions. We believe that we have clearly documented the advantages of more collaborative union–management relationships, and this approach could be extended more widely to include governments, industry, employees, and community organizations. Pressing issues such as income security, removing barriers to job-related training, flexible work designs, workplace equity and diversity, and sustainable economic development along with environmental sustainability, must be addressed in such multi-stakeholder forums. In a sense, the ingredients of what used to be called *industrial policy*—economic growth, innovation, productivity—are now tightly intertwined with *social policy* concerns about equity, opportunity, and human development.

Robert Reich, commenting on American society 15 years ago but with observations equally relevant to Canada today, alerted us to the connection between economic globalization and polarization within national economies. In his view, if things are left as they are, we will see further inequality between nations and within industrialized capitalist nations. In his words:

> The choice is ours to make. We are no more slaves to present trends than to vestiges of the past. We can, if we choose, assert that our mutual obligations as citizens extend beyond our economic usefulness to one another, and act accordingly. (Reich 1991: 313)

We agree.

REFERENCES

Abella, Irving 1974 "Oshawa 1937." In Irving Abella, ed., *On Strike: Six Key Labour Struggles in Canada 1919–1949.* Toronto: James Lewis and Samuel.

Abu-Laban, Yasmeen, and Christina Gabriel 2002 *Selling Diversity: Immigration, Multiculturalism, Employment Equity and Globalization.* Peterborough, ON: Broadview Press.

Acker, Joan 1990 "Hierarchies, jobs, bodies: A theory of gendered organizations." *Gender & Society.* 4(2): 139–58.

Adamchak, Donald J., Shuo Chen, and Jiangtao Li 1999 "Occupations, work units and work rewards in urban China." *International Sociology* 14(4): 423–41.

Adams, Roy J. 1995 "Canadian industrial relations in comparative perspective." In Morley Gunderson and Allen Ponak, eds., *Union–Management Relations in Canada.* 3rd ed. Don Mills, ON: Addison-Wesley Publishers.

———— **2000** "Canadian industrial relations at the dawn of the 21st century— prospects for reform." *Workplace Gazette: An Industrial Relations Quarterly* 3(4): 109–15.

Adams, Roy J., and Margaret Hallock 2001 "The anti-sweatshop movement and corporate codes of conduct." *Perspectives on Work* 5(1): 15–18.

Adams, Roy J., and Parbudyal Singh 1997 "Worker rights under NAFTA: Experience with the North American Agreement on Labor Cooperation." In Rick Chaykowski, Paul-André Lapointe, Guylaine Vallée, and Anil Verma, eds., *Worker Representation in the Era of Trade and Deregulation. Selected Papers from the 33rd Annual Canadian Industrial Relations Association Conference, Brock University, St. Catharines, ON.* CIRA.

Adams, Tracy L. 1998 "Gender and women's employment in the male-dominated profession of dentistry: 1867–1917." *Canadian Review of Sociology and Anthropology* 35(1): 21–42.

———— **2000** *A Dentist and a Gentleman: Gender and the Rise of Dentistry in Ontario.* Toronto: University of Toronto Press.

———— **2003** "Professionalization, gender and female-dominated professions in Ontario." *The Canadian Review of Sociology and Anthropology* 40(3): 267–89.

Advisory Committee on Labour Management Relations in the Federal Public Service 2001 *Working Together in the Public Interest. Final Report.* Ottawa: Treasury Board of Canada.

Aguayo, Rachael 1990 *Dr. Deming: The American Who Taught the Japanese About Quality.* New York: Simon and Schuster.

Aguiar, Luís L. M. 2001 "Doing cleaning work 'scientifically': the reorganization of work in the contract building cleaning industry." *Economic and Industrial Democracy* 22(2): 239–69.

Aguren, Stefan, Christer Bredbacka, Reine Hansson, Kurt Ihregren, and K. G. Karlson 1985 *Volvo Kalmar Revisited: Ten Years of Experience.* Stockholm: Efficiency and Participation Development Council.

Ainley, Pat 1993 *Class and Skill: Changing Divisions of Knowledge and Labour.* London: Cassell Educational Limited.

Akyeampong, Ernest B. 1992 "Discouraged workers—where have they gone?" *Perspectives on Labour and Income* (Winter): 38–44.

———— **1993** "Flextime work arrangements." *Perspectives on Labour and Income* (Autumn): 17–30.

——— 1997 "Work arrangements: 1995 overview." *Perspectives on Labour and Income* (Spring): 48–52.

——— 2002 "Unionization and fringe benefits." *Perspectives on Labour and Income* 14(Autumn): 42–46.

——— 2004 "The union movement in transition." *Perspectives on Labour and Income* 16(Autumn): 39–47.

Alexander, Karl L., Doris R. Entwisle, and Carrie S. Horsey 1998 "From first grade forward: Early foundations of high school dropout." *Sociology of Education* 70(2): 87–107.

Aligisakis, Maximos 1997 "Labour disputes in Western Europe: Typology and tendencies." *International Labour Review* 136(1): 73–94.

Allen, Mary, and Chantal Vaillancourt 2004 "Class of 2000—student loans." *Canadian Social Trends (Autumn)*: 18–21.

Altman, Morris 2001 *Worker Satisfaction and Economic Performance: Microfoundations of Success and Failure.* Armonk, NY: M. E. Sharpe.

Alvesson, Mats, and Stefan Sveningsson 2003 "Managers doing leadership: The extra-ordinarization of the mundane." *Human Relations* 56(12): 1435–59.

American Federation of Labor (AFL–CIO) 1995 *What's Wrong With Right-to-Work: A Tale of Two Nations.* Rev. ed. Washington, DC: AFL–CIO.

Anand, Vikas, Blake E. Ashforth, and Mahendra Joshi 2004 "Business as usual: The acceptance and perpetuation of corruption in organizations." *Academy of Management Executive* 18(2): 39–53.

Anderson, Kay J. 1991 *Vancouver's Chinatown: Racial Discourse in Canada, 1875–1980.* Montreal and Kingston: McGill–Queen's University Press.

Anderson-Connolly, Richard, L. Grunberg, E. S. Greenberg, and S. Moore 2002 "Is lean mean? Workplace transformation and employee well-being." *Work, Employment & Society* 16(3): 389–413.

Andres, Lesley 1993 "Life trajectories, action, and negotiating the transition from high school." In Paul Anisef and Paul Axelrod, eds., *Transitions: Schooling and Employment in Canada.* Toronto: Thompson Educational Publishing.

Andres, Lesley, Paul Anisef, Harvey Krahn, Dianne Looker, and Victor Thiessen 1999 "The persistence of social structure: Cohort, class and gender effects on the occupational aspirations and expectations of Canadian youth." *Journal of Youth Studies* 2(3): 261–82.

Andres, Lesley, and Harvey Krahn 1999 "Youth pathways in articulated postsecondary systems: Enrolment and completion patterns of urban young women and men." *Canadian Journal of Higher Education* 29(1): 47–82.

Andres, Lesley, and E. Dianne Looker 2001 "Rurality and capital: Educational expectations and attainments of rural, urban/rural and metropolitan youth." *Canadian Journal of Higher Education* 31(2): 1–46.

Andrew, Caroline, Pat Armstrong, Hugh Armstrong, Wallace Clement, and Leah Vosko, eds. 2003 *Studies in Political Economy: Developments in Feminism.* Toronto: Women's Press.

Angus, Charlie, and Brit Griffin 1996 *We Lived a Life and Then Some: The Life, Death, and Life of a Mining Town.* Toronto: Between the Lines.

Angus Reid Group 1996 "The public's agenda over 1996." *The Angus Reid Report* 11(6 Nov/Dec).

Anisef, Paul, Paul Axelrod, Etta Baichman-Anisef, Carl James, and Anton Turrittin 2000 *Opportunity and Uncertainty: Life Course Experiences of the Class of '73.* Toronto: University of Toronto Press.

Anthony, P. D. 1977 *The Ideology of Work.* London: Tavistock.

References

Antoniou, Andreas, and Robin Rowley 1986 "The ownership structure of the largest Canadian corporations, 1979." *Canadian Journal of Sociology* 11: 253–68.

Antos, Joseph R., Mark Chandler, and Wesley Mellow 1980 "Sex differences in union membership." *Industrial and Labor Relations Review* 33: 162–69.

Aoyama, Yuko, and Manuel Castells 2002 "An empirical assessment of the informational society: Employment and occupational structures in G-7 countries." *International Labor Review* 141(1/2): 123–59.

Apostle, Richard, and Gene Barrett 1992 *Emptying Their Nets: Small Capital and Rural Industrialization in the Nova Scotia Fishing Industry.* Toronto: University of Toronto Press.

Apostle, Richard, D. Clairmont, and L. Osberg 1985 "Segmentation and wage determination." *Canadian Review of Sociology and Anthropology* 22: 30–56.

Appelbaum, Eileen, Thomas Bailey, Peter Berg, and Arne L. Kalleberg 2000 *Manufacturing Advantage: Why High-Performance Work Systems Pay Off.* Ithaca, NY: ILR Press.

Appelbaum, Eileen, and Rosemary Batt 1994 *The New American Workplace: Transforming Work Systems in the United States.* Ithaca, NY: ILR Press.

Arai, Bruce A. 1997 "The road not taken: The transition from unemployment to self-employment in Canada, 1961–1994." *Canadian Journal of Sociology* 22(3): 365–82.

———— **2000** "Self-employment as a response to the double day for women and men in Canada." *Canadian Review of Sociology and Anthropology* 37(2): 125–42.

Archibald, W. Peter 1978 *Social Psychology as Political Economy.* Toronto: McGraw-Hill Ryerson.

Argote, Linda 1999 *Organizational Learning: Creating, Retaining, and Transferring Knowledge.* Boston: Kluwer Academic.

Argyris, Chris 1999 "The next challenge for TQM—taking the offensive on defensive reasoning." *Journal for Quality and Participation* 22(6): 41–43.

Armstrong, Pat, and Hugh Armstrong 1983 *A Working Majority: What Women Must Do for Pay.* Ottawa: Canadian Advisory Council on the Status of Women.

———— **1990** *Theorizing Women's Work.* Toronto: Garamond Press.

———— **1994** *The Double Ghetto: Canadian Women and Their Segregated Work.* 3rd ed. Toronto: McClelland and Stewart.

Armstrong, Pat, Jacqueline Choiniere, and Elaine Day 1993 *Vital Signs: Nursing in Transition.* Toronto: Garamond Press.

Armstrong, Pat, and Mary Cornish 1997 "Restructuring pay equity for a restructured work force: Canadian perspectives." *Gender, Work and Organization* 4(2): 67–85.

Armstrong, Robin 1999 "Mapping the conditions of First Nations communities." *Canadian Social Trends* (Winter): 14–18.

Aronowitz, S. 1973 *False Promises: The Shaping of American Working Class Consciousness.* New York: McGraw-Hill.

Arsen, David D., Mark I. Wilson, and Jonas Zoninsein 1996 "Trends in manufacturing employment in the NAFTA region: Evidence of a giant sucking sound?" In Karen Roberts and Mark I. Wilson, eds., *Policy Choices: Free Trade Among NAFTA Nations,* East Lansing: Michigan State University Press.

Ashton, David 1986 *Unemployment Under Capitalism: The Sociology of British and American Labour Markets.* Brighton, England: Wheatsheaf Books.

Ashton, David N., and Johnny Sung 2002 *Supporting Workplace Learning for High Performance Working.* Geneva: International Labour Office.

References

Ashwin, Sarah 2004 "Social partnership or a complete sellout? Russian trade unions' responses to conflict." *British Journal of Industrial Relations* 42(1): 23–46.

Askenazy, Philippe 2001 "Innovative workplace practices and occupational injuries and illnesses in the United States." *Economic and Industrial Democracy* 22(4): 485–516.

Aston, T. H., and C. H. E. Philpin, eds. 1985 *The Brenner Debate: Agrarian Class Structure and Economic Development in Pre-Industrial Europe.* Cambridge: Cambridge University Press.

Atkinson, J. 1985 "Flexibility: Planning for an uncertain future." *Manpower Policy and Practice* 1(Summer): 26–29.

Attewell, Paul 1987 "The deskilling controversy." *Work and Occupations* 14: 323–46.

Avery, Donald H. 1995 *Reluctant Host: Canada's Response to Immigrant Workers, 1896–1994.* Toronto: McClelland and Stewart.

Babcock, Robert H. 1974 *Gompers in Canada: A Study in American Continentalism Before the First World War.* Toronto: University of Toronto Press.

Bachmann, Kimberley 2000 *More Than Just Hard Hats: Creating Healthier Work Environments.* Ottawa: Conference Board of Canada.

Bacon, Nicholas, Mike Wright, and Natalia Demina 2004 "Management buyouts and human resource management." *British Journal of Industrial Relations* 42(2): 325–47.

Baer, Douglas 2004 "Educational credentials and the changing occupational structure." In James Curtis, Edward Grabb, and Neil Guppy, eds., *Social Inequality in Canada: Patterns, Problems, and Policies.* 4th ed. Toronto: Pearson/Prentice Hall.

Baer, Douglas E., Edward Grabb, and William A. Johnston 1991 "Class, crisis and political ideology in Canada: Recent trends." *Canadian Review of Sociology and Anthropology* 24: 1–22.

Bailey, Thomas, and Annette D. Bernhard 1997 "In search of the high road in a low-wage industry." *Politics and Society* 25(2): 179–201.

Bailyn, Lotte, and John T. Lynch 1983 "Engineering as a life-long career: Its meaning, its satisfactions, its difficulties." *Journal of Occupational Behaviour* 4: 263–83.

Bain, George S. 1978 *Union Growth and Public Policy in Canada.* Ottawa: Labour Canada.

Bain, George S., and H. A. Clegg 1974 "A strategy for industrial relations research in Britain." *British Journal of Industrial Relations* 12: 91–113.

Bakan, Abigail, and Audrey Kobayashi 2000 *Employment Equity Policy in Canada: An Interprovincial Comparison.* Ottawa: Status of Women Canada.

Baldi, Stéphane, and Debra Branch McBrier 1997 "Do the determinants of promotion differ for blacks and whites? Evidence from the U.S. labor market." *Work and Occupations* 24(4): 478–97.

Bamber, Greg J., and Russell D. Lansbury 1998 "An introduction to international and comparative employment relations." In Greg J. Bamber and Russell D. Lansbury, eds., *International and Comparative Employment Relations.* 3rd ed. Thousand Oaks, CA: Sage Publications.

Bamber, Greg J., Russell D. Landsbury, and Nick Wailes, eds. 2004 *International and Comparative Employment Relations: Globalisation and the Developed Market Economies.* London: Sage.

Barling, Julian, Clive Fullagar, and E. K. Kelloway 1992 *The Union and Its Members: A Psychological Approach.* New York: Oxford University Press.

Barling, Julian, E. Kevin Kelloway, and Michael Frone, eds. 2005 *Handbook of Work Stress.* Thousand Oaks, CA: Sage.

References

Barlow, Maude, and Heather-Jane Robertson 1994 *Class Warfare: The Assault on Canada's Schools.* Toronto: Key Porter Books.

Barnard, Chester 1938 *The Functions of the Executive.* Cambridge, MA: Harvard University Press.

Barnet, Richard J., and John Cavanagh 1994 *Global Dreams: Imperial Corporations and the New World Order.* New York: Touchstone.

Barnett, William P., James N. Baron, and Toby E. Stuart 2000 "Avenues of attainment: Occupational demography and organizational careers in the California civil service." *American Journal of Sociology* 106: 88–144.

Baron, James N., and Michael T. Hannan 2001 "Labor pains: Change in organizational models and employee turnover in young, high-tech firms." *American Journal of Sociology* 4: 960–1012.

Basu, Kaushik, ed. 2004 *India's Emerging Economy: Performance and Prospects in the 1990s and Beyond.* London and Cambridge, MA: MIT Press.

Batstone, Eric, Ian Boraston, and Stephen Frenkel 1978 *The Social Organization of Strikes.* Oxford: Basil Blackwell.

Batt, Rosemary 2000 "Strategic segmentation in front-line services: Matching customers, employees and human resource systems." *International Journal of Human Resource Management* 11(3): 540–61.

Battle, Ken, and Sherri Torjman, eds. 1999 *Employment Policy Options.* Ottawa: Caledon Institute of Social Policy.

Bauder, Harold 2001 "Employment, ethnicity and metropolitan context: The case of young Canadian immigrants." *Journal of International Migration and Integration* 2(3): 315–41.

Baumol, William J., Alan S. Blinder, and Edward N. Wolff 2003 *Downsizing in America: Reality, Causes, and Consequences.* New York: Russell Sage Foundation.

Baureiss, Gunter 1987 "Chinese immigration, Chinese stereotypes, and Chinese labour." *Canadian Ethnic Studies* 19: 15–34.

Beaud, Michel 1983 *A History of Capitalism 1500–1980.* New York: Monthly Review.

Beaujot, Roderic 1997 "Parental preferences for work and childcare." *Canadian Public Policy* 23(3): 275–88.

Beaumont, P. B. 1995 *The Future of Employment Relations.* London: Sage.

Beck, J. Helen, Jeffrey G. Reitz, and Nan Weiner 2002 "Addressing systemic racial discrimination in employment: The Health Canada case and implications of legislative change." *Canadian Public Policy* 28(3): 373–94.

Becker, Brian E., Mark A. Huselid, and Dave Ulrich 2001 *The HR Scorecard: Linking People, Strategy, and Performance.* Boston: Harvard Business School Press.

Becker, Gary S. 1975 *Human Capital: A Theoretical and Empirical Analysis with Special Reference to Education.* 2nd ed. Chicago: University of Chicago Press.

Beirne, Martin, Kathleen Riach, and Fiona Wilson 2004 "Controlling business? Agency and constraint in call centre working." *New Technology, Work and Employment* 19(2): 96–109.

Bélanger, Jacques 2000 *The Influence of Employee Involvement on Productivity: A Review of Research.* Research Paper R-00-4E. Hull, QC: Applied Research Branch, Human Resources Development Canada.

Bell, Daniel 1973 *The Coming of Post-Industrial Society.* New York: Basic Books.

Bellin, Seymour S., and S. M. Miller 1990 "The split society." In Kai Erikson and Steven Peter Vallas, eds., *The Nature of Work: Sociological Perspectives.* New Haven, CT: American Sociological Association and Yale University Press.

Belt, Vicki, Ranald Richardson, and Juliet Webster 2002 "Women, social skill and interactive service work in telephone call centres." *New Technology, Work and Employment* 17(1): 20–34.

Bender, Keith A., and Peter J. Sloane 1998 "Job satisfaction, trade unions, and exit-voice revisited." *Industrial and Labor Relations Review* 51(2): 222–40.

Bendix, Reinhard 1974 *Work and Authority in Industry.* Berkeley: University of California Press.

Bennis, Warren 1999 "The end of leadership: Exemplary leadership is impossible without full inclusion, initiatives, and cooperation of followers." *Organizational Dynamics* 28(1): 71–80.

Bercuson, David J. 1974 *Confrontation at Winnipeg: Labour, Industrial Relations, and the General Strike.* Montreal and Kingston: McGill–Queen's University Press.

Berg, Maxine 1988 "Women's work, mechanization and the early phases of industrialization in England." In R. E. Pahl, ed., *On Work: Historical, Comparative and Theoretical Approaches.* Oxford: Basil Blackwell.

Berg, Peter 1999 "The effects of high performance work practices on job satisfaction in the United States steel industry." *Relations industrielles/Industrial Relations* 54(1): 111–34.

Berle, Adolf A., and Gardiner C. Means 1968 *The Modern Corporation and Private Property.* Rev. ed. New York: Harcourt, Brace and World. (Orig. pub. 1932.)

Bernhardt, Annette, Martina Morris, and Mark S. Handcock 1995 "Women's gains or men's losses? A closer look at the shrinking gender gap in earnings." *American Journal of Sociology* 101: 302–28.

Bernstein, Jared, and Lawrence Mishel 1997 "Has wage inequality stopped growing?" *Monthly Labor Review* (December): 3–16.

Bernstein, Paul 1997 *American Work Values: Their Origin and Development.* Albany: State University of New York Press.

Betcherman, Gordon 2000 "Structural unemployment: How important are labour market policies and institutions?" *Canadian Public Policy* (July Supplement): S131–40.

Betcherman, Gordon, and Norm Leckie 1997 *Youth Employment and Education Trends in the 1980s and 1990s.* Ottawa: Canadian Policy Research Networks (CPRN) Working Paper No. W03.

Betcherman, Gordon, and Graham Lowe 1997 *The Future of Work in Canada: A Synthesis Report.* Ottawa: Canadian Policy Research Networks.

Betcherman, Gordon, Kathryn McMullen, and Katie Davidman 1998 *Training for the New Economy: A Synthesis Report.* Ottawa: Canadian Policy Research Networks.

Betcherman, Gordon, Kathryn McMullen, Norm Leckie, and Christina Caron 1994 *The Canadian Workplace in Transition.* Kingston, ON: IRC Press.

Beynon, H. 1984 *Working for Ford.* 2nd ed. Harmondsworth, England: Penguin.

Bielby, Denise B., and William T. Bielby 1988 "She works hard for her money: Household responsibilities and the allocation of work effort." *American Journal of Sociology* 93: 1031–59.

Birdi, Kamal, Peter Warr, and Andrew Oswald 1995 "Age differences in three components of employee well-being." *Applied Psychology: An International Review* 44: 345–73.

Birdsall, Nancy, and Carol Graham, eds. 2000 *New Markets, New Opportunities? Economic and Social Mobility in a Changing World.* Washington, DC: Brookings Institution Press.

Black, Julie, and Yvonne Stanford 2005 "When Martha and Henry are poor: The poverty of Alberta's social assistance programs." In Trevor Harrison, ed., *The Return of the Trojan Horse: Alberta and the New World (Dis)Order.* Montreal: Black Rose Books.

References

Black, Sandra E., and Lisa M. Lynch 2000
*What's Driving the New Economy? The
Benefits of Workplace Innovation.*
Cambridge, MA: National Bureau of
Economic Research. NBER Working
Paper No. W7479.

**Blackburn, Robert M., Jennifer Jarman,
and Janet Siltanen 1993** "The analysis of
occupational gender segregation over time
and place: Considerations of measurement
and some new evidence." *Work,
Employment & Society* 7: 335–62.

**Blackburn, R. M., and Michael Mann
1979** *The Working Class in the Labour
Market.* London: Macmillan.

**Blau, Francine D., and Marianne A.
Ferber 1986** *The Economics of Women,
Men and Work.* Englewood Cliffs, NJ:
Prentice-Hall.

Blau, Peter M. 1963 *The Dynamics of
Bureaucracy.* 2nd ed. Chicago: University
of Chicago Press.

**Blau, Peter M., and Otis Dudley Duncan
1967** *The American Occupational
Structure.* New York: John Wiley and Sons.

Blau, Peter M., and W. Richard Scott 1963
*Formal Organizations: A Comparative
Approach.* London: Routledge and Kegan
Paul.

Blauner, Robert 1964 *Alienation and
Freedom: The Factory Worker and His
Industry.* Chicago: University of Chicago
Press.

Bleasdale, Ruth 1981 "Class conflict on the
canals of Upper Canada in the 1840s."
Labour/Le Travail 7: 9–39.

**Blishen, Bernard R., W. K. Carroll, and C.
Moore 1987** "The 1981 socioeconomic
index for occupations in Canada."
*Canadian Review of Sociology and
Anthropology* 24: 465–88.

Block, Fred 1990 *Postindustrial Possibilities:
A Critique of Economic Discourse.*
Berkeley: University of California Press.

**Block, Richard N., and Karen Roberts
2000** "A comparison of labour standards
in the United States and Canada."
Relations industrielles/Industrial Relations
55(2): 273–306.

**Bluestone, Barry, and Bennett Harrison
1982** *The Deindustrialization of America.*
New York: Basic Books.

———— **2000** *Growing Prosperity: The Battle
for Growth with Equity in the Twenty-first
Century.* Boston: Houghton.

**Bognanno, Mario F., and Kathryn J.
Ready, eds. 1993** *The North American
Free Trade Agreement: Labor, Industry, and
Government Perspectives.* Westport, CT:
Praeger.

Boivin, Jean, and Esther Déom 1995
"Labour–management relations in
Quebec." In Morley Gunderson and Allen
Ponak, eds., *Union-Management Relations
in Canada.* 3rd ed. Don Mills, ON:
Addison-Wesley Publishers.

Booth, Patricia 1997 *Contingent Work:
Trends, Issues and Challenges for Employers.*
Report 192-97. Ottawa: Conference
Board of Canada.

Boulet, Jac-André, and Laval Lavallée 1984
The Changing Economic Status of Women.
Ottawa: Supply and Services Canada
(Economic Council of Canada).

Bourdieu, Pierre 1986 "The forms of capital." In J. C. Richardson, ed., *Handbook
of Theory and Research for the Sociology of
Education.* New York: Greenwood
Press.

Bowen, David E., and Cheri Ostroff 2004
"Understanding HRM-firm performance
linkages: The role of the 'strength' of the
HRM system." *Academy of Management
Review* 29(2): 203–21.

**Bowes-Sperry, Lynn, and Jasmine Tata
1999** "A multiperspective framework of
sexual harassment." In Gary Powell, ed.,
Handbook of Gender and Work. Thousand
Oaks, CA: Sage.

Bowlby, Geoff 2000 "The school-to-work
transition." *Perspectives on Labour and
Income* 12(Spring): 43–48.

References

———— 2001 "The labour market: Year-end review." *Perspectives on Labour and Income* 13(Spring): 9–19.

———— 2002 "Farmers leaving the field." *Perspectives on Labour and Income* 14(Spring): 23–28.

Boyd, Monica 1985 "Revising the stereotype: Variations in female labour force interruptions." Paper presented at the annual meetings of the Canadian Sociology and Anthropology Association and the Canadian Population Society, Montreal.

Boyd, Monica, and Doug Norris 1999 "The crowded nest: Young adults at home." *Canadian Social Trends* (Spring): 2–5.

Boyd, Monica, and Michael Vickers 2000 "100 years of immigration in Canada." *Canadian Social Trends* (Autumn): 2–12.

Boyer, Robert, and Daniel Drache, eds. 1996 *States Against Markets: The Limits of Globalization*. London: Routledge.

Boyett, Joseph H., and Henry P. Conn 1991 *Workplace 2000: The Revolution Reshaping American Business*. New York: Dutton.

Bradbury, Bettina 1993 *Working Families: Age, Gender, and Daily Survival in Industrializing Montreal*. Toronto: McClelland and Stewart.

Bradley, Harriet 1989 *Men's Work, Women's Work: A Sociological History of the Sexual Division of Labour in Employment*. Cambridge: Polity Press.

Bradwin, Edmund 1972 *The Bunkhouse Man: A Study of Work and Pay in the Camps of Canada*. Toronto: University of Toronto Press. (Orig. pub. 1928.)

Braverman, Harry 1974 *Labor and Monopoly Capital: The Degradation of Work in the Twentieth Century*. New York: Monthly Review Press.

Brenner, Mark D., David Farris, and John Ruser 2004 "'Flexible' work practices and occupational safety and health: Exploring the relationship between cumulative trauma disorders and workplace transformation." *Industrial Relations* 43(1): 232–66.

Brisbois, Richard 2003 *How Canada Stacks Up: The Quality of Work—An International Perspective*. Ottawa: Canadian Policy Research Networks, CPRN Research Paper No. W/23. (http://www.cprn.com/en/doc.cfm)

Briskin, Linda, and Patricia McDermott 1993 *Women Challenging Unions: Feminism, Democracy, and Militancy*. Toronto: University of Toronto Press.

Broad, Dave, and Wayne Antony 1999 *Citizens or Consumers? Social Policy in a Market Society*. Halifax: Fernwood Publishing.

Brockman, Joan 2001 *Gender in the Legal Profession: Fitting or Breaking the Mould*. Vancouver: University of British Columbia Press.

Brohawn, Dawn K., ed. 1997 *Journey to an Ownership Culture: Insights from the ESOP Community*. Washington, DC: The ESOP Association and Scarecrow Press.

Brown, Amanda, and Jim Stanford 2000 *Flying Without a Net: The "Economic Freedom" of Working Canadians in 2000*. Ottawa: Canadian Centre for Policy Alternatives.

Brown, Leslie H. 1997 "Organizations for the 21st century? Co-operatives and 'new' forms of organization." *Canadian Journal of Sociology* 22: 65–93.

Brown, Lorne 1987 *When Freedom Was Lost: The Unemployed, The Agitator, and the State*. Montreal: Black Rose Books.

Brown, Megan 2003 "Survival at work: Flexibility and adaptability in American corporate culture." *Cultural Studies* 17(5): 713–33.

Brown, Shona L., and Kathleen M. Eisenhardt 1998 *Competing on the Edge: Strategy as Structured Chaos*. Boston: Harvard Business School Press.

References

Bruno, Robert 1998 "Property rights or entitlements: How ESOPS influence what workers value about their unions." *Labor Studies Journal* 23(3): 55–83.

Brym, Robert 1996 "The third Rome and the end of history: Notes on Russia's second communist revolution." *Canadian Review of Sociology and Anthropology* 33: 391–406.

Bryson, Alex, Lorenzo Cappellari, and Claudio Lucifora 2004 "Does union membership really reduce job satisfaction?" *British Journal of Industrial Relations* 42(3): 439–59.

Buchanan, Ruth 2002 "Lives on the line: Low-wage work in the teleservice economy." In Frank Munger, ed., *Laboring Below the Line: The New Ethnography of Poverty, Low-Wage Work, and Survival in the Global Economy* (pp. 45–72). New York: Russell Sage.

Buchanan, Ruth, and Sarah Koch-Schulte 2000 *Gender on the Line: Technology, Restructuring and the Reorganization of Work in the Call-Centre Industry.* Ottawa: Status of Women Canada. Cat. no. SW21-44/2000E. (http://www.swc-cfc .gc.ca/pubs /index_e.html)

Buckingham, Alan 1999 "Is there an underclass in Britain?" *British Journal of Sociology* 50(1): 49–75.

Budd, John W. 2004 "Non-wage forms of compensation." *Journal of Labor Research* 25(4): 597–622.

Bunting, Madeline 2004 *Willing Slaves: How the Overwork Culture Is Ruling Our Lives.* London: HarperCollins.

Burawoy, Michael 1979 *Manufacturing Consent: Changes in the Labor Process under Monopoly Capitalism.* Chicago: University of Chicago Press.

——— **1984** "Karl Marx and the satanic mills: Factory politics under early capitalism in England, the United States, and Russia." *American Journal of Sociology* 90: 247–82.

Burchell, Brendan, David Ladipo, and Frank Wilkinson, eds. 2002 *Job Insecurity and Work Intensification.* London and New York: Routledge.

Burchell, Brendan, and Jill Rubery 1990 "An empirical investigation into the segmentation of the labour supply." *Work, Employment & Society* 4: 551–75.

Burke, Ronald J., and Mary C. Mattis 2000 *Women on Corporate Boards of Directors: International Challenges and Opportunities.* London: Kluwer Academic Publishers.

Burman, Patrick 1988 *Killing Time, Losing Ground: Experiences of Unemployment.* Toronto: Wall & Thompson.

——— **1996** *Poverty's Bonds: Power and Agency in the Social Relations of Welfare.* Toronto: Thompson Educational Publishing.

Burnham, J. 1941 *The Managerial Revolution.* Harmondsworth, England: Penguin.

Burns, T., and G. M. Stalker 1961 *The Management of Innovation.* London: Tavistock Publications.

Burris, Beverly H. 1998 "Computerization of the workplace." *Annual Review of Sociology* 24: 141–57.

Burstein, M., N. Tienharra, P. Hewson, and B. Warrander 1975 *Canadian Work Values: Findings of a Work Ethic Survey and a Job Satisfaction Survey.* Ottawa: Information Canada.

Busby, Nicole, and Sam Middlemiss 2001 "The equality deficit: Protection against discrimination on the grounds of sexual orientation in employment." *Gender, Work and Organization* 8(4): 387–410.

Byrne, Edmund F. 1990 *Work, Inc.: A Philosophical Inquiry.* Philadelphia. Temple University Press.

Calás, Marta B., and Linda Smircich 1996 "From 'the woman's point of view':

References

Feminist approaches to organization studies." In Stewart R. Clegg, Cynthia Hardy, and Walter R. Nord, eds., *Handbook of Organization Studies* (pp. 218–57). London: Sage.

Callaghan, George, and Paul Thompson **2001** "Edwards revisited: Technical control and call centres." *Economic and Industrial Democracy* 22(1): 13–37.

Calliste, Agnes 1987 "Sleeping car porters in Canada: An ethnically submerged split labour market." *Canadian Ethnic Studies* 19: 1–20.

Cameron, David, and Janice Gross Stein **2000** "Globalization, culture and society: The state as place amidst shifting spaces." *Canadian Public Policy* 24 (2nd Supplement): S15–34.

Canada 1969 *Canadian Industrial Relations: The Report of the Task Force on Labour Relations.* Ottawa: Queen's Printer.

—— **1984** *Report of the Commission on Equality in Employment* [The Abella Report]. Ottawa: Supply and Services.

—— **1985** *Employment Equity Act.* Chapter 23, 2nd Supplement, Revised Statutes of Canada.

—— **1992** *A Matter of Fairness. Report of the Social Committee on the Review of the Employment Equity Act.* Ottawa: House of Commons.

—— **2004** *Pay Equity: A New Approach to a Fundamental Right. Final Report, Pay Equity Task Force.* Ottawa: Department of Justice.

Canada, Department of Labour 1958 *Survey of Married Women Working for Pay in Eight Canadian Cities.* Ottawa: Queen's Printer.

Canadian Committee on Women in Engineering 1992 *More than Just Numbers: Report of the Canadian Committee on Women in Engineering.* Fredericton: Faculty of Engineering, University of New Brunswick.

Canadian Conference of Catholic Bishops **1991** "The crisis of work." Canadian Conference of Catholic Bishops, Work and Solidarity Project.

Canadian Council on Social Development **2001** *The Progress of Canada's Children 2001.* Ottawa: Canadian Council on Social Development. (http://www.ccsd.ca/pubs/2001/pcc2001/)

Canadian Labour Congress (CLC) 1997 *Women's Work: A Report.* Ottawa: Canadian Labour Congress.

Canadian Social Trends 1997 "Canadian children in the 1990s: Selected findings of the National Longitudinal Survey of Children and Youth." *Canadian Social Trends* 44 (Spring): 2–9.

Cant, Sarah, and Ursala Sharma 1995 "The reluctant profession: Homeopathy and the search for legitimacy." *Work, Employment & Society* 9: 743–62.

Cappelli, Peter 1992 "Is the 'skills gap' really about attitudes?" Philadelphia: National Center on the Educational Quality of the Workforce, University of Pennsylvania.

Caragata, Warren 1979 *Alberta Labour: A Heritage Untold.* Toronto: James Lorimer.

Carayon, Pascale, and Fred Zijlstra 1999 "Relationship between job control, work pressure and strain: Studies in the USA and in the Netherlands." *Work and Stress* 13(1): 32–48.

Card, David, Thomas Lemieux, and W. Craig Riddell 2004 "Unions and wage inequality." *Journal of Labor Research* 25(4): 519–62.

Carey, Alex 1967 "The Hawthorne studies: A radical criticism." *American Sociological Review* 32: 403–16.

Carley, Mark 2005 *Industrial Relations in the EU, Japan and USA, 2003–4* European Industrial Relations Observatory. (http://www.eiro.eurofound.ie)

Carré, Françoise, Marianne A. Ferber, Lonnie Golden, and Stephen A. Herzenberg, eds. 2000 *Nonstandard*

Work: The Nature and Challenge of Changing Employment Relationships. Champaign, IL: Industrial Relations Research Association.

Carroll, William K. 2004 *Corporate Power in a Globalizing World: A Study of Elite Social Organizations.* Toronto: Oxford University Press.

Carson, Paula P., Patricia A. Lanier, Kerry D. Carson, and Brandi N. Guidry 2000 "Clearing a path through the management fashion jungle: Some preliminary trailblazing." *Academy of Management Journal* 43(6): 1143–58.

Carter, Chris, and Frank Mueller 2002 "The 'long march' of the management modernizers." *Human Relations* 55(11): 1325–54.

Carter, Donald D. 1998 "Employment benefits for same sex couples: The expanding entitlement." *Canadian Public Policy* 14(1): 107–17.

Cartwright, Susan, and Cary L. Cooper 1997 *Managing Workplace Stress.* Thousand Oaks, CA: Sage.

Castells, Manuel 1996 *The Information Age: Economy, Society and Culture. Vol. I: The Rise of the Network Society.* Oxford: Blackwell.

———— **1998** *The Information Age: Economy, Society and Culture. Vol. III: End of Millennium.* Oxford: Blackwell.

Catalyst 2004 *Catalyst Census of Women Corporate Officers and Top Earners in Canada.* New York and Toronto: Catalyst.

CAW–Canada Research Group on CAMI 1993 *The CAMI Report: Lean Production in a Unionized Auto Plant.* Willowdale, ON: Canadian Auto Workers Research Department.

Centre for Studies of Aging 1996 *Issues of an Aging Workforce in a Changing Society: Cases and Comparisons.* Toronto: University of Toronto, Centre for Studies of Aging.

Chaison, Gary 2004 "Union mergers in the U.S. and abroad." *Journal of Labor Research* 25(1): 97–115.

Chaison, Gary N., and Joseph B. Rose 1990 "New directions and divergent paths: The North American labor movement in troubled times." *Proceedings of the Spring Meeting of the Industrial Relations Research Association.* Madison, WI: IRRA.

Chan, Tak W. 2000 "Revolving doors reexamined: Occupational sex segregation over the life course." *American Sociological Review* 64: 86–96.

Chandler, Alfred D., Jr. 1962 *Strategy and Structure: Chapters in the History of the American Industrial Enterprise.* Cambridge, MA: MIT Press.

———— **1977** *The Visible Hand: The Managerial Revolution in American Business.* Cambridge, MA: Harvard University Press.

Chang, Clara, and Constance Sorrentino 1991 "Union membership statistics in 12 countries." *Monthly Labor Review* (December): 46–53.

Chawla, Raj K. 1992 "The changing profile of dual-earner families." *Perspectives on Labour and Income* (Summer): 22–29.

———— **2004** "Wealth inequality by province." *Perspectives on Labour and Income* 16(Winter): 15–22.

Chaykowski, Richard 2005 *Non-Standard Work and Economic Vulnerability.* Ottawa: Canadian Policy Research Networks.

Chemers, Martin M., Stuart Oskamp, and Mark A. Costanzo, eds. 1995 *Diversity in Organizations: New Perspectives for a Changing Workplace.* Thousand Oaks, CA: Sage.

Cherns, A. 1976 "The principles of sociotechnical design." *Human Relations* 29: 783–92.

Child, John 1972 "Organizational structure, environment and performance: The role of strategic choice." *Sociology* 6: 2–22.

References

——— 1985 "Managerial strategies, new technology and the labour process." In David Knights, Hugh Willmott, and David Collison, eds., *Job Redesign: Critical Perspectives on the Labour Process.* Aldershot, England: Gower.

Chirot, Daniel 1986 *Social Change in the Modern Era.* San Diego: Harcourt Brace Jovanovich.

Chiu, Warren C. K., Andy W. Chan, Ed Snape, and Tom Redman 2001 "Age stereotypes and discriminatory attitudes towards older workers: An East–West comparison." *Human Relations* 54(5): 629–61.

Chui, Tina, Kelly Tran, and John Flanders 2005 "Chinese Canadians: Enriching the cultural mosaic." *Canadian Social Trends* (Spring): 24–32.

Chui, Tina, and Danielle Zietsma 2003 "Earnings of immigrants in the 1990s." *Canadian Social Trends* (Autumn): 24–28.

Chung, Lucy 2004 "Low-paid workers: How many live in low-income families?" *Perspectives on Labour and Income* 16(Winter): 23–32.

Church, Elizabeth 2001 "CEOs reap big bonuses as investors take losses." *The Globe and Mail* (14 June): A1.

——— 2005 "Market recovery delivers executive payout bonanza." *The Globe and Mail* (4 May): B1.

Citizenship and Immigration Canada 2005 *The Monitor* (Spring). Ottawa. (http://www.cic.gc.ca/english/monitor/issue09/05-overview.html)

Clark, Andrew, Andrew Oswald, and Peter Warr 1996 "Is job satisfaction U-shaped in age?" *Journal of Occupational and Organizational Psychology* 69: 57–81.

Clark, R. D. 1982 "Worker participation in health and safety in Canada." *International Labour Review* 121: 199–206.

Clark, Warren 1999 "Search for success: Finding work after graduation." *Canadian Social Trends* (Summer): 10–15.

——— 2000 "100 years of education." *Canadian Social Trends* (Winter): 3–7.

Clarke, Jan 2000 "Skill as a complex gendered concept: A qualitative study of women, information work, and technology." *Knowledge and Society* 12: 209–28.

Clarke, Louise, and Larry Haiven 1999 "Workplace change and continuous bargaining: Saskatoon Chemicals then and now." *Relations industrielles/Industrial Relations* 54(1): 168–91.

Clarke, Simon, Lee Chang-Hee, and Li Qi 2004 "Collective consultation and industrial relations in China." *British Journal of Industrial Relations* 42(2): 235–54.

Clarke, Thomas, and Stewart Clegg 1998 *Changing Paradigms: The Transformation of Management Knowledge for the 21st Century.* London: Harper Collins Business.

Clawson, Dan, and Mary A. Clawson 1999 "What has happened to the US labor movement?" *Annual Review of Sociology* 25: 95–119.

Clegg, Stewart R. 1990 *Modern Organizations: Organization Studies in the Postmodern World.* London: Sage.

Clegg, Stewart R., Cynthia Hardy, and Walter R. Nord, eds. 1996 *Handbook of Organization Studies.* Thousand Oaks, CA: Sage.

Clemenson, Heather A. 1992 "Are single industry towns diversifying? A look at fishing, mining and wood-based communities." *Perspectives on Labour and Income* (Spring): 31–43.

Clement, Wallace 1981 *Hardrock Mining: Industrial Relations and Technological Changes at Inco.* Toronto: McClelland and Stewart.

——— 1986 *The Struggle to Organize: Resistance in Canada's Fishery.* Toronto: McClelland and Stewart.

Clement, Wallace, and John Myles 1994 *Relations of Ruling: Class and Gender in*

Postindustrial Societies. Montreal and Kingston: McGill–Queen's University Press.

Cleveland, Gordon, Morley Gunderson, and Douglas Hyatt 2003 "Union effects in low-wage services: Evidence from Canadian childcare." *Industrial and Labor Relations Review* 56(2): 295–305.

Coates, Mary Lou, David Arrowsmith, and Melanie Courchene 1989 *The Canadian Industrial Relations Scene in Canada 1989: The Labour Movement and Trade Unionism Reference Tables.* Kingston, ON: Industrial Relations Centre, Queen's University.

Cobb, Clifford, Ted Halstead, and Jonathan Rowe 1995 "If the GDP is up, why is America down?" *Atlantic Monthly* (October): 60–74.

Cockburn, Cynthia 1991 *In the Way of Women: Men's Resistance to Sex Equality in Organizations.* Ithaca, NY: ILR Press.

Cohen, Marjorie Griffen 1988 *Women's Work, Markets and Economic Development in Nineteenth Century Ontario.* Toronto: University of Toronto Press.

Colling, Herb 1995 *Ninety-Nine Days: The Ford Strike in Windsor, 1945.* Toronto: NC Press.

Collins, James C. 2001 *Good to Great: Why Some Companies Make the Leap and Others Don't.* New York: HarperBusiness.

Collins, Randall 1990 "Market closure and the conflict theory of professions." In Michael Burrage and Rolf Torstendahl, eds., *Professions in Theory and History: Rethinking the Study of the Professions.* London: Sage.

Collinson, David, and Stephen Ackroyd 2005 "Resistance, misbehaviour, and dissent." In Stephen Ackroyd, R. Batt, P. Thompson, and P. S. Tolbert, eds., *The Oxford Handbook of Work and Organization.* Oxford: Oxford University Press.

Coltrane, Scott 1996 *Family Man: Fatherhood, Housework, and Gender Equity.* New York: Oxford University Press.

——— **2000** "Research on household labor: Modeling and measuring the social embeddedness of routine family work." *Journal of Marriage and the Family* 62(4): 1208–33.

Comish, Shaun 1993 *The Westray Tragedy: A Miner's Story.* Halifax: Fernwood Publishing.

Commission for Labor Cooperation 2000 *Standard and Advanced Practices in the North American Garment Industry.* Washington, DC: Secretariat of the Commission for Labor Cooperation.

——— **2003** *North American Labor Markets: Main Changes Since NAFTA.* Washington, DC: Secretariat of the Commission for Labor Cooperation.

Conference Board of Canada 1993 *Employability Skills Profile.* Ottawa: Conference Board of Canada.

——— **2000** *What To Do Before The Well Runs Dry: Managing Scarce Skills.* Ottawa: Conference Board of Canada.

Conley, James R. 1988 "More theory, less fact? Social reproduction and class conflict in a sociological approach to working-class history." *Canadian Journal of Sociology* 13: 73–102.

Conrad, Peter 1987 "Wellness in the workplace: Potentials and pitfalls of work-site health promotion." *Milbank Quarterly* 65: 255–75.

Conway, Hugh, and Jens Svenson 1998 "Occupational injury and illness rates, 1992–96: Why they fell." *Monthly Labor Review* (November): 36–58.

Cooke-Reynolds, Melissa, and Nancy Zukewich 2004 "The feminization of work." *Canadian Social Trends* (Spring): 24–29.

Co-operatives Secretariat 2000 *Worker Co-operative Success Stories.* Ottawa: Government of Canada, Co-operatives Secretariat.

Copp, Terry 1974 *The Anatomy of Poverty: The Condition of the Working Class in Montreal, 1897–1929.* Toronto: McClelland and Stewart.

Corak, Miles, ed. 1998 *Labour Markets, Social Institutions, and the Future of Canada's Children.* Ottawa: Statistics Canada and Human Resources Development Canada.

Cornfield, Daniel B. 2001 "Shifts in public approval of labor unions in the United States, 1936–1999." *Guest Scholar Poll Review.* The Gallup Organization. (http://www.gallup.com)

Coser, Lewis A. 1971 *Masters of Sociological Thought: Ideas in Historical and Social Context.* New York: Harcourt Brace Jovanovich.

Côté, James 2000 *Arrested Adulthood: The Changing Nature of Maturity and Identity.* New York: New York University Press.

Coupland, Douglas 1993 *Generation X: Tales for an Accelerated Culture.* New York: St. Martin's Press.

Coverman, Shelley 1988 "Sociological explanations of the male–female wage gap: Individual and structuralist theories." In Ann Helton Stromberg and Shirley Harkess, eds., *Women Working: Theories and Facts in Perspective.* 2nd ed. Mountain View, CA: Mayfield.

Cranford, Cynthia J. 2004 "Gendered resistance: Organizing justice for janitors in Los Angeles." In Jim Stanford and Leah F. Vosko, eds., *Challenging the Market: Struggles to Regulate Work and Income* (pp. 309–29). Montreal and Kingston: McGill–Queen's University Press.

Cranswick, Kelly 2003 *General Social Survey Cycle 16: Caring for an Aging Society.* Ottawa: Statistics Canada. Cat. no. 89-582-XIE.

Craven, Paul 1980 *An Impartial Umpire: Industrial Relations and the Canadian State, 1900–1911.* Toronto: University of Toronto Press.

Creese, Gillian 1988–89 "Exclusion or solidarity? Vancouver workers confront the oriental problem." *B.C. Studies* 80: 24–51.

———— **1999** *Contracting Masculinity: Gender, Class, and Race in a White-Collar Union, 1944–1994.* Don Mills, ON: Oxford University Press.

Creese, Gillian, Neil Guppy, and Martin Meissner 1991 *Ups and Downs on the Ladder of Success: Social Mobility in Canada.* Ottawa: Statistics Canada, General Social Survey Analysis Series 5. Cat. no. 11-612E, no. 5.

Creese, Gillian, and Edith Ngene Kambere 2003 "What colour is your English." *The Canadian Review of Sociology and Anthropology* 40(5): 565–73.

Crocker, Diane, and Valery Kalemba 1999 "The incidence and impact of women's experiences of sexual harassment in Canadian workplaces." *Canadian Review of Sociology and Anthropology* 36(4): 541–55.

Crompton, Susan 1994 "Left behind: Lone mothers in the labour market." *Perspectives on Labour and Income* (Summer): 23–28.

———— **2002** "I still feel over-qualified for my job." *Canadian Social Trends* (Winter): 23–26.

Crompton, Susan, and Leslie Geran 1995 "Women as main wage-earners." *Perspectives on Labour and Income* (Winter): 26–29.

Crompton, Susan, and Michael Vickers 2000 "One hundred years of labour force." *Canadian Social Trends* (Summer): 2–14.

Crouch, Colin 1982 *Trade Unions: The Logic of Collective Action.* Glasgow: Fontana.

Crozier, Michel 1964 *The Bureaucratic Phenomenon.* Chicago: University of Chicago Press.

Culbert, Samuel A., and Scott J. Schroeder 2003 "Getting hierarchy to work." In Subir Chowdhury, ed., *Organization 21C: Someday All Organizations Will Lead This*

Way (pp. 105–23). Upper Saddle River, NJ: Financial Times Prentice Hall.

Daft, Richard L. 1995 *Organization Theory and Design.* 5th ed. Minneapolis/St. Paul, MN: West Publishing.

Dalton, Melville 1959 *Men Who Manage: Fusions of Feeling and Theory in Administration.* New York: John Wiley and Sons.

Danford, Andy, M. Richardson, P. Stewart, S. Tailby, and M. Upchurch 2004 "High performance work systems and workplace partnership: A case study of aerospace workers." *New Technology, Work and Employment* 19(3): 14–29.

Danysk, Cecilia 1995 *Hired Hands: Labour and the Development of Prairie Agriculture, 1880 to 1930.* Don Mills, ON: Oxford University Press.

Das Gupta, Tania 1996 *Racism and Paid Work.* Toronto: Garamond Press.

Davies, Lorraine, and Patricia J. Carrier 1999 "The importance of power relations for the division of household labour." *Canadian Journal of Sociology* 24(1): 35–51.

Davies, Scott 2004 "Stubborn disparities: Explaining class inequalities in schooling." In James Curtis, Edward Grabb, and Neil Guppy, eds., *Social Inequality in Canada: Patterns, Problems, and Policies.* 4th ed. Toronto: Pearson/Prentice Hall.

Davies, Scott, and Neil Guppy 1997 "Fields of study, college selectivity, and student inequalities in higher education." *Social Forces* 75(4): 1417–38.

———— **1998** "Race and Canadian education." In Vic Satzewich, ed., *Racism and Social Inequality in Canada: Concepts, Controversies and Strategies of Resistance.* Toronto: Thompson Educational Publishing.

Davies, Scott, Clayton Mosher, and Bill O'Grady 1996 "Educating women: Gender inequalities among Canadian university graduates." *Canadian Review of Sociology and Anthropology* 33: 125–42.

Davis, Bob 2000 *Skills Mania: Snake Oil in Our Schools?* Toronto: Between the Lines.

Davis, Kingsley, and Wilbert E. Moore 1945 "Some principles of stratification." *American Sociological Review* 10: 242–49.

Davis, Louis E., and Charles Sullivan 1980 "A labour–management contract and quality of working life." *Journal of Occupational Behaviour* 1: 29–41.

de Broucker, Patrice, and Laval Lavallée 1998 "Getting ahead in life: Does your parents' education count?" *Canadian Social Trends* (Summer): 11–15.

Deitch, Elizabeth A., A. Barsky, R. M. Butz, S. Chan, A. P. Brief, and J. C. Bradley 2003 "Subtle yet significant: The existence and impact of everyday racial discrimination in the workplace." *Human Relations* 56(1): 1299–1324.

de Jonge, Jan, Hans Bosma, Richard Peter, and Johannes Siegrist 2000 "Job strain, effort-reward imbalance and employee well-being: A large-scale cross-sectional study." *Social Science and Medicine* 50(9): 1317–27.

DeLong, David W. 2004 *Lost Knowledge: Confronting the Threat of an Aging Workforce.* New York: Oxford University Press.

De Santis, Solange 1999 *Life on the Line: One Woman's Tale of Work, Sweat, and Survival.* New York: Doubleday.

de Silva, Arnold 1999 "Wage discrimination against natives." *Canadian Public Policy* 25(1): 65–83.

Devane, Tom 2004 *Integrating Lean Six Sigma and High-Performance Organizations: Leading the Charge Toward Dramatic, Rapid and Sustainable Improvement.* San Francisco: Pfeiffer.

de Vaus, David, and Ian McAllister 1991 "Gender and work orientation: Values and satisfaction in Western Europe." *Work and Occupations* 18: 72–93.

De Witte, Hans, and Katharina Näswall 2003 "'Objective' vs 'subjective' job insecurity: Consequences of temporary work for job satisfaction and organizational commitment in four European countries." *Economic and Industrial Democracy* 24(2): 149–88.

de Witte, Marco, and Bram Steijn 2000 "Automation, job content, and underemployment." *Work, Employment & Society* 14(2): 245–64.

Dickson, Tony, and Hugh V. McLachlan 1989 "In search of the spirit of capitalism: Weber's misinterpretation of Franklin." *Sociology* 23: 81–89.

DiGiacomo, Gordon 1999 "Aggression and violence in the workplace." *Workplace Gazette: An Industrial Relations Quarterly* 2(2): 72–85.

Diprete, Thomas A., and Whitman T. Soule 1988 "Gender and promotion in segmented job ladder systems." *American Sociological Review* 53: 26–40.

Dixon, Nancy M. 2000 *Common Knowledge: How Companies Thrive by Sharing What They Know.* Boston: Harvard Business School Press.

Dohm, Arlene 2000 "Gauging the labour force effects of retiring baby boomers." *Monthly Labor Review* (July): 17–25.

Donaldson, Lex 1995 *In Defence of Organization Theory: A Reply to Critics.* Cambridge: Cambridge University Press.

—— **1996** *For Positivist Organization Theory.* London: Sage.

Doucet, Andrea 2001 "You see the need perhaps more clearly than I have: Exploring gendered processes of domestic responsibility." *Journal of Family Issues* 22(3): 328–57.

Doucouliagos, Chris 1995 "Worker participation and productivity in labor-managed and participatory capitalist firms: A meta-analysis." *Industrial and Labor Relations Review* 49(1): 58–77.

Downie, Bryan, and Mary L. Coates 1995 "Barriers, challenges, and future directions." In Bryan Downie and Mary

L. Coates, eds., *Managing Human Resources in the 1990s and Beyond: Is the Workplace Being Transformed?* Kingston, ON: IRC Press.

Doyle, Joyce 1980 *Our Family History.* Unpublished family history.

Drache, Daniel 1994 "Lean production in Japanese auto transplants in Canada." *Canadian Business Economics* (Spring): 45–59.

Drache, Daniel, and Meric S. Gertler 1991 "The world economy and the nation-state: The new international order." In D. Drache and M. S. Gertler, eds., *The New Era of Global Competition: State Policy and Market Power.* Montreal and Kingston: McGill–Queen's University Press.

Drolet, Marie 2002a "The male-female wage gap." *Perspectives on Labour and Income* 14(Spring): 29–37.

—— **2002b** "New evidence on gender-pay differentials: Does measurement matter?" *Canadian Public Policy* 28(1): 1–16.

Drolet, Marie, and René Morissette 2002 "Better jobs in the new economy?" *Perspectives on Labour and Income* 14(Autumn): 47–58.

Drost, Helmar 1996 "Joblessness among Canada's Aboriginal peoples." In Brian MacLean and Lars Osberg, eds., *The Unemployment Crisis: All for Nought?* Montreal and Kingston: McGill–Queen's University Press.

Drucker, Peter F. 1993 *Post-Capitalist Society.* New York: HarperBusiness.

—— **1999** "Knowledge-worker productivity: The biggest challenge." *California Management Review* 41(2): 79–94.

Dubé, Vincent 2004 "Sidelined in the labour market." *Perspectives on Labour and Income* 16(Summer): 25–31.

Duchesne, Doreen 2004 "More seniors at work." *Perspectives on Labour and Income* 16(Spring): 55–67.

Dudley, Kathryn Marie 1994 *The End of the Line: Lost Jobs, New Lives in*

Postindustrial America. Chicago: University of Chicago Press.

Duffy, Ann, and Norene Pupo 1992 *Part-Time Paradox: Connecting Gender, Work and Family*. Toronto: McClelland and Stewart.

Duncan, Greg J., Martha S. Hill, and Saul D. Hoffman 1988 "Welfare dependence within and across generations." *Science* (January): 467–71.

Duncan, Greg. J., W. Jean Yeung, Jeanne Brooks-Gunn, and Judith R. Smith 1998 "How much does childhood poverty affect the life chances of children?" *American Sociological Review* 63: 406–23.

Dunlop, John T. 1971 *Industrial Relations Systems*. Carbondale: Southern Illinois University Press. (Orig. pub. 1958.)

Dunne, Gillian 1996 *Lesbian Lifestyles: Women's Work and the Politics of Sexuality*. Toronto: University of Toronto Press.

Durkheim, Émile 1960 *The Division of Labour in Society*. New York: Free Press. (Orig. pub. 1897.)

Duxbury, Linda, Lorraine Dyke, and Natalie Lam 2000 *Managing High Technology Employees*. Scarborough, ON: Carswell.

Duxbury, Linda, and Christopher Higgins 2001 *Work–Life Balance in the New Millennium*. Ottawa: Canadian Policy Research Networks.

Easterby-Smith, Mark, John Burgoyne, and Luis Araujo, eds. 2001 *Organizational Learning and the Learning Organization: Developments in Theory and Practice*. London: Sage.

Easterlin, Richard 1980 *Birth and Fortune: The Impact of Numbers on Personal Welfare*. New York: Basic Books.

Economic Council of Canada 1990 *Good Jobs, Bad Jobs: Employment in the Service Economy*. Ottawa: Supply and Services Canada.

————— **1992** *Pulling Together: Productivity, Innovation and Trade*. Ottawa: Supply and Services Canada.

Economist, The 1996 "Cultural explanations: The man in the Baghdad café." (9 November): 23–26.

————— **2005** "The tiger in front—India." (3 March): 3–5.

Edwards, Richard C. 1979 *Contested Terrain: The Transformation of the Workplace in the Twentieth Century*. New York: Basic Books.

Ehrenreich, Barbara 2001 *Nickel and Dimed: On (Not) Getting By in America*. New York: Henry Holt and Company.

Elliot, James R., and Ryan A. Smith 2004 "Race, gender and workplace power." *American Sociological Review* 69: 365–86.

Emery, F. E., and Einar Thorsrud 1969 *Form and Content in Industrial Democracy*. London: Tavistock Publications.

Engels, Friedrich 1971 *The Condition of the Working Class in England*. Oxford: Basil Blackwell. (Orig. pub. 1845.)

Erikson, Kai 1990 "On work and alienation." In Kai Erikson and Steven Peter Vallas, eds., *The Nature of Work: Sociological Perspectives*. New Haven, CT: American Sociological Association and Yale University Press.

Esping-Anderson, Gøsta 2004 "Unequal opportunities and the mechanisms of social inheritance." In Miles Corak, ed., *Generational Income Mobility in North America and Europe*. Cambridge: Cambridge University Press.

Etzioni, Amitai 1975 *A Comparative Analysis of Complex Organizations*. 2nd. ed. New York: Free Press.

Evans, Peter, and James E. Rauch 1999 "Bureaucracy and growth: A cross-national analysis of the efforts of 'Weberian' state structures on economic growth." *American Sociological Review* 64: 748–65.

References

Evetts, Julia 2003 "The sociological analysis of professionalism: Occupational change in the modern world." *International Sociology* 18(2): 395–415.

Ewan, Stuart 1976 *Captains of Consciousness: Advertising and the Social Roots of the Consumer Culture.* New York: McGraw-Hill.

Expert Panel on Skills 2000 *Stepping Up: Skills and Opportunities in the Knowledge Economy.* Ottawa: Advisory Council on Science and Technology. Cat. no. C2-467/2000. (http://acst-ccst.gc.ca)

Fairris, David, and Edward Levine 2004 "Declining union density in Mexico, 1984–2000." *Monthly Labor Review* 127(9): 10–17.

Farber, Henry S., and Daniel H. Saks 1980 "Why workers want unions: The role of relative wages and job characteristics." *Journal of Political Economy* 88: 349–69.

Fast, Janet E., and Judith A. Frederick 1996 "Working arrangements and time stress." *Canadian Social Trends* 43 (Winter): 14–19.

Feather, Norman T. 1990 *The Psychological Impact of Unemployment.* New York: Springer-Verlag.

Federal, Provincial and Territorial Ministers Responsible for Social Services 2000 *In Unison 2000: Persons with Disabilities in Canada.* Ottawa: Human Resources Development Canada. Cat. no. SP-182-01-01E.

Feldberg, Roslyn, and Evelyn Nakano Glenn 1979 "Male and female: Job versus gender models in the sociology of work." *Social Problems* 26: 524–38.

Feldman, Daniel C., Helen I. Doerpinghaus, and William H. Turnley 1994 "Managing temporary workers: A permanent HR challenge." *Organizational Dynamics* 23: 49–63.

Felstead, Alan, Nick Jewson, and Sally Walters 2003 "Managerial control of employees working at home." *British*
Journal of Industrial Relations* 41(2): 241–64.

Felstead, Alan, Harvey Krahn, and Marcus Powell 1999 "Young and old at risk: Comparative trends in 'non-standard' patterns of employment in Canada and the United Kingdom." *International Journal of Manpower* 20(5): 277–96.

Felt, Lawrence, and Peter Sinclair 1992 "Everybody does it: Unpaid work in a rural peripheral region." *Work, Employment & Society* 6: 43–64.

Ferber, Marianne A., and Jane Waldfogel 1998 "The long-term consequences of nontraditional employment." *Monthly Labor Review* (May): 3–12.

Fernie, Sue, and David Metcalf 1998 *(Not) Hanging on the Telephone: Payment Systems in the New Sweatshop.* London: Centre for Economic Performance, London School of Economics.

Feuchtwang, Stephen 1982 "Occupational ghettos." *Economy and Society* 11: 251–91.

Figart, Deborah M., and June Lapidus 1996 "The impact of comparable worth on earnings inequality." *Work and Occupations* 23: 297–318.

Finnie, Ross 1999 "Patterns of inter-provincial migration, 1982–95." *Canadian Economic Observer* (December): 3.1–3.12.

Firebaugh, Glenn, and Brian Harley 1995 "Trends in job satisfaction in the United States by race, gender, and type of occupation." *Research in Sociology of Work* 5: 87–104.

Fisher, Cynthia D. 2003 "Why do lay people believe that satisfaction and performance are correlated? Possible sources of a commonsense theory." *Journal of Organizational Behavior* 24(6): 753–77.

Fisher, E. G. 1982 "Strike activity and wildcat strikes in British Columbia: 1945–1975." *Relations industrielles/ Industrial Relations* 37: 284–312.

References

Fisher, Susan R., and Margaret A. White 2000 "Downsizing in a learning organization: Are there hidden costs?" *Academy of Management Review* 25(1): 244–51.

Fishman, Ted 2005 *China Inc: How the Rise of the Next Superpower Challenges America and the World.* New York: Scribner.

Flanders, Allan 1970 *Management and Unions: The Theory and Reform of Industrial Relations.* London: Faber and Faber.

Florida, Richard 2002 *The Rise of the Creative Class: And How It's Transforming Work, Leisure, Community and Everyday Life.* New York: Basic Books.

———— **2005** *The Flight of the Creative Class: The New Global Competition for Talent.* New York: Harper Collins.

Foot, David K., and Jeanne C. Li 1986 "Youth employment in Canada: A misplaced priority?" *Canadian Public Policy* 12(3): 499–506.

Foot, David K., with Daniel Stoffman 1998 *Boom, Bust and Echo 2000: Profiting from the Demographic Shift in the New Millennium.* Toronto: Macfarlane Walter & Ross.

Foot, David K., and Rosemary A. Venne 1990 "Population pyramids and promotional prospects." *Canadian Public Policy* 16: 387–98.

Forester, Tom 1989 *Computers in the Human Context: Information Technology, Productivity, and People.* Cambridge, MA: MIT Press.

Form, William 1983 "Sociological research on the American working class." *Sociological Quarterly* 24: 163–84.

———— **1987** "On the degradation of skills." *Annual Review of Sociology* 13: 29–47.

Forrest, Anne 2000 "Pay equity: The state of the debate." In Yonatan Reshef, Colette Bernier, Denis Harrison, and Terry H. Wagar, eds., *Industrial Relations in a New Millennium. Selected Papers from the 37th Annual Canadian Industrial Relations Association Conference, May 25–27, 2000, Edmonton, Alberta* (pp. 65–76). CIRA.

Fortin, Pierre 1996 "The unbearable lightness of zero-inflation optimism." In Brian K. MacLean and Lars Osberg, eds., *The Unemployment Crisis: All for Nought?* (pp. 14–38). Montreal and Kingston: McGill–Queen's University Press.

Foucault, Michel 1977 *Discipline and Punish: The Birth of the Prison.* Harmondsworth, England: Penguin.

Fox, Bonnie, and Pamela Sugiman 1999 "Flexible work, flexible workers: The restructuring of clerical work in a large telecommunications company." *Studies in Political Economy* 60(Autumn): 59–84.

Frager, Ruth A. 1992 *Sweatshop Strife: Class, Ethnicity, and Gender in the Jewish Labour Movement of Toronto 1900–1939.* Toronto: University of Toronto Press.

Frank, David 1986 "Contested terrain: Workers' control in the Cape Breton coal mines in the 1920s." In Craig Heron and Robert Storey, eds., *On the Job: Confronting the Labour Process in Canada.* Montreal and Kingston: McGill–Queen's University Press.

Franke, Richard Herbert, and James D. Kaul 1978 "The Hawthorne experiments: First statistical interpretation." *American Sociological Review* 43: 623–43.

Frederick, Judith A., and Janet E. Fast 1999 "Eldercare in Canada: Who does how much?" *Canadian Social Trends* 54(Autumn): 26–30.

Freedman, Marcia 1976 *Labor Markets: Segments and Shelters.* Montclair, NJ: Allanheld, Osmun and Co.

Freeman, Bill 1982 *1005: Political Life in a Union Local.* Toronto: James Lorimer.

Freeman, Richard B. 1995 "The future for unions in decentralized collective bargaining systems: US and UK unionism in an era of crisis." *British Journal of Industrial Relations* 33: 519–36.

Freeman, Richard B., and J. L. Medoff 1984 *What Do Unions Do?* New York: Basic Books.

Freeman, Richard B., and Joel Rogers 1999 *What Workers Want.* Ithaca, NY: Cornell University Press.

Frege, Carola 2005 "The discourse of industrial democracy: Germany and the US revisited." *Economic and Industrial Democracy* 26(1): 151–75.

Freidson, Eliot 1986 *Professional Powers: A Study of the Institutionalization of Formal Knowledge.* Chicago: University of Chicago Press.

Frenette, Marc 2001 "Overqualified? Recent graduates, employer needs." *Perspectives on Labour and Income* 13(Spring): 45–53.

Frenkel, Steve, Marek Korczynski, Leigh Donoghue, and Karen Shire 1995 "Reconstituting work: Trends towards knowledge work and info-normative control." *Work, Employment & Society* 9: 773–96.

Friedman, Andrew L. 1977 *Industry and Labour: Class Struggle at Work and Monopoly Capitalism.* London: Macmillan.

Friedman, Thomas L. 2000 *The Lexus and the Olive Tree.* New York: Anchor Books.

—— **2005** *The World Is Flat: A Brief History of the 21st Century.* New York: Farrar, Strauss, Giroux.

Friendly, Martha, and Jane Beach 2005 *Early Childhood Education and Care in Canada 2004.* 6th ed. Toronto: Childcare Resource and Research Unit, University of Toronto.

Fudge, Judy, and Patricia McDermott, eds. 1991 *Just Wages: A Feminist Assessment of Pay Equity.* Toronto: University of Toronto Press.

Fudge, Judy, and Leah F. Vosko 2001 "Gender, segmentation and the standard employment relationship in Canadian labour law, legislation and policy." *Economic and Industrial Democracy* 22(2): 271–310.

Fullerton, Howard N., Jr. 1997 "Labour force 2006: Slowing down and changing composition." *Monthly Labor Review* (November): 23–38.

Furnham, Adrian 1990 *The Protestant Work Ethic: The Psychology of Work-Related Beliefs and Behaviours.* London: Routledge.

Galarneau, Diane 2005 "Earnings of temporary versus permanent employees." *Perspectives on Labour and Income* 17(Spring): 40–53.

Galarneau, Diane, and Louise Earl 1999 "Women's earnings / men's earnings." *Perspectives on Labour and Income* (Winter): 20–26.

Galarneau, Diane, Jean-Pierre Maynard, and Jin Lee 2005 "Whither the workweek?" *Perspectives on Labour and Income* 17(June): 5–17.

Galarneau, Diane, and René Morissette 2004 "Immigrants: Settling for less?" *Perspectives on Labour and Income* 16(Autumn): 7–18.

Gallagher, Daniel G., and Magnus Sverke 2005 "Contingent employment contracts: Are existing employment theories still relevant?" *Economic and Industrial Democracy* 26(2): 181–203.

Gallie, Duncan 1978 *In Search of the New Working Class: Automation and Social Integration within the Capitalist Enterprise.* Cambridge: Cambridge University Press.

—— **1983** *Social Inequality and Class Radicalism in France and Britain.* Cambridge: Cambridge University Press.

—— **1991** "Patterns of skill change: Upskilling, deskilling or the polarization of skills?" *Work, Employment & Society* 5: 319–51.

Gallie, Duncan, and Helen Russell 1998 "Unemployment and life satisfaction: A cross-cultural comparison." *Archives Européennes de Sociologie* 39(2): 248–80.

References

Gallie, Duncan, Michael White, Yuan Cheng, and Mark Tomlinson 1998 *Restructuring the Employment Relationship.* Oxford: Clarendon Press.

Gannagé, Charlene 1986 *Double Day, Double Bind: Women Garment Workers.* Toronto: Women's Press.

———— **1995** "Union women in the garment industry respond to new managerial strategies." *Canadian Journal of Sociology* 20(4): 469–95.

Gardell, B. 1977 "Autonomy and participation at work." *Human Relations* 30: 515–33.

———— **1982** "Scandinavian research on stress in working life." *International Journal of Health Services* 12: 31–41.

Gardell, B., and B. Gustavsen 1980 "Work environment research and social change: Current developments in Scandinavia." *Journal of Occupational Behavior* 1: 3–17.

Garnsey, E., J. Rubery, and F. Wilkinson 1985 "Labour market structure and work-force divisions." In R. Deem and G. Salaman, eds., *Work, Culture and Society.* Milton Keynes and Philadelphia: Open University Press.

Garvin, David A. 2000 *Learning in Action: A Guide to Putting the Learning Organization to Work.* Boston: Harvard Business School Press.

Gaskell, Jane 1992 *Gender Matters from School to Work.* Milton Keynes, England: Open University Press.

Gendell, Murray 1998 "Trends in retirement age in four countries, 1965–95." *Monthly Labor Review* (August): 20–30.

George, Jennifer M., and Gareth R. Jones 1997 "Experiencing work values, attitudes, and moods." *Human Relations* 50(4): 393–416.

Gephart, Martha A., Victoria J. Marsick, Mark E. Van Buren, and Michelle S. Spiro 1996 "Learning organizations come alive." *Training & Development* (December): 35–45.

Gera, Surendra, ed. 1991 *Canadian Unemployment: Lessons from the 80s and Challenges for the 90s.* Ottawa: Economic Council of Canada.

Gerber, Theodore P., and Michael Hout 1998 "More shock than therapy: Market transition, employment, and income in Russia, 1991–1995." *American Journal of Sociology* 104(1): 1–50.

Gerson, Kathleen 1993 *No Man's Land: Men's Changing Commitments to Family and Work.* New York: Basic Books.

Gherardi, Silvia 1999 "Learning as problem-driven or learning in the face of mystery?" *Organisation Studies* 20(1): 101–23.

Gilbert, Sid, L. Barr, W. Clark, M. Blue, and D. Sunter 1993 *Leaving School: Results from a National Survey Comparing School Leavers and High School Graduates 18 to 20 Years of Age.* Ottawa: Minister of Supply and Services.

Gilbert, Sid, and Jeff Frank 1998 "Skill deficits among the young." *Canadian Social Trends* (Winter): 17–20.

Giles, Anthony 1996 "Globalization and industrial relations." In Anthony Giles, Anthony E. Smith, and Gilles Trudeau, eds., *The Globalization of the Economy and the Worker. Selected papers from the 32nd Annual Canadian Industrial Relations Association Conference, Laval, QC.* CIRA.

———— **2000** "Industrial relations in the millennium: Beyond employment?" *Labour/Le Travail* 46(Fall): 37–67.

Gillespie, Richard 1991 *Manufacturing Knowledge: A History of the Hawthorne Experiments.* New York: Cambridge University Press.

Gindin, Sam 1995 *The Canadian Auto Workers: The Birth and Transformation of a Union.* Toronto: James Lorimer.

Gingras, Yves, and Richard Roy 2000 "Is there a skills gap in Canada?" *Canadian Public Policy* (July Supplement): S159–74.

References

Gitter, Robert J., and Markus Scheuer 1997 "U.S. and German youths: Unemployment and the transition from school to work." *Monthly Labor Review* (March): 16–20.

——— **1998** "Low unemployment in the Czech Republic: 'Miracle' or 'mirage'?" *Monthly Labor Review* (August): 31–37.

Godard, John 1994 *Industrial Relations: The Economy and Society.* Toronto: McGraw-Hill Ryerson.

——— **1997** "Managerial strategies, labour and employment relations and the state: The Canadian case and beyond." *British Journal of Industrial Relations* 35(3): 399–426.

——— **2001** "Beyond the high performance paradigm? An analysis of variation in Canadian managerial perceptions of reform programme effectiveness." *British Journal of Industrial Relations* 39(1): 25–52.

——— **2004** "A critical assessment of the high-performance paradigm." *British Journal of Industrial Relations* 42(2): 349–78.

Goldfield, Michael 1987 *The Decline of Organized Labor in the United States.* Chicago: University of Chicago Press.

Goldthorpe, John H., D. Lockwood, F. Bechhofer, and J. Platt 1969 *The Affluent Worker in the Class Structure.* Cambridge: Cambridge University Press.

Goleman, Daniel 2004 "What makes a leader?" *Harvard Business Review* (January): 82–101. (Orig. pub. in same journal 1998.)

Gordon, Brett R. 1994 "Employee involvement in the enforcement of the occupational safety and health laws of Canada and the United States." *Comparative Labor Law Journal* 15: 527–60.

Gordon, David M., R. Edwards, and M. Reich 1982 *Segmented Work, Divided Workers: The Historical Transformation of Labor in the United States.* New York: Cambridge University Press.

Gordon, Michael E., and Angelo S. Denisi 1995 "A re-examination of the relationship between union membership and job satisfaction." *Industrial and Labor Relations Review* 48: 222–36.

Gorham, Deborah 1994 "'No longer an invisible minority': Women physicians and medical practice in late twentieth-century North America." In Dianne Dodd and Deborah Gorham, eds., *Caring and Curing: Historical Perspectives on Women and Healing in Canada* (pp. 183–211). Ottawa: University of Ottawa Press.

Gorz, Andre 1982 *Farewell to the Working Class: An Essay on Post-Industrial Socialism.* London: Pluto Press.

——— **1999** *Reclaiming Work: Beyond the Wage-Based Society.* Malden, MA: Polity Press.

Gouldner, Alvin W. 1954 *Patterns of Industrial Bureaucracy.* New York: Free Press.

——— **1955** *Wildcat Strike.* New York: Free Press.

Goyder, John 2005 "The dynamics of occupational prestige: 1975–2000." *Canadian Review of Sociology and Anthropology* 42(1): 1–23.

Goyder, John, Neil Guppy, and Mary Thompson 2003 "The allocation of male and female occupational prestige in an Ontario urban area: A quarter-century replication." *Canadian Review of Sociology and Anthropology* 40(4): 417–39.

Grabb, Edward G. 2002 *Theories of Social Inequality.* 4th ed. Toronto: Harcourt Canada.

——— **2004** "Economic power in Canada: Corporate concentration, foreign ownership and state involvement." In James Curtis, Edward Grabb, and Neil Guppy, eds., *Social Inequality in Canada: Patterns, Problems, and Policies.* 4th ed. Toronto: Pearson/Prentice Hall.

Graham, Laurie 1995 *On the Line at Subaru–Isuzu: The Japanese Model and the American Worker.* Ithaca, NY: ILR Press.

Grayson, Paul J. 1994 "Perceptions of workplace hazards." *Perspectives on Labour and Income* (Spring): 41–47.

Green, Francis 2001 "It's been a hard day's night: The concentration and intensification of work in late twentieth-century Britain." *British Journal of Industrial Relations* 39(1): 53–80.

Green, Francis, and David Ashton 1992 "Skill shortages and skill deficiency: A critique." *Work, Employment & Society* 6: 287–301.

Green, William C., and Ernest J. Yanarella, eds. 1996 *North American Auto Unions in Crisis: Lean Production as Contested Terrain.* Albany: State University of New York Press.

Greenhaus, Jeffrey H., and Saroj Parasuraman 1999 "Research on work, family, and gender: Current status and future directions." In Gary Powell, ed., *Handbook of Gender and Work.* Thousand Oaks, CA: Sage Publications.

Grenon, Lee 1997 "Northern earnings and income." *Perspectives on Labour and Income* (Spring): 28–32.

Grenon, Lee, and Barbara Chun 1997 "Non-permanent paid work." *Perspectives on Labour and Income* (Autumn): 21–31.

Grimshaw, Damian, F. L. Cooke, I. Grugulis, and S. Vincent 2002 "New technology and changing organizational forms: Implications for managerial control and skills." *New Technology, Work and Employment* 14(7): 186–203.

Grint, Keith 1991 *The Sociology of Work: An Introduction.* Cambridge: Polity Press.

Gunderson, Morley 1994 *Comparable Worth and Gender Discrimination: An International Perspective.* Geneva: International Labour Office.

——— 1998 "Harmonization of labour policies under trade liberalization." *Relations industrielles/Industrial Relations* 53(1): 24–52.

Gunderson, Morley, Bob Hebdon, Douglas Hyatt, and Allen Ponak 2005 "Strikes and dispute resolution." In Morley Gunderson, Allen Ponak, and Daphne Gottlieb Taras, eds., *Union–Management Relations in Canada* (pp. 332–70). 5th ed. Toronto: Pearson Addison Wesley.

Gunderson, Morley, and Douglas Hyatt 2005 "Union impact on compensation, productivity, and management of the organization." In Morley Gunderson, Allen Ponak, and Daphne Gottlieb Taras, eds., *Union–Management Relations in Canada* (pp. 394–413). 5th ed. Toronto: Pearson Addison Wesley.

Gunderson, Morley, Douglas Hyatt, and Allen Ponak 1995 "Strikes and dispute resolution." In Morley Gunderson and Allen Ponak, eds., *Union–Management Relations in Canada.* 3rd ed. Don Mills, ON: Addison-Wesley Publishers.

Gunderson, Morley, and W. Craig Riddell 2000 "The changing nature of work: Implications for public policy." In W. Craig Riddell and France St-Hilaire, eds., *Adapting Public Policy to a Labour Market in Transition* (pp. 1–37). Montreal: Institute for Research on Public Policy.

Gunderson, Morley, Jeffrey Sack, James McCartney, David Wakely, and Jonathan Eaton 1995 "Employee buyouts in Canada." *British Journal of Industrial Relations* 33: 417–42.

Gunderson, Morley, Andrew Sharpe, and Steven Wald 2000 "Youth unemployment in Canada, 1976–1998." *Canadian Public Policy* (July Supplement): S85–S100.

Gustafsson, Bjorn, and Mats Johansson 1999 "In search of smoking guns: What makes income inequality vary over time in different countries?" *American Sociological Review* 64: 585–605.

Guzda, Henry P. 1993 "Workplace partnerships in the United States and Europe."

Monthly Labor Review (October): 67–72.

Haber, Samuel 1964 *Efficiency and Uplift: Scientific Management in the Progressive Era, 1890–1920.* Chicago: University of Chicago Press.

Habtu, Roman 2003 "Information technology workers." *Perspectives on Labour and Income* 15(Autumn): 15–21.

Hachen, David S., Jr. 1988 "Gender differences in job mobility rates in the United States." *Social Science Research* 17: 93–116.

Hagan, John, and Fiona Kay 1995 *Gender in Practice: A Study of Lawyers' Lives.* New York: Oxford University Press.

Hage, J. T. 1999 "Organizational 'innovation' and organizational change." *Annual Review of Sociology* 25: 597–622.

Hakim, Catherine 2002 "Lifestyle preferences as determinants of women's differentiated labor market careers." *Work and Occupations* 29(4): 428–59.

Hale, Thomas W., Howard V. Hayghe, and John M. McNell 1998 "Persons with disabilities: Labor market activity, 1994." *Monthly Labor Review* (September): 3–12.

Hall, Karen 1999 "Hours polarization at the end of the 1990s." *Perspectives on Labour and Income* (Summer): 28–37.

Hall, Michael, Larry McKeown, and Karen Roberts 2001 *Caring Canadians, Involved Canadians: Highlights from the 2000 National Survey of Giving, Volunteering and Participating.* Ottawa: Statistics Canada. Cat. no. 71-542-XPE.

Halpern, Norman 1984 "Sociotechnical systems design: The Shell Sarnia experience." In J. B. Cunningham and T. H. White, eds., *Quality of Working Life: Contemporary Cases.* Ottawa: Labour Canada.

Hamblin, Heather 1995 "Employees' perspectives on the dimension of labour flexibility: Working at a distance." *Work, Employment & Society* 9(3): 473–98.

Hamilton, Gary G., and Nicole Woolsey Biggart 1988 "Market, culture, and authority: A comparative analysis of management and organization in the Far East." *American Journal of Sociology* 94: S52–S94.

Hamilton, Richard F., and James D. Wright 1986 *The State of the Masses.* New York: Aldine.

Hammer, Michael, and James Champy 1993 *Reengineering the Corporation: A Manifesto for Business Revolution.* New York: HarperBusiness.

Hammer, Tove Helland, and Ariel Avgar 2005 "The impact of unions on job satisfaction, organizational commitment, and turnover." *Journal of Labor Research* 26(2): 241–66.

Hamper, Ben 1986 *Rivethead: Tales from the Assembly Line.* New York: Warner Books.

Hampson, Ian 1999 "Lean production and the Toyota production system—or, the case of the forgotten production concepts." *Economic and Industrial Democracy* 20(3): 369–91.

Handel, Michael J. 2003 "Skills mismatch in the labor market." *Annual Review of Sociology* 29: 135–65.

Hansen, Susan 2004 "From 'common observation' to behavioural risk management: Workplace surveillance and employee assistance 1914–2003." *International Sociology* 19(2): 151–71.

Harchaoui, Tarek M., F. Tarkani, C. Jackson, and P. Armstrong 2002 "Information technology and economic growth in Canada and the U.S." *Monthly Labor Review* 125(10): 3–12.

Harpaz, Itzhak, and Raphael Snir 2003 "Workaholism: Its definition and nature." *Human Relations* 56(3): 291–319.

Harrington, Charles C., and Susan K. Boardman 1997 *Paths to Success: Beating the Odds in American Society.* Cambridge, MA: Harvard University Press.

Harris, Lloyd C., and Emmanuel Ogbonna 1999 "Developing a market oriented culture: A critical evaluation." *Journal of Management Studies* 36(2): 177–96.

Harrison, Bennett 1997 *Lean and Mean: Why Large Corporations Will Continue to Dominate the Global Economy.* New York: Guilford Press.

Harrison, Trevor 2005 *The Return of the Trojan Horse: Alberta and the New World (Dis)Order.* Montreal: Black Rose Books.

Hartmann, Heidi 1976 "Capitalism, patriarchy, and job segregation by sex." *Signs* 1: 137–69.

Harvey, Andrew S., Katherine Marshall, and Judith A. Frederick 1991 *Where Does Time Go?* Ottawa: Statistics Canada, General Social Survey Analysis Series 4. Cat. no. 11-612E, no. 4.

Hauser, Robert M., John Robert Warren, Min-Hsiung Huang, and Wendy C. Carter 2000 "Occupational status, education, and social mobility in the meritocracy." In Kenneth Arrow, Samuel Bowles, and Steven Durlauf, eds., *Meritocracy and Economic Inequality* (pp. 179–229). Princeton, NJ: Princeton University Press.

Hayden, Anders 1999 *Sharing the Work, Sparing the Planet: Work Time, Consumption, and Ecology.* Toronto: Between the Lines.

Hayes, Robert H., and Ramchandran Jaikumar 1988 "Manufacturing's crisis: New technologies, obsolete organizations." *Harvard Business Review* (September/October): 77–85.

Health Canada 2004 "Social inequality and health." In James Curtis, Edward Grabb, and Neil Guppy, eds., *Social Inequality in Canada: Patterns, Problems, and Policies.* 4th ed. Toronto: Pearson/Prentice Hall.

Hearn, Jeff, Deborah L. Sheppard, Peta Tancred-Sheriff, and Gibson Burrell, eds. 1989 *The Sexuality of Organization.* London: Sage.

Heil, Gary, Warren Bennis, and Deborah C. Stephens 2000 *Douglas McGregor, Revisited: Managing the Human Side of the Enterprise.* New York: John Wiley & Sons.

Heinzl, John 1997 "Nike's hockey plans put Bauer on thin ice." *The Globe and Mail* (2 July): B2.

Heisig, Ulrich, and Wolfgang Littek 1995 "Trust as a basis for work organization." In Wolfgang Littek and Tony Charles, eds., *The New Division of Labour: Emerging Forms of Work Organisation in International Perspective.* Berlin: Walter de Gruyter.

Heisz, Andrew 2002 *The Evolution of Job Stability in Canada: Trends and Comparisons to U.S. Results.* Statistics Canada. Cat. no. 11F0019MIE–No. 162.

Hellgren, Johnny, and Magnus Sverke 2001 "Unionized employees' perceptions of role stress and fairness during organizational downsizing: Consequences for job satisfaction, union satisfaction, and well-being." *Economic and Industrial Democracy* 22(4): 543–67.

Helling, Jan 1985 *Innovations in Work Practices at Saab–Scania* (Saab–Scania Personnel Division). Paper delivered at the U.S.–Japan Automotive Industry Conference, Ann Arbor, MI, March 5–6.

Hemingway, Fred, and Kathryn McMullen 2004 *A Family Affair: The Impact of Paying for College or University. A Literature Review and Gap Analysis.* Montreal: The Canadian Millennium Scholarship Foundation.

Hendry, John 1999 "Cultural theory and contemporary management organization." *Human Relations* 52(5): 557–77.

Henry, Manon 2004 "Union membership in Canada—2004." *Workplace Gazette* 7(3): 42–48.

Henson, Kevin D. 1996 *Just a Temp.* Philadelphia: Temple University Press.

Heron, Craig 1980 "The crisis of the craftsmen: Hamilton's metal workers in the early twentieth century." *Labour/Le Travail* 6: 7–48.

References

———— 1989 *The Canadian Labour Movement: A Short History.* Toronto: James Lorimer.

Heron, Craig, and Bryan Palmer 1977 "Through the prism of the strike: Industrial conflict in southern Ontario, 1910–14." *Canadian Historical Review* 58: 423–58.

Herzberg, Frederick 1966 *Work and the Nature of Man.* New York: World.

———— 1968 "One more time: How do you motivate employees?" *Harvard Business Review* 46: 53–62.

Hessing, Melody 1991 "Talking shop(ping): Office conversations and women's dual labour." *Canadian Journal of Sociology* 16: 23–50.

Hickson, David J. 1987 "Decision making at the top of organizations." *Annual Review of Sociology* 13: 165–92.

Hill, Stephen 1981 *Competition and Control at Work.* London: Heinemann.

———— 1988 "Technology and organizational culture: The human imperative in integrating new technology into organization design." *Technology in Society* 10: 233–53.

Hilton, Rodney, ed. 1976 *The Transition from Feudalism to Capitalism.* London: New Left Books.

Hinings, C. R., and Royston Greenwood 1988 *The Dynamics of Strategic Change.* Oxford: Basil Blackwell.

Hirschman, A. O. 1970 *Exit, Voice, and Loyalty.* Cambridge, MA: Harvard University Press.

Hirst, Paul, and Grahame Thompson 1995 "Globalization and the future of the nation state." *Economy and Society* 24: 408–42.

Hochschild, Arlie 1989 *The Second Shift: Working Parents and the Revolution at Home.* New York: Viking Penguin.

———— 1997 *The Time Bind: When Work Becomes Home and Home Becomes Work.* New York: Henry Holt and Company.

Hodson, Randy 1996 "Dignity in the workplace under participative management: Alienation and freedom revisited." *American Sociological Review* 61: 719–38.

———— 2001 *Dignity at Work.* Cambridge: Cambridge University Press.

———— 2004 "Work life and social fulfillment: Does social affiliation at work reflect a carrot or a stick?" *Social Science Quarterly* 85(2): 221–39.

Hodson, Randy, and Robert L. Kaufman 1982 "Economic dualism: A critical review." *American Sociological Review* 47: 727–39.

Hofstede, Geert 1984 *Culture's Consequences: International Differences in Work-Related Values.* Beverly Hills, CA: Sage.

Hofstede, Geert, Bram Neuijen, Denise Daval Ohayv, and Geert Sanders 1990 "Measuring organizational cultures: A qualitative and quantitative study across twenty cases." *Administrative Science Quarterly* 35: 286–316.

Holzer, Boris 2000 "Miracles with a system: The economic rise of East Asia and the role of sociocultural patterns." *International Sociology* 15(3): 455–78.

Holzer, Harry J. 1996 *What Employers Want: Job Prospects for Less-Educated Workers.* New York: Russell Sage Foundation.

Homans, George C. 1950 *The Human Group.* New York: Harcourt, Brace and World.

Hoque, Kim, and Ian Kirkpatrick 2003 "Non-standard employment in the management and professional workforce: Training, consultation and gender implications." *Work, Employment and Society* 17(4): 667–89.

Horrell, Sara, J. Rubery, and B. Burchell 1990 "Gender and skills." *Work, Employment & Society* 4: 189–216.

House, J. D. 1980 *The Last of the Free Enterprisers: The Oilmen of Calgary.* Toronto: Macmillan.

References

Hsiung, Ping-Chun 1996 *Living Rooms as Factories: Class, Gender, and the Satellite Factory System in Taiwan.* Philadelphia: Temple University Press.

Huberman, Michael, and Paul Lanoie 2000 "Changing attitudes toward work-sharing: Evidence from Quebec." *Canadian Public Policy* 26(2): 141–55.

Hughes, Everett C. 1943 *French Canada in Transition.* Chicago: University of Chicago Press.

Hughes, Karen D. 1989 "Office automation: A review of the literature." *Relations industrielles/Industrial Relations* 44: 654–79.

——— **1995** "Women in non-traditional occupations." *Perspectives on Labour and Income* (Autumn): 14–19.

——— **1996** "Transformed by technology? The changing nature of women's 'traditional' and 'non-traditional' white-collar work." *Work, Employment & Society* 10: 227–50.

——— **1999** *Gender and Self-Employment in Canada: Assessing Trends and Policy Implications.* CPRN Study No. W04. Ottawa: Canadian Policy Research Networks.

——— **2000a** *Women and Corporate Directorships in Canada: Trends and Issues.* CPRN Discussion Paper No. CPRN-01. Ottawa: Canadian Policy Research Networks.

——— **2000b** "Restructuring work, restructuring gender: Women's movement into non-traditional occupations in Canada." In Victor W. Marshall et al., eds., *Restructuring Work and the Life Course.* Toronto: University of Toronto Press.

——— **2005** *Female Enterprise in the New Economy.* Toronto: University of Toronto Press.

Hughes, Karen D., and Graham S. Lowe 2000 "Surveying the 'post industrial' landscape: Information technologies and labour market polarization in Canada."

Canadian Review of Sociology and Anthropology 37(1): 29–53.

Hughes, Karen D., and Vela Tadic 1998 "'Something to deal with': Customer sexual harassment and women's retail work in Canada." *Gender, Work and Organizations* 5(4): 207–19.

Hulin, Charles 1991 "Adaptation, persistence, and commitment in organizations." In Marvin D. Dunnette and Leaetta M. Hough, eds., *Handbook of Industrial and Organizational Psychology.* 2nd ed. Vol. 2. Palo Alto, CA: Consulting Psychologists Press.

Hum, Derek, and Wayne Simpson 1996 "Canadians with disabilities and the labour market." *Canadian Public Policy* 22: 285–97.

——— **1999a** "Training and unemployment." In Ken Battle and Sherri Torjman, eds., *Employment Policy Options.* Ottawa: Caledon Institute of Social Policy.

——— **1999b** "Wage opportunities for visible minorities in Canada." *Canadian Public Policy* 25(3): 379–94.

——— **2002** "Do immigrants catch up economically?" *Policy Options* 23(2): 47–50.

Human Resources Development Canada 1994 *Report of the Advisory Committee on Working Time and the Distribution of Work.* Ottawa: Supply and Services Canada.

——— **2000a** *Statistical Analysis: Occupational Injuries and Fatalities Canada.* Hull, QC: HRDC.

——— **2000b** *Occupational Injuries and Their Cost in Canada 1993–1997.* Hull, QC: HRDC.

——— **2003** *Disability in Canada: A 2001 Profile.* Ottawa: Human Resources Development Canada.

Human Resources Development Canada, and Statistics Canada 2002 *Results from the Survey of Self-Employment in Canada.* Ottawa: Applied Research Branch, Human Resources Development Canada.

References

Human Resources Development Canada, Workplace Information Directorate 2000 *Directory of Labour Organizations in Canada.* Hull, QC: HRDC.

Human Resources and Skills Development Canada 2004 *Annual Report— Employment Equity Act.* Ottawa: HRSCD. (http://www.hrsdc.gc.ca)

Humphrey, James H. 1998 *Job Stress.* Boston: Allyn and Bacon.

Humphries, Jane 1977 "Class struggle and the persistence of the working class family." *Cambridge Journal of Economics* 1: 241–58.

Hunter, Alfred A., and Michael C. Manley 1986 "On the task content of work." *Canadian Review of Sociology and Anthropology* 23: 47–71.

Hunter, Larry W., and John J. Lafkas 2003 "Opening the box: Information technology, work practices, and wages." *Industrial and Labor Relations Review* 56(2): 224–43.

Huxley, Christopher 1979 "The state, collective bargaining and the shape of strikes in Canada." *Canadian Journal of Sociology* 4: 223–39.

Hyman, Richard 1975 *Industrial Relations: A Marxist Introduction.* London: Macmillan.

———— **1978** *Strikes.* 2nd ed. Glasgow: Fontana.

Iaffaldano, Michelle T., and Paul M. Muchinsky 1985 "Job satisfaction and job performance: A meta-analysis." *Psychological Bulletin* 97: 251–73.

Idle, Thomas R., and Arthur J. Cordell 1994 "Automating work." *Society* 36 (September–October): 65–71.

Ilg, Randy E. 1995 "The changing face of farm employment." *Monthly Labor Review* (April): 3–12.

International Centre for Human Rights and Democratic Development 1997 *Commerce with Conscience? Human Rights and Corporate Codes of Conduct.* Montreal: ICHRDD.

International Labour Office (ILO) 1997 *World Employment Report 1997–98: Industrial Relations, Democracy and Stability.* Geneva: ILO.

———— **1999** *Decent Work.* Geneva: ILO.

Jackson, Andrew 2005 *Work and Labour in Canada: Critical Issues.* Toronto: Canadian Scholars' Press.

Jackson, Andrew, David Robinson, Bob Baldwin, and Cindy Wiggins 2000 *Falling Behind: The State of Working Canada, 2000.* Ottawa: Canadian Centre for Policy Alternatives.

Jacoby, Sanford M. 1985 *Employing Bureaucracy: Managers, Unions, and the Transformation of Work in American Industry, 1900–1945.* New York: Columbia University Press.

———— **1997** *Modern Manners: Welfare Capitalism Since the New Deal.* Princeton, NJ: Princeton University Press.

———— **2005** *The Embedded Corporation: Corporate Governance and Employment Relations in Japan and the United States.* Princeton, NJ: Princeton University Press.

Jahoda, M. 1982 *Employment and Unemployment: A Social-Psychological Approach.* Cambridge: Cambridge University Press.

Jain, Harish C., and Mohammed Al-Waqfi 2000 "Racial discrimination in employment in Canada." In Yonatan Reshef et al., eds., *Industrial Relations in a New Millennium. Selected Papers from the 37th Annual Canadian Industrial Relations Association Conference, Quebec.* CIRA.

Jain, Hem C. 1990 "Worker participation in Canada: Current developments and challenges." *Economic and Industrial Democracy* 11: 279–90.

Jamieson, Stuart Marshall 1971 *Times of Trouble: Labour Unrest and Industrial Conflict in Canada, 1900–66.* Ottawa: Queen's Printer.

Jaynes, Gerald D. 2000 "Identity and economic performance." *Annals of the*

References

American Academy of Political and Social Science 568(March): 128–39.

Jenness, Diamond 1977 *Indians of Canada.* 7th ed. Toronto: University of Toronto Press.

Jenson, Jane 1989 "The talents of women, the skills of men: Flexible specialization and women." In Stephen Wood, ed., *The Transformation of Work: Skill, Flexibility and the Labour Process.* London: Unwin Hyman.

Jenson, Jane, and Sherry Thompson 1999 *Comparative Family Policy: Six Provincial Stories.* CPRN Study No. F-08. Ottawa: Canadian Policy Research Networks.

Jermier, John M. 1998 "Introduction: Critical perspectives on organizational control." *Administrative Science Quarterly* 43(2): 235–56.

Jex, Steve 1998 *Stress and Job Performance: Theory, Research, and Implications for Managerial Practice.* Thousand Oaks, CA: Sage.

Jha, Prem Shankar 2002 *The Perilous Road to the Market: The Political Economy of Reform in Russia, India and China.* London: Pluto Press.

Jiang, Shanhe, Richard H. Hall, Karyn L. Loscocco, and John Allen 1995 "Job satisfaction theories and job satisfaction: A China and U.S. comparison." *Research in the Sociology of Work* 5: 161–78.

Johnson, Carol 1996 "Does capitalism really need patriarchy? Some old issues reconsidered." *Women's Studies International Forum* 19: 193–202.

Johnson, Gloria Jones, and Roy W. Johnson 1995 "Subjective underemployment and job satisfaction." *International Review of Modern Sociology* 25: 73–84.

Johnson, Holly 1994 "Work-related sexual harassment." *Perspectives on Labour and Income* (Winter): 9–12.

Johnson, Jeffrey V. 1989 "Control, collectivity and the psychosocial work environment." In S. L. Sauter, J. J. Hurrell Jr.,

and C. L. Cooper, eds., *Job Control and Worker Health.* New York: Wiley.

Jones, Charles, Lorna Marsden, and Lorne Tepperman 1990 *Lives of Their Own: The Individualization of Women's Lives.* Don Mills, ON: Oxford University Press.

Jones, Frank E. 1999 "Seniors who volunteer." *Perspectives on Labour and Income* (Autumn): 9–17.

——— 2000 "Youth volunteering on the rise." *Perspectives on Labour and Income* 12(Spring): 36–42.

Jones, Oswald 2000 "Scientific management, culture and control: A first-hand account of Taylorism in practice." *Human Relations* 53(5): 631–53.

Jones, Stephen R. G. 1990 "Worker interdependence and output: The Hawthorne studies reevaluated." *American Sociological Review* 55: 176–90.

Jonsson, Berth 1980 "The Volvo experiences of new job design and new production technology." *Working Life in Sweden* 18 (September).

Judge, Timothy A., J. E. Bono, C. J. Thoresen, and G. K. Patton 2001 "The job satisfaction–job performance relationship: A quantitative and qualitative review." *Journal of Applied Psychological Bulletin* 127(3): 376–407.

Kalleberg, Arne L. 1977 "Work values and job rewards: A theory of job satisfaction." *American Sociological Review* 42: 124–43.

——— 1988 "Comparative perspectives on work structures and inequality." *Annual Review of Sociology* 14: 203–25.

——— 2001 "Organizing flexibility: The flexible firm in a new century." *British Journal of Industrial Relations* 39(4): 479–504.

Kalleberg, Arne, and Karyn A. Loscocco 1983 "Aging, values and rewards: Explaining age differences in job satisfaction." *American Sociological Review* 48: 78–90.

Kalleberg, Arne, Barbara F. Reskin, and Ken Hudson 2000 "Bad jobs in America:

Standard and nonstandard employment relations and job quality in the United States." *American Sociological Review* 65: 256–78.

Kalleberg, Arne L., and Mark E. Van Buren 1996 "Is bigger better? Explaining the relationship between organization size and job rewards." *American Sociological Review* 61: 47–66.

Kamata, Satoshi 1983 *Japan in the Passing Lane: An Insider's Account of Life in a Japanese Factory.* New York: Pantheon.

Kanter, Rosabeth M. 1977 *Men and Women of the Corporation.* New York: Basic Books.

——— 1989 *When Giants Learn to Dance: Mastering the Challenges of Strategy, Management, and Careers in the 1990s.* New York: Simon and Shuster.

——— 1995 *World Class: Thriving Locally in the Global Economy.* New York: Simon & Shuster.

Kaplan, Robert S., and David P. Norton 2001 *The Strategy-Focused Organization: How Balanced Scorecard Companies Thrive in the New Business Environment.* Boston: Harvard Business School Press.

Kapsalis, Constantine 1998 "An international comparison of employee training." *Perspectives on Labour and Income* (Spring): 23–28.

Kapsalis, Constantine, René Morissette, and Garnett Picot 1999 "The returns to education, and the increasing wage gap between younger and older workers." Ottawa: Statistics Canada, Analytic Studies Branch, Research Paper no. 131.

Kapstein, Ethan B. 1996 "Workers and the world economy." *Foreign Affairs* 75: 16–37.

Karasek, Robert 1979 "Job demands, job decision latitude and mental health implications for job redesign." *Administrative Science Quarterly* 24: 285–308.

Karasek, Robert, and Töres Theorell 1990 *Healthy Work: Stress, Productivity, and the*

Reconstruction of Working Life. New York: Basic Books.

Kazemipur, Abdolmohammad, and Shiva S. Halli 2001 "The changing colour of poverty in Canada." *Canadian Review of Sociology and Anthropology* 38(2): 217–38.

Kealey, Gregory S. 1980 *Toronto Workers Respond to Industrial Capitalism, 1867–1892.* Toronto: University of Toronto Press.

——— 1981 "The bonds of unity: The Knights of Labour in Ontario, 1880–1900." *Histoire sociale/Social History* 14: 369–411.

——— 1986 "Work control, the labour process, and nineteenth-century Canadian printers." In Craig Heron and Robert Storey, eds., *On the Job: Confronting the Labour Process in Canada.* Montreal and Kingston: McGill–Queen's University Press.

——— 1995 *Workers and Canadian History.* Montreal and Kingston: McGill–Queen's University Press.

Keddie, V. 1980 "Class identification and party preference among manual workers." *Canadian Review of Sociology and Anthropology* 17: 24–36.

Keenoy, Tom 1985 *Invitation to Industrial Relations.* Oxford: Basil Blackwell.

Kelly, John E. 1982 *Scientific Management, Job Redesign and Work Performance.* London: Academic Press.

Kelly, Karen, Linda Howatson-Leo, and Warren Clark 1997 "'I feel overqualified for my job . . .'" *Canadian Social Trends* (Winter): 11–16.

Kemeny, Anna 2002 "Driven to excel: A portrait of Canada's workaholics." *Canadian Social Trends* (Spring): 2–7.

Kerr, Clark, J. T. Dunlop, F. H. Harbison, and C. A. Myers 1973 *Industrialization and Industrial Man.* London: Penguin.

Kerr, Clark, and Abraham Siegel 1954 "The interindustry propensity to strike:

An international comparison." In Arthur Kornhauser et al., eds., *Industrial Conflict*. New York: McGraw-Hill.

Kettler, David, James Struthers, and Christopher Huxley 1990 "Unionization and labour regimes in Canada and the United States." *Labour/Le Travail* 25: 161–87.

Kimmel, Jean, and Lisa M. Powell 1999 "Moonlighting trends and related policy issues in Canada and the United States." *Canadian Public Policy* 25(2): 207–31.

King, W. L. Mackenzie 1918 *Industry and Humanity: A Study in the Principles Underlying Industrial Reconstruction*. Toronto: Thomas Allen.

Kleemann, Frank, and Ingo Matuschek 2002 "Between job and satisfaction: Motivations and career orientations of German 'high quality' call center employees." *Electronic Journal of Sociology* 6: 2. (http://www.sociology.org)

Knight, Rolf 1978 *Indians at Work: An Informal History of Native Indian Labour in British Columbia 1858–1930*. Vancouver: New Star Books.

Knighton, Tamara, and Sheba Mirza 2002 "Postsecondary participation: The effects of parents' education and household income." *Education Quarterly Review* 8(3): 25–32.

Knights, David, and Darren McCabe 2000 "'Ain't misbehavin'? Opportunities for resistance under new forms of 'quality' management." *Sociology* 34(3): 421–36.

Knights, David, and Glenn Morgan 1990 "The concept of strategy in sociology: A note of dissent." *Sociology* 24: 475–83.

Knights, David, Hugh Willmott, and David Collison, eds. 1985 *Job Redesign: Critical Perspectives on the Labour Process*. Aldershot, England: Gower.

Kochan, Thomas A. 1979 "How American workers view labor unions." *Monthly Labor Review* 102(April): 23–31.

Kochan, Thomas A., Harry C. Katz, and Robert B. McKersie 1986 *The Transformation of American Industrial Relations*. New York: Basic Books.

Kochan, Thomas, Russell Lansbury, and John Paul MacDuffle, eds. 1997 *After Lean Production: Evolving Practices in the World Auto Industry*. Ithaca, NY: ILR Press.

Kochan, Thomas A., and Paul Osterman 1994 *The Mutual Gains Enterprise: Forging a Winning Partnership Among Labor, Management, and Government*. Boston: Harvard Business School Press.

Kohn, Melvin L. 1990 "Unresolved issues in the relationship between work and personality." In Kai Erikson and Steven Peter Vallas, eds., *The Nature of Work: Sociological Perspectives*. New Haven, CT: American Sociological Association and Yale University Press.

Kohn, Melvin L., and Carmi Schooler 1983 *Work and Personality: An Inquiry into the Impact of Social Stratification*. Norwood, NJ: Ablex.

Kompier, M., E. Degier, P. Smulders, and D. Draasisma 1994 "Regulations, policies and practices concerning work stress in five European countries." *Work and Stress* 8: 296–318.

Kopinak, Kathryn 1996 *Desert Capitalism: Maquiladoras in North America's Western Industrial Corridor*. Tucson: University of Arizona Press.

Korczynski, Marek, Karen Shire, Stephen Frenkel, and May Tam 2000 "Service work in consumer capitalism: Customers, control and contradictions." *Work, Employment & Society* 14(4): 669–87.

Krahn, Harvey 1992 *Quality of Work in the Service Sector*. Ottawa: Statistics Canada, General Social Survey Analysis Series 6. Cat. no. 11 612F, No. 6.

——— 1995 "Non-standard work on the rise." *Perspectives on Labour and Income* (Winter): 35–42.

References

——— **1997** "On the permanence of human capital: Use it or lose it." *Policy Options* 18(6): 17–21.

——— **2004** "Choose your parents carefully: Social class, post-secondary participation, and occupational outcomes." In James Curtis, Edward Grabb, and Neil Guppy, eds., *Social Inequality in Canada: Patterns, Problems, and Policies.* 4th ed. Toronto: Pearson/Prentice Hall.

Krahn, Harvey, Tracey Derwing, Marlene Mulder, and Lori Wilkinson 2000 "Educated and underemployed: Refugee integration into the Canadian labour market." *Journal of International Migration and Integration* 1(Winter): 59–84.

Krahn, Harvey, and Trevor Harrison 1992 "Self-referenced relative deprivation and economic beliefs: The effects of the recession in Alberta." *Canadian Review of Sociology and Anthropology* 29: 191–209.

Krahn, Harvey, and Graham S. Lowe 1984 "Public attitudes towards unions: Some Canadian evidence." *Journal of Labor Research* 5: 149–64.

——— **1997** *Literacy Utilization in Canadian Workplaces.* Ottawa: Human Resources Development Canada.

——— **1999** "Literacy in the workplace." *Perspectives on Labour and Income* (Summer): 38–44.

Krahn, Harvey, Norma Wolowyk, and Alison Yacyshyn 1998 *Metis Settlements General Council Census 1998. Report I: Enumeration Results and Settlement Profiles.* Edmonton: Population Research Laboratory, University of Alberta.

Krogman, Naomi, and Tom Beckley 2002 "Corporate 'bail-outs' and local 'buyouts': Pathways to community forestry." *Society and Natural Resources* 15(2): 109–27.

Kruse, Douglas L. 1998 "Persons with disabilities: Demographic, income, and health care characteristics, 1993." *Monthly Labor Review* (September): 13–22.

Kuhn, Thomas S. 1970 *The Structure of Scientific Revolutions.* 2nd ed. Chicago: University of Chicago Press.

Kumar, Krishnan 1995 *From Post-Industrial to Post-Modern Society: New Theories of the Contemporary World.* Oxford: Blackwell.

Kumar, Pradeep 1995 *Unions and Workplace Change in Canada.* Kingston, ON: IRC Press.

Kumar, Pradeep, and Lynn Acri 1991 "Women's issues and collective bargaining." In Donald Carter, ed., *Women and Industrial Relations. Proceedings of the 28th Conference of the Canadian Industrial Relations Association, June 2–4, 1991, Kingston, ON* (pp. 581–95). CIRA.

Kumar, Pradeep, David Arrowsmith, and Mary Lou Coates 1991 *Canadian Labour Relations: An Information Manual.* Kingston, ON: Industrial Relations Centre, Queen's University.

Kumar, Pradeep, Graham S. Lowe, and Grant Schellenberg 2001 "Workers' experience of job quality and the willingness to join a union." *Canadian Industrial Relations Association Annual Conference, May 27, 2001, Laval University, Quebec City.* CIRA.

Kunda, Gideon, and Galit Ailon-Souday 2005 "Managers, markets, and ideologies: Design and devotion revisited." In Stephen Ackroyd, R. Batt, P. Thompson, and P. S. Tolbert, eds., *The Oxford Handbook of Work and Organization.* Oxford: Oxford University Press.

Kunz, Jean Lock, Anne Milan, and Sylvain Schetagne 2000 *Unequal Access: A Canadian Profile of Racial Differences in Education, Employment and Income.* Canadian Race Relations Foundation.

Labour Canada 1986 *Women in the Labour Force, 1985–86 Edition.* Ottawa: Labour Canada, Women's Bureau.

Labour Market Ministers 2000 *Profile of Canadian Youth in the Labour Market: Second Annual Report to the Forum of*

Labour Market Ministers. Hull, QC: HRDC. Cat. no. RH61-1/2000E.

Lacroix, R. 1986 "Strike activity in Canada." In W. Craig Riddell, ed., *Canadian Labour Relations.* Toronto: University of Toronto Press.

Lait, Jana, and Jean E. Wallace 2002 "Stress at work: A study of organizational-professional conflict and unmet expecta-tions." *Relations industrielles/Industrial Relations* 57(3): 463–87.

Laliberte, Ron, and Vic Satzewich 1999 "Native migrant labour in the southern Alberta sugar-beet industry: Coercion and paternalism in the recruitment of labour." *Canadian Review of Sociology and Anthropology* 36(1): 65–85.

Lam, Helen, and Yonatan Reshef 1999 "Are quality improvement and downsizing compatible? A human resources perspec-tive." *Relations industrielles/Industrial Relations* 54(4): 727–47.

Lamba, Navjot 2003 "The employment experiences of Canadian refugees: Measuring the impact of human and social capital on employment out-comes." *Canadian Review of Sociology and Anthropology* 40(1): 45–64.

Land, Hillary 1980 "The family wage." *Feminist Review* 6: 55–77.

Langford, Tom 1996 "Effects of strike participation on the political conscious-ness of Canadian postal workers." *Relations industrielles/Industrial Relations* 51(3): 651–82.

Laroche, Mireille 1998 "In and out of low income." *Canadian Social Trends* (Autumn): 20–24.

Laroche, Mireille, Marcel Merette, and G. C. Ruggeri 1999 "On the concept and dimensions of human capital in a knowledge-based economy context." *Canadian Public Policy* 25(1): 87–100.

Lash, Scott 1984 *The Militant Worker: Class and Radicalism in France and America.* London: Heinemann.

——— 1990 *Sociology of Postmodernism.* London: Routledge.

Lawler, Edward E., III, and Susan Albers Mohrman 2003 *Creating a Strategic Human Resources Organization: An Assessment of Trends and New Directions.* Stanford, CA: Stanford University Press.

Laxer, Gordon 1989 *Open for Business: The Roots of Foreign Ownership in Canada.* Don Mills, ON: Oxford University Press.

——— 1995 "Social solidarity, democracy and global capitalism." *Canadian Review of Sociology and Anthropology* 32: 287–313.

Leadbeater, David, and Peter Suschnigg 1997 "Training as the principal focus of adjustment policy: A critical view from northern Ontario." *Canadian Public Policy* 23: 2–19.

Leck, Joanne 2002 "Making employment equity programs work for women." *Canadian Public Policy* 28(S1): S85–S100.

Leckie, Norm, André Léonard, Julie Turcotte, and David Wallace 2001 *Employer and Employee Perspectives on Human Resource Practices.* Ottawa: Statistics Canada and Human Resources Development Canada. Cat. no. 71-584-MPE, No. 1, The Evolving Workplace Series.

Lehmann, Wolfgang 2000 "Is Germany's dual system still a model for Canadian youth apprenticeship initiatives?" *Canadian Public Policy* 26(2): 225–40.

Leicht, Kevin T., and Mary L. Fennell 2001 *Professional Work: A Sociological Approach.* Oxford: Blackwell.

Leiter, Michael P. 1991 "The dream denied: Professional burnout and the constraints of human service organizations." *Canadian Psychology* 32: 547–55.

Lendvay-Zwickl, Judy 2004 *The Canadian Industrial Relations System: Current Challenges and Future Options.* Ottawa: Conference Board of Canada.

References

Lenski, Gerhard 1966 *Power and Privilege: A Theory of Social Stratification.* New York: McGraw-Hill.

Leonard, Madeleine 1998 "The long-term unemployed, informal economic activity, and the 'underclass' in Belfast: Rejecting or reinstating the work ethic." *International Journal of Urban and Regional Research* 22(1): 42–59.

Lerner, Gerda 1986 *The Creation of Patriarchy.* New York: Oxford University Press.

Lettau, Michael K., and Thomas C. Buchmueller 1999 "Comparing benefit costs for full- and part-time workers." *Monthly Labor Review* (March): 30–35.

Levine, David I. 1995 *Reinventing the Workplace: How Business and Employees Can Both Win.* Washington, DC: Brookings Institution.

Levinson, Klas 2000 "Codetermination in Sweden: Myth and reality." *Economic and Industrial Democracy* 21(4): 457–73.

Levitan, Sar A., and Clifford M. Johnson 1982 *Second Thoughts on Work.* Kalamazoo, MI: W. E. Upjohn Institute for Employment Research.

Lewchuk, Wayne, A. Leslie Robb, and Vivienne Walters 1996 "The effectiveness of Bill 70 and joint health and safety committees in reducing injuries in the workplace: The case of Ontario." *Canadian Public Policy* 22: 225–43.

Lewicki, Roy J., Daniel J. McAllister, and Robert J. Bies 1998 "Trust and distrust: New relationships and realities." *Academy of Management Review* 23(3): 438–58.

Li, Peter S. 1982 "Chinese immigrants on the Canadian prairie, 1919–47." *Canadian Review of Sociology and Anthropology* 19: 527–40.

——— 2001 "The market worth of immigrants' educational credentials." *Canadian Public Policy* 27(1): 23–38.

——— 2003 "Initial earnings and catch-up capacity of immigrants." *Canadian Public Policy* 29(3): 319–37.

Lian, Jason Z., and David Ralph Matthews 1998 "Does the vertical mosaic still exist? Ethnicity and income in Canada, 1991." *Canadian Review of Sociology and Anthropology* 35(4): 461–81.

Liker, Jeffrey K. 2004 *The Toyota Way: 14 Management Principles from the World's Greatest Manufacturer.* New York: McGraw-Hill.

Liker, Jeffrey K., Carol J. Haddad, and Jennifer Karlin 1999 "Perspectives on technology and work organization." *Annual Review of Sociology* 25: 575–96.

Lillie, Nathan 2005 "Union networks and global unionism in maritime shipping." *Relations industrielles/Industrial Relations* 60(1): 88–111.

Lincoln, James R. 1990 "Japanese organization and organization theory." *Research in Organizational Behavior* 12: 255–94.

Lincoln, James R., and Arne L. Kalleberg 1990 *Culture, Control, and Commitment: A Study of Work Organization and Work Attitudes in the United States and Japan.* Cambridge: Cambridge University Press.

Lincoln, James R., and Kerry McBride 1987 "Japanese industrial organization in comparative perspective." *Annual Review of Sociology* 13: 289–312.

Lindenfeld, Fran, and Pamela Wynn 1997 "Success and failure of worker co-ops: The role of internal and external environmental factors." *Humanity and Society* 21(2): 148–61.

Lindsay, Colin 1999 "Seniors: A diverse group aging well." *Canadian Social Trends* (Spring): 24–26.

Linhart, Robert 1981 *The Assembly Line.* London: John Calder.

Lipset, Seymour Martin 1990 *Continental Divide: The Values and Institutions of the*

United States and Canada. New York: Routledge.

Lipset, Seymour Martin, and Reinhard Bendix 1959 *Social Mobility in Industrial Society.* Berkeley: University of California Press.

Lipset, Seymour M., and Noah M. Meltz 1997 "Canadian and American attitudes toward work and institutions." *Perspectives on Work* 1(3):14–19.

Lipset, Seymour Martin, Martin Trow, and James Coleman 1956 *Union Democracy: The Internal Politics of the International Typographical Union.* Garden City, NJ: Anchor Books.

Lipshitz, Raanan 2000 "Chic, mystique, and misconception: Argyris and Schon and the rhetoric of organizational learning." *Journal of Applied Behavioral Science* 36(4): 456–73.

Lipsig-Mummé, Carla 1987 "Organizing women in the clothing trades: Homework and the 1983 garment strike in Canada." *Studies in Political Economy* 22: 41–71.

——— **2005** "Trade unions and labour-relations systems in comparative perspective." In Morley Gunderson, Allen Ponak, and Daphne Gottlieb Taras, eds., *Union–Management Relations in Canada* (pp. 476–93). 5th ed. Toronto: Pearson Addison Wesley.

Little, Don 1997 "Financing universities: Why are students paying more?" *Education Quarterly Review* 4(2): 10–26.

Littler, Craig R. 1982 *The Development of the Labour Process in Capitalist Societies.* London: Heinemann.

Littler, Craig R., and Peter Innes 2003 "Downsizing and deknowledging the firm." *Work, Employment & Society* 17(1): 73–100.

Livingstone, David W. 1999 *The Education-Jobs Gap: Underemployment or Economic Democracy.* Toronto: Garamond Press.

——— **2005** *Basic Findings of the 2004 Canadian Learning and Work Survey.* (http://www.wallnetwork.ca/updates/ WALLBasicSummJune05.pdf)

Livingstone, D. W., and Meg Luxton 1996 "Gender consciousness at work: Modification of the male breadwinner norm." In David W. Livingstone and J. M. Mangan, eds., *Recast Dreams: Class and Gender Consciousness in Steeltown* (pp. 100–29). Toronto: Garamond Press.

Livingstone, David W., and J. Marshall Mangan, eds. 1996 *Recast Dreams: Class and Gender Consciousness in Steeltown.* Toronto: Garamond Press.

Living Wage 2005 Living Wage Resource Center. (http://www.livingwagecampaign .org)

Lockwood, David 1966 "Sources of variation in working class images of society." *Sociological Review* 14: 249–67.

Logue, John 1981 "Saab/Trollhattan: Reforming work life on the shop floor." *Working Life in Sweden* 23(June).

Logue, John, and Jacquelyn S. Yates 1999 "Worker ownership American style: Pluralism, participation and performance." *Economic and Industrial Democracy* 20(2): 225–52.

Long, Richard J. 1989 "Patterns of workplace innovation in Canada." *Relations industrielles/Industrial Relations* 44: 805–26.

——— **1995** "Employee buyouts: The Canadian experience." *Canadian Business Economics* (Summer): 28–41.

Looise, Jan Kees, and Michiel Drucker 2003 "Dutch works councils in times of transition: The effects of changes in society, organizations and work on the position of works councils." *Economic and Industrial Democracy* 24(3): 379–409.

Looker, Dianne 1994 "Active capital: The impact of parents on youths' educational performance and plans." In Lorna Erwin and David MacLennan, eds., *Sociology of*

References

Education in Canada: Critical Perspectives on Theory, Research and Practice. Toronto: Copp Clark Longman.

——— 1997 "In search of credentials: Factors affecting young adults' participation in postsecondary education." *Canadian Journal of Higher Education* 27(2/3): 1–36.

Looker, E. Dianne, and Peter Dwyer 1998 "Education and negotiated reality: Complexities facing rural youth in the 1990s." *Journal of Youth Studies* 1(1): 5–22.

Looker, E. Dianne, and Victor Thiessen 1999 "Images of work: Women's work, men's work, housework." *Canadian Journal of Sociology* 24(2): 225–54.

Lowe, Graham S. 1981 "Causes of unionization in Canadian banks." *Relations industrielles/Industrial Relations* 36: 865–92.

——— 1984 "The rise of modern management in Canada." In G. S. Lowe and H. J. Krahn, eds., *Working Canadians.* Toronto: Methuen.

——— 1987 *Women in the Administrative Revolution: The Feminization of Clerical Work.* Toronto: University of Toronto Press.

——— 1997 "Computers in the workplace." *Perspectives on Labour and Income* (Summer): 29–36.

——— 1998 "The future of work: Implications for unions." *Relations industrielles/Industrial Relations* 53(2): 235–57.

——— 2000 *The Quality of Work: A People-Centred Agenda.* Don Mills, ON: Oxford University Press.

——— 2001 *Employer of Choice? Workplace Innovation in Government: A Synthesis Report.* Ottawa: Canadian Policy Research Networks.

——— 2003 *Healthy Workplaces and Productivity: A Discussion Paper.* Ottawa: Health Canada. (http://www.grahamlowe.ca/documents/93)

Lowe, Graham S., and Harvey Krahn 1985 "Where wives work: The relative effects of situational and attitudinal factors." *Canadian Journal of Sociology* 10: 1–22.

——— 1989 "Recent trends in public support for unions in Canada." *Journal of Labor Research* 10: 391–410.

——— 1999 "Reconceptualizing youth unemployment." In Julian Barling and E. Kevin Kelloway, eds., *Young Workers: Varieties of Experience.* Washington, DC: American Psychological Association.

——— 2000 "Work aspirations and attitudes in an era of labour market restructuring: A comparison of two Canadian youth cohorts." *Work, Employment & Society* 14(1): 1–22.

Lowe, Graham S., Harvey Krahn, and Jeff Bowlby 1997 *1996 Alberta High School Graduate Survey: Report of Research Findings.* Edmonton: Population Research Laboratory, University of Alberta.

Lowe, Graham S., and Herbert C. Northcott 1986 *Under Pressure: A Study of Job Stress.* Toronto: Garamond Press.

Lowe, Graham S., and Sandra Rastin 2000 "Organizing the next generation: Influences on young workers' willingness to join unions in Canada." *British Journal of Industrial Relations* 38(2): 203–22.

Lowe, Graham S., and Grant Schellenberg 2001 *What's a Good Job? The Importance of Employment Relationships.* CPRN Study W-05. Ottawa: Canadian Policy Research Networks.

Lucas, Rex A. 1971 *Minetown, Milltown, Railtown.* Toronto: University of Toronto Press.

Luce, Stephanie 2004 *Fighting for a Living Wage.* Ithaca, NY: Cornell University Press.

Luffman, Jacqueline 2003 "Taking stock of equity compensation." *Perspectives on Labour and Income* 15(Summer): 26–33.

Luxton, Meg 1980 *More Than a Labour of Love.* Toronto: Women's Press.

References

Luxton, Meg, and Leah Vosko 1998 "Where women's efforts count: The 1996 census campaign and 'family politics' in Canada." *Studies in Political Economy* 56(Summer): 49–81.

Macarov, David 1982 *Worker Productivity: Myths and Reality.* Beverly Hills, CA: Sage.

Macdonald, Martha 1991 "Post-Fordism and the flexibility debate." *Studies in Political Economy* 36: 177–201.

MacDonald, Robert 1994 "Fiddly jobs, undeclared working and the something for nothing society." *Work, Employment & Society* 8: 507–30.

Maclarkey, Robert L. 1997 "Information technology and the workplace: An empirical investigation of anti-technology theory." *International Review of Modern Sociology* 27(2): 33–44.

MacLean, Brian K., and Lars Osberg, eds. 1996 *The Unemployment Crisis: All for Nought?* Montreal and Kingston: McGill–Queen's University Press.

Macleod, Gus 1997 *From Mondragon to America: Experiments in Community Economic Development.* Sydney, NS: University College of Cape Breton Press.

Mahon, Rianne 1984 *The Politics of Industrial Restructuring: Canadian Textiles.* Toronto: University of Toronto Press.

——— **1987** "From Fordism to ? New technology, labour markets and unions." *Economic and Industrial Democracy* 8: 5–60.

Mahon, Rianne, with Sonya Michel 2002 *Child Care Policy at the Crossroads: Gender and Welfare State Restructuring.* London: Routledge.

Malone, Thomas W. 2004 *The Future of Work: How the New Order of Business Will Shape Your Organization, Your Management Style, and Your Life.* Boston: Harvard Business School Press.

Mann, Michael 1970 "The social cohesion of liberal democracy." *American Sociological Review* 35: 423–39.

Mann, Sandi, and Lynn Holdsworth 2003 "The psychological impact of teleworking: Stress, emotions and health." *New Technology, Work and Employment* 18(3): 196–211.

Manser, Marilyn E., and Garnett Picot 1999 "Self-employment in Canada and the United States." *Perspectives on Labour and Income* (Autumn): 37–44.

Marchak, M. Patricia 1981 *Ideological Perspectives on Canada.* 2nd ed. Toronto: McGraw-Hill Ryerson.

Marjoribanks, Kevin 2002 *Family and School Capital: Towards a Context Theory of Students' School Outcomes.* Dordrecht, The Netherlands: Kluwer Academic Publishers.

Marquardt, Richard 1998 *Enter at Your Own Risk: Canadian Youth and the Labour Market.* Toronto: Between the Lines.

Mars, Gerald 1982 *Cheats at Work: An Anthropology of Workplace Crime.* London: Unwin Paperbacks.

Marsh, Catherine, R. McAuley, and S. Penlington 1990 "The road to recovery? Some evidence from vacancies in one labour market." *Work, Employment & Society* 4: 31–58.

Marshall, Katherine 1989 "Women in the professional occupations: Progress in the 1980s." *Canadian Social Trends* (Spring): 13–16.

——— **1993** "Dual earners: Who's responsible for housework?" *Canadian Social Trends* (Winter): 11–14.

——— **1996** "A job to die for." *Perspectives on Labour and Income* (Summer): 26–31.

——— **1997** "Job sharing." *Perspectives on Labour and Income* (Summer): 6–10.

——— **1998** "Stay-at-home dads." *Perspectives on Labour and Income* (Spring): 9–15.

——— **1999** "Seasonality in employment." *Perspectives on Labour and Income* (Spring): 16–22.

References

———— **2000** "Incomes of young retired women: The past 30 years." *Perspectives on Labour and Income* 12(Winter): 9–17.

———— **2001a** "Part-time by choice." *Perspectives on Labour and Income* 13(Spring): 20–27.

———— **2001b** "Working with computers." *Perspectives on Labour and Income* 13(Summer): 9–15.

———— **2002** "Duration of multiple job-holding." *Perspectives on Labour and Income* 14(Summer): 17–24.

———— **2003** "Benefits of the job." *Perspectives on Labour and Income* 15(Summer): 7–14.

Marshall, Victor W., and Margaret M. Mueller 2002 *Rethinking Social Policy for An Aging Society: Insights from the Life Course Perspective.* Ottawa: Canadian Policy Research Networks. (http://www.cprn.org)

Martin, D'Arcy 1995 *Thinking Union: Activism and Education in Canada's Labour Movement.* Toronto: Between the Lines Press.

Martin, Gary 2000 "Employment and unemployment in Mexico in the 1990s." *Monthly Labor Review* (November): 3–18.

Martin, Jack K., and Paul M. Roman 1996 "Job satisfaction, job reward characteristics, and employees' problem drinking behaviors." *Work and Occupations* 23: 4–25.

Martin, Jack K., and Constance L. Shehan 1989 "Education and job satisfaction: The influences of gender, wage-earning status, and job values." *Work and Occupations* 16: 184–99.

Martin, Joanne, Kathleen Knopoff, and Christine Beckman 1998 "An alternative to bureaucratic impersonality and emotional labor: Bounded emotionality at The Body Shop." *Administrative Science Quarterly* 43(2): 429–69.

Maslach, Christina, and Michael P. Leiter 1997 *The Truth about Burnout: How Organizations Cause Personal Stress and What To Do About It.* San Francisco: Jossey-Bass.

Matthews, Roy A. 1985 *Structural Change and Industrial Policy: The Redeployment of Canadian Manufacturing, 1960–80.* Ottawa: Supply and Services Canada.

Maxim, Paul S., Jerry P. White, Dan Beavon, and Paul C. Whitehead 2001 "Dispersion and polarization of income among Aboriginal and non-Aboriginal Canadians." *The Canadian Review of Sociology and Anthropology* 38(4): 465–76.

Maxim, Paul S., Jerry P. White, Paul C. Whitehead, and Daniel Beavon 2000 *An Analysis of Wage and Income Inequality: Dispersion and Polarization of Income among Aboriginal and Non Aboriginal Canadians.* London, ON: Population Studies Centre, Discussion Paper no. 00-9.

Maximova, Katerina, and Harvey Krahn 2005 "Does race matter? Earnings of visible minority graduates from Alberta universities." *Canadian Journal of Higher Education* 35(1): 85–110.

Maynard, Rona 1987 "How do you like your job?" *Report on Business Magazine* (November): 112–22.

Mayo, Elton 1945 *The Social Problems of an Industrial Civilization.* Cambridge, MA: Harvard University Press.

McBrier, Debra Branch, and George Wilson 2004 "Going down: Race and downward occupational mobility for white-collar workers in the 1990s." *Work and Occupations* 31(4): 283–322.

McCabe, Darren 1999 "Total quality management: Antiunion Trojan horse or management albatross." *Work, Employment & Society* 13(4): 665–91.

McCormack, A. Ross 1978 *Reformers, Rebels and Revolutionaries: The Western Canadian Radical Movement, 1899–1919.* Toronto: University of Toronto Press.

References

McCormick, Chris, ed. 1998 *The Westray Chronicles: A Case Study of an Occupational Disaster.* Halifax: Fernwood Publishing.

McDonald, Judith A., and Robert J. Thornton 1998 "Private-sector experience with pay equity in Ontario." *Canadian Public Policy* 24(2):185–208.

McFarlane, Seth, Roderic Beaujot, and Tony Haddad 2000 "Time constraints and relative resources as determinants of the sexual division of domestic work." *Canadian Journal of Sociology* 25(1): 61–82.

McGregor, Douglas 1960 *The Human Side of Enterprise.* New York: McGraw-Hill.

McIlwee, Judith S., and J. Gregg Robinson 1992 *Women in Engineering: Gender, Power, and Workplace Culture.* Albany: State University of New York Press.

McKay, Shona 1991 "Willing and able." *Report on Business Magazine* (October): 58–63.

————— **1996** "You're (still) hired." *Report on Business Magazine* (December): 54–60.

McKinlay, Alan, and Ken Starkey, eds. 1998 *Foucault, Management and Organization Theory: From Panopticon to Technologies of Self.* Thousand Oaks, CA: Sage.

McLuhan, Marshall 1964 *Understanding Media: The Extensions of Man.* New York: McGraw-Hill.

McRae, Susan 2003 "Constraints and choices in mothers' employment careers: A consideration of Hakim's preference theory." *British Journal of Sociology* 54(3): 317–38.

Meissner, Martin 1971 "The long arm of the job: A study of work and leisure." *Industrial Relations* 10: 239–60.

Meissner, Martin, E. W. Humphreys, S. M. Meis, and W. J. Scheu 1975 "No exit for wives: Sexual division of labour and the cumulation of household demands." *Canadian Review of Sociology and Anthropology* 12: 424–39.

Menzies, Heather 1996 *Whose Brave New World? The Information Highway and the New Economy.* Toronto: Between the Lines.

Merton, Robert K. 1952 "Bureaucratic structure and personality." In Robert K. Merton, A. P. Gray, B. Hockey, and H. C. Selvin, eds., *Reader in Bureaucracy.* New York: Free Press.

Metcalfe, Beverly, and Alison Linstead 2003 "Gendering teamwork: Re-writing the feminine." *Gender, Work & Organization* 10(1): 94–119.

Meunier, Dominique, Paul Bernard, and Johanne Boisjoly 1998 "Eternal youth? Changes in the living arrangements of young people." In Miles Corak, ed., *Labour Markets, Social Institutions, and the Future of Canada's Children.* Ottawa: Statistics Canada.

Meyer, Stephen 1981 *The Five Dollar Day: Labor Management and Social Control in the Ford Motor Company, 1908–1921.* Albany: State University of New York Press.

Michels, Robert 1959 *Political Parties: A Sociological Study of the Oligarchical Tendencies of Modern Democracy.* New York: Dover Publications. (Orig. pub. 1915.)

Micklethwait, John, and Adrian Wooldridge 1996 *The Witch Doctors: Making Sense of the Management Gurus.* New York: Times Books.

Middleton, Chris 1988 "The familiar fate of the *famulae*: Gender divisions in the history of wage labour." In R. E. Pahl, ed., *On Work: Historical, Comparative and Theoretical Approaches.* Oxford: Basil Blackwell.

Milan, Anne, and Kelly Tran 2004 "Blacks in Canada: A long history." *Canadian Social Trends* (Spring): 2 7.

Miles, Raymond E. 1965 "Human relations or human resources." *Harvard Business Review* 43 (July–August): 148–63.

Milkman, Ruth 1991 *Japan's California Factories: Labor Relations and Economic Globalization.* Los Angeles: Institute of Industrial Relations, University of California at Los Angeles.

———— **1997** *Farewell to the Factory: Auto Workers in the Late Twentieth Century.* Berkeley: University of California Press.

Milkman, Ruth, and Cydney Pullman 1991 "Technological change in an auto assembly plant: The impact on workers' tasks and skills." *Work and Occupations* 18: 123–47.

Miller, Gloria 2004 "Frontier masculinity in the oil industry: The experience of women engineers." *Gender Work and Organization* 11(1): 47–73.

Miller, Joanne, C. Schooler, M. L. Kohn, and K. A. Miller 1979 "Women and work: The psychological effects of occupational conditions." *American Journal of Sociology* 85: 66–94.

Miller, Karen A., Melvin L. Kohn, and Carmi Schooler 1985 "Educational self-determination and the cognitive functioning of students." *Social Forces* 63(4): 923–44.

Mills, C. Wright 1948 *The New Men of Power.* New York: Harcourt-Brace.

———— **1956** *White Collar: The American Middle Classes.* New York: Oxford University Press.

Mills, Terry L., Craig A. Boylstein, and Sandra Lorean 2001 "'Doing' organizational culture in the Saturn corporation." *Organisation Studies* 22(1): 117–43.

Milner, Henry 1989 *Sweden: Social Democracy in Practice.* Oxford: Oxford University Press.

Miniti, Maria, Pia Arenius, and Nan Langowitz 2004 *Global Entrepreneurship Monitor: 2004 Report on Women and Entrepreneurship.* Available at http://www.gemconsortium.org.

Mintzberg, Henry 1989 *Mintzberg on Management: Inside Our Strange World of Organizations.* New York: Free Press.

———— **1994** "The fall and rise of strategic planning." *Harvard Business Review* 72: 107–14.

———— **1998** "Covert leadership: Notes on managing professionals." *Harvard Business Review* 76(6): 140–47.

Mintzberg, Henry, Bruce Ahlstrand, and Joseph Lampel 1998 *Strategy Safari: A Guided Tour Through the Wilds of Strategic Management.* New York: The Free Press.

Mirchandani, Kiran 1999 "Legitimizing work: Telework and the gendered reification of the work-nonwork dichotomy." *Canadian Review of Sociology and Anthropology* 36(1): 87–107.

Mirus, Rolf, Roger J. Smith, and Vladimir Karoleff 1994 "Canadian underground economy revisited: Update and critique." *Canadian Public Policy* 20: 235–52.

Mishel, Lawrence, and Jared Bernstein 2003 "Wage inequality and the new economy in the US: Does IT-led growth generate wage inequality?" *Canadian Public Policy* 29(S1): S203–S221.

Mishra, Aneil K., and Gretchen M. Spreitzer 1998 "Explaining how survivors respond to downsizing: The roles of trust, empowerment, justice, and work redesign." *Academy of Management Review* 23(3): 567–88.

Mitchell, Alanna 1997 "The poor fare worst in schools." *The Globe and Mail* (18 April): A1.

Mitchell, Barbara A., Andrew Wister, and Ellen Gee 2004 "The ethnic and family nexus of homeleaving and returning among Canadian young adults." *Canadian Journal of Sociology* 29(4): 543–76.

Mitchell, Jared 1999 "The pink ceiling." *Report on Business Magazine* (June): 78–84.

References

Mittelstaedt, Martin 2004 "Dying for a living." *The Globe and Mail* (13 March): F1.

Molgat, Marc 2002 "Leaving home in Quebec: Theoretical and social implications of (im)mobility among youth." *Journal of Youth Studies* 5(2): 135–52.

Morgan, Gareth 1997 *Images of Organization.* Thousand Oaks, CA: Sage Publications.

Morissette, René 1997 "Declining earnings of young men." *Canadian Social Trends* (Autumn): 8–12.

——— **2002a** "Families on the financial edge." *Perspectives on Labour and Income* 14(Autumn): 9–20.

——— **2002b** "Cumulative earnings among young workers." *Perspectives on Labour and Income* 14(Winter): 33–40.

——— **2004** "Permanent layoff rates." *Perspectives on Labour and Income* 16(Summer): 15–24.

Morissette, René, Grant Schellenberg, and Cynthia Silver 2004 "Retaining older workers." *Perspectives on Labour and Income* 16(Winter): 33–38.

Morissette, René, and Xuelin Zhang 2004 "Retirement plan awareness." *Perspectives on Labour and Income* 16(Spring): 37–44.

——— **2005** "Escaping low earnings." *Perspectives on Labour and Income* 17(Summer): 37–44.

Morissette, René, Xuelin Zhang, and Marie Drolet 2002 "Wealth inequality." *Perspectives on Labour and Income* 14(Spring): 15–22.

Morita, Masaya 2001 "Have the seeds of Japanese teamworking taken root abroad?" *New Technology, Work and Employment* 16(3): 178–90.

Morris, Lydia, and Sarah Irwin 1992 "Employment histories and the concept of the underclass." *Sociology* 26: 401–20.

Morton, Desmond 1989 "The history of the Canadian labour movement." In John C.

Anderson, Morley Gunderson, and Allen Ponak, eds., *Union–Management Relations in Canada.* 2nd ed. Don Mills, ON: Addison-Wesley.

Moulton, David 1974 "Ford Windsor 1945." In Irving Abella, ed., *On Strike: Six Key Labour Struggles in Canada 1919–1949.* Toronto: James Lewis and Samuel.

Muhlhausen, David B. 2005 "Do job programs work: A review article." *Journal of Labor Research* 26(2): 299–322.

Murray, Gregor 2001 "Unions: Membership, structures, actions, and challenges." In Morley Gunderson, Allen Ponak, and Daphne G. Taras, eds., *Union–Management Relations in Canada.* 4th ed. Toronto: Addison Wesley Longman.

Mustard, Cam, John N. Lavis, and Aleck Ostry 2005 "Work and health: New evidence and enhanced understandings." In Jody Heymann, C. Hertzman, M. Barer, and R. Evans, eds., *Creating Healthier Societies: From Analysis to Action.* Oxford: Oxford University Press.

Muszynski, Alicja 1996 *Cheap Wage Labour: Race and Gender in the Fisheries of British Columbia.* Montreal and Kingston: McGill–Queen's University Press.

Myles, John 1988 "The expanding middle: Some Canadian evidence on the deskilling debate." *Canadian Review of Sociology and Anthropology* 25: 335–64.

——— **1996** "Public policy in a world of market failure." *Policy Options* (July–August): 14–19.

——— **2000** "Incomes of seniors." *Perspectives on Labour and Income* 12(Winter): 23–32.

Myles, John, and Jill Quadagno, eds. 2005 *States, Labour Markets and the Future of Old Age Policy.* Philadelphia: Temple University Press.

Nadwodny, Richard 1996 "Canadians working at home." *Canadian Social Trends* (Spring): 16–20.

References

National Center on Education and the Economy (NCEE) 1990 *America's Choice: High Skills or Low Wages! Report of the Commission on the Skills of the American Workforce.* Rochester, NY: NCEE.

National Council of Welfare 2004 *Poverty Profile 2001.* Ottawa: National Council of Welfare. (http://www.ncwcnbes.net)

———— **2005** *Welfare Incomes 2004.* Ottawa: National Council of Welfare.

Neale, Deborah 1992 "Will Bob Rae deliver on his promise?" *Policy Options* (January–February): 25–28.

Nee, Victor, and Rebecca Matthews 1996 "Market transformation and societal transformation in reforming state socialism." *Annual Review of Sociology* 22: 401–35.

Neis, Barbara 1991 "Flexible specialization: What's that got to do with the price of fish?" *Studies in Political Economy* 36: 145–76.

Nelson, Daniel 1980 *Frederick W. Taylor and the Rise of Scientific Management.* Madison: University of Wisconsin Press.

Nelson, Joel I. 1995 *Post-Industrial Capitalism: Exploring Economic Inequality in America.* Thousand Oaks, CA: Sage.

Nevitte, Neil 2000 "Value change and reorientations in citizen–state relations." *Canadian Public Policy* 24(2 Supplement): S73–S94.

Newman, Katherine S. 1989 *Falling from Grace: The Experience of Downward Mobility in the American Middle Class.* New York: Vintage Books.

Nichols, Donald, and Chandra Subramaniam 2001 "Executive compensation: Excessive or equitable?" *Journal of Business Ethics* 29: 339–51.

Niezen, Ronald 1993 "Power and dignity: The social consequences of hydro-electric development for the James Bay Cree." *Canadian Review of Sociology and Anthropology* 30: 510–29.

Nilsson, Tommy 1996 "Lean production and white collar work: The case of Sweden." *Economic and Industrial Democracy* 17: 447–72.

Nishman, Robert F. 1995 *Worker Ownership and the Restructuring of Algoma Steel in the 1990s.* Kingston, ON: IRC Press.

Nissen, Bruce 1997 *Unions and Workplace Reorganization.* Detroit, MI: Wayne State University Press.

Noble, David 1995 *Progress without People: New Technology, Unemployment, and the Message of Resistance.* Toronto: Between the Lines.

Noble, Douglas D. 1994 "Let them eat skills." *The Review of Education/Pedagogy/ Cultural Studies* 16(1): 15–29.

Noël, Alain, and Keith Gardner 1990 "The Gainers strike: Capitalist offensive, militancy, and the politics of industrial relations in Canada." *Studies in Political Economy* 31: 31–72.

Nolan, Peter, and P. K. Edwards 1984 "Homogenise, divide and rule: An essay on *Segmented Work, Divided Workers.*" *Cambridge Journal of Economics* 8(2): 197–215.

Nollen, Stanley, and Helen Axel 1996 *Managing Contingent Workers: How to Reap the Benefits and Reduce the Risks.* New York: American Management Association.

Northcott, Herbert C., and Graham S. Lowe 1987 "Job and gender influences in the subjective experience of work." *Canadian Review of Sociology and Anthropology* 24: 117–31.

Nussbaum, Karen, and Virginia duRivage 1986 "Computer monitoring: Mismanagement by remote control." *Business and Society Review* 56(Winter): 16–20.

Oakes, Leslie S., Barbara Townley, and David J. Cooper 1998 "Business planning as pedagogy: Language and control in a changing institutional field." *Administrative Science Quarterly* 43: 257–92.

References

Oakeshott, Robert 1978 *The Case for Workers' Co-ops.* London: Routledge and Kegan Paul.

O'Donnell, Vivian, and Heather Tait 2004 "Well-being of the non-reserve Aboriginal population." *Canadian Social Trends* (Spring): 19–23.

O'Grady, John 2000 "Joint health and safety committees: Finding a balance." In Terrance Sullivan, ed., *Injury and the New World of Work.* Vancouver and Toronto: UBC Press.

O'Hara, Bruce 1993 *Working Harder Isn't Working.* Vancouver: New Star Books.

Ollivier, Michèle 2000 "'Too much money off other people's backs': Status in late modern societies." *Canadian Journal of Sociology* 25(4): 441–70.

Olsen, Gregg 1999 "Half empty or half full? The Swedish welfare state in transition." *Canadian Review of Sociology and Anthropology* 36(2): 241–67.

Olson, Mancur 1965 *The Logic of Collective Action.* Cambridge, MA: Harvard University Press.

O'Neill, Jeff 1991 "Changing occupational structure." *Canadian Social Trends* (Winter): 8–12.

Ontario Ministry of Education and Training 1996 *Excellence, Accessibility, Responsibility: Report of the Advisory Panel on Future Directions for Postsecondary Education.* Toronto: Minister of Education and Training.

O'Reilly, Charles A., III, and Jeffrey Pfeffer 2000 *Hidden Value: How Great Companies Achieve Extraordinary Results With Ordinary People.* Boston: Harvard Business School Press.

Organisation for Economic Co-operation and Development (OECD) 1994 *Employment Outlook, July 1994.* Paris: OECD.

——— **1996** *Lifelong Learning for All.* Paris: OECD.

——— **2003** *Corporate Data Environment (Labour Database).* Available online at http://www1.oecd.org/scripts/cde/default .asp.

——— **2004** *OECD Employment Outlook 2004.* Paris: OECD. IXBN 92-64-01045-9.

——— **2005** *OECD Employment Outlook 2005.* Paris: OECD. IXBN 92-64-01045-9.

Organisation for Economic Co-operation and Development, Human Resources Development Canada, and Statistics Canada 1998 *Literacy Skills for the Knowledge Society: Further Results of the International Adult Literacy Survey.* Paris and Ottawa: OECD, HRDC and Statistics Canada. Cat. no. 89-556-XPE.

Orme, W. A., Jr. 1996 *Understanding NAFTA: Mexico, Free Trade, and the New North America.* Austin: University of Texas Press.

Orr, Julian E. 1996 *Talking about Machines: An Ethnography of a Modern Job.* Ithaca, NY: ILR Press.

Osberg, Lars, and Pierre Fortin, eds. 1996 *Unnecessary Debt.* Toronto: Lorimer.

Osberg, Lars, and Zhengxi Lin 2000 "How much of Canada's unemployment is structural?" *Canadian Public Policy* (July Supplement): S141–S158.

Osberg, Lars, and Andrew Sharpe 1998 "An index of economic well-being for Canada." Ottawa: Centre for the Study of Living Standards. (http://www.csls.ca)

Osberg, Lars, Fred Wein, and Jan Grude 1995 *Vanishing Jobs: Canada's Changing Workplace.* Toronto: Lorimer.

Ospina, Sonia 1996 *Illusions of Opportunity: Employee Expectations and Workplace Inequality.* Ithaca and London: Cornell University Press.

Osterman, Paul 1999 *Securing Prosperity: The American Labor Market. How It Has Changed and What to Do About It.* Princeton, NJ: Princeton University Press.

References

——— 2000 "Work reorganization in an era of restructuring: Trends in diffusion and effects on employee welfare." *Industrial and Labor Relations Review* 53(2): 179–96.

Osterman, Paul, and M. Diane Burton 2005 "Ports and ladders: The nature and relevance of internal labor markets in a changing world." In Stephen Ackroyd, R. Batt, P. Thompson, and P. S. Tolbert, eds., *The Oxford Handbook of Work and Organization.* Oxford: Oxford University Press.

Ostroff, Frank 1999 *The Horizontal Organization: What the Organization of the Future Actually Looks Like and How it Delivers Value to Customers.* New York: Oxford University Press.

Ostry, Sylvia 1968 *The Female Worker in Canada.* Ottawa: Queen's Printer.

O'Toole, James, ed. 1977 *Work, Learning and the American Future.* San Francisco: Jossey-Bass.

Ouchi, William 1981 *Theory Z: How American Business Can Meet the Japanese Challenge.* Reading, MA: Addison-Wesley.

Owram, Doug 1996 *Born at the Right Time: A History of the Baby Boom Generation.* Toronto: University of Toronto Press.

Page, Marianne E. 2004 "New evidence on the intergenerational correlation in welfare participation." In Miles Corak, ed., *Generational Income Mobility in North America and Europe.* Cambridge: Cambridge University Press.

Palameta, Boris 2004 "Low income among immigrants and visible minorities." *Perspectives on Labour and Income* 16(Summer): 32–37.

Palmer, Bryan 1975 "Class, conception and conflict: The thrust for efficiency, managerial views of labor and the working class rebellion, 1902–22." *Radical Review of Political Economics* 7: 31–49.

——— **1979** *A Culture in Conflict: Skilled Workers and Industrial Capitalism in Hamilton, Ontario, 1860–1914.* Montreal and Kingston: McGill–Queen's University Press.

——— **1986** *The Character of Class Struggle: Essays in Canadian Working Class History, 1850–1985.* Toronto: McClelland and Stewart.

——— **1992** *Working Class Experience: Rethinking the History of Canadian Labour, 1800–1991.* 2nd ed. Toronto: McClelland and Stewart.

Palmer, Craig, and Peter Sinclair 1997 *When the Fish are Gone: Ecological Disaster and Fishers in Northwestern Newfoundland.* Halifax: Fernwood Publishing.

Panitch, Leo, and Donald Swartz 1993 *The Assault on Trade Union Freedoms: From Wage Controls to Social Contract.* Toronto: Garamond.

Paris, Hélène 1989 *The Corporate Response to Workers with Family Responsibilities.* Ottawa: Conference Board of Canada, Report 43-89.

Parr, Joy 1990 *The Gender of Breadwinners: Women, Men and Change in Two Industrial Towns 1880–1950.* Toronto: University of Toronto Press.

Parthasarathy, Balaji 2004 "India's Silicon Valley or Silicon Valley's India? Socially embedding the computer software industry in Bangalore." *International Journal of Urban and Regional Research* 28(3): 664–85.

Patterson, E. Palmer, II 1972 *The Canadian Indian: A History Since 1500.* Don Mills, ON: Collier-Macmillan.

Pauly, Edward, H. Kopp, and J. Haimson 1995 *Homegrown Lessons: Innovative Programs Linking School and Work.* San Francisco, CA: Jossey-Bass.

Payette, Suzanne 2000 "What's new in workplace innovations?" *Workplace Gazette* 3(1): 110–19.

Payne, Joan, and Clive Payne 1994 "Recession, restructuring and the fate of the unemployed: Evidence in the underclass debate." *Sociology* 28: 1–19.

References

Pendleton, Andrew, John McDonald, Andrew Robinson, and Nicholas Wilson 1996 "Employee participation and corporate governance in employee-owned firms." *Work, Employment & Society* 10: 205–26.

Pentland, H. Claire 1979 "The Canadian industrial relations system: Some formative factors." *Labour/Le Travail* 4: 9–23.

———— **1981** *Labour and Capital in Canada, 1650–1860.* Toronto: James Lorimer.

Perlman, Selig 1928 *A Theory of the Labor Movement.* New York: Macmillan.

Perrow, Charles 1986 *Complex Organizations: A Critical Essay.* 3rd ed. New York: Random House.

***Perspectives on Labour and Income* 2004** "Unionization." *Perspectives on Labour and Income* 16(Autumn): 58–65.

Peters, Suzanne 1995 *Exploring Canadian Values: Foundations for Well-Being.* Ottawa: Canadian Policy Research Networks Study Inc.

Peters, Tom 1987 *Thriving on Chaos: Handbook for a Management Revolution.* New York: Alfred A. Knopf.

Peters, Thomas J., and Robert H. Waterman Jr. 1982 *In Search of Excellence.* New York: Warner.

Peters, Valerie 2004 *Working and Training: First Results of the 2003 Adult Education and Training Survey.* Statistics Canada and Human Resources and Skills Development Canada. Cat. no. 81-595-MIE No. 015.

Petersen, Trond, Ishak Saporta, and Marc-David L. Seidel 2000 "Offering a job: Meritocracy and social networks." *American Journal of Sociology* 106: 763–816.

Petty, M. M., G. McGee, and J. Cavender 1984 "A meta-analysis of the relationship between individual job satisfaction and individual performance." *Academy of Management Review* 9: 712–21.

Pfeffer, Jeffrey 1994 *Competitive Advantage through People: Unleashing the Power of the Workforce.* Boston: Harvard University Press.

———— **1997** *New Directions in Organization Theory: Problems and Prospects.* New York: Oxford University Press.

Pfeffer, Jeffrey, and Robert I. Sutton 2000 *The Knowing–Doing Gap: How Smart Companies Turn Knowledge into Action.* Boston: Harvard Business School Press.

Phelan, Jo 1994 "The paradox of the contented female worker: An assessment of alternative explanations." *Social Psychology Quarterly* 57: 95–107.

Phillips, Paul, and Erin Phillips 1993 *Women & Work: Inequality in the Canadian Labour Market.* Toronto: James Lorimer & Company.

Picot, Garnett 1987 "The changing industrial mix of employment, 1951–1985." *Canadian Social Trends* (Spring): 8–11.

———— **1998** "What is happening to earnings inequality and youth wages in the 1990s?" *Canadian Economic Observer* (September): 3.1–3.18.

Picot, Garnett, and Andrew Heisz 2000 "The performance of the 1990s Canadian labour market." *Canadian Public Policy* 26(Supplement): 7–25.

Piore, Michael J., and Charles F. Sabel 1984 *The Second Industrial Divide: Possibilities for Prosperity.* New York: Basic Books.

Pitts, Gordon 2005 "CEO's harmonious-society plan? Fire 14,000 staff." *The Globe and Mail* (6 July): A1.

Piva, Michael J. 1979 *The Condition of the Working Class in Toronto, 1900–1921.* Ottawa: University of Ottawa Press.

Podolny, Joel M., and Karen L. Page 1998 "Network forms of organization." *Annual Review of Sociology* 24: 57–64.

Poggio, Barbara 2000 "Between bytes and bricks: Gender cultures in work contexts." *Economic and Industrial Democracy* 21(3): 381–402.

References

Polanyi, Karl 1957 *The Great Transformation.* Boston: Beacon Press.

Pollert, Anna 1988 "The flexible firm: Fixation or fact?" *Work, Employment & Society* 2: 281–316.

Pollitt, Christopher 1995 "Management techniques for the public service: Pulpit and practice." In B. G. Peters and D. J. Savoie, eds., *Governance in a Changing Environment.* Montreal and Kingston: McGill–Queen's University Press.

Ponak, Allen, and Larry F. Moore 1981 "Canadian bank unionism: Perspectives and issues." *Relations industrielles/ Industrial Relations* 36: 3–30.

Ponak, Allen, and Daphne Taras 1995 *Right-to-Work.* Submission to Alberta Economic Development Authority Joint Review Committee.

Poole, Michael 1981 *Theories of Trade Unionism.* London: Routledge and Kegan Paul.

Popper, Micha, and Raanan Lipshitz 2000 "Organizational learning: Mechanisms, culture, and feasibility." *Management Learning* 31(2): 181–96.

Poutsma, Erik, John Hendrickx, and Fred Huijgen 2003 "Employee participation in Europe: In search of the participative workplace." *Economic and Industrial Democracy* 24(1): 45–76.

Powell, Gary N. 1993 *Women and Men in Management.* 2nd ed. Newbury Park, CA: Sage.

Preuss, Gil A., and Brenda A. Lautsch 2002 "The effect of formal versus informal job security on employee involvement programs." *Relations industrielles/Industrial Relations* 57(3): 517–39.

Prince, Michael J. 2004 "Canadian disability policy: Still a hit-and-miss affair." *Canadian Journal of Sociology* 29(1): 59–82.

Pringle, Rosemary 1989 "Bureaucracy, rationality and sexuality: The case of secretaries." In Jeff Hearn, Deborah L. Sheppard, Peta Tancred-Sheriff, and Gibson Burrell, eds., *The Sexuality of Organization.* London: Sage.

Pruijt, Hans 2003 "Teams between neo-Taylorism and anti-Taylorism." *Economic and Industrial Democracy* 24(1): 77–101.

Pugh, D. S., D. J. Hickson, and C. R. Hinings 1985 *Writers on Organizations.* Beverly Hills, CA: Sage.

Pupo, Norene 1997 "Always working, never done: The expansion of the double day." In Ann Duffy, Daniel Glenday, and Norene Pupo, eds., *Good Jobs, Bad Jobs, No Jobs: The Transformation of Work in the 21st Century.* Toronto: Harcourt-Brace Canada.

Purcell, Kate 1979 "Militancy and acquiescence amongst women workers." In Sandra Burman, ed., *Fit Work for Women.* London: Croom Helm.

Purcell, Patrick J. 2000 "Older workers: Retirement and retirement trends." *Monthly Labor Review* (October): 19–30.

Quarter, Jack 1992 *Canada's Social Economy: Co-operatives, Non-Profits, and Other Community Enterprises.* Toronto: Lorimer.

Randle, Keith 1996 "The white-coated worker: Professional autonomy in a period of change." *Work, Employment & Society* 10: 737–53.

Rank, Mark, and Thomas A. Hirschl 1999 "The likelihood of poverty across the American adult life span." *Social Work* 44(3): 201–16.

Rankin, Tom 1990 *New Forms of Work Organization: The Challenge for North American Unions.* Toronto: University of Toronto Press.

Ranson, Gillian 1998 "Education, work and family decision-making: Finding the 'right time' to have a baby." *Canadian Review of Sociology and Anthropology* 35(4): 517–33.

Rashid, Abdul 1994 "High income families." *Perspectives on Labour and Income* (Winter): 46–57.

References

———— 1999 "Family income: 25 years of stability and change." *Perspectives on Labour and Income* (Spring): 9–15.

Ray, Carol Axtell 1986 "Corporate culture: The last frontier of control?" *Journal of Management Studies* 23: 287–97.

Rayman, Paula M. 2001 *Beyond the Bottom Line: The Search for Dignity at Work.* New York: Palgrave.

Reich, Robert B. 1991 *The Work of Nations: Preparing Ourselves for 21st-Century Capitalism.* New York: Alfred A. Knopf.

———— 2000 *The Future of Success.* New York: Alfred Knopf Inc.

Reid, Frank 1982 "Wage-and-price controls in Canada." In John Anderson and Morley Gunderson, eds., *Union–Management Relations in Canada.* Don Mills, ON: Addison-Wesley.

Reimer, Neil 1979 "Oil, Chemical and Atomic Workers International Union and the quality of working life: A union perspective." *Quality of Working Life: The Canadian Scene* (Winter): 5–7.

Reiter, Ester 1991 *Making Fast Food: From the Frying Pan into the Fryer.* Montreal and Kingston: McGill–Queen's University Press.

Reitz, Jeffrey G. 1988 "Less racial discrimination in Canada, or simply less racial conflict? Implications of comparisons with Britain." *Canadian Public Policy* 14: 424–41.

———— 2001 "Immigrant skill utilization in the Canadian labour market: Implications of human capital research." *Journal of International Migration and Integration* 2(3): 347–78.

Reitz, Jeffery G., and Anil Verma 2004 "Immigration, race, and labor: Unionization and wages in the Canadian labor market." *Industrial Relations* 43(4): 835–54.

Reshef, Yonatan 1990 "Union decline: A view from Canada." *Journal of Labor Research* 9: 25–39.

Reshef, Yonatan, and Sandra Rastin 2003 *Unions in the Time of Revolution: Government Restructuring in Alberta and Ontario.* Toronto: University of Toronto Press.

Reskin, Barbara 1998 *The Realities of Affirmative Action in Employment.* Washington, DC: American Sociological Association.

———— 2000 "Getting it right: Sex and race inequality in work organizations." *Annual Review of Sociology* 26: 707–9.

Reskin, Barbara F., and Debra B. McBrier 2000 "Why not ascription? Organizations' employment of male and female managers." *American Sociological Review* 65: 210–33.

Reskin, Barbara, and Irene Padavic 2002 *Men and Women at Work.* 2nd ed. Thousand Oaks, CA: Pine Forge Press.

Reynolds, John R. 1997 "The effects of industrial employment conditions on job-related distress." *Journal of Health and Social Behavior* 38(2): 105–16.

Richards, John, Aidan Vining, David M. Brown, Michael Krashinsky, William J. Milne, Ernie S. Lightman, and Shirley Hoy 1995 *Helping the Poor: A Qualified Case for "Workfare."* Ottawa: C. D. Howe Institute.

Richardson, Charley 1996 "Computers don't kill jobs, people do: Technology and power in the workplace." *Annals of the American Academy of Political and Social Science* (March): 167–79.

Riddell, W. Craig 1985 "Work and pay: The Canadian labour market: An overview." In W. Craig Riddell, ed., *Work and Pay: The Canadian Labour Market.* Toronto: University of Toronto Press.

———— 1986 "Canadian labour relations: An overview." In W. Craig Riddell, ed., *Canadian Labour Relations.* Toronto: University of Toronto Press.

Riddell, W. Craig, and France St-Hilaire, eds. 2000 *Adapting Public Policy to a Labour Market in Transition.* Montreal: Institute for Research on Public Policy.

Riddell, W. Craig, and Andrew Sharpe 1998 "The Canada–US unemployment rate gap: An introduction and overview." *Canadian Public Policy* (February Supplement): S1–S37.

Riddell, W. Craig, and Arthur Sweetman 2000 "Human capital formation in a period of rapid change." In W. Craig Riddell and France St-Hilaire, eds., *Adapting Public Policy to a Labour Market in Transition* (pp. 85–141). Montreal: Institute for Research on Public Policy.

Rifkin, Jeremy 1995 *The End of Work: The Decline of the Global Labor Force and the Dawn of the Post-Market Era.* New York: Putnam.

Rinehart, James 1978 "Contradictions of work-related attitudes and behaviour: An interpretation." *Canadian Review of Sociology and Anthropology* 15: 1–15.

———— **1984** "Appropriating workers' knowledge: Quality control circles at a General Motors plant." *Studies in Political Economy* 14: 75–97.

———— **2006** *The Tyranny of Work: Alienation and the Labour Process.* 5th ed. Toronto: Thomson Nelson.

Rinehart, James, Christopher Huxley, and David Robertson 1997 *Just Another Car Factory? Lean Production and its Discontents.* Ithaca, NY: ILR Press.

Rioux, Marcia H. 1985 "Labelled disabled and wanting to work." In *Research Studies of the Commission on Equality in Employment* [Abella Commission]. Ottawa: Supply and Services Canada.

Roberts, J. Timmons, and John E. Baugher 1995 "Hazardous workplaces and job strain: Evidence from an eleven nation study." *International Journal of Contemporary Sociology* 32: 235–49.

Roberts, Karen, Doug Hyatt, and Peter Dorman 1996 "The effect of free trade on contingent work in Michigan." In Karen Roberts and Mark I. Wilson, eds., *Policy Choices: Free Trade Among NAFTA Nations.* East Lansing: Michigan State University Press.

Roberts, Wayne 1990 *Cracking the Canadian Formula: The Making of the Energy and Chemical Workers Union.* Toronto: Between the Lines.

Robertson, David, and Jeff Wareham 1987 *Technological Change in the Auto Industry.* Willowdale, ON: Canadian Auto Workers (CAW).

———— **1989** *Changing Technology and Work: Northern Telecom.* Willowdale, ON: Canadian Auto Workers (CAW).

Robin, Martin 1968 *Radical Politics and Canadian Labour: 1880–1930.* Kingston, ON: Industrial Relations Centre, Queen's University.

Robinson, Allan 2005 "Layoffs dim glow of consumer stocks." *The Globe and Mail* (28 June): B16.

Rodrigues, Suzana B., and David J. Hickson 1995 "Success in decision making: Different organizations, differing reasons for success." *Journal of Management Studies* 32(5): 655–78.

Roethlisberger, F. J., and W. J. Dickson 1939 *Management and the Worker.* Cambridge, MA: Harvard University Press.

Rollings-Magnusson, Sandra 2000 "Canada's most wanted: Pioneer women on the western prairies." *Canadian Review of Sociology and Anthropology* 37(2): 223–38.

Rollings-Magnusson, Sandra, Harvey Krahn, and Graham S. Lowe 1999 *Does a Decade Make a Difference? Education and Work among 1985 and 1996 University Graduates.* Edmonton: Population Research Laboratory, University of Alberta.

References

Rónas-Tas, Ákos 1994 "The first shall be last? Entrepreneurship and Communist cadres in the transition from socialism." *American Journal of Sociology* 100: 40–69.

Rones, Philip L., Randy E. Ilg, and Jennifer M. Gardner 1997 "Trends in hours of work since the mid-1970s." *Monthly Labor Review* (April): 3–14.

Rose, Joseph 2001 "From softball to hardball: The transition in labour–management relations in the Ontario public service." In Gene Swimmer, ed., *Public-Sector Labour Relations in an Era of Restraint and Restructuring.* Toronto: Oxford University Press.

Rose, Joseph B., and Gary N. Chaison 2001 "Unionism in Canada and the United States in the 21st century: The prospects for revival." *Relations industrielles/Industrial Relations* 56(1): 34–62.

Rosenberg, Samuel, ed. 1989 *The State and the Labor Market.* New York: Plenum.

Rosenthal, Patrice, Stephen Hill, and Riccardo Peccei 1997 "Checking out service: Evaluating excellence, HRM and TQM in retailing." *Work, Employment and Society* 11(3): 481–503.

Ross, Becki L. 2000 "Bumping and grinding on the line: Making nudity pay." *Labour/Le Travail* 46(Fall): 221–50.

Ross, Peter, Greg J. Bamber, and Gillian Whitehouse 1998 "Employment, economics and industrial relations: Comparative statistics." In Greg J. Bamber and Russell D. Lansbury, eds., *International and Comparative Employment Relations.* 3rd ed. Thousand Oaks, CA: Sage Publications.

Rossides, Daniel W. 1998 *Professions and Disciplines: Functional and Conflict Perspectives.* Upper Saddle River, NJ: Prentice Hall.

Rowe, Reba, and William E. Snizek 1995 "Gender differences in work values: Perpetuating the myth." *Work and Occupations* 22: 215–29.

Rowlinson, Michael, and Stephen Procter 1999 "Organizational culture and business history." *Organisation Studies* 20(3): 369–96.

Roy, Arun S., and Ging Wong 2000 "Direct job creation programs: Evaluation lessons on cost-effectiveness." *Canadian Public Policy* 26(2): 157–69.

Roy, Donald 1952 "Quota restriction and goldbricking in a machine shop." *American Journal of Sociology* 57: 427–42.

——— **1959–60** "'Bananatime': Job satisfaction and informal interaction." *Human Organization* 18: 158–68.

Rubery, Jill 1988 "Employers and the labour market." In Duncan Gallie, ed., *Employment in Britain.* Oxford: Basil Blackwell.

Rubery, Jill, and Colette Fagan 1995 "Gender segregation in societal context." *Work, Employment & Society* 9: 213–40.

Rubery, Jill, and Damian Grimshaw 2001 "ICTs and employment: The problem of job quality." *International Labour Review* 140(2): 165–92.

Russell, Bob 1990 *Back to Work? Labour, State, and Industrial Relations in Canada.* Scarborough, ON: Nelson.

——— **1997** "Rival paradigms at work: Work reorganization and labour force impacts in a staple industry." *Canadian Review of Sociology and Anthropology* 34: 25–52.

——— **1999** *More with Less: Work Reorganization in the Canadian Mining Industry.* Toronto: University of Toronto Press.

——— **2004** "Are all call centres the same?" *Labour & Industry* 14(3): 91–109.

Ryerson, Stanley B. 1968 *Unequal Union: Confederation and the Roots of Conflict in the Canadas, 1815–1873.* Toronto: Progress Books.

Sallaz, Jeffrey J. 2004 "Manufacturing concessions: Attritionary outsourcing at GM's Lordstown, USA assembly plant." *Work, Employment & Society* 18(4): 687–708.

References

Sandberg, Ake 1994 "'Volvoism's at the end of the road?" *Studies in Political Economy* 45: 170–82.

Sangster, Joan 1978 "The 1907 Bell telephone strike: Organizing women workers." *Labour/Le Travail* 3: 109–30.

———— **1995** "Doing two jobs: The wage-earning mother, 1945–70." In Joy Parr, ed., *A Diversity of Women: Ontario, 1945–1980.* Toronto: University of Toronto Press.

Sargent, Timothy C. 2000 "Structural unemployment and technological change in Canada, 1990–1999." *Canadian Public Policy* 26(S1): 109–23.

Sass, Robert 1986 "Workplace health and safety: Report from Canada." *International Journal of Health Services* 16: 565–82.

———— **1995** "A conversation about the work environment." *International Journal of Health Services* 25: 117–28.

———— **1996** "A message to the labor movement: Stop and think!" *International Journal of Health Services* 26: 595–609.

Sassen, Saskia 2002 "Global cities and survival circuits." In Barbara Ehrenreich and Arlie Hochschild, eds., *Global Woman: Nannies, Maids and Sex Workers in the New Economy.* New York: Metropolitan Books.

Satzewich, Vic, ed. 1998 *Racism and Social Inequality in Canada: Concepts, Controversies, and Strategies of Resistance.* Toronto: Thompson Educational Publishing.

Saul, John R. 1995 *The Unconscious Civilization.* Toronto: Anansi.

Saunders, Ron 2005 *Lifting the Boats: Policies to Make Work Pay.* Ottawa: Canadian Policy Research Networks.

Sauter, Steven L., J. J. Hurrell Jr., and C. L. Cooper 1989 *Job Control and Worker Health.* New York: Wiley.

Schecter, Stephen, and Bernard Paquet 1999 "Contested approaches in the study of poverty: The Canadian case and the argument for inclusion." *Current Sociology* 47(3): 43–64.

Schein, Edgar H. 2004 *Organizational Culture and Leadership.* 3rd ed. San Francisco: Jossey-Bass.

Schellenberg, Grant, and David P. Ross 1997 *Left Poor by the Market: A Look at Family Poverty and Earnings.* Ottawa: Canadian Council on Social Development.

Schellenberg, Grant, and Cynthia Silver 2004 "You can't always get what you want: Retirement preferences and experiences." *Canadian Social Trends* (Winter): 2–7.

Schissel, Bernard, and Terry Wotherspoon 2003 *The Legacy of School for Aboriginal People: Education, Oppression, and Emancipation.* Don Mills, ON: Oxford University Press.

Schmidt, Sarah 2005 "Six out of 10 new MDs are women." *National Post* (July 28): A4.

Schooler, Carmi 1984 "Psychological effects of complex environments during the life span: A review and theory." *Intelligence* 8: 259–81.

———— **1996** "Cultural and social-structural explanations of cross-national psychological differences." *Annual Review of Sociology* 22: 323–49.

Schor, Juliet 1991 *The Overworked American: The Unexpected Decline of Leisure.* New York: Basic Books.

———— **1998** *The Overspent American.* New York: Basic Books.

Schouteten, Roel, and Jos Benders 2004 "Lean production assessed by Karasek's job demand–job control model." *Economic and Industrial Democracy* 25(3): 347–73.

Schwartz, Barry 1999 "Capitalism, the market, the 'underclass,' and the future." *Society* 37(1): 33–42.

Scott, John 1988 "Ownership and employer control." In Duncan Gallie, ed., *Employment in Britain.* Oxford: Basil Blackwell.

References

Seager, Allen 1985 "Socialists and workers: The Western Canadian coal miners, 1900–21." *Labour/Le Travail* 16: 25–39.

Sedivy-Glasgow, Marie 1992 "Nursing in Canada." *Canadian Social Trends* (Spring): 27–29.

Seeman, Melvin 1967 "On the personal consequences of alienation in work." *American Sociological Review* 32: 273–85.

——— **1975** "Alienation studies." *Annual Review of Sociology* 1: 91–125.

Semler, Ricardo 1993 "Workers' paradise?" *Report on Business Magazine* (December): 39–48.

Senge, Peter M. 1990 *The Fifth Discipline: The Art and Practice of the Learning Organization.* New York: Doubleday.

Senge, Peter, Art Kleiner, Charlotte Roberts, Richard B. Ross, and Bryan J. Smith 1994 *The Fifth Discipline Fieldbook: Strategies and Tools for Building a Learning Organization.* New York: Currency Doubleday.

Sennett, Richard 1998 *The Corrosion of Character: The Personal Consequences of Work in the New Capitalism.* New York: W.W. Norton.

Sennett, Richard, and Jonathan Cobb 1972 *The Hidden Injuries of Class.* New York: Knopf.

Serwer, Andy 2005 "Bruised in Bentonville." *Fortune* (April 18): 84–89.

Sev'er, Aysan 1999 *Special Issue: Sexual Harassment. Canadian Review of Sociology and Anthropology* 36(4).

Sewell, Graham 1998 "The discipline of teams: The control of team-based industrial work through electronic and peer surveillance." *Administrative Science Quarterly* 43: 397–428.

Shain, Alan 1995 "Employment of people with disabilities." *Canadian Social Trends* (Autumn): 8–13.

Shalla, Vivian 2002 "Jettisoned by design? The truncated employment relationship of customer sales and service agents under airline restructuring." *Canadian Journal of Sociology* 27(1): 1–32.

——— **2004** "Time warped: The flexibilization and maximization of flight attendant working time." *The Canadian Review of Sociology and Anthropology* 41(3): 345–68.

Shapiro, Daniel M., and Morton Stelcner 1997 "Language and earnings in Quebec: Trends over twenty years, 1970–1990." *Canadian Public Policy* 23(2): 115–40.

Sharpe, Andrew 1999 "The nature and causes of unemployment in Canada." In Ken Battle and Sherri Torjman, eds., *Employment Policy Options.* Ottawa: Caledon Institute of Social Policy.

Sharpe, Dennis B., and Gerald White 1993 *Educational Pathways and Experiences of Newfoundland Youth.* St. John's: Centre for Educational Research and Development, Memorial University.

Sherman, Barrie, and Phil Judkins 1995 *Licensed to Work.* London: Cassel.

Shewell, Hugh 2004 *"Enough to Keep Them Alive": Indian Welfare in Canada, 1873–1965.* Toronto: University of Toronto Press.

Shieh, G. S. 1992 *"Boss" Island: The Subcontracting Network and Micro-Entrepreneurship in Taiwan's Development.* New York: Peter Lang.

Shields, Margot 2000 "Long working hours and health." *Perspectives on Labour and Income* 12(Spring): 49–56.

——— **2003** "The health of Canada's shift workers." *Canadian Social Trends* (Summer): 21–25.

Shorter, Edward, and Charles Tilly 1974 *Strikes in France, 1830–1968.* Cambridge, MA: Cambridge University Press.

Siltanen, Janet 1994 *Locating Gender: Occupational Segregation, Wages and Domestic Responsibilities.* London: UCL Press.

Sinclair, Peter R., and Lawrence F. Felt 1992 "Separate worlds: Gender and

References

domestic labour in an isolated fishing region." *Canadian Review of Sociology and Anthropology* 29: 55–71.

Skrypnek, Berna J., and Janet E. Fast 1996 "Work and family policy in Canada." *Journal of Family Issues* 17: 793–812.

Smith, Adam 1976 *The Wealth of Nations.* Chicago: University of Chicago Press. (Orig. pub. 1776.)

Smith, Michael R. 1978 "The effects of strikes on workers: A critical analysis." *Canadian Journal of Sociology* 3: 457–72.

————— 1999 "The production of flexible attitudes in the Canadian pulp and paper industry." *Relations industrielles/Industrial Relations* 54(3): 581–608.

————— 2001 "Technological change, the demand for skills, and the adequacy of their supply." *Canadian Public Policy* 27(1): 1–22.

Smith, Michael R., Anthony C. Masi, Axel van den Berg, and Joseph Smucker 1995 "External flexibility in Sweden and Canada: A three industry comparison." *Work, Employment & Society* 9: 689–718.

————— 1997 "Insecurity, labour relations, and flexibility in two process industries: A Canada/Sweden comparison." *Canadian Journal of Sociology* 22: 31–63.

Smith, Vicki 1994 "Braverman's legacy: The labour process tradition at 20." *Work and Occupations* 21: 403–21.

————— 1997 "New forms of work organization." *Annual Review of Sociology* 23: 315–39.

Smucker, Joseph 1980 *Industrialization in Canada.* Scarborough, ON: Prentice-Hall.

Smucker, Joseph, Axel van den Berg, Michael R. Smith, and Anthony C. Masi 1998 "Labour deployment in plants in Canada and Sweden: A three-industry comparison." *Relations industrielles/Industrial Relations* 53(3): 430–56.

Soderfeldt, Bjorn, Marie Soderfeldt, Kelvyn Jones, and Patricia O'Camp 1997 "Does organization matter? A multilevel analysis of the demand-control model applied to human services." *Social Science and Medicine* 44(4): 527–34.

Sonnenfeld, Jeffrey A. 1985 "Shedding light on the Hawthorne studies." *Journal of Occupational Behaviour* 6: 111–30.

Sorrentino, Constance 1995 "International unemployment indicators, 1983–93." *Monthly Labor Review* (August): 31–50.

Sosteric, Mike 1996 "Subjectivity and the labour process: A case study in the restaurant industry." *Work, Employment & Society* 10: 297–318.

Spain, Daphne, and Suzanne M. Bianchi 1996 *Balancing Act: Motherhood, Marriage, and Employment among American Women.* New York: Russell Sage Foundation.

Spenner, Kenneth I. 1983 "Deciphering Prometheus: Temporal change in the skill level of work." *American Sociological Review* 48: 824–37.

————— 1990 "Skill: Meanings, methods, and measures." *Work and Occupations* 14: 399–421.

Spilerman, Seymour 2000 "Wealth and stratification processes." *Annual Review of Sociology* 26: 497–524.

Staber, Udo 1993 "Worker cooperatives and the business cycle: Are cooperatives the answer to unemployment?" *The American Journal of Economics and Sociology* 52: 129–43.

Stabler, Jack C., and Eric C. Howe 1990 "Native participation in northern development: The impending crisis in the NWT." *Canadian Public Policy* 16: 262–83.

Stasiulis, Daiva, and Abigail B. Bakan 2003 *Negotiating Citizenship: Migrant Women in Canada and the Global System.* New York: Palgrave MacMillan.

Statistics Canada 1995 "Moving with the times: Introducing change to the LFS"

References

The Labour Force (December): C–2–C–19.

——— **2000a** "Stress and well-being." *Health Reports* (Winter): 21–32.

——— **2000b** *Women in Canada 2000: A Gender-Based Statistical Report.* Ottawa: Statistics Canada. Cat. no. 89-503-XPE.

——— **2001** *Population Projections for Canada, Provinces and Territories, 2000–2026.* Cat. no. 91-520-XIB.

——— **2002** *Analysis of Income in Canada.* Ottawa: Statistics Canada. Cat. no. 75-203-XIE.

——— **2003a** *The Changing Profile of Canada's Labour Force.* 2001 Census Analysis Series. (http://www.statcan.ca)

——— **2003b** *Canada's Ethnocultural Portrait: The Changing Mosaic.* 2001 Census Analysis Series. (http://www.statcan.ca)

——— **2003c** *Aboriginal Peoples of Canada: A Demographic Profile.* 2001 Census Analysis Series. (http://www.statcan.ca)

——— **2003d** *Education, Employment and Income of Adults with and without Disabilities–Tables: 2001 Participation and Activity Limitation Survey.* Ottawa: Statistics Canada, Housing, Family and Social Statistics Division. Cat. no. 89-587-XIE.

——— **2003e** *Women in Canada: Work Chapter Updates.* Ottawa: Statistics Canada. Cat. no. 89F0133XIE.

——— **2003f** *Education in Canada: Raising the Standard.* Ottawa: Statistics Canada. Cat. no. 96F0030XIE2001012.

——— **2004a** "Childcare Arrangements." *Perspectives on Labour and Income* 16(Summer): 54–58.

——— **2004b** "University Enrolment." *The Daily* July 30.

——— **2004c** *The Canadian Labour Market at a Glance 2003.* Cat. no. 71-222-XIE. Available at http://www.statcan.ca.

——— **2004d** "The near-retirement rate." *Perspectives on Labour and Income* 16(Spring): 68–72.

——— **2005a** *Aboriginal Conditions in Census Metropolitan Areas, 1981–2001.* Ottawa. Cat. no. 89-613-MIE Paper 008. Available at http://www.statcan.ca.

——— **2005b** *A Guide to the Labour Force Survey.* Ottawa. Cat. no. 71-543-GIE. Available at http://www.statcan.ca.

——— **2005c** *Population Projections of Visible Minority Groups, Canada, Provinces and Territories.* Ottawa. Cat. no. 91-541-XIE. Available at http://www.statcan.ca.

——— **2005d** *Improvements in 2005 to the Labour Force Survey (LFS).* Ottawa. Cat. no. 71F0031XIE-No. 002. Available at http://www.statcan.ca.

Statistics Canada and Human Resources Development Canada 1996 *Reading the Future: A Portrait of Literacy in Canada.* Cat. no. 89–551.

——— **1998** *The Evolving Workplace: Findings from the Pilot Workplace and Employee Survey.* Ottawa: Statistics Canada and HRDC. Cat. no. 71-583-XPE.

Steinberg, Ronnie J. 1990 "Social construction of skill: Gender, power, and comparable worth." *Work and Occupations* 17: 449–82.

Stern, David, N. Finkelstein, J. R. Rose III, J. Latting, and C. Dornsife 1995 *School to Work: Research on Programs in the United States.* Washington, DC: Falmer.

Stern, Robert N. 1976 "Intermetropolitan pattern of strike frequency." *Industrial and Labor Relations Review* 29: 218–35.

Stewart, A., and R. M. Blackburn 1975 "The stability of structured inequality." *Sociological Review* 23: 481–508.

Stinson, John F., Jr. 1997 "New data on multiple jobholding available from the CPS," *Monthly Labor Review* (March): 3–8.

Storey, Robert 1983 "Unionization versus corporate welfare: The Dofasco way." *Labour/Le Travail* 12: 7–42.

References

Storey, Robert, and Wayne Lewchuk 2000 "From dust to DUST to dust: Asbestos and the struggle for worker health and safety at Bendix Automotive." *Labour/Le Travail* 45(Spring): 103–40.

Strong-Boag, Veronica 1988 *The New Day Recalled: Lives of Girls and Women in English Canada, 1919–1939.* Markham, ON: Penguin Books.

Strong-Boag, Veronica, and A. Fellman, eds. 1997 *Rethinking Canada: The Promise of Women's History.* 3rd ed. Don Mills, ON: Oxford University Press.

Stubblefield, Al 2005 *The Baptist Health Care Journey to Excellence: Creating a Culture that WOWs!* Hoboken, NJ: John Wiley & Sons.

Sufrin, Eileen 1982 *The Eaton Drive: The Campaign to Organize Canada's Largest Department Store, 1948 to 1952.* Toronto: Fitzhenry and Whiteside.

Sugiman, Pamela 1992 "That wall's comin' down: Gendered strategies of worker resistance in the UAW Canadian region (1963–1970)." *Canadian Journal of Sociology* 17: 24–27.

Sullivan, Maureen 1996 "Rozzie and Harriet? Gender and family patterns of lesbian coparents." *Gender and Society* 10(6): 747–67.

Sullivan, Terrance, ed. 2000 *Injury and the New World of Work.* Vancouver and Toronto: UBC Press.

Sunter, Deborah 2001 "Demography and the labour market." *Perspectives on Labour and Income* 13(Spring): 28–39.

Sunter, Deborah, and Geoff Bowlby 1998 "Labour force participation in the 1990s." *Perspectives on Labour and Income* (Autumn): 15–21.

Sunter, Deborah, and René Morissette 1994 "The hours people work." *Perspectives on Labour and Income* (Autumn): 8–13.

Super, Donald E., and Branimir Šverko, eds. 1995 *Life Roles, Values, and Careers: International Findings of the Work Importance Study.* San Francisco: Jossey-Bass.

Suplee, Curt 1997 "Robot revolution." *National Geographic* (July): 76–93.

Sussman, Deborah 1998 "Moonlighting: A growing way of life." *Perspectives on Labour and Income* (Summer): 24–31.

——— **2000** "Unemployment kaleidoscope." *Perspectives on Labour and Income* 12(Autumn): 9–15.

——— **2002** "Barriers to job-related training." *Perspectives on Labour and Income* 14(Summer): 25–32.

Sussman, Deborah, and Martin Tabi 2004 "Minimum wage workers." *Perspectives on Labour and Income* 16(Summer): 5–14.

Swanson, Jean 2001 *Poor-Bashing: The Politics of Exclusion.* Toronto: Between the Lines.

Swift, Jamie 1995 *Wheel of Fortune: Work and Life in the Age of Falling Expectations.* Toronto: Between the Lines.

Swimmer, Gene, ed. 2001 *Public-Sector Labour Relations in an Era of Restraint and Restructuring.* Toronto: Oxford University Press.

Szelényi, Iván, and Eric Kostello 1996 "The market transition debated: Toward a synthesis?" *American Journal of Sociology* 10: 1082–96.

Tanner, Julian 1984 "Skill levels of manual workers and beliefs about work, management, and industry: A comparison of craft and non-craft workers in Edmonton." *Canadian Journal of Sociology* 9: 303–18.

Tanner, Julian, and Rhonda Cockerill 1986 "In search of working-class ideology: A test of two perspectives." *Sociological Quarterly* 27: 389–402.

Tanner, Julian, Rhonda Cockerill, Jan Barnsley, and A. P. Williams 1999 "Flight paths and revolving doors: A case study of gender desegregation in pharmacy." *Work, Employment & Society* 13(2): 275–93.

Tanner, Julian, Harvey Krahn, and Timothy F. Hartnagel 1995 *Fractured Transitions from School to Work: Revisiting the Dropout Problem.* Don Mills, ON: Oxford University Press.

Tapscott, Don 1996 *The Digital Economy: Promise and Peril in the Age of Networked Intelligence.* San Francisco, CA: McGraw-Hill.

Taras, Daphne G. 2002 "Alternative forms of employee representation and labour policy." *Canadian Public Policy* 28(1): 105–16.

Taras, Daphne G., Allen Ponak, and Morley Gunderson 2005 "Introduction to Canadian industrial relations." In Morley Gunderson, Allen Ponak, and Daphne Gottlieb Taras, eds., *Union–Management Relations in Canada* (pp. 1–22). 5th ed. Toronto: Pearson Addison Wesley.

Taylor, Alison 2001 *The Politics of Educational Reform in Alberta.* Toronto: University of Toronto Press.

Taylor, Philip, and Peter Bain 2001 "Trade unions, workers' rights and the frontier of control in UK call centres." *Economic and Industrial Democracy* 22(1): 39–66.

——— **2003** "'Subterranean worksick blues': Humour as subversion in two call centres." *Organization Studies* 24(9): 1487–1509.

Taylor, Philip, Jeff Hyman, Gareth Mulvey, and Peter Bain 2002 "Work organization, control and the experience of work in call centres." *Work, Employment & Society* 16(1): 133–50.

Teeple, Gary 1972 "Land, labour and capital in pre-Confederation Canada." In Gary Teeple, ed., *Capitalism and the National Question in Canada.* Toronto: University of Toronto Press.

Terkel, Studs 1971 *Working.* New York: Pantheon.

Therrien, Pierre, and André Léonard 2003 *Empowering Employees: A Route to Innovation.* Ottawa: Statistics Canada and Human Resources Development Canada. Cat. No 71-584-MEI, No. 8, The Evolving Workplace Series.

Thomas, Alan Berkeley 1988 "Does leadership make a difference to organizational performance?" *Administrative Science Quarterly* 33: 388–400.

Thomason, Terry, and Silvana Pozzebon 2002 "Determinants of firm workplace health and safety and claims management practices." *Industrial and Labor Relations Review* 55(2): 286–307.

Thompson, Paul, and Chris Smith 2000/1 "Follow the redbrick road." *International Studies of Management & Organization* 30(4): 40–67.

Thorsrud, Einar 1975 "Collaborative action research to enhance the quality of working life." In L. E. Davis and A. B. Cherns, eds., *The Quality of Working Life.* Vol. 1. New York: Free Press.

Tilly, Charles 1979 *From Mobilization to Revolution.* Reading, MA: Addison-Wesley.

——— **1998** *Durable Inequality.* Berkeley: University of California Press.

Timpson, Annis May 2001 *Driven Apart: Women's Employment Equality and Child Care in Canadian Public Policy.* Vancouver and Toronto: University of British Columbia Press.

Tindale, Joseph A. 1991 *Older Workers in an Aging Workforce.* Ottawa: National Advisory Council on Aging.

Tobin, James 1996 "Business cycles and economic growth: Current controversies about theory and policy." In Lars Osberg and Pierre Fortin, eds., *Unnecessary Debts.* Toronto: Lorimer.

Toffler, Alvin 1980 *The Third Wave.* New York: Bantam.

Tran, Kelly 2004 "Visible minorities in the labour force: 20 years of change." *Canadian Social Trends* (Summer): 7–11.

References

Trice, Harrison M., and Janice M. Beyer 1993 *The Cultures of Work Organizations.* Englewood Cliffs, NJ: Prentice-Hall.

Trist, E. L., and K. W. Bamforth 1951 "Some social and psychological consequences of the longwall method of coal-getting." *Human Relations* 4: 3–38.

Tucker, Eric 1992 "Worker participation in health and safety regulation: Lessons from Sweden." *Studies in Political Economy* 37: 95–127.

——— 2003 "Diverging trends in worker health and safety protection and participation in Canada, 1985–2000." *Relations industrielles/Industrial Relations* 58(3): 395–424.

Turk, James L., ed. 2000 *The Corporate Campus: Commercialization and the Dangers to Canada's Colleges and Universities.* Toronto: Lorimer.

Tushman, Michael L., and Philip Anderson, eds. 2004 *Managing Strategic Innovation and Change.* 2nd ed. New York: Oxford University Press.

Ulrich, Dave 1998 "Intellectual capital = competence × commitment." *Sloan Management Review* 39(2):15–26.

Vahtera, Jussi, Mika Kivimäki, Jaana Pentti, Anne Linna, Marianna Virtanen, Pekka Virtanen, and Jane E. Ferrie 2004 "Organisational downsizing, sickness absence, and mortality: 10-town prospective cohort study." *British Medical Journal* 328(7439): 555. (http://www.bmj.com)

Vallas, Steven Peter 1990 "The concept of skill." *Work and Occupations* 14: 379–98.

——— 2003 "Why teamwork fails: Obstacles to workplace change in four manufacturing plants." *American Sociological Review* 68: 223–50.

van den Berg, Axel, and Joseph Smucker, eds. 1997 *The Sociology of Labour Markets: Efficiency, Equity, Security.* Scarborough, ON: Prentice-Hall.

Van Houten, Donald R. 1990 "The political economy and technical control of work humanization in Sweden during the 1970s and 1980s." *Work and Occupations* 14: 483–513.

Van Kirk, Sylvia 1980 *Many Tender Ties: Women in Fur-Trade Society, 1670–1870.* Winnipeg: Watson and Dwyer.

Vaughan, Diane 1999 "The dark side of organizations: Mistake, misconduct, and disaster." *Annual Review of Sociology* 25: 271–305.

Veltmeyer, Henry 1983 "The development of capitalism and the capitalist world system." In J. Paul Grayson, ed., *Introduction to Sociology: An Alternative Approach.* Toronto: Gage.

Vosko, Leah F. 1998 "Regulating precariousness? The temporary employment relationship under the NAFTA and the EC treaty." *Relations industrielles/Industrial Relations* 53(1): 123–53.

——— 2000 *Temporary Work: The Gendered Rise of a Precarious Employment Relationship.* Toronto: University of Toronto Press.

Vosko, Leah, ed. 2005 *Precarious Employment: Understanding Labour Market Insecurity in Canada.* Montreal and Kingston: McGill–Queen's University Press.

Vosko, Leah, Nancy Zukewich, and Cynthia Cranford 2003 "Precarious jobs: Towards a new typology." *Perspectives on Labour and Income* 15(Winter): 39–49.

Voss, Kim, and Rachel Sherman 2000 "Breaking the iron law of oligarchy: Union revitalization in the American labor movement." *American Journal of Sociology* 106(2): 303–49.

Wajcman, Judy 1991 "Patriarchy, technology, and conceptions of skill." *Work and Occupations* 18: 29–45.

Walby, Sylvia 1990 *Theorizing Patriarchy.* Oxford: Basil Blackwell.

References

Waldie, Paul 2005 "How health costs hurt the Big Three." *The Globe and Mail* (22 March): B1.

Waldram, James B. 1987 "Native employment and hydroelectric development in northern Manitoba." *Journal of Canadian Studies* 22: 62–76.

Walker, C. R., and R. H. Guest 1952 *Man on the Assembly Line.* Cambridge, MA: Harvard University Press.

Wallace, Jean E. 1995a "Corporatist control and organizational commitment among professionals: The case of lawyers working in law firms." *Social Forces* 73: 811–39.

––––––– **1995b** "Organizational and professional commitment in professional and nonprofessional organizations." *Administrative Science Quarterly* 40: 228–55.

Wallace, Jean E., and Merlin B. Brinkerhoff 1991 "The measurement of burnout revisited." *Journal of Social Service Research* 14: 85–111.

Wallace, Jo-Ann 1995 "'Fit and qualified': The equity debate at the University of Alberta." In Stephen Richer and Lorna Weir, eds., *Beyond Political Correctness: Toward the Inclusive University.* Toronto: University of Toronto Press.

Walmsley, Ann 1992 "Trading places." *Report on Business Magazine* (March): 17–27.

Walsh, Janet 1997 "Employment systems in transition? A comparative analysis of Britain and Australia." *Work, Employment & Society* 11: 1–25.

Walters, Vivienne 1985 "The politics of occupational health and safety: Interviews with workers' health and safety representatives and company doctors." *Canadian Review of Sociology and Anthropology* 22: 57–79.

Walters, Vivienne, and Ted Haines 1988 "Workers' use and knowledge of the internal responsibility system: Limits to participation in occupational health and safety." *Canadian Public Policy* 14: 411–23.

Wanner, Richard A. 1999 "Expansion and ascription: Trends in educational opportunity in Canada, 1920–1944." *Canadian Review of Sociology and Anthropology* 36(3): 409–42.

––––––– **2000** "A matter of degree(s): Twentieth-century trends in occupational status returns to educational credentials in Canada." *Canadian Review of Sociology and Anthropology* 37(3): 313–43.

––––––– **2004** "Social mobility in Canada: Concepts, patterns, and trends." In James Curtis, Edward Grabb, and Neil Guppy, eds., *Social Inequality in Canada: Patterns, Problems, and Policies.* 4th ed. Toronto: Pearson/Prentice Hall.

Warhurst, Christopher 1998 "Recognizing the possible: The organization and control of a socialist labor process." *Administrative Science Quarterly* 43(2): 470–97.

Watkins, Mel 1991 "A staples theory of economic growth." In Gordon Laxer, ed., *Perspectives on Canadian Economic Development.* Toronto: Oxford University Press.

Webb, Sidney, and Beatrice Webb 1894/1911 *The History of Trade Unionism.* London: Longmans, Green.

Weber, Max 1946 "Bureaucracy." In H. H. Gerth and C. Wright Mills, eds., *From Max Weber.* New York: Oxford University Press.

––––––– **1958** *The Protestant Ethic and the Spirit of Capitalism.* New York: Scribner.

––––––– **1964** *The Theory of Social and Economic Organization.* New York: Free Press.

Weiss, David 2003 *In Search of the Eighteenth Camel: Discovering a Mutual Gains Oasis for Unions and Management.* Kingston, ON: Industrial Relations Centre, Queen's University.

References

Weiss, Donald D. 1976 "Marx versus Smith on the division of labour." *Monthly Review* 28: 104–18.

Weitzel, William, and Ellen Jonsson 1989 "Decline in organizations: A literature integration and extension." *Administrative Science Quarterly* 34: 91–109.

Wells, Donald M. 1986 *Soft Sell: Quality of Working Life Programs and the Productivity Race.* Ottawa: Canadian Centre for Policy Alternatives.

——— **1997** "When push comes to shove: Competitiveness, job security and labour–management cooperation in Canada." *Economic and Industrial Democracy* 18(2): 167–200.

Wetzel, Kurt, Daniel G. Gallagher, and Donna E. Soloshy 1991 "Union commitment: Is there a gender gap?" *Relations industrielles/Industrial Relations* 46: 564–83.

White, Jerry P. 1990 *Hospital Strike: Women, Unions, and Public Sector Conflict.* Toronto: Thompson Educational Publishing.

White, Jerry, Paul Maxim, and Stephen Obeng Gyimah 2003 "Labour force activity of women in Canada: A comparative analysis of Aboriginal and non-Aboriginal women." *The Canadian Review of Sociology and Anthropology* 40(4): 391–415.

White, Julie 1990 *Male and Female: Women and the Canadian Union of Postal Workers.* Toronto: Thompson Educational Publishing.

White, Lynn, Jr. 1962 *Medieval Technology and Social Change.* Oxford: Oxford University Press.

White, Michael, Stephen Hill, Patrick McGovern, Colin Mills, and Deborah Smeaton 2003 "'High-performance' management practices, working hours and work-life balance." *British Journal of Industrial Relations* 41(2): 175–95.

Whitehead, T. N. 1936 *Leadership in a Free Society.* Cambridge, MA: Harvard University Press.

Whitley, Richard 1992 *Business Systems in East Asia: Firms, Markets and Societies.* London: Sage.

Whittaker, D. H. 1990 "The end of Japanese-style employment?" *Work, Employment & Society* 4: 321–47.

Whyman, Philip 2004 "An analysis of wage-earner funds in Sweden: Distinguishing myth from reality." *Economic and Industrial Democracy* 25(3): 411–45.

Whyte, William Foote, and Kathleen King Whyte 1988 *Making Mondragon: The Growth and Dynamics of the Worker Cooperative Complex.* Ithaca, NY: ILR Press.

Wilensky, Jeanne L., and Harold L. Wilensky 1951 "Personnel counseling: The Hawthorne case." *American Journal of Sociology* 57: 265–80.

Wilgosh, Lorraine R., and Deborah Skaret 1987 "Employer attitudes toward hiring individuals with disabilities: A review of the recent literature." *Canadian Journal of Rehabilitation* 1: 89–98.

Wilkins, Kathryn, and Marie P. Beaudet 1998 "Work stress and health." *Health Reports* (Winter): 47–62.

Williams, Cara 2003 "Sources of workplace stress." *Perspectives on Labour and Income* 15(Autumn): 23–30.

——— **2004** "The sandwich generation." *Perspectives on Labour and Income* 16(Winter): 7–14.

Williams, Christine L. 1989 *Gender Differences at Work: Women and Men in Nontraditional Occupations.* Berkeley: University of California Press.

——— **1995** *Still a Man's World: Men Who Do "Women's Work."* Berkeley: University of California Press.

Wilson, John 2000 "Volunteering." *Annual Review of Sociology* (26): 215–40.

Wilson, William J. 1997 *When Work Disappears: The World of the New Urban Poor.* New York: Alfred A. Knopf.

References

Winson, Anthony 1996 "In search of the part-time capitalist farmer: Labour use and farm structure in central Canada." *Canadian Review of Sociology and Anthropology* 33: 89–110.

Winson, Anthony, and Belinda Leach 2002 *Contingent Work, Disrupted Lives: Labour and Community in the New Rural Economy.* Toronto: University of Toronto Press.

Wirth, Linda 2004 *Breaking Through the Glass Ceiling: Women in Management: Update.* Geneva: International Labour Office.

Wolfson, Michael C., and Brian B. Murphy 1998 "New views on inequality trends in Canada and the United States." *Monthly Labor Review* (April): 3–23.

Womack, James P., Daniel T. Jones, and Daniel Roos 1990 *The Machine That Changed the World.* New York: Harper Perennial.

Wood, Stephen 1989a "The Japanese management model." *Work and Occupations* 16: 446–60.

——— **1989b** *The Transformation of Work? Skill, Flexibility and the Labour Process.* London: Unwin Hyman.

——— **1989c** "New wave management?" *Work, Employment & Society* 3: 379–402.

Woodward, Joan 1980 *Industrial Organization: Theory and Practice.* 2nd ed. Oxford: Oxford University Press.

World Bank 1993 *The East Asian Miracle: Economic Growth and Public Policy.* New York: Oxford University Press.

——— **2001** *Rethinking the East Asian Miracle.* New York: Oxford University Press and the World Bank.

——— **2005a** *World Development Indicators 2005.* (http://www.worldbank.org/data/wdi2005/wditext/Cover.htm)

——— **2005b** *East Asia Update 2005.* (http://siteresources.worldbank.org/INTEAPHALFYEARLYUPDATE/Resources/eapupdate.pdf)

World Commission on Environment and Development 1987 *Our Common Future.* Oxford: Oxford University Press.

Wotherspoon, Terry 1995 "Transforming Canada's education system: The impact on educational inequalities, opportunities, and benefits." In B. Singh Bolaria, ed., *Social Issues and Contradictions in Canadian Society.* 2nd ed. Toronto: Harcourt Brace Jovanovich.

Wrege, Charles D., and Richard M. Hodgetts 2000 "Frederick W. Taylor's 1899 pig iron observations: Examining fact, fiction, and lessons for the new millennium." *Academy of Management Journal* 43(6): 1283–91.

Wright, Erik Olin 1997 *Class Counts: Comparative Studies in Class Analysis.* Cambridge: Cambridge University Press.

Wright, Erik Olin, C. Costello, D. Hachen, and J. Sprague 1982 "The American class structure." *American Sociological Review* 47: 709–26.

Wright, James D., and Richard F. Hamilton 1979 "Education and job attitudes among blue-collar workers." *Sociology of Work and Occupations* 6: 59–83.

Wright, Susan 1993 "Blaming the victim, blaming society, or blaming the discipline: Fixing responsibility for homelessness." *Sociological Quarterly* 34: 1–16.

Wu, Xiaogang, and Yu Xie 2002 "Does the market pay off? Earnings returns to education in urban China." *American Sociological Review* 68: 425–42.

Wyatt Canada 1991 *The New National Benchmark on Worker Attitudes.* Vancouver: The Wyatt Company.

Wylie, William N. T. 1983 "Poverty, distress, and disease: Labour and the construction of the Rideau Canal, 1826–1832." *Labour/Le Travail* 11: 7–30.

Yakabuski, Konrad 2001 "The extended family." *Report on Business Magazine* (July): 55–56.

References

Yalnizyan, Armine, T. Ran Ide, and Arthur J. Cordell 1994 *Shifting Time: Social Policy and the Future of Work.* Toronto: Between the Lines Press.

Yammarino, Frances J., Fred Dansereau, and Christina J. Kennedy 2001 "A multiple-level multidimensional approach to leadership: Viewing leadership through an elephant's eye." *Organizational Dynamics* 29(3): 149–63.

Yanow, Dvora 2000 "Seeing organizational learning: A 'cultural' view." *Organization* 7(2): 247–68.

Yates, Charlotte 1990 "The internal dynamics of union power: Explaining Canadian autoworkers' militancy in the 1980s." *Studies in Political Economy* 31: 73–105.

——— 1998 "Unity and diversity: Challenges to an expanding Canadian autoworkers' union." *Canadian Review of Sociology and Anthropology* 35(1): 93–118.

——— 2000 "Staying the decline in union membership: Union organizing in Ontario, 1985–1999." *Relations industrielles/Industrial Relations* 55(4): 640–71.

Yates, Charlotte, Wayne Lewchuk, and Paul Stewart 2001 "Empowerment as a Trojan horse: New systems of work organization in the North American automobile industry." *Economic and Industrial Democracy* 22(4): 517–41.

Yoder, Janice D. 1991 "Rethinking tokenism: Looking beyond numbers." *Gender & Society* 5: 178–92.

Zaleznik, Abraham 2004 "Managers and leaders: Are they different?" *Harvard Business Review* (January): 74–81. (Orig. pub. in same journal 1977.)

Zbaracki, Mark J. 1998 "The rhetoric and reality of total quality management." *Administrative Science Quarterly* 43(3): 602–36.

Zeitlin, Irving M. 1968 *Ideology and the Development of Sociological Theory.* Englewood Cliffs, NJ: Prentice-Hall.

Zeitlin, M. 1974 "Corporate ownership and control." *American Journal of Sociology* 79: 1073–1119.

Zetka, James R., Jr. 1992 "Work organization and wildcat strikes in the U.S. automobile industry, 1946–1963." *American Sociological Review* 57: 214–26.

Zeytinoglu, Isik U., and Gordon B. Cooke 2005 "Non-standard work and benefits: Has anything changed since the Wallace Report?" *Relations industrielles/Industrial Relations* 60(1): 29–63.

Zeytinoglu, Isik U., and Jacinta K. Muteshi 2000 "Gender, race and class dimensions of nonstandard work." *Industrial Relations* 55(1): 133–65.

Zhao, John, Doug Drew, and T. Scott Murray 2000 "Knowledge workers on the move." *Perspectives on Labour and Income* 12(Summer): 32–46.

Zuboff, Shoshana 1988 *In the Age of the Smart Machine: The Future of Work and Power.* New York: Basic Books.

INDEX

Index

Index

Index

Index

Index

Index